Smart Nanomaterials Technology

Nanotechnology is a rapidly growing scientific field and has attracted a great interest over the last few years because of its abundant applications in different fields like biology, physics and chemistry. This science deals with the production of minute particles called nanomaterials having dimensions between 1 and 100 nm which may serve as building blocks for various physical and biological systems. On the other hand, there is the class of smart materials where the material that can stimuli by external factors and results a new kind of functional properties. The combination of these two classes forms a new class of smart nanomaterials, which produces unique functional material properties and a great opportunity to larger span of application. Smart nanomaterials have been employed by researchers to use it effectively in agricultural production, soil improvement, disease management, energy and environment, medical science, pharmaceuticals, engineering, food, animal husbandry and forestry sectors.

This book series in Smart Nanomaterials Technology aims to comprehensively cover topics in the fabrication, synthesis and application of these materials for applications in the following fields:

- Energy Systems—Renewable energy, energy storage (supercapacitors and electrochemical cells), hydrogen storage, photocatalytic water splitting for hydrogen production
- Biomedical—controlled release of drugs, treatment of various diseases, biosensors,
- Agricultural—agricultural production, soil improvement, disease management, animal feed, egg, milk and meat production/processing,
- Forestry—wood preservation, protection, disease management
- Environment—wastewater treatment, separation of hazardous contaminants from wastewater, indoor air filters.

Azamal Husen
Editor

Nanobiotechnology for Abiotic Stress Adaptation and Mitigation in Agricultural Crops

Physiological and Molecular Intervention

Springer

Editor
Azamal Husen
Wolaita Sodo University
Wolaita, Ethiopia

Graphic Era (Deemed to be University)
Dehradun, India

ISSN 3004-8273 ISSN 3004-8281 (electronic)
Smart Nanomaterials Technology
ISBN 978-981-96-9297-2 ISBN 978-981-96-9298-9 (eBook)
https://doi.org/10.1007/978-981-96-9298-9

This Springer imprint is published by the registered company Springer Nature Singapore Pte Ltd.
The registered company address is: 152 Beach Road, #21-01/04 Gateway East, Singapore 189721, Singapore

Preface

The Green Revolution of the mid-twentieth century played a key role in increasing global food production, allowing us to feed a population that is now approaching 8 billion. To meet the needs of a growing global population by 2050, food production will need to rise significantly by 60–80%. However, climate change has amplified various challenges, such as land degradation, water scarcity, and extreme temperatures, all of which negatively impact crop yields. In this connection, a range of abiotic stresses, including salinity, drought, heavy metals, extreme temperatures, and air pollution, further hinder crop growth and productivity. To alleviate these negative impacts, various strategies have been employed to increase plant resistance, including genetic improvement, extensive breeding programs, plant-soil optimization, and substantial financial investments. Recently, nanobiotechnology has emerged as an innovative and rapidly advancing field that combines the progress of nanotechnology with biological sciences, especially cellular and molecular biology. In this context, several nanoparticles (NPs), including zinc oxide, silicon oxide, magnesium oxide, iron oxide, aluminum oxide, copper oxide, and silver NPs, have shown great potential in boosting the performance of agricultural crops under challenging conditions. NPs enhance plant resistance to stress by strengthening physical barriers, facilitating the uptake and transport of essential minerals and nutrients, improving photosynthetic efficiency, increasing the synthesis of anti-stress compounds, and modulating hormone balance, mainly auxins and gibberellins, thereby promoting plant tolerance to several abiotic stresses. Due to their high surface energy and large surface-to-volume ratio, NPs exhibit superior bioavailability and bioactivity compared to their bulk counterparts. They easily penetrate plant cells, are readily absorbed, and influence crucial stages in the plant life cycle, including seed germination, seedling growth, root formation, photosynthesis, flowering, and yield. However, while NPs offer significant benefits under abiotic stress, they can also be toxic to crop plants at higher concentrations. Thus, this situation needs to be examined in different crop plant species.

NPs may mitigate oxidative damage caused by abiotic stress by scavenging reactive oxygen species and activating antioxidant defense mechanisms, particularly at lower concentrations in certain crop plants. This book provides valuable insights into the application of NPs as a promising approach to enhancing the resilience of agricultural crop plants and ensuring food security in a changing environment. Despite the positive outcomes observed in improving plant performance under abiotic stress, the molecular mechanisms underlying the beneficial effects of NPs are still being investigated. Furthermore, the book also explores the long-term effects, safety concerns, and optimal application methods of NPs in agricultural practices. This book offers readers a thorough, up-to-date overview of how NPs impact crop responses at the molecular, cellular, metabolic, biochemical, and physiological levels, particularly under stressful conditions. In terms of physiology, NPs enhance stomatal behavior, boost photosynthesis, improve water use efficiency, and promote seed germination and growth. Biochemically, they alter metabolite profiles, reduce the accumulation of reactive oxygen species, and increase the activity of various antioxidant enzymes. NPs also improve lipid, protein, and glucose metabolism, as well as better regulate secondary metabolites like flavonoids and phenolics. At the cellular level, NPs help maintain osmotic balance, ion homeostasis, and membrane stability. On a molecular level, they affect stress-responsive transcription factors, influence gene expression, and impact signaling pathways such as calcium signaling. However, NPs can also have negative phytotoxic effects, such as inhibiting germination, stunting plant growth, disrupting cellular metabolism through oxidative stress, and damaging proteins and DNA, ultimately leading to reduced crop yield and performance. These adverse effects are typically observed at higher NP concentrations, though they can vary by plant species. Therefore, it is essential to standardize different NP formulations, concentrations, and delivery methods for various crop plants.

This book is an invaluable resource for advanced students, faculty, scientists, and researchers in fields such as nanobiotechnology, plant biology, plant biochemistry, plant physiology, plant stress physiology, plant microbiology, soil microbiology, agricultural sciences, botanical and crop sciences, environmental botany, and related disciplines. I am deeply thankful to all the contributors for accepting my invitation, sharing their expertise in specialized research areas, and incorporating feedback to improve their contributions. Lastly, I would like to offer special thanks to Shagufta Yasmeen, Zaara Ajmal, Mehwish Ajmal, and Huzaifa Ajmal for their time, dedication, and unwavering support in bringing this project together.

Wolaita, Ethiopia/Dehradun, India Azamal Husen

About This Book

Climate change has amplified various challenges, such as land degradation, water scarcity, and extreme temperatures, all of which negatively impact crop yields. In this connection, a range of abiotic stresses hinder crop growth and productivity. To alleviate these adverse impacts, various strategies have been employed to enhance plant resistance, including genetic improvement, extensive breeding programs, plant-soil optimization, and substantial financial investments. In this connection, nanobiotechnology has emerged as an innovative and rapidly advancing field that combines the progress of nanotechnology with biological sciences, especially cellular and molecular biology. Several metal and metal-oxide nanoparticles (NPs) have shown great potential in boosting the performance of agricultural crops under challenging conditions. NPs enhance plant resistance to stress by strengthening physical barriers, facilitating the uptake and transport of essential minerals and nutrients, improving photosynthetic efficiency, increasing the synthesis of anti-stress compounds, and modulating hormone balance, mainly auxins and gibberellins, thereby promoting plant tolerance to various abiotic stresses. Due to their high surface energy and large surface-to-volume ratio, NPs exhibit superior bioavailability and bioactivity compared to their bulk counterparts. This book offers readers a thorough, up-to-date overview of how NPs impact crop responses at the molecular, cellular, metabolic, biochemical, and physiological levels, particularly under stressful conditions. In terms of physiology, NPs enhance stomatal behavior, boost photosynthesis, improve water use efficiency, and promote seed germination and growth. Biochemically, they alter metabolite profiles, reduce the accumulation of reactive oxygen species, and increase the activity of various antioxidant enzymes. NPs also improve lipid, protein, and glucose metabolism, as well as better regulate secondary metabolites like flavonoids and phenolics. At the cellular level, NPs help maintain osmotic balance, ion homeostasis, and membrane stability. At the molecular level, they affect stress-responsive transcription factors, influence gene expression, and impact signaling pathways such as calcium signaling. However, NPs can also have negative phytotoxic effects, such as inhibiting germination, stunting plant growth, disrupting cellular metabolism through oxidative stress, and damaging proteins and DNA, ultimately leading to reduced crop yield and performance. These adverse effects are typically observed at higher NP concentrations, though they can vary by plant species. Therefore, it is essential to standardizing different NP formulations, concentrations, and delivery methods for various crop plants. This book is an invaluable resource for advanced students, faculty, scientists, and researchers in fields such as nanobiotechnology, plant biology, plant biochemistry, plant physiology, plant stress physiology, plant microbiology, soil microbiology, agricultural sciences, botanical and crop sciences, environmental botany, and related disciplines.

Contents

About the Editor

Azamal Husen serves as a Full Professor and Head of the Department of Biology at the University of Gondar, Ethiopia. He is currently a Foreign Delegate at Wolaita Sodo University, Wolaita, Ethiopia, and also holds the position of Adjunct Professor at Graphic Era (Deemed to be University) in Dehradun, Uttarakhand, India. Previously, he was a Full Professor at Sankalchand Patel University in Visnagar, India, and a Visiting Faculty member at the Forest Research Institute and Doon College of Agriculture and Forest in Dehra Dun, India. With 25 years of research and teaching experience, Husen specializes in Nanobiotechnology, Plant Physiology, Environmental Botany, Medicinal Plants, and Plant Propagation. He has conducted research supported by organizations such as the World Bank, the National Agricultural Technology Project, the Indian Council of Agricultural Research, the Indian Council of Forest Research Education, and the Japan Bank for International Cooperation. Husen has published extensively, with over 350 works, including research papers, books, and book chapters. He has also served on the Editorial Boards and as a reviewer for respected journals published by Elsevier, Frontiers Media, Taylor & Francis, Springer Nature, RSC, Oxford University Press, Sciendo, the Royal Society, CSIRO, PLOS, MDPI, John Wiley & Sons, and UPM Journals. He is a member of the advisory board of Cambridge Scholars Publishing, UK. He is a fellow of the Plantae group of the American Society of Plant Biologists and belongs to several professional organizations, including the International Society of Root Research, the Asian Council of Science Editors, the International Natural Product Sciences, and so on. He has delivered numerous talks and presentations at conferences and workshops. Additionally, Husen has conducted various orientation courses for academic staff, developed educational modules, and contributed to the creation of B.Sc., M.Sc., and Ph.D. curricula. In recognition of his decade of service in teaching, research, and community engagement, he received an award from the University of Gondar, Ethiopia. He is the Editor-in-Chief of the *American Journal of Plant Physiology* and serves as a Book Series Editor for *Exploring Medicinal Plants* (Taylor & Francis Group, USA), *Plant Biology, Sustainability, and Climate Change* (Elsevier, USA), and *Smart Nanomaterials Technology* (Springer Nature, Singapore). He is also the Handbook Series Editor for *Nanotechnology and Its Applications* (Springer Nature, Singapore). Husen has earned the distinguished honor of being recognized as one of the "World's Top 2% Scientists" from 2022 onwards by Stanford University, USA. This prestigious recognition has been prominently featured in the Elsevier Data Repository. Additionally, he has served as a guest editor for special issues on various topics, including *Advances in Plant Metabolomics: Insights, Applications, and Integration* (Plant Physiology and Biochemistry), *Exploring Nanotechnology in Medicinal Plants* (Journal of Herbal Medicine), and *Integrating Nanotechnology with Plant Compounds: Enhancing Abiotic Stress Resilience and Crop Improvement via Signaling Molecules* (Frontiers in Plant Science), among others.

Significance of Nanoparticles to Modulate Crop Responses at Physiological, Biochemical, Metabolic, Cellular, and Molecular Levels Under Abiotic Stress Conditions: An Overview

Smita Parekh, Megha Barot, Arpan Modi, Yesha Master, Ayushi Agrawal, Noreen Zahra, and Azamal Husen

Abstract In the recent era of agriculture science, nanoparticles (NPs) have shown great promise in reducing the damaging effects of abiotic stress on crops. This chapter emphasizes how NPs affect crop responses at the molecular, cellular, metabolic, biochemical, and physiological levels. In terms of physiology, NP exposure improves stomatal behavior, increases photosynthesis and water usage efficiency, and encourages seed germination and development. At the biochemical level, they change the metabolite profiles linked to stress signaling and defence pathways, decrease the accumulation of reactive oxygen species, and increase the activity of various antioxidant enzymes. Improved lipid, protein, and glucose metabolism as well as better

S. Parekh
The Mandvi Education Society Science College, VNSGU, Surat, Gujarat, India

M. Barot
Research and Development Cell, Parul University, Vadodara, Gujarat, India

Department of Environmental Science, Parul Institute of Applied Sciences, Parul University, Vadodara, Gujarat, India

A. Modi
Axenic Agritech Private Limited, Ahmedabad, Gujarat, India

Y. Master (✉)
Department of Life Science, GSFC University, Vadodara, Gujarat, India
e-mail: yeshamaster.91@gmail.com

A. Agrawal
Department of Biotechnology, VNSGU, Surat, Gujarat, India

N. Zahra
Department of Botany, Government College Women University, Faisalabad, Pakistan

A. Husen
Wolaita Sodo University, Wolaita Sodo, Ethiopia

Department of Biotechnology, Graphic Era (Deemed to be University), Dehradun, Uttarakhand, India

A. Husen (ed.), *Nanobiotechnology for Abiotic Stress Adaptation and Mitigation in Agricultural Crops*, Smart Nanomaterials Technology,
https://doi.org/10.1007/978-981-96-9298-9_1

control of secondary metabolites like flavonoids and phenolics are changes brought about by NPs. Changes at the cellular level include osmotic balance, ion homeostasis maintenance, and membrane stabilization. At the molecular level, NPs modify stress-responsive transcription factors, impact gene expression, and also impact signaling pathways like calcium signaling and so on. NPs have also been shown to cause adverse phytotoxic effects, including inhibition of germination, reduced plant growth, disruption of cellular metabolism through the generation of oxidative stress, and damage to proteins and DNA, ultimately leading to a decline in crop yield and overall performance. These responses are noticed usually at higher concentration of NP treatments, although it varies from one plant species to another. Thus, various NP formulations, concentrations, and delivery methods for diverse crop plants are required to be standardized. Further, due to NP exposure, the ecotoxicological hazards (in plant-soil-water-human systems) must be examined. Overall, this chapter provides an overview of the role of NPs in enhancing abiotic stress tolerance in crop plants, focusing on their effects at the physiological, biochemical, metabolic, cellular, and molecular levels.

Keywords Abiotic stress tolerance · Ecotoxicological hazard · Food security · Nanoparticles · Sustainability

Abbreviations

ACO	ACC Oxidase
ACS	Acetyl CoA-Synthetase
AgNPs	Silver Nanoparticles
Al_2O_3	Aluminum Oxide
AP2/ERF	APETALA2/Ethylene Response Factor
AuNPs	Gold Nanoparticles
CuNPs	Copper Nanoparticles
CuO	Copper Oxide
ERF1	Ethylene Response Factor 1
ETR2	Ethylene Response 2
Fe_3O_4	Iron Oxide
H_2O_2	Hydrogen Peroxide
MAPKs	Mitogen-Activated Protein Kinases
MDA	Malondialdehyde
MDHA	Monodehydroascorbate
MgO	Magnesium Oxide
NPs	Nanoparticles
PI3K	Phosphatidylinositol 3-kinase
ROS	Reactive Oxygen Species
SeNPs	Selenium Nanoparticles
SiNPs	Silica Nanoparticles

SiO_2	Silicon Dioxide
TiO_2	Titanium Dioxide
ZnO	Zinc Oxide

1 Introduction

Nanotechnology is the branch of science and engineering that manipulates materials at the nanoscale (1–100 nm) to create structures, devices, or systems with novel properties and functions as compared to bulk materials. Nanoparticles (NPs) are the key products of this technology [105] and have numerous applications in science and technology [57, 58, 65, 71, 99, 102].

1.1 NPs in Agriculture

NPs are particles with < 100 nm dimension and unique physicochemical properties such as high surface area-to-volume ratio, increased reactivity, and higher rate of adsorption [31]. In agriculture, NPs have several applications in nanofertilizers, nano-pesticides, nano-sensors, and disease management [90, 91, 66, 63, 68, 133]. They are used as micro- and macro-nutrient fertilizers to provide various nutrients to plants for enhanced growth and improved productivity [76]. These are used to make plants resistant to target and non-target plant pests, and improve plant health by managing plant diseases [31]. They detect phytopathogens and monitor crop growth and soil conditions, entry of agrochemical residues, nutrient deficiency, soil and plant health, and environmental pollution, for sustainable agriculture [132].

1.2 Overview of Abiotic Stress

Abiotic stressors include drought, salt, temperature extremes, various atmospheric pollution, and the accumulation of heavy metals which cause morphological, physiological, and biochemical alterations in plants [31, 41, 51, 54–61, 159, 166, 167]. Figure 1 illustrates the impact of abiotic stresses on plants through various stress-induced changes, highlighting their detrimental effects on plant growth and development. Drought limits water availability to plants leading to reduced photosynthesis, decreases turgor pressure hindering cell growth, protoplasm may become rigid altering the cellular metabolism, and decreases leaf area and plant height leading to a reduction in plant growth and development [159]. High salt concentrations alter the plant's hormonal balance, delay seed germination by increasing germination

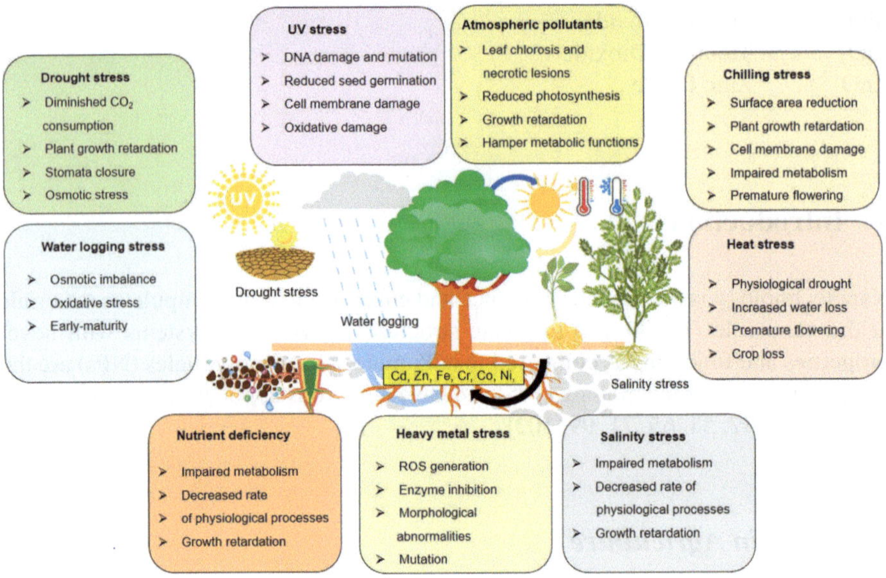

Fig. 1 Impact of abiotic stresses on plant growth and development

time, disrupt ion balance and water uptake causing toxicity, and reduce photosynthetic activities [67]. Accumulation of heavy metals such as cadmium and lead at high concentrations becomes highly toxic to plants such as degradation of protein structures, disruption of cytoplasmic membrane integrity, suppression of photosynthesis and enzymatic activities, and also causes DNA damage in plants [31, 48]. Extreme temperatures both high and low temperatures can impair enzymatic activities and cellular structures. Cold stress leads to cell and tissue dehydration, cellular water crystallization, reduced membrane conductivity, increased water viscosity, and increased leakage of electrolytes [115]. High-temperature stress increases water loss by transpiration, leading to drought conditions. It decreases photosynthesis, causes protein damage, inhibits the synthesis of proteins, inactivates enzymes, and damages membranes [146, 159].

1.3 *Mechanisms of Stress Tolerance*

In agriculture, NPs have gained significant attention because of their potential to reduce the effects of abiotic stresses (drought, salinity, temperature extremes, and heavy metal toxicity) by interacting with plants at cellular and molecular levels to enhance stress tolerance [52, 53, 132, 136, 137, 143, 144, 145] (Table 1). Abiotic stresses cause oxidative stress in plants by generating excessive reactive oxygen species (ROS). The increased ROS damage membranes, proteins, and DNA, and lead to lipid peroxidation in plants [9, 22]. NPs suppress the levels of ROS in plants in several ways. For example, zinc oxide (ZnO) NPs boost the activity of antioxidant enzymes such as superoxide dismutase (SOD), catalase (CAT), ascorbate peroxidase (APX), and glutathione peroxidase (GPX) [31], Cerium oxide (CeO_2) NP surface lattice has vacant oxygen sites, which allows them to scavenge O_2 radicals, thus reducing the levels of ROS [79], and aluminum oxide (Al_2O_3) NPs scavenge ROS by elevating the ascorbate/glutathione pathway [9]. NPs mitigate the toxicity of heavy metals like lead and cadmium in soil. Titanium dioxide (TiO_2) NPs reduce cadmium heavy metal toxicity by reducing cadmium-induced oxidative stress [31]. The high surface area-to-volume ratio of NPs allows them to bind with toxic metal ions and makes them less bioavailable [69]. Silicon dioxide (SiO_2) NPs bind active heavy metal ions and induce detoxifying enzymes that help plants tolerate heavy metal stress [31]. Under stressful conditions, NPs enhance the availability of nutrients to plants. Under salt stress conditions, iron oxide (Fe_3O_4) NPs improve iron uptake and TiO_2 NPs enhance nutrient uptake in plants [9]. Nanofertilizers ensure sufficient nutrient supply during stress conditions by releasing nutrients more efficiently [69].

2 Physiological Modulation by NPs Under Abiotic Stress

Plant development and growth, fruit yield, and productivity are negatively affected by abiotic stress factors like drought, salinity, high temperature, toxicity of heavy metals, and pollution. Using their distinct physicochemical characteristics, NPs provide a novel means of reducing these kinds of stresses through physiological control. NPs can affect physiological functions such as photosynthesis, water absorption, nutrient assimilation, gaseous exchange, and oxidative stress management by interacting with plants at the cellular and molecular levels. In this section, a summary of the effect of NPs on different physiological parameters has been discussed. Photosynthetic processes, which are frequently hampered by abiotic stress, are made more efficient by NPs. For example, under drought stress, TiO_2 NPs prevent photosynthetic pigments from degrading and improve chlorophyll stability [96]. By controlling stomatal conductance, minimizing excessive water loss, and enhancing root electric conductivity, NPs can improve water use efficiency (WUE) in water-limiting environments. For example, silicon NPs assist plants in maintaining cellular hydration by increasing root water intake and inducing osmolyte synthesis [123]. ROS, which are produced in excess as a result of abiotic stress, oxidatively damage cellular constituents. Strong antioxidants, such as CeO_2 NPs scavenge ROS and preserve cellular redox equilibrium [31]. By lowering DNA damage, protein oxidation, and lipid peroxidation, this modulation increases a plant's ability to withstand stress. Abiotic stress frequently interferes with the metabolism and absorption of nutrients. NPs aid in the restoration of metabolic balance by increasing soil nutrient availability and promoting nutrient transport. For instance, it has been demonstrated that iron oxide NPs enhance iron absorption, which is necessary to sustain electron transport and chlorophyll production under stressful situations [9]. NPs'

Table 1 Major abiotic stress, important agricultural crops, used different nanoparticles, type of exposure, various concentrations, studied parameters, and overall impact on the studied parameters

Abiotic stress	Crop plants	Nanoparticles	Mode of exposure	Concentration	Studied parameters	Impact	Key references
Drought stress	*Spinacia oleracea* L.	Green synthesized iron nanoparticles	Foliar spray	40 mg/L	Photosynthesis efficiency, growth parameters, gas exchange rate, total soluble proteins, inorganic ions	Recovered plant growth, photosynthesis efficiency, gas exchange attributes, total soluble proteins and inorganic ions when iron nanoparticles were sprayed under drought conditions	Naseem et al. [108]
Water deficit (Drought)	Potato	Silicon dioxide and zinc oxide nanoparticles	Exogenous application	Zinc oxide nanoparticles (0, 50, 100 ppm); silicon dioxide nanoparticles (0, 25, 50 ppm)	Relative water content of leaves, leaf gas exchange traits, and proline content, photosynthesis rate, intercellular carbon dioxide concentration, water use efficiency, stomatal conductance	Increase in leaf gas exchange, photosynthesis rate, improved stomatal conductance, relative water contents (LRWC), proline content, photosynthetic pigments, and polysaccharide crude nanoparticles; leaf green index; total yield and harvest index; and water consumption efficiency	Al-Selwey et al. [10]

(continued)

Table 1 (continued)

Abiotic stress	Crop plants	Nanoparticles	Mode of exposure	Concentration	Studied parameters	Impact	Key references
Drought stress	Wheat	Silicon nanoparticles	Seed priming, soil and foliar application	0, 300, 600, and 900 mg/L	Plant height, relative water content, number of grains per spike, spike length, 1000-grain weight, organic yield per plant, yield per plant, peroxidase and superoxide dismutase activity, nitrogen, phosphorous and potassium content, chlorophyll content, ear length and hydrogen peroxide content	Increases plant height, overall yield, relative water content, stomatal conductance, chlorophyll content, superoxide dismutase and peroxidase activity, nitrogen, phosphorous, potassium, content. Reduced ear length and hydrogen peroxide content	Raza et al. [124]
Drought stress	Pea	Boron nanoparticles	Foliar application	12.5, 25, and 50 ppm	Plant height, root- shoot ratio, yield, superoxide dismutase, catalase, ascorbate peroxidase and glutathione peroxidase activity, malonaldehyde and hydrogen peroxide content	Increased plant height, root-shoot ratio, yield superoxide dismutase, catalase, ascorbate peroxidase and glutathione peroxidase activity. Decreased malonaldehyde and hydrogen peroxide content	Sutuliene et al. [150]

(continued)

Table 1 (continued)

Abiotic stress	Crop plants	Nanoparticles	Mode of exposure	Concentration	Studied parameters	Impact	Key references
Drought stress	Maize	Copper nanoparticles	Foliar application	3.333, 4.444, 5.556 mg/L	Seed number and dry weight, shoot height and biomass, anthocyanin content, relative leaf water content, chlorophyll content, leaf protein content, carotenoid contents, reactive oxygen species scavenging enzyme activities	Increases average yield per plant and number of seed, shoot height and biomass, leaf protein content, relative water content, anthocyanin content, chlorophyll content, carotenoid contents, enhancement of reactive oxygen species scavenging activities of ascorbate peroxidase and superoxide dismutase. Decreases reactive oxygen species accumulation	Nguyen et al. [109]

(continued)

Table 1 (continued)

Abiotic stress	Crop plants	Nanoparticles	Mode of exposure	Concentration	Studied parameters	Impact	Key references
Drought stress	Tomato	Green zinc oxide nanoparticles	Foliar application	25, 50, and 100 mg/L	Biomass, shoot dry weight, shoot fresh weight, root dry weight, and root fresh weight, free phenols, malonaldehyde and hydrogen peroxide concentration, catalase, ascorbate peroxidase, and superoxide dismutase activity	At lower concentration, significantly increases biomass, free phenols content, superoxide dismutase, ascorbate peroxidase, and catalase activity and decreases hydrogen peroxide and malonaldehyde content but at higher concentration (100 mg/L) nanoparticles showed adverse result in mitigating drought stress	El-Zohri et al. [32]

(continued)

Table 1 (continued)

Abiotic stress	Crop plants	Nanoparticles	Mode of exposure	Concentration	Studied parameters	Impact	Key references
Drought stress	Eggplant (*Solanum melongena* L.)	Zinc oxide nanoparticles	Foliar Application	50 and 100 ppm	Growth characteristics include plant height, leaf count per plant, stem diameter, fresh and dry plant shoot weights, and total leaf area, tissue water status, water productivity, membrane stability index, yield, relative chlorophyll content, relative water content (RWC), photosynthetic efficiency, and nutrient content (Iron, Phosphorous, Potassium, Nitrogen, and Manganese)	Increased growth traits, photosynthetic efficiency and chlorophyll content, photosynthetic efficiency, water productivity, relative water content and membrane stability index, nutrient content (Iron, Phosphorous, Potassium, Nitrogen, and Manganese). Increased photosynthetic efficiency and better stem and leaf anatomical features	Semida et al. [131]

(continued)

Table 1 (continued)

Abiotic stress	Crop plants	Nanoparticles	Mode of exposure	Concentration	Studied parameters	Impact	Key references
Cadmium and drought stress	Wheat	Iron nanoparticles	Drenching	0, 25, 50, and 100 mg/kg	Spike length, plant height, dry biomass yields of wheat shoot, root, and grains, ear length, hydrogen peroxide malonaldehyde content, peroxidase activity, cadmium and iron concentration in plant	Increases plant height, spike length, biomass, chlorophyll content, Fe concentrations. Improves superoxide dismutase and peroxidase activities. Reduces ear length, hydrogen peroxide, and malonaldehyde contents, cadmium concentration in plant tissues	Adrees et al. [5]
Drought stress	Strawberry	Selenium and silicon dioxide nanoparticles	Foliar application	Selenium/silicon nanoparticles (50 and 100 mg/L) and selenium nanoparticles (25 mg/L) silicon nanoparticles (125 mg/L)	Dry and fresh weight of shoot and root, total chlorophyll and carotenoid, leaf total soluble relative water content, carbohydrates, proline, leaf malonaldehyde and hydrogen peroxide contents, number of inflorescences, flower and fruit, and total phenolic compounds, fruit anthocyanin, antioxidant, and ascorbic acid content	Increases fresh and dry weight of roots and shoots, chlorophyll, carotenoid, carbohydrate, RWC contents. Decreases MDA and H_2O_2 content, number of inflorescence, flower and fruit and fruit anthocyanin, total phenolic compounds, ascorbic acid content and antioxidant was also increased significantly	Zahedi et al. [165]

(continued)

Table 1 (continued)

Abiotic stress	Crop plants	Nanoparticles	Mode of exposure	Concentration	Studied parameters	Impact	Key references
Drought stress	Wheat	SiO_2 nanoparticles	Soil, root and foliar application	0, 30, 60, and 90 ppm	Increase plant height, biomass, and yield, content of proline, leaf area, greenness, relative water content and superoxide dismutase activity	Increase leaf area, chlorophyll and relative water content, proline content, superoxide dismutase activity, productivity, plant height and biomass	Behboudi et al. [16]
Drought and salinity stress	Eggplant	Polysaccharide crude nanoparticles	Exogenous application	15, 35, and 65 mM	Chlorophyll content determination, proline content, total soluble protein, total soluble sugar, activities of antioxidant enzymes (glutathione reductase, superoxide dismutase, peroxidase, and catalase) activity	Increases in the activity of antioxidant enzymes (glutathione reductase, superoxide dismutase, peroxidase, and catalase), chlorophyll concentration, chlorophyll fluorescence, and chlorophyll stability index. Decreased malonaldehyde content, lipid peroxidation, proline content, total soluble proteins, and total soluble sugar content	Alabdallah et al. [11]

(continued)

Table 1 (continued)

Abiotic stress	Crop plants	Nanoparticles	Mode of exposure	Concentration	Studied parameters	Impact	Key references
Drought and salt stress	*Elymus sibiricus*	Silicon nanoparticles	Seed priming	50, 100, 150, 200, 250 mg/L	Chlorophyll content, osmolyte content, activities of superoxide dismutase, catalase, peroxidase, ascorbate peroxidase	Increased accumulation of chlorophyll, total proteins and osmolytes. Increased activities of superoxide dismutase, catalase, peroxidase, ascorbate peroxidase under both the stresses	Khan et al. [83]
Salinity stress	Rice	Zinc oxide nanoparticles	Hydroponic culture media	50 mg/L	Root fresh weight, root dry weight, root length, root potash content, root sodium content, ascorbate peroxidase, catalase, malonaldehyde content, superoxide dismutase activity	Enhances root dry weight, root fresh weight, root length root potassium content, and increases the concentration level of catalase, superoxide dismutase, and ascorbate peroxidase activity in root. Reduction in the sodium concentration of the roots, which improved the potassium/sodium ratio in the root system	Singh et al. [140]

(continued)

Table 1 (continued)

Abiotic stress	Crop plants	Nanoparticles	Mode of exposure	Concentration	Studied parameters	Impact	Key references
Salinity stress	Rice	Zinc oxide nanoparticles	Foliar spray	50 mg/L	Potassium level and photosynthetic pigment, morphology of trichome and stomata and antioxidant enzymes catalase, ascorbate peroxidase, glutathione reductase, and superoxide dismutase, activity, level of malonaldehyde, proline, and hydrogen peroxide	Enhances the morphology of the stomata and trichomes, raises the potassium level and accumulation of photosynthetic pigment, and increases the activity of the antioxidant enzymes catalase, ascorbate peroxidase, glutathione reductase, and superoxide dismutase. Reduces the level of malonaldehyde, proline, and hydrogen peroxides	Singh et al. [139]

(continued)

Table 1 (continued)

Abiotic stress	Crop plants	Nanoparticles	Mode of exposure	Concentration	Studied parameters	Impact	Key references
Salinity stress	Wheat	Green titanium oxide nanoparticles	Foliar application	20, 40, 60, and 80 mg/L)	Yield per pot, spike length, number of grains per spike, hundred grains weight, osmotic potential, water potential, carotenoid contents, production of osmolytes, soluble proteins and amino acids, superoxide dismutase activity, malonaldehyde content, total flavonoid and total phenolic contents	Increases carotenoid, germination qualities, osmotic and water potential, total flavonoid and total phenolic contents soluble proteins and sugar, amino acid and proline content, superoxide dismutase activity, and reduces malonaldehyde content	Mustafa et al. [107]
Salinity stress	Wheat	Selenium nanoparticles	Foliar application	0.01, 0.05, and 0.1%	Plant height, biomass/plant, yield/plant, 1000 grain weight, total chlorophyll, superoxide dismutase, peroxidase, total free amino acids, malonaldehyde, hydrogen peroxide	Increases plant height, biomass/plant, yield/plant, 1000 grain weight. Total chlorophyll, superoxide dismutase and peroxidase activity, total free amino acid content. Decreases malonaldehyde, hydrogen peroxide content	Zafar et al. [163]

(continued)

Table 1 (continued)

Abiotic stress	Crop plants	Nanoparticles	Mode of exposure	Concentration	Studied parameters	Impact	Key references
Salinity stress (in vitro)	Alfalfa	Calcium oxide, copper oxide, zinc oxide nanoparticles	Media supplement	0.8 ppm	Hydrogen peroxide, peroxidase activity, microscopic examination	Reduced activity of peroxidase and hydrogen peroxide content in copper oxide nanoparticle treated callus as compared to others under 50 mM salt stress conditions	Akçay et al. [8]
Salinity stress	Strawberry	Selenium nanoparticles	Foliar application	10 and 20 mg/L	Shoot dry weight and root dry weight, relative water content, chlorophyll and total carotenoid contents, total soluble carbohydrate and free proline content, malonaldehyde and hydrogen peroxide contents, superoxide dismutase activity, abscisic acid, and indole-3-acetic acid contents, malic acid, citric acid, succinic acid content	Increases shoot dry weight and root dry weight, relative water content, chlorophyll and total carotenoid contents, and free proline content, superoxide dismutase activity indole-3-acetic acid, and abscisic acid contents. Also increases malic acid, citric acid, succinic acid content, total soluble carbohydrate in fruits. Decreases malonaldehyde and hydrogen peroxide content	Zahedi et al. [164]

(continued)

Table 1 (continued)

Abiotic stress	Crop plants	Nanoparticles	Mode of exposure	Concentration	Studied parameters	Impact	Key references
Salinity stress	Chili pepper	Spermine (Spm), chitosan nanoparticles encapsulated spermine (CSpm NPs), chitosan nanoparticles (C-NPs)	Exogenous application	Spermine (0.5 mM), chitosan nanoparticles encapsulated spermine (0.5, 1, 1.5 mM), chitosan nanoparticles (0.5 mM)	Leaf area, number of leaves, shoot dry weight, shoot fresh weigh, yield, chlorophyll, carotenoid content, bioactive compounds, and capsaicinoids enzymes activity (superoxide dismutase and catalase)	Higher amounts of bioactive substances, capsicinoids, growth, yield, and antioxidant enzyme activity (superoxide dismutase and catalase)	Ramadan et al. [122]
Salinity stress	Pepper	Nano-silicon (nSi)	Foliar application	0, 0.5, 1, 2, and 3 mM	Photosynthetic pigments, total carotenoid content, malondialdehyde content, antioxidative enzyme activity (ascorbate peroxidase, catalase, glutathione reductase, and superoxide dismutase), malonaldehyde content	Considerably increases chlorophyll content, activities of superoxide dismutase, catalase, ascorbate peroxidase and glutathione reductase, and total carotenoid content. Decreased malonaldehyde content	Kusvuran et al. [92]

(continued)

Table 1 (continued)

Abiotic stress	Crop plants	Nanoparticles	Mode of exposure	Concentration	Studied parameters	Impact	Key references
Salinity stress	Bell pepper fruits	Selenium, silicon and copper	Soil drenching	Copper nanoparticles 100 and 500 mg/L, Selenium nanoparticles 10 and 50 mg/L, Silicon nanoparticles 200 and 1000 mg/L	Chlorophyll content, phenols, flavonoids, lycopene, β-carotene carotenoids and glutathione content, glutathione peroxidase, catalase, ascorbate peroxidase and phenylalanine ammonia lyase activity, and growth	Elevated levels of lycopene, glutathione peroxidase, chlorophylls, and the enzymes ascorbate peroxidase, glutathione peroxidase, catalase, and phenylalanine ammonia lyase in the leaves	González García et al. [40]
Salinity stress	Potato	Silicone nanoparticles	Foliar application	Sodium silicate nanoparticles (400 ppm)	Stomatal conductance, transpiration rate, radical scavenging activity, quantum yield of PSII, leaf transpiration	Increases- radical scavenging activity. Decreased potato leaf stomatal conductance, salinity level, and leaf transpiration	Kafi et al. [74]

(continued)

Table 1 (continued)

Abiotic stress	Crop plants	Nanoparticles	Mode of exposure	Concentration	Studied parameters	Impact	Key references
Cadmium and lead toxicity	Rice	Selenium and silicon nanoparticles	Foliar application	$SeSi_2$ (5:10 mg/L), SeSi (5:5 mg/L), $SeSi_3$ (5:20 mg/L), Se_2Si_2 (10:10 mg/L), Se_3Si (20:5 mg/L), Se_2Si (10:5 mg/L), Se_2Si_3 (10:20 mg/L), Se_3Si_3 (20:20 mg/L), Se_3Si_2 (20:10 mg/L)	Concentration of heavy metal in grounded shoots, roots, grains and husk, grains yield, biomass, protein content and phytic acid in rice grains	Silicon and selenium combinations meaningfully reduced the cadmium and lead concentration in plant tissue. The amount of phytic acid in rice grains was reduced by either Si or Se alone or in combination. elevated Se levels, biomass, grain yield, and protein content in rice grains	Hussain et al. [66]
Cold stress	Chili	Selenium and zinc oxide nanoparticles	Foliar application		Number of leaves, height of plant, fresh and dry weight of plant, leaf area, photosynthesis rate, transpiration rate, capsaicin, ascorbic acid, macro- and micro-nutrient content, peroxidase, abscisic acid, malonaldehyde, hydrogen peroxide content	Significantly increases plant height, plant weight, number of leaves, leaf area, capsaicin as well as ascorbic acid content, macro- and micro-nutrient content, photosynthesis rate, transpiration rate, peroxidase and glutathione dehydrogenase. Decreases abscisic acid, malonaldehyde, hydrogen peroxide content	Sayed et al. [130]

(continued)

Table 1 (continued)

Abiotic stress	Crop plants	Nanoparticles	Mode of exposure	Concentration	Studied parameters	Impact	Key references
Oxidative stress	Maize	Silver nanoparticles	Hydroponics	0, 100, 200, and 500 and 1000 mg/L	Silver accumulation in plant tissues, root and shoot lengths, biomass, activities of antioxidant enzymes catalase, superoxide dismutase, peroxidase, hydrogen peroxide, ascorbate peroxidase, production level and lipid peroxidation estimation by leaves thiobarbituric acid reactive substances content	Silver accumulation in plant tissues, increases root and shoot length at a lower concentration of silver nanoparticles (100 mg/L) but with 500 and 1000 mg/L treatments, biomass, shoot and root lengths reduced significantly. Increases the activities of superoxide dismutase, catalase, ascorbate peroxidase, peroxidase. Also increases thiobarbituric acid reactive substances content and hydrogen peroxide production level	Abbas et al. [1]
Different stress condition	*Nicotiana tabacum* L. cv. xanthi	Gold nanoparticles	Exogenous treated	100, 200, 300, 400, and 500 ppm	Seed germination, shoot and root length, reactive oxygen species scavenging activity, capacity of scavenging free radical, total chlorophyll content	Increased—seed germination, photosynthetic parameters such as total chlorophyll content, ion leakage percentage shoot and root length, reactive oxygen species scavenging activity, growth when treated with low concentration	Jalil et al. [68]

high surface area-to-volume ratio improves their ability to interact with plant cells, promoting biochemical processes and nutrition delivery [96]. NPs can alter gene expression, increasing the production of proteins and enzymes linked to stress by stress-responsive pathway induction mechanism [31]. Certain NPs are incorporated into cell membranes, altering their fluidity and permeability to enhance the transfer of water and nutrients [123].

2.1 Impact of NPs on the Process of Photosynthesis and Water Use Efficiency (WUE)

Photosynthesis and WUE are two of the most important physiological processes that are being affected by NPs. The general health, development, and production of the plant depend on these activities, particularly in harsh climatic circumstances like drought, salt, and extremely high or low temperatures. By increasing the amount of chlorophyll, promoting enzyme activity, and raising the efficiency of the photosynthetic machinery, NPs have been demonstrated in several studies to have a favorable impact on photosynthesis.

For instance, it has been observed that NPs, particularly those based on silicon (Si), titanium dioxide (TiO_2), and zinc oxide (ZnO), enhance the concentration of chlorophyll a and b, improving light absorption and facilitating effective energy conversion [96]. Because plants frequently suffer from oxidative stress brought on by elements like dryness or excessive salt, this impact is especially significant when abiotic stress conditions are present.

Furthermore, by boosting photophosphorylation efficiency and photosystem II (PSII) activity, NPs can influence the electron transport chain [9]. ROS and oxidative damage are better balanced as a result, which is important for preserving cellular activities under stress. Research indicates that by shielding photosynthetic pigments from deterioration, TiO_2 NPs can increase chlorophyll stability in drought-stressed environments. By changing the optical characteristics of leaves, NPs can improve the effectiveness of light harvesting [26, 27]. NPs like CeO_2 scavenge ROS and shield photosynthetic equipment by acting as antioxidants. ZnO NPs can boost carbon fixation rates by increasing the activity of ribulose-1,5-bisphosphate carboxylase/oxygenase (RuBisCO) [26, 27].

The capacity of plants to convert absorbed water into biomass is known as water use efficiency, or WUE. This is a crucial characteristic for plants that are growing in water-limited environments. By altering plant water relations and improving stomatal control, NPs can increase WUE. It has been shown that some NPs, including those made of carbon, increase plants' ability to retain water, enabling them to better handle the drought effect [123]. Additionally, NPs' function in controlling transpiration and stomatal conductance has drawn attention. By influencing stomata's opening and shutting, NPs can reduce water loss due to the transpiration process [20]. Water conservation is aided by this technique, especially when environmental stress levels are high. Furthermore, research has demonstrated that NPs may trigger the synthesis of Osmoprotectants, which protect against dehydration by stabilizing proteins and enzymes and preserving cellular turgor. These osmo-protectants include proline and soluble sugars [31]. Lowering the total amount of water needed to sustain metabolic activities improves WUE.

2.2 Effect of NPs on Stomatal Regulation

Plants use stomata, which are tiny openings on the leaf surface, to control transpiration and gas exchange. In order for plants to adapt to abiotic stress like drought, salt, and severe temperature fluctuation, their stomata must be able to react to environmental stimuli. Because of the minuscule size, special surface characteristics, and biological activity, NPs can affect stomatal control and increase plant resistance to stress.

By adjusting the stomatal opening, NPs can maximize the equilibrium between water loss through transpiration and carbon dioxide (CO_2) intake for photosynthesis. SiNPs can improve the mechanical strength of guard cells, which helps them control stomatal opening under stress [96]. TiO_2 NPs can promote abscisic acid (ABA) signaling, a crucial hormonal mechanism involved in stress response, to increase stomatal closure during drought. When it comes to stomatal accessibility, ROS are crucial secondary mediators. Excessive ROS generation during abiotic stress might cause stomatal impairment. As antioxidants, NPs like ZnO and CeO_2 scavenge ROS and preserve ideal levels for stomatal function [9, 127]. It has been demonstrated that NPs control aquaporins, which are proteins that help water pass through cell membranes. Improved guard cell turgor due to increased aquaporin activity allows for accurate stomatal control. Carbon-based nanoparticles, such as graphene oxide, enhance stomatal responsiveness by protecting cell hydration and improving water transport [19]. The hormonal mechanisms that regulate stomata can be impacted by NPs. AgNPs, in situations of water deprivation, facilitate quick stomatal closure by promoting the production of stress-related hormones like ABA [9]. Fe_3O_4 NPs increase cytokinin levels, which can postpone stomatal closure brought on by stress and preserve photosynthetic activity in the early phases of stress [50]. By lowering excessive water loss, preserving guard cell turgidity, and guaranteeing improved control over stomatal dynamics, NPs like silica and alumina have been shown to enhance leaf water potential [90].

2.3 Impact of NPs on Seed Germination and Growth Parameters

The use of NPs in agriculture has attracted a lot of interest because of their potential to increase plant growth metrics and seed germination. This novel method, which is sometimes called "seed nano-priming," entails applying nanoparticle substances to seeds to improve their ability to thrive in both ideal and stressful situations [44]. Mechanisms of how NPs positively affect seed germination are mentioned below. Because NPs may pass through seed coverings, they can activate metabolic pathways necessary for germination and facilitate water intake. For example, ZnO NPs have been demonstrated to increase seed metabolic rates, vigor index, and seedling traits, resulting in improved germination and vigor [29]. Seeds treated with NPs show improved germination and growth in spite of abiotic challenges like dehydration and salt stress. According to reports, carbon-based NPs (CNPs) greatly enhance the growth of seeds in high-temperature and salt-exposed environments, underscoring their function in stress tolerance [152]. NPs can even improve the accessibility and absorption of vital nutrients. For instance, by improving nutrient absorption efficiency, silica NPs have been employed in seed priming to enhance germination characteristics and subsequent seedling development [44]. Some NPs affect the hormonal equilibrium in seeds, which encourages development. Levels of growth hormone have been found to be impacted by AgNPs, which increase sprouting of seeds and overall growth [4]. Research showed that triploid seeds treated with AgNPs had a noticeably greater rate of seedling emergence at 14 days post sowing the seeds treated with other treatments, suggesting that AgNPs are effective in encouraging early seedling growth [4]. Studies on tomato seeds primed with Se NPs revealed that a concentration of 10 ppm considerably lengthened the overall length of tomato seedlings primed with Se NPs, indicating an ideal dosage for growth promotion [37]. The potential of silica NPs in seed treatment procedures was highlighted when stevia seedlings were nano-primed with silica NPs, which resulted in a 106% rise in germination percentage and improvements in other germination metrics [44].

Although using NPs for seed priming has several advantages, it is important to take into account variables such as nanoparticle concentration, size, and crop type to prevent phytotoxicity. To clarify the underlying mechanisms and provide standardized procedures for the secure and efficient application of nanoparticles in agriculture, more investigation is required.

3 Biochemical Responses Induced by NPs

NPs have proven important tools for increasing plants' tolerance to abiotic stress by modifying several stress-related metabolite biosynthesis and accumulation.

3.1 Antioxidant Enzyme Activity Modulation by NPs

NPs can enhance the expression of gene coding for antioxidant enzymes through several mechanisms including induction of signal transduction pathways, generation of controlled ROS levels, and epigenetic modifications. ZnO NPs were found to induce the activation of stress signaling pathways, such as the PI3K/MAPK dual pathway [147]. These pathways upregulate the activation of transcription factors such as AP2/ERF and NAM, ATAF, and CUC (NAC), which bind to the promoter region of genes encoding antioxidant enzymes. Thus, the expression of enzymes increases and leads to enhanced ROS scavenging [170]. A study conducted on *Arabidopsis thaliana* reported that AgNP treatment upregulates the expression of gene coding for SOD and peroxidases [3].

Engineered NPs such as ZnO and SiO_2 can produce low levels of ROS that act as signaling molecules, stimulating antioxidant defense systems in plants [25]. The generated ROS activates MAPK signaling cascades and cellular antioxidant mechanisms [6]. Also, at low concentrations, ROS participate in cell growth, survival, immune responses, metabolic regulation, and cell signaling processes. This indicates that NP-induced ROS can act as signaling molecules, participating in various biological functions [23, 168]. NPs diffuse into the nucleus, influence epigenetic modifications, and modulate chromatin structure and function, potentially influencing gene expression. Nanomaterials and NP exposure alters DNA methylation and histone modifications (acetylation and methylation), regulating the expression of gene coding for antioxidant enzymes and are involved in oxidative stress tolerance [43, 116, 135].

3.2 Reduction of ROS Accumulation

Under stress conditions, the overproduction and accumulation of ROS in plants causes oxidative stress, leading to the disintegration of membrane lipids and biomolecules, and leakage of electrolytes [143, 144]. According to Fig. 2, the mechanisms by which NPs mitigate ROS accumulation in plants are categorized into direct ROS scavenging, enhancement of antioxidant systems, and regulation of ROS-related gene expression.

NPs such as nanoceria (CeO_2 NPs) protect the plant's photosynthetic system by catalytically scavenging ROS. Nanoceria have ROS scavenging properties as they can switch between Ce^{+3} and Ce^{+4} oxidation states. This protects plants from the adverse effects of ROS accumulation under abiotic stress conditions [158]. NPs such as SiO_2, ZnO, TiO_2, and AgNPs, upregulate the expression and activity of antioxidant enzymes (SOD, GPX, APX, and CAT) during stress periods. These nanoparticle-induced increased enzymatic activities help plants tolerate oxidative stress. As depicted in Fig. 2, this enhancement of enzymatic activities by NPs plays a crucial role in mitigating oxidative stress, illustrating their pivotal role in plant

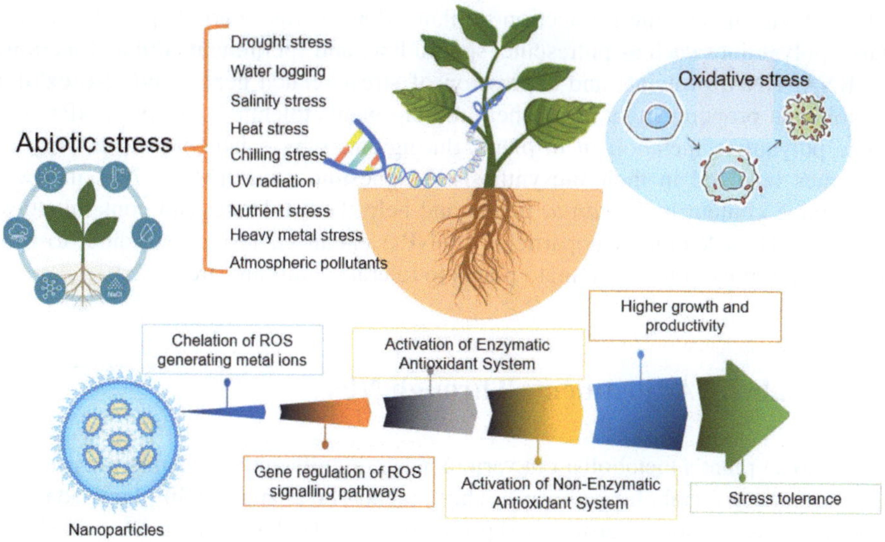

Fig. 2 Mechanisms of ROS reduction in plants as mediated by nanoparticles

stress tolerance mechanisms [31, 143, 144]. NPs enhance the level of nitric oxide (NO), which acts as a stress-signaling molecule and activates the expression of ROS-related genes, ultimately activating the plant's defense system against stress [143, 144]. Similarly, PVP-coated AgNPs upregulate the expression of genes encoding antioxidant enzymes that regulate ROS levels [69]. These genetic modulations by NPs help plants survive during stress periods, ensuring plant growth and tolerance to stress conditions [31].

3.3 *Changes in Stress-Related Metabolites*

Osmolytes (stress-related metabolites) such as polyamines, proline, and glycine-betaine, as well as secondary metabolites, help plants survive abiotic stress conditions [142]. NPs help to maintain the accumulation and utilization of these metabolites, enhancing plants' tolerance to stress [9]. In case of osmolytes, proline and glycine-betaine play an important role in maintaining osmotic balance and stabilizing plant cellular structures during stress [13]. NPs such as ZnO NPs and Copper Oxide (CuO NPs enhance the biosynthesis of proline and glycine-betaine by upregulating nitrogen metabolism in plants [38, 79]. In a similar way, secondary metabolites such as flavonoids and phenolic compounds help plants tolerate stress conditions by scavenging free radicals, strong antioxidant activity, and inhibiting hydrolytic and oxidative enzymes [70]. SiO_2 NPs and TiO_2 NPs are found to increase the amount of secondary metabolite production in plants during stress periods [17, 39]. Similarly, polyamines such as putrescine, spermidine, and spermine modulate functions of RNA, DNA, proteins, and expression of stress-related genes, and also exhibit antioxidant properties that allow them to help plants tolerate stress [93]. NPs alleviate polyamine metabolism in plants during stress by enhancing the activity of enzymes involved in their biosynthesis. Magnesium oxide (MgO) NPs increase polyamine content in *D. carota* plants and help them tolerate lead contamination in soil [34]. Selenium nanoparticles (SeNPs) enhanced levels and shoot-to-root transport of polyamines in tomato plants to tolerate cadmium stress [75].

4 Metabolic Adjustments Through NPs

NPs impact plant's metabolism in various ways, significantly adjusting the primary and secondary metabolic pathways. These metabolic changes help plants adapt to abiotic stresses, improve growth and productivity, and enhance stress tolerance [28].

4.1 Influence on Primary Metabolism (Carbohydrates, Proteins, and Lipids)

Primary metabolism involves pathways such as carbohydrate metabolism, protein synthesis, and lipid metabolism that are essential for plant growth and development. In the case of carbohydrates, NPs such as ZnO and TiO_2 enhance photosynthetic efficiency and sugar accumulation in plants during stress. These NPs enable increased production of soluble sugars that act as osmoprotectants during stress by enhancing light absorption and reducing photo-oxidative damage. A study found that TiO_2 NPs increased the photosynthetic activity and plant immunity, leading to a 30% increase in the yield of radish plants [155]. Similarly, ZnO NPs were found to increase photosynthesis and antioxidant enzyme activity in tomato plants [35]. In addition to their role in carbohydrate metabolism, NPs also influence protein regulation by stabilizing ribosomal activity and reducing oxidative damage to protein structures, thereby ensuring efficient protein synthesis. AgNPs enhance the synthesis of stress-response proteins, such as heat shock proteins (HSPs), which help properly fold proteins during stress conditions. AgNPs also enhance seed germination and plant growth and improve photosynthetic efficiency [82, 103]. Additionally, NPs modulate the lipid content of membranes by upregulating the synthesis of unsaturated fatty acids, thus enhancing membrane integrity and fluidity. This modulation helps plants maintain cell structures during extreme temperatures or salinity stress [126, 129].

4.2 *Regulation of Secondary Metabolites (Phenolics, Flavonoids, and Alkaloids)*

Secondary metabolites, such as phenolics, flavonoids, and alkaloids, play an important role in enhancing a plant's tolerance against abiotic stress due to their antioxidant activity and ability to act as signaling compounds by stimulating the resistance system in plants [128]. Among these, phenolics are secondary metabolites having strong antioxidant properties that help in ROS scavenging and also stabilize cellular structures, thus decreasing the effects of oxidative damage caused by abiotic stresses. NPs such as CeO_2 and CuO enhance the synthesis of phenolic compounds in stressed plants [42]. These NPs increase the activation of enzymes such as phenylalanine ammonia-lyase (PAL) that produce precursors for phenolic compound synthesis [24]. Flavonoids, on the other hand, are multifaceted secondary metabolites that act as powerful antioxidants, helping protect plants from UV and oxidative stress. Flavonoids regulate auxin transport and signaling pathways and neutralize ROS. Zinc oxide (ZnO) NPs enhance flavonoid synthesis by upregulating the expression of related genes. This leads to a higher accumulation of flavonoids, which provides better stress tolerance capacity to plants [18, 33, 38]. Alkaloids, which are nitrogen-containing metabolites, also play a crucial role in plant defense. They serve as effective compounds against pathogens and help plants tolerate environmental stress. Some alkaloids, such as atropine, morphine, and nicotine, act as antioxidants and help in the detoxification of ROS generated by different stresses [104]. AgNPs induce alkaloid synthesis, especially in medicinal plants, by triggering a signaling pathway that involves hydrogen peroxide and MAPKs and upregulation of related biosynthetic genes. This improves survival, adaptation, and growth of plants under stressful conditions [36].

4.3 *Impact on Hormonal Balance*

NPs influence the hormonal balance in plants either through direct interactions or by influencing their biosynthesis and signaling pathways under abiotic stress conditions. These regulations of hormones in plants help them to better tolerate stress and sustain growth in stressful conditions [138]. Auxins, in particular, control growth processes such as cell elongation by altering cell wall plasticity, root development, and phototropism in plants [112]. Several studies have been done on the role of different NPs in influencing the synthesis of auxin and supporting plant resilience under stress. In a study on *Eruca sativa*, gold nanoparticles (AuNPs) enhanced root growth and expression of genes that regulate the release of nitrile, a precursor of auxin biosynthesis. Similarly, TiO_2 NP treatment in *A. thaliana* plants increased auxin accumulation in root tips. However, copper oxide (CuO) NPs were found to have negative feedback regulation in auxin signaling by downregulating auxin signaling F-box proteins [153, 156]. In contrast, cytokinins induce cell division and chlorophyll biosynthesis and also help delay senescence in plants [112]. TiO_2 NPs and nanoceria were found to upregulate the cytokinin-mediated signaling pathway in *A. thaliana.* ZnO and SiO_2 NPs regulate the expression level of cytokinin signaling pathway-related genes (*ARR7* and *ARR15*), beneficially affecting plant growth under stress [156]. Similarly, gibberellins are important in seed germination, stem elongation, inducing maleness in dioecious flowers (sex expression), and causing parthenocarpic fruit development [112]. AgNPs and CuNPs significantly increased GA3 amount in the *Annona muricata* plant, while TiO_2 NPs regulated GA biosynthesis and signaling at the genetic level [72, 156]. In addition, abscisic acid plays a role in stomatal closure and stress signaling [112]. AgNPs modulate the production of ABA and the function of antioxidant machinery in salt stress in plants [26, 27]. Another hormone, ethylene is involved in fruit ripening, leaf senescence, and stress response in plants [112]. ZnO NPs enhance the expression of ethylene-related genes (*ACS* and *ACO* genes for ethylene biosynthesis; and *ETR2* and *ERF1* genes for ethylene signaling pathway) in abiotic stress conditions in plants [14].

5 Cellular-Level Changes Triggered by NPs

NPs induce significant cellular-level changes that enhance plants' resistance to abiotic stress by improving membrane stability, ion homeostasis, osmotic adjustments, and cell wall remodeling.

5.1 NPs and Membrane Stability Under Stress

Cellular membrane stability is an important factor for plant survival under stress conditions. Abiotic stresses induce oxidative damage by excess ROS production, leading to lipid peroxidation and disrupting membrane integrity [9]. NPs function to reduce the effects of oxidative damage on cellular membranes. NPs such as ZnO, Al_2O_3, and TiO_2 reduce oxidative stress by scavenging ROS, upregulating the activity of antioxidant enzymes, and enhancing the expression of stress-related genes, thus preventing lipid peroxidation and sustaining the phospholipids membrane integrity [143, 158]. For instance, TiO_2 NPs were reported to reduce lipid peroxidation by reducing H_2O_2 and MDA (a marker of lipid peroxidation) contents, reducing electrolyte leakage, and enhancing antioxidant activity in plants [2]. It has been noticed that the TiO_2 nanoparticles significantly reduce H_2O_2 and MDA contents, thereby decreasing electrolyte leakage and oxidative damage. Similarly, AgNPs regulate the expression of genes that code for antioxidant enzymes, thus helping in scavenging ROS [69] and ZnO NPs increase the photosynthetic pigment amount and enhance the antioxidant defense system in plants during stressful conditions which helps to control ROS production [31].

5.2 *Ion Homeostasis and Osmotic Adjustment*

NPs play a critical role in maintaining ion homeostasis and facilitating osmotic adjustments during stress. They regulate the uptake and distribution of ions such as potassium (K^+), calcium (Ca^{+2}), and magnesium (Mg^{+2}) while reducing the accumulation of toxic elements like sodium (Na^+). SiNPs were found to decrease the Na/K ratio and concentration of Na by regulating polyamine metabolism in rice plants under salinity stress. SiNPs reduce Na^+ influx and increase K^+ accumulation in plant roots [160, 162]. In addition, osmolytes such as proline, glycine-betaine, and soluble sugars maintain osmotic balance and reduce oxidative stress damage in plants. NPs such as ZnO, TiO_2, and CuO were found to increase the biosynthesis and accumulation of proline and glycine-betaine by enhancing nitrogen metabolism and increase the accumulation of soluble sugars by enhancing light absorption [13, 38, 155].

5.3 *Role in Cell Wall Remodeling*

Cell wall remodeling is a critical plant response to abiotic stress, ensuring structural support, flexibility, and defense against mechanical and oxidative stresses. NPs play an important role in modulating cell wall composition and properties by influencing the deposition of key cell wall components and regulating the activity of cell wall-modifying enzymes, thus enhancing plants' resistance against abiotic stress conditions [75]. Specifically, CeO_2 and SiO_2 NPs promote the synthesis and deposition of lignin, cellulose, and hemicellulose, thus strengthening the cell wall. This enhances the plants' ability to tolerate mechanical stress and osmotic fluctuations associated with abiotic stresses [134, 169]. Additionally, NPs modulate the activity of enzymes involved in cell wall remodeling, such as peroxidases and cellulases. Peroxidases are essential enzymes in the lignin biosynthetic pathway, catalyzing the polymerization of lignin monomers, which strengthen cell walls under stress conditions [95]. NPs also enhance cell wall plasticity by influencing their composition and structure, allowing for better water transport and cell expansion during stress. This is important for promoting internalization and maintaining cellular functions and overall plant growth under stressful conditions [113].

6 Molecular Mechanisms of Stress Modulation by NPs

Physiological, biochemical, metabolic, and cellular level changes, induced by NPs, have one thing in common, which is the molecular level regulation. NP-modulated responses of plants can be seen as the changes in the types and synthesis of stress-responsive proteins or enzymes. These include the production of transcription factors, ROS-scavenging enzymes, plant hormones (Abscisic acid, Salicylic acid, Jasmonic acid), calcium signaling, and epigenetic modifications [153]. It has been observed that NPs modulate the genetic network by switching on and off stress-responsive genes and transcription factors in a variety of plant species (Table 2).

6.1 NPs in Gene Expression and Stress-Responsive Transcription Factors

Transcription factors are proteins mainly associated with the regulatory elements of the genes. These transcription factors are modulated by a variety of molecules and signals. Plants contain stress-responsive genes, which are triggered to increase resistance or tolerance once the signal is generated. It is now well-evident that transcription factors are primarily involved in the number of biological processes, including stress responses. At the molecular level they are directly involved in the start of transcription by interacting with regulatory elements of the gene. Major transcription factors engaged in the abiotic stress responses are ethylene-responsive-element-binding factor (ERF); basic-domain leucine zipper (bZIP); WRKY transcription factors; and dehydration-responsive element binding factor (DREB). Their synthesis and level of regulation are thoroughly reviewed [141]. It has been observed that 40 nm sized NPs have been taken up by roots. These NPs are then transported to different parts of the plant via vascular system. Effects of different nanomaterials on transcriptome of model organisms have been studied in detail by [94] and found limited overlap in gene expression patterns. After being penetrated through cell wall, NPs bind to the specific receptors present in the plasma membrane, and a signal for the stress-responsive genes is generated. These signals correspond to the production of specific proteins (transcription factors), which in turn initiate the stress-responsive mechanism. Moreover, these NPs also interact with cell cycle process at the chromatin level and can interfere in the methylation and chromatin remodeling. These in turn introduce the changes in gene expression patterns that may or may not be inherited. Such changes are known as epigenetic modifications [106].

6.2 Modulation of Signaling Pathways

In their entire life cycles, plants are exposed to two different phases, viz., growth and stress. Plant growth regulators (or phytohormones) play a significant role during the life cycle of plants, starting from seed germination to seed dispersion. Major plant growth regulators are auxins, cytokinins, gibberellic acid, ethylene, abscisic acid, jasmonic acid, salicylic acid, and brassinosteroids. Among them auxins, cytokinins, and gibberellins help in plant growth and development, whereas the remaining help in stress conditions. On the other hand, signaling pathways include metabolism (synthesis and degradation) of plant growth regulators, calcium signaling, and ROS

Table 2 Genetic and epigenetic regulation of stress-responsive genes through the application of NPs in different plant species

Nanoparticle	Plant species	Stress	Target gene(s)	Regulation	Key references
Zinc oxide NPs	*Medicago sativa* L.	Heavy metal (Cd)	Program cell death related genes	Reduced the Cd toxicity and regulated the genes involved in program cell death Enhanced the activity of antioxidant enzymes	Chen et al. [21]
Carbon NPs	*Oryza sativa* L.	Drought	*OsSNRK1*, *OsSNRK2*, aquaporin genes, *OsPHYB*	Increased relative water content, lower stomatal conductance, downregulation of phytochrome B gene Downregulation of *OsSNRKs*	Kumar et al. [89]
Iron oxide NPs	*Brassica napus*	Heavy metal (Cd)	Oxidative stress-related genes, chlorophyll biosynthesis genes, and Cd transporters	Differential regulation of genes associated with superoxide dismutase, peroxidase, catalase, glutathione reductase	He et al. [46]
Selenium NP (lower dose; < 10 mg/L)	*Momordica charantia*	In vitro environment	*WRKY1*; phenylalanine ammonia-lyase (*PAL*); 4-coumarate CoA-ligase (*4CL*)	Upregulation of all the genes studied Increased activities of catalase and peroxidase enzymes DNA cytosine methylation (epigenetic modifications)	Rajaee Behbahani et al. [119]
CuO and ZnO NPs	*Triticum aestivum* L.	Drought; metal stress	*LEA*, *DREB*, *Dehydrin*	Upregulation of genes associated with drought tolerance Consistent expression of genes related to metal stress	Yang et al. [161]
Chitosan NPs	*Catharanthus roseus*	Drought	Strictosidine synthase (*STR*); deacetylvindoline-4-O-acetyltransferase (*DAT*); peroxidase 1 (*PRX1*) and geissoschizine synthase (*GS*)	More than fivefold-change increase in the expression genes associated with secondary metabolite production NPs effectively mitigated the negative effects caused by drought stress	Ali et al. [12]
Fe, Cu, Co, and Zn oxide NPs	*Glycine max* L.	Drought	*GmDREB32*; *GmNAC11*; *GmMYB118*; *GmWRKY27*; *GmERD1*	Expression of all the genes were found to be upregulated in shoots and roots of all the NP-treated plants as compared to control under drought Iron NPs were found to be most effective among all the NPs tested *GmWRKY* and *GmMYB* expression were highest among all the genes	Linh et al. [97]
Cu- NPs	*Solanum lycopersicum* L.	Salt	Genes of SOD, CAT, GPX, PR1, and JA biosynthesis	Irrespective of stress condition, the CuNPs along with chitosan showed increased expression of SOD, CAT, and GPX genes	Hernández-Hernández et al. [47]
Cerium oxide NPs	*Oryza sativa* L.	Salt	Nitrate reductase	Enhanced nitric oxide production resulted in salt stress tolerance *nia2* gene expression was induced in treated plants as compared to control in stressed condition	Zhou et al. [171]
TiO_2 NPs	*Cicer arietinum* L.	Cold stress	*CaLRubisco*, *CaSRubisco* and *Cachlorophyll a/b*-binding protein genes	Increased expression of genes involved in photosynthesis process Significantly decreased levels of H_2O_2	Hasanpour et al. [45]

(continued)

Table 2 (continued)

Nanoparticle	Plant species	Stress	Target gene(s)	Regulation	Key references
ZnO NPs	*Oryza sativa* L.	Chilling stress	Genes involved in antioxidative defense system and transcription factors	Upregulation of *OsCu/ZnSOD1*, *OsCu/ZnSOD2*, *OsCu/ZnSOD3*, *OsPRX11*, *OsPRX65*, *OsPRX89*, *OsCATA*, and *OsCATB* and chilling responsive transcription factors *OsbZIP52*, *OsMYB4*, *OsMYB30*, *OsNAC5*, *OsWRKY76*, and *OsWRKY94* Reduced levels of H_2O_2, MDA and Proline	Song et al. [148]
Iron oxide NPs and hydrogel NPs	*Oryza sativa* L.	Heavy metal stress	Cadmium transporter genes (*OsHMA2*; *OsHMA3*; *OsLCT1*)	Downregulation of genes involved in Cd transport from root to the storage organs in treated plants as compared to control Increased biomass, increased content of antioxidant enzymes and increased efficiency of photosynthesis in cadmium-stressed plants treated with iron oxide NPs and hydrogel NPs	Ahmed et al. [7]

generation during stress conditions. Nanoparticle-mediated modulation of signaling pathways is done either through activation and deactivation of genes involved in phytohormone biosynthesis, transport, and degradation or through interaction of NPs with existing plant growth regulators to limit or delimit their bioavailability [153]. NPs effectively modulate the expression of *AUX1* (AUXIN TRANSPORTER CARRIER1/AUXIN TRANS-PORTER-LIKE PROTEINS), auxin receptor-related genes (*AFB, TIR*), auxin carrier genes (*PIN1, PIN2, PIN3*), and auxin biosynthesis gene *YUC8* for auxin metabolism; *ARR7* and *ARR15* for cytokinins; REPRESSOR OF GIBBERELLIC ACID 1 (*RGA1*) for gibberellins; *NCED* (ABA synthesis) and *RD22* (ABA-mediated dehydration responsive protein) for ABA; and *ACC, ACO, ACS, ERF1*, and *ETR2* involved in ethylene synthesis and response [14, 88, 151, 157, 164]. It is also noteworthy that the modulation of signaling pathways by NPs greatly depends on the size, type, plant species, and experimental conditions [153].

7 Advantages and Limitations of Using NPs in Agriculture

7.1 Advantages of NPs in Agriculture

NPs increase nutrient absorption efficiency by facilitating slow and targeted nutrient release, minimizing losses through leaching and volatilization. ZnO NPs enhance zinc uptake and growth in maize [149]. Nano-pesticides provide targeted pest control, reducing chemical use. They can penetrate pest cuticles to deliver active ingredients effectively. ZnO NPs reduce the need for several chemical products by acting as an antibacterial and fertilizer [78]. SiO_2 NPs are effective against pests like aphids and mites [118]. They can improve plant tolerance to abiotic stress, such as drought and salinity, through modulation of stress-responsive pathways [121]. NPs serve as carriers for DNA, RNA, or proteins, enabling more efficient and less invasive crop genetic engineering. In chemical processes, NPs are extremely effective catalysts that use less material while producing faster reaction rates than bulk catalysts. Controlled release made possible by NPs guarantees sustained action and lessens the need for repeated treatments [81, 80]. Agrochemicals are encapsulated by nanocarriers and released gradually, increasing the effectiveness of pest management and lowering environmental pollution [120].

7.2 Ecotoxicological Issues, Risk Evaluations, Prolonged Effects, and Drawbacks of NPs Compared to Traditional Techniques

NPs are well known for their special qualities and benefits above traditional techniques in a variety of industries, including environmental remediation, agriculture, and medicine. However, there are a number of obstacles to their use, such as ecotoxicological issues, use dangers, long-term effects, and other restrictions [77]. A thorough comprehension of these elements is necessary to guarantee the long-term application of NPs.

Ecotoxicological Issues: Although NPs have many advantages, there are some ecological and toxicological hazards due to their tiny size and strong reactivity. NPs have the ability to interact with aquatic life, soil microorganisms, and other non-target species, upsetting the natural equilibrium [84]. Through soil, water, or air, NPs can make their way into food chains. Long-term exposure may cause organisms to bioaccumulate, which might be harmful at higher trophic levels [114, 125]. According to Jośko et al. [73], metal oxide NPs such as ZnO and TiO_2 can change the microbial diversity, nutrient cycling, and soil structure.

Lack of Standardization and Economic: Inconsistent synthesis and characterization of NPs lead to unpredictable efficacy and safety. High production costs can limit access for small-scale farmers. NPs can change the toxicity of other chemicals in the environment or produce secondary pollutants through their interactions. Although these interactions are crucial, little is known about them. It takes sophisticated instruments to detect and measure NPs in intricate environmental matrices like soil and water, which makes evaluations resource-intensive.

Long-Term Effects: It is yet unclear how NPs may affect the environment and living things in the long run. Concerns over cumulative environmental effects are raised by the great stability of certain NPs, particularly those with coatings, which can remain in ecosystems for a decade [110]. Excessive usage of antimicrobials based on nanoparticles, such as copper and AgNPs, may result in the emergence of resistant bacteria strains, reducing the effectiveness of these treatments. Long-term exposure to NPs, whether directly (by ingestion or inhalation) or indirectly (via food chains), may have unidentified health effects. According to research, NPs have the ability to penetrate biological barriers, such as the blood–brain barrier, which may result in oxidative stress or inflammation [111].

NP Restrictions in Contrast to Traditional Methods: Notwithstanding their benefits, NPs have drawbacks that should be taken into account. The cost of producing NPs with exact properties may prevent them from being widely used, particularly in small-scale enterprises and agriculture. The laws governing NPs are still being developed. Their safe and appropriate use is hampered by the absence of thorough instructions. Depending on environmental factors including pH, salinity, and organic matter content, NPs may exhibit unpredictable behavior that produces erratic results [98]. The environmental advantages of some nanoparticle production techniques may be outweighed by their high energy requirements. The application of nanotechnology, particularly in food and health, may be impacted by ethical and societal considerations.

7.3 Obstacles to Expanding Nanoparticle-Based Strategies for Crop Plant Development

Nanoparticles (NPs) hold significant promise for advancing crop improvement through genetic editing, enhanced stress tolerance, and targeted pesticide delivery; however, multiple challenges continue to limit their broader application in agriculture. Nonetheless, despite the considerable promise of nanotechnology, numerous challenges must be addressed before its advancements can be effectively translated into large-scale agricultural applications. A thorough examination of the difficulties in expanding crop enhancement techniques based on NPs may be found below. Expensive and energy-intensive advanced methods are needed to synthesize NPs with specific properties (such as size, surface charge, and functionality). For example, large-scale manufacturing is less practical because of the high costs of techniques like chemical vapor deposition and laser ablation [120]. Furthermore, compared to seed investments, investors may make less money with NPs. The production of NPs in the lab is sometimes quite regulated, requiring particular circumstances in order to get the desired results. Variability in raw materials, equipment, and climatic conditions might make it difficult to replicate these operations on an industrial scale [98].

When NPs are used extensively, they may accumulate in soil and water habitats, which might have an impact on creatures that are not intended targets, such as pollinators, aquatic animals, and beneficial soil bacteria. Some NPs, such as metal oxides (like ZnO and TiO_2), are persistent in the environment, which raises questions regarding potential long-term ecological effects [73]. It's crucial to making sure that NPs are stable in harsh field environments including high humidity, temperature swings, and soil pH fluctuations. In these circumstances, a lot of NPs agglomerate or break down, which lessens their effectiveness. It's still difficult to create effective NP delivery methods that guarantee crop plants properly absorb them. In order to minimize losses from leaching or runoff, NPs must pass through plant barriers such as the cell wall and cuticle. Certain NPs are harmful to plant cells, preventing them from growing, germinating, or performing metabolic processes. For instance, it has been demonstrated that high levels of silver NPs have a detrimental effect on wheat (*Triticum aestivum*) root elongation and seed germination [141]. It might be challenging to generalize results and create standardized applications since various crop species—and even variants within the same species—respond to NPs in different ways.

Moreover, regulatory bodies frequently don't have clear rules for approving agricultural goods based on NPs. Large-scale commercialization investments are discouraged by this regulatory uncertainty. The majority of research on the use of NPs to enhance crops is carried out in greenhouses or controlled laboratories. There are a few field tests, which are crucial for comprehending the viability and difficulties in the actual world [87]. The rational design of NPs is hampered by an incomplete understanding of the molecular and cellular processes via which they interact with plants. Hence, it is very essential to considering all the above-mentioned parameters before using NPs in agricultural practice.

8 Recent Advances and Future Perspectives

Abiotic stress including drought, salt, heavy metals, and extreme temperatures are serious threats to agriculture. Present advances of NPs include their significance in alleviating stress, as nanofertilizers, nano-pesticides, and nanocarriers have been recognized. Metal and metal oxide NPs such as ZnO NPs have shown significant potential against drought and salinity stress, and underlying mechanisms are increasing antioxidant enzyme activity, reducing oxidative damage, and improving photosynthetic activity [38]. FeO NPs have been reported for seed priming, resulting in increased germination rates and vigor index of seedlings under stressful conditions [100]. Carbon nanotubes (CNTs) and graphene oxide are reported for improving water and nutrient intake during drought and salt stress conditions, respectively, which resulted in increased root growth and photosynthesis [15, 101]. NPs are also recently used as nanofertilizers such as phosphorus-based nanofertilizers have been reported for good absorption efficiency due to their controlled delivery of nutrients and hence stress reduction. Copper-based nano-pesticides have shown their pest control activity, which allows less usage of chemical-based pesticides. However, careful research is to be conducted for its formulation to balance efficacy and cost [30]. Very recently, the development of smartly designed nanocarriers for accurate pesticide delivery has revolutionized farming. The specific feature of these smart nanocarriers is that they can deliver their payload when triggered by certain environmental signals, such as pH, temperature, or moisture levels. This phenomenon can guarantee precise action and reduce waste [49]. Furthermore, NPs increase plant stress tolerance through a variety of ways, including activating the antioxidant defense systems to decrease reactive oxygen species (ROS), improving photosynthetic efficiency and chlorophyll content, regulating genes and proteins that respond to stress, and modulating plant hormone levels for stress adaptation [31].

Future Perspectives: Future research must focus on green synthesis methods that use plant extracts or microorganisms in order to generate promising biodegradable and ecologically friendly NPs [154]. Further research is needed to understand the molecular mechanisms underlying NP-plant interactions. This knowledge can be used to build NPs that are specific to crops and stress circumstances [81, 80]. Commercialization of nanoparticle-based agricultural products necessitates strong regulatory frameworks to assure their safety for humans and the environment [77, 91]. Moreover, integrating nanotechnology with precision agriculture equipment like drones and sensors allows for real-time monitoring and management of crop health and stress situations [117]. Overall, the successful integration of nanotechnology into mainstream agriculture will necessitate addressing safety, scalability, and regulatory issues. With sustained study and innovation, NPs have the potential to play a critical role in sustainable agriculture and global food security.

9 Conclusion

NPs are found to be a promising facet in sustainable agricultural practice, by enhancing plant growth, productivity, and resilience under abiotic stress conditions. The application of NPs as nanofertilizers, nano-pesticides, and nano-sensors has demonstrated significant improvements in nutrient delivery, pest management, and environmental monitoring. NPs fortify plants against unfavorable environmental conditions by enhancing antioxidant defense mechanisms, improved nutrient uptake, and efficient detoxification processes. NPs help plants mitigate stress caused by salinity, heavy metals, droughts, and temperature extremes. Along with enhanced crop productivity, challenges impose various difficulties including financial, technical, and safety-related concerns and finally gaps in regulatory frameworks and research. With further research for filling up the current gaps in using NPs in agricultural practices along with safety concerns, NPs can play a pivotal role in improving agricultural yield and resilience in the face of global climatic and environmental challenges.

References

1. Q Abbas B Yousaf H Ullah MU Ali M Zia-ur-Rehman M Rizwan J Rinklebe 2020 Biochar-induced immobilization and transformation of silver-nanoparticles affect growth, intracellular-radicles generation and nutrients assimilation by reducing oxidative stress in maize J Hazard Mater 390 121976
2. Abdalla H, Adarosy MH, Hegazy HS, Abdelhameed RE (2022) Potential of green synthesized titanium dioxide nanoparticles for enhancing seedling emergence, vigor and tolerance indices and DPPH free radical scavenging in two varieties of soybean under salinity stress. BMC Plant Biol 22(1):560. https://doi.org/10.1186/s12870-022-03945-7
3. Abideen Z, Hanif M, Munir N, Nielsen BL (2022) Impact of nanomaterials on the regulation of gene expression and metabolomics of plants under salt stress. Plants 11(5):691. https://doi.org/10.3390/plants11050691
4. Acharya P, Jayaprakasha GK, Crosby KM, Jifon JL, Patil BS (2020) Nanoparticle-mediated seed priming improves germination, growth, yield, and quality of watermelons (*Citrullus lanatus*) at multi-locations in Texas. Sci Rep 10(1). https://doi.org/10.1038/s41598-020-61696-7
5. M Adrees ZS Khan S Ali M Hafeez S Khalid MZ Rehman A Hussain K Hussain SA Chatha M Rizwan 2020 Simultaneous mitigation of cadmium and drought stress in wheat by soil application of iron nanoparticles Chemosphere 238 124681
6. Ahmad P, Sarwat M, Sharma S (2008) Reactive oxygen species, antioxidants and signaling in plants. J Plant Biol 51(3)
7. Ahmed T, Noman M, Manzoor N, Shahid M, Abdullah M, Ali L, Wang G, Hashem A, Al-Arjani ABF, Alqarawi AA, Abd_Allah EF (2021) Nanoparticle-based amelioration of drought stress and cadmium toxicity in rice via triggering the stress responsive genetic mechanisms and nutrient acquisition. Ecotoxicol Environ Saf 209:111829
8. Akçay M, Geyik MS, Yazicilar B, Boke F, Nadaroglu H, Atıcı O, Bezirganoğlu, I (2025) Assessment of the effects of newly fabricated CaO, CuO, ZnO nanoparticles on callus formation maintenance of Alfalfa (*Medicago sativa* L.) under in vitro salt stress. Agronomy 15(1):180

9. Al-Khayri JM, Rashmi R, Surya Ulhas R, Sudheer WN, Banadka A, Nagella P, Aldaej MI, Rezk AAS, Shehata WF, Almaghasla MI (2023) The role of nanoparticles in response of plants to abiotic stress at physiological, biochemical, and molecular levels. Plants 12(2):292. https://doi.org/10.3390/plants12020292
10. WA Al-Selwey AA Alsadon AA Ibrahim JP Labis MF Seleiman 2023 Effects of zinc oxide and silicon dioxide nanoparticles on physiological, yield, and water use efficiency traits of potato grown under water deficit Plants 12 1 218
11. NM Alabdallah MM Hasan I Hammami AI Alghamdi D Alshehri HA Alatawi 2021 Green synthesized metal oxide nanoparticles mediate growth regulation and physiology of crop plants under drought stress Plants 10 1730 https://doi.org/10.3390/plants10081730
12. EF Ali AM El-Shehawi OHM Ibrahim EY Abdul-Hafeez MM Moussa FAS Hassan 2021 A vital role of chitosan nanoparticles in improvisation of drought stress tolerance in *Catharanthus roseus* (L.) through biochemical and gene expression modulation Plant Physiol Biochem 161 166 175
13. M Ashraf MR Foolad 2007 Roles of glycine betaine and proline in improving plant abiotic stress resistance Environ Exp Bot 59 2 206 216 https://doi.org/10.1016/j.envexpbot.2005.12.006
14. Azhar BJ, Noor A, Zulfiqar A, Zeenat A, Ahmad S, Chishti I, Abbas Z, Shakeel SN (2021) Effect of ZnO, SiO_2 and composite nanoparticles on *Arabidopsis thaliana* and involvement of ethylene and cytokinin signaling pathways. Pak J Bot 53(2):437–446. https://doi.org/10.30848/PJB2021-2(40)
15. Bakry BA, Sadak MS, Al Ashkar NM, Ibrahim OM, Okla MK, El-Tahan AM (2024) The role of carbon nanotubes in improving drought tolerance via upregulation of the physiological processes of peanut plants grown in sandy soils. Agronomy 14(3):611. https://doi.org/10.3390/agronomy14030611
16. F Behboudi Z Tahmasebi Sarvestani MZ Kassaee SAM Modares Sanavi A Sorooshzadeh 2018 Improving growth and yield of wheat under drought stress via application of SiO_2 nanoparticles J Agric Sci Technol 20 7 1479 1492
17. D Bhatt G Sharma 2018 Role of silicon in counteracting abiotic and biotic plant stresses Int J Chem Stud 6 2 1434 1442
18. Brunetti C, Fini A, Sebastiani F, Gori A, Tattini M (2018) Modulation of phytohormone signaling: a primary function of flavonoids in plant–environment interactions. Front Plant Sci 9. https://doi.org/10.3389/fpls.2018.01042
19. Byrt CS, Zhang RY, Magrath I, Chan KX, De Rosa A, McGaughey S (2023) Exploring aquaporin functions during changes in leaf water potential. Front Plant Sci 14. https://doi.org/10.3389/fpls.2023.1213454
20. Chandrashekar HK, Singh G, Kaniyassery A, Thorat SA, Nayak R, Murali TS, Muthusamy A (2023) Nanoparticle-mediated amelioration of drought stress in plants: a systematic review. 3 Biotech 13(10). https://doi.org/10.1007/s13205-023-03751-4
21. Chen Z, Feng Y, Guo Z, Han M, Yan X (2024) Zinc oxide nanoparticles alleviate cadmium toxicity and promote tolerance by modulating programmed cell death in alfalfa (*Medicago sativa* L.). J Hazard Mater 469:133917
22. A Choudhary S Sharma H Kaur N Sharma MM Gadewar S Mehta A Husen 2024 Plant system, abiotic stress resilience, reactive oxygen species, and coordination of engineered nanomaterials: a review S Afr J Bot 171 45 59 https://doi.org/10.1016/j.sajb.2024.05.053
23. Dai Y, Guo Y, Tang W, Chen D, Xue L, Chen Y, Guo Y, Wei S, Wu M, Dai J, Wang S (2024) Reactive oxygen species-scavenging nanomaterials for the prevention and treatment of age-related diseases. J Nanobiotechnol 22(1). https://doi.org/10.1186/s12951-024-02501-9
24. M Darbahani MR Ghiyasi M Rahaie 2024 Nanoparticles as new elicitors for the production of bioactive and phytochemicals in vitro and in vivo plant culture Phytochem Rev https://doi.org/10.1007/s11101-024-10022-4
25. Dayem AA, Hossain MK, Lee SB, Kim K, Saha SK, Yang GM, Choi HY, Cho SG (2017) The role of reactive oxygen species (ROS) in the biological activities of metallic nanoparticles. Int J Mol Sci 18(1):120. https://doi.org/10.3390/ijms18010120

26. F Dilnawaz MH Kalaji AN Misra 2023 Nanotechnology in improving photosynthesis under adverse climatic conditions: cell to canopy action Plant Nano Biol 4 100035https://doi.org/10.1016/j.plana.2023.100035
27. Dilnawaz F, Misra AN, Apostolova E (2023) Involvement of nanoparticles in mitigating plant's abiotic stress. Plant Stress 10. https://doi.org/10.1016/j.stress.2023.100280
28. Djanaguiraman M, Anbazhagan V, Dhankher OP, Prasad PVV (2024) Uptake, translocation, toxicity, and impact of nanoparticles on plant physiological processes. Plants 13(22):3137. https://doi.org/10.3390/plants13223137
29. Donia DT, Carbone M (2023) Seed priming with zinc oxide nanoparticles to enhance crop tolerance to environmental stresses. Int J Mol Sci 24(24):17612. https://doi.org/10.3390/ijms242417612
30. El-Saadony MT, Abd El-Hack ME, Taha AE, Fouda MMG, Ajarem JS, Maodaa SN, Allam AA, Elshaer N (2020) Ecofriendly synthesis and insecticidal application of copper nanoparticles against the storage pest *Tribolium castaneum*. Nanomaterials 10(3):587. https://doi.org/10.3390/nano10030587
31. El-Saadony MT, Saad AM, Soliman SM, Salem HM, Desoky ESM, Babalghith AO, El-Tahan AM, Ibrahim OM, Ebrahim AAM, Abd El-Mageed TA, Elrys AS, Elbadawi AA, El-Tarabily KA, AbuQamar SF (2022) Role of nanoparticles in enhancing crop tolerance to abiotic stress: a comprehensive review. Front Plant Sci 13. https://doi.org/10.3389/fpls.2022.946717
32. M El-Zohri NA Al-Wadaani SO Bafeel 2021 Foliar sprayed green zinc oxide nanoparticles mitigate drought-induced oxidative stress in tomato Plants 10 11 2400
33. H Etesami H Fatemi M Rizwan 2021 Interactions of nanoparticles and salinity stress at physiological, biochemical and molecular levels in plants: a review Ecotoxicol Environ Saf 225 112769https://doi.org/10.1016/j.ecoenv.2021.112769
34. S Faiz NA Yasin WU Khan AA Shah W Akram A Ahmad A Ali NH Naveed L Riaz 2022 Role of magnesium oxide nanoparticles in the mitigation of lead-induced stress in *Daucus carota*: modulation in polyamines and antioxidant enzymes Int J Phytoremediation 24 4 364 372 https://doi.org/10.1080/15226514.2021.1949263
35. M Faizan A Faraz M Yusuf ST Khan S Hayat 2018 Zinc oxide nanoparticle-mediated changes in photosynthetic efficiency and antioxidant system of tomato plants Photosynthetica 56 2 678 686 https://doi.org/10.1007/s11099-017-0717-0
36. Fouad A, Hegazy AE, Azab E, Khojah E, Kapiel T (2021) Boosting of antioxidants and alkaloids in *Catharanthus roseus* suspension cultures using silver nanoparticles with expression of *CrMPK3* and *STR* genes. Plants 10(10). https://doi.org/10.3390/plants10102202
37. García-Locascio E, Valenzuela EI, Cervantes-Avilés P (2024) Impact of seed priming with selenium nanoparticles on germination and seedlings growth of tomato. Sci Rep 14(1). https://doi.org/10.1038/s41598-024-57049-3
38. Ghani MI, Saleem S, Rather SA, Rehmani MS, Alamri S, Rajput VD, Kalaji HM, Saleem N, Sial TA, Liu M (2022) Foliar application of zinc oxide nanoparticles: an effective strategy to mitigate drought stress in cucumber seedling by modulating antioxidant defense system and osmolytes accumulation. Chemosphere 289. https://doi.org/10.1016/j.chemosphere.2021.133202
39. M Ghorbanpour 2015 Major essential oil constituents, total phenolics and flavonoids content and antioxidant activity of *Salvia officinalis* plant in response to nano-titanium dioxide Indian J Plant Physiol 20 249 256 https://doi.org/10.1007/s40502-015-0170-7
40. Y González-García C Cárdenas-Álvarez G Cadenas-Pliego A Benavides-Mendoza M Cabrera-De-La-Fuente A Sandoval-Rangel A Juárez-Maldonado 2021 Effect of three nanoparticles (Se, Si and Cu) on the bioactive compounds of bell pepper fruits under saline stress Plants 10 2 217
41. MB Hafeez A Ghaffar N Zahra N Ahmad WH Al-Qahtani P Sharma A Husen J Li 2024 Exploring the efficacy of plant growth promoters in nutrient acquisition and antioxidant defense responses of late sown wheat under saline conditions S Afr J Bot 171 812 824 https://doi.org/10.1016/j.sajb.2024.06.042

42. Haghmadad Milani M, Mohammadi A, Panahirad S, Farhadi H, Labib P, Kulak M, Gohari G, Fotopoulos V, Vita F (2024) Cerium oxide nanoparticles (CeO_2 NPs) enhance salt tolerance in spearmint (*Mentha spicata* L.) by boosting the antioxidant system and increasing essential oil composition. Plants 13:20. https://doi.org/10.3390/plants13202934
43. Haliloğlu K, Türko A, Balpınar Ö, Nadaro H, Alaylı A, Poczai P (2022) Effects of zinc, copper and iron oxide nanoparticles on induced DNA methylation, genomic instability and LTR retrotransposon polymorphism in wheat (*Triticum aestivum* L.). Plants. https://doi.org/10.3390/plants
44. Hasanaklou NT, Mohagheghi V, Hasanaklou HT, Ma'mani L, Malekmohammadi M, Moradi F, Dalvand Y (2023) Seed nano-priming using silica nanoparticles: effects in seed germination and physiological properties of *Stevia rebaudiana* Bertoni. Chem Biol Technol Agric 10:1. https://doi.org/10.1186/s40538-023-00445-0
45. H Hasanpour R Maali-Amir H Zeinali 2015 Effect of TiO_2 nanoparticles on metabolic limitations to photosynthesis under cold in chickpea Russ J Plant Physiol 62 779 787
46. He D, Kaleem Z, Ali S, Shahbaz H, Zhang K, Li J, Sheteiwy MS, Ulhassan Z, Zhou W (2025) Impact of iron oxide nanoparticles on cadmium toxicity mitigation in *Brassica napus*. Plant Physiol Biochem 109500
47. H Hernández-Hernández A Juárez-Maldonado A Benavides-Mendoza H Ortega-Ortiz G Cadenas-Pliego D Sánchez-Aspeytia S González-Morales 2018 Chitosan-PVA and copper nanoparticles improve growth and overexpress the SOD and JA genes in tomato plants under salt stress Agronomy 8 175
48. MA Hossain P Piyatida JAT Silva da M Fujita 2012 Molecular mechanism of heavy metal toxicity and tolerance in plants: central role of glutathione in detoxification of reactive oxygen species and methylglyoxal and in heavy metal chelation J Bot 2012 1 37 https://doi.org/10.1155/2012/872875
49. Q Hou H Zhang L Bao Z Song C Liu Z Jiang Y Zheng 2021 Ncs-delivered pesticides: a promising candidate in smart agriculture Int J Mol Sci 22 23 https://doi.org/10.3390/ijms222313043
50. Hudeček M, Nožková V, Plíhalová L, Plíhal O (2023) Plant hormone cytokinin at the crossroads of stress priming and control of photosynthesis. Front Plant Sci 13. https://doi.org/10.3389/fpls.2022.1103088
51. A Husen 2010 Growth characteristics, physiological and metabolic responses of teak (*Tectona grandis* Linn. f.) clones differing in rejuvenation capacity subjected to drought stress Silvae Genetica 59 124 136
52. Husen A (2020) Interactions of metal and metal-oxide nanomaterials with agricultural crops: an overview. In: Husen, A, Jawaid M (eds) Nanomaterials for agriculture and forestry applications. Elsevier Inc., pp 167–197
53. Husen A (2020) Carbon-based nanomaterials and their interactions with agricultural crops. In: Husen, A, Jawaid M (eds) Nanomaterials for agriculture and forestry applications. Elsevier Inc., pp 199–218
54. Husen A (2021) The harsh environment and resilient plants: an overview. In: Husen A (ed) Harsh environment and plant resilience. Springer International Publishing AG, Cham, pp 1–23 https://doi.org/10.1007/978-3-030-65912-7_1
55. Husen A (2021) Morpho-anatomical, physiological, biochemical and molecular responses of plants to air pollution. In: Husen A (ed) Harsh environment and plant resilience. Springer International Publishing AG, Cham, pp 203–234. https://doi.org/10.1007/978-3-030-65912-7_9
56. Husen A (2021) Cross talk between autophagy and hormones for abiotic stress tolerance in plants. In: Husen A (eds) Plant performance under environmental stress. Springer International Publishing AG, Cham, pp 1–15. https://doi.org/10.1007/978-3-030-78521-5_1
57. Husen A (2022) Engineered nanomaterials for sustainable agricultural production, soil improvement and stress management. Elsevier Inc. https://doi.org/10.1016/C2021-0-00054-7
58. Husen A (2022) Environmental pollution and medicinal plants. CRC Press. https://doi.org/10.1201/9781003178866

59. Husen A (2023) Plants and their interaction to environmental pollution (Damage detection, adaptation, tolerance, physiological and molecular responses). Elsevier Inc. https://doi.org/10.1016/C2020-0-03319-0
60. Husen A (2024) Current understanding of biostimulants in plant protection and performance. Elsevier Inc.
61. Husen A (2024) Essential minerals in plant-soil system (Coordination, Signalling, and Interaction under Adverse Situations). Elsevier Inc.
62. Husen A (2024) Plant response to gold nanoparticles. Springer Nature Singapore, Singapore. https://doi.org/10.1007/978-981-97-7174-5
63. Husen A (2025) Plant response to silver nanoparticles. Springer Nature Singapore, Singapore. https://doi.org/10.1007/978-981-97-7352-7
64. Husen A (2025) Emerging carbon nanomaterials for sustainable agricultural practices (Synthesis, Plant Growth, Performance, Production and Protection). Springer Nature Singapore, Singapore
65. Husen A, Siddiqi KS (2023) Advances in smart nanomaterials and their applications. Elsevier Inc. https://doi.org/10.1016/C2021-0-02202-1
66. B Hussain Q Lin Y Hamid M Sanaullah L Di MB Khan X Yang 2020 Foliage application of selenium and silicon nanoparticles alleviates Cd and Pb toxicity in rice (*Oryza sativa* L) Sci Total Environ 712 136497
67. EA Ibrahim 2016 Seed priming to alleviate salinity stress in germinating seeds J Plant Physiol 192 38 46 https://doi.org/10.1016/j.jplph.2015.12.011
68. SU Jalil M Zahera MS Khan MI Ansari 2019 Biochemical synthesis of gold nanoparticles from *Nicotiana tabacum* L cv xanthi leaf protein and their physiological, developmental, and ROS scavenging effects on tobacco plants under stress IET Nanobiotechnol 13 1 23 29
69. Jalil SU, Ansari MI (2019) Nanoparticles and abiotic stress tolerance in plants: synthesis, action, and signaling mechanisms. Plant Signal Mol 549–561. https://doi.org/10.1016/B978-0-12-816451-8.00034-4
70. Jia Z, Tang M, Wu J (1999) The determination of flavonoid contents. Food Chem 555–559
71. Jin-Chul K, Madhusudhan A, Husen A (2021) Smart nanomaterials in biomedical applications. Springer Nature Switzerland AG, Cham, Switzerland. https://doi.org/10.1007/978-3-030-84262-8
72. Jonapá-Hernández F, Gutiérrez-Miceli F, Santos-Espinosa A, Ruíz-Lau N, Ruíz-Valdiviezo V, Valdez-Salas B, González-Mendoza D (2020) Foliar application of green nanoparticles in *Annona muricata* L. plants and their effects in physiological and biochemical parameters. Biocatal Agric Biotechnol 28:101751. https://doi.org/10.1016/j.bcab.2020.101751
73. I Jośko M Kusiak P Oleszczuk 2021 The chronic effects of CuO and ZnO nanoparticles on *Eisenia fetida* in relation to the bioavailability in aged soils Chemosphere 266 128982https://doi.org/10.1016/j.chemosphere.2020.128982
74. M Kafi J Nabati B Saadatian A Oskoueian J Shabahang 2019 Potato response to silicone compounds (micro and nanoparticles) and potassium as affected by salinity stress Ital J Agron 14 3 162 169
75. Y Kang H Qin G Wang B Lei X Yang M Zhong 2024 Selenium nanoparticles mitigate cadmium stress in tomato through enhanced accumulation and transport of sulfate/selenite and polyamines J Agric Food Chem 72 1473 1486 https://doi.org/10.1021/acs.jafc.3c07504
76. AG Kaningini AM Nelwamondo S Azizi M Maaza KC Mohale 2022 Metal nanoparticles in agriculture: a review of possible use Coatings 12 10 https://doi.org/10.3390/coatings12101586
77. Kapoor D, Sharma MMM, Yadav S, Sharma P, Husen A (2025) Environmental safety and ethical issues related to use of silver nanoparticles in plants and soil system. In: Husen A (ed) Plant response to silver nanoparticles. Smart nanomaterials technology. Springer, Singapore. https://doi.org/10.1007/978-981-97-7352-7_18
78. PL Kashyap X Xiang P Heiden 2015 Chitosan nanoparticle-based delivery systems for sustainable agriculture Int J Biol Macromol 77 36 51 https://doi.org/10.1016/j.ijbiomac.2015.02.039

79. MF Khalid R Iqbal Khan MZ Jawaid W Shafqat S Hussain T Ahmed M Rizwan S Ercisli OL Pop R Alina Marc 2022 Nanoparticles: the plant saviour under abiotic stresses Nanomaterials 12 21 https://doi.org/10.3390/nano12213915
80. MR Khan V Adam TF Rizvi B Zhang F Ahamad I Jośko Y Zhu M Yang C Mao 2019 Nanoparticle–plant interactions: two-way traffic Small 15 37 https://doi.org/10.1002/smll.201901794
81. I Khan K Saeed I Khan 2019 Nanoparticles: properties, applications and toxicities Arab J Chem 12 908 931 https://doi.org/10.1016/j.arabjc.2017.05.011
82. S Khan M Zahoor R Sher Khan M Ikram NU Islam 2023 The impact of silver nanoparticles on the growth of plants: the agriculture applications Heliyon 9 6 https://doi.org/10.1016/j.heliyon.2023.e16928
83. Khan I, Awan SA, Rizwan M, Huizhi W, Ulhassan Z, Xie W (2024) Silicon nanoparticles improved the osmolyte production, antioxidant defense system, and phytohormone regulation in *Elymus sibiricus* (L.) under drought and salt stress. Environ Sci Pollut Res 31(6):8985–8999
84. SJ Klaine PJJ Alvarez GE Batley TF Fernandes RD Handy DY Lyon S Mahendra MJ McLaughlin JR Lead 2008 Nanomaterials in the environment: behavior, fate, bioavailability, and effects Environ Toxicol Chem 27 1825 1851 https://doi.org/10.1897/08-090.1
85. A Kumar A Choudhary H Kaur S Mehta A Husen 2021 Metal-based nanoparticles, sensors and their multifaceted application in food packaging J Nanobiotechnol 19 256 1 25 https://doi.org/10.1186/s12951-021-00996-0
86. A Kumar A Choudhary H Kaur S Mehta A Husen 2021 Smart nanomaterial and nanocomposite with advanced agrochemical activities Nanoscale Res Lett 16 156 1 26 https://doi.org/10.1186/s11671-021-03612-0
87. S Kumar M Nehra N Dilbaghi G Marrazza AA Hassan K-H Kim 2019 Nano-based smart pesticide formulations: emerging opportunities for agriculture J Control Release 294 131 153 https://doi.org/10.1016/j.jconrel.2018.12.012
88. A Kumar A Singh M Panigrahy PK Sahoo KC Panigrahi 2018 Carbon nanoparticles influence photomorphogenesis and flowering time in *Arabidopsis thaliana* Plant Cell Rep 37 901 912
89. Kumar A, Manasa LS, Kashyap J, Mishra G, Mandal S, Rath J, Rout GR, Panigrahi KC, Panigrahy M (2024) Carbon nanoparticle exposure strengthens water-relation parameters by stimulating abscisic acid pathway and aquaporins genes in rice. Plant Stress 13:100540
90. VV Kumari P Banerjee VC Verma S Sukumaran MAS Chandran KA Gopinath G Venkatesh SK Yadav VK Singh NK Awasthi 2022 Plant nutrition: an effective way to alleviate abiotic stress in agricultural crops Int J Mol Sci 23 15 https://doi.org/10.3390/ijms23158519
91. Kumari R, Suman K, Karmakar S, Mishra V, Lakra SG, Saurav GK, Mahto BK (2023) Regulation and safety measures for nanotechnology-based agri-products. Front Genome Ed 5. https://doi.org/10.3389/fgeed.2023.1200987
92. S Kusvuran B Cengil F Mutlu 2024 Effect of nano-silicon application on salt tolerance of pepper (*Capsicum annuum* L) Proc Bulg Acad Sci 77 3 467 474
93. Kuznetsov VV, Shevyakova NI (2007) Polyamines and stress tolerance of plants. Plant Stress 50–71
94. P Landa R Vankova J Andrlova J Hodek P Marsik H Storchova JC White T Vanek 2012 Nanoparticle-specific changes in *Arabidopsis thaliana* gene expression after exposure to ZnO, TiO_2, and fullerene soot J Hazard Mater 241 55 62
95. BR Lee KY Kim WJ Jung JC Avice A Ourry TH Kim 2007 Peroxidases and lignification in relation to the intensity of water-deficit stress in white clover (*Trifolium repens* L.) J Exp Bot 58 1271 1279 https://doi.org/10.1093/jxb/erl280
96. Y Liang H Liu Y Fu P Li S Li Y Gao 2023 Regulatory effects of silicon nanoparticles on the growth and photosynthesis of cotton seedlings under salt and low-temperature dual stress BMC Plant Biol 23 1 https://doi.org/10.1186/s12870-023-04509-z
97. TM Linh NC Mai PT Hoe LQ Lien NK Ban LT Hien NH Chau NT Van 2020 Metal-based nanoparticles enhance drought tolerance in soybean J Nanomater 2020 4056563
98. GV Lowry KB Gregory SC Apte JR Lead 2012 Transformations of nanomaterials in the environment Environ Sci Technol 46 6893 6899 https://doi.org/10.1021/es300839e

99. Madhusudhan A, Purohit SD, Prasad R, Husen A (2024) Functional smart nanomaterials and their theranostics approaches. Springer Nature Singapore, Singapore
100. Madusanka HKS, Chathurika Sri JAS, Aruggoda AGB, Chathurika JAS, Weerakoon SR (2024) Effect of seed priming of green synthesized iron oxide magnetic nanoparticles using *Salvinia molesta* plant extract on seed germination and seedlings growth of tomato (*Solanum lycopersicum*). Int J Adv Res Innov 9(6):2669–2686. https://www.researchgate.net/publication/378292438
101. Mahmoud NE, Abdelhameed RM (2021) Superiority of modified graphene oxide for enhancing the growth, yield, and antioxidant potential of pearl millet (*Pennisetum glaucum* L.) under salt stress. Plant Stress 2. https://doi.org/10.1016/j.stress.2021.100025
102. Mandal AK, Ghorai S, Husen A (2023) Functionalized smart nanomaterials for point-of-care testing. Springer Nature Singapore, Singapore
103. Martínez-Cisterna D, Rubilar O, Tortella G, Chen L, Chacón-Fuentes M, Lizama M, Parra P, Bardehle L (2024) Silver nanoparticles as a potent nanopesticide: toxic effects and action mechanisms on pest insects of agricultural importance—a review. Molecules 29(23):5520. https://doi.org/10.3390/molecules29235520
104. Matsuura HN, Fett-Neto AG (2015) Plant alkaloids: main features, toxicity, and mechanisms of action. In: Plant toxins. Springer Netherlands, pp 1–15. https://doi.org/10.1007/978-94-007-6728-7_2-1
105. SE McNeil 2005 Nanotechnology for the biologist J Leukoc Biol 78 3 585 594 https://doi.org/10.1189/jlb.0205074
106. N Munir W Gulzar Z Abideen M Hasanuzzaman A El-Keblawy F Zhao 2023 Plant-nanoparticle interactions: transcriptomic and proteomic insights Agronomy 13 8 2112
107. Mustafa N, Raja NI, Ilyas N, Abasi F, AhmadMS, Ehsan M, Próchów J (2022) Exogenous application of green titanium dioxide nanoparticles (TiO_2 NPs) to improve the germination, physiochemical, and yield parameters of wheat plants under salinity stress. Molecules 27(15):4884
108. Naseem J, Shah AA, Usman S, Ahmed S, Gatasheh MK, Shaffique S, Javad S (2025) Green synthesized FeNPs ameliorate drought stress in *Spinacia oleracea* L. through improved photosynthetic capacity, redox balance, and antioxidant defense. Sci Rep 15(1):1782
109. Nguyen QN, Sidorova A, Torres R (2022) User interactions with chatbot interfaces vs. Menu-based interfaces: an empirical study. Comput Hum Behav 128. https://doi.org/10.1016/j.chb.2021.107093
110. B Nowack JF Ranville S Diamond JA Gallego-Urrea C Metcalfe J Rose N Horne AA Koelmans SJ Klaine 2012 Potential scenarios for nanomaterial release and subsequent alteration in the environment Environ Toxicol Chem 31 1 50 59 https://doi.org/10.1002/etc.726
111. G Oberdörster E Oberdörster J Oberdörster 2005 Nanotoxicology: an emerging discipline evolving from studies of ultrafine particles Environ Health Perspect 113 7 823 839 https://doi.org/10.1289/ehp.7339
112. Ogunyale OG, Fawibe OO, Ajiboye AA, Agboola DA (2014) A review of plant growth substances: their forms, structures, synthesis and functions. J Adv Lab Res Biol 5(4):152–168. https://e-journal.sospublication.co.in
113. HC Oliveira AB Seabra S Kondak OP Adedokun Z Kolbert 2023 Multilevel approach to plant–nanomaterial relationships: from cells to living ecosystems J Exp Bot 74 12 3406 3424 https://doi.org/10.1093/jxb/erad107
114. C Parisi M Vigani E Rodríguez-Cerezo 2015 Agricultural nanotechnologies: what are the current possibilities? Nano Today 10 2 124 127 https://doi.org/10.1016/j.nantod.2014.09.009
115. Pearce RS (2001) Plant freezing and damage. Ann Bot 87(4):417–424. https://doi.org/10.1006/anbo.2000.1352
116. Pogribna M, Hammons G (2021) Epigenetic effects of nanomaterials and nanoparticles. J Nanobiotechnol 19(1). https://doi.org/10.1186/s12951-020-00740-0
117. A Pratama S Rahayu 2024 Biosensors enabled by nanotechnology: advancing precision agriculture through IoT monitoring J Int Connect Emerg Technol 9 1 25 39

118. M Rai AP Ingle J Trzcińska-Wencel M Wypij S Bonde A Yadav G Kratošová P Golińska 2021 Biogenic silver nanoparticles: what we know and what do we need to know? Nanomaterials 11 11 2901 https://doi.org/10.3390/nano11112901
119. S Rajaee Behbahani A Iranbakhsh M Ebadi A Majd ZO Ardebili 2020 Red elemental selenium nanoparticles mediated substantial variations in growth, tissue differentiation, metabolism, gene transcription, epigenetic cytosine DNA methylation, and callogenesis in bittermelon (*Momordica charantia*); an in vitro experiment PLoS ONE 15 7 e0235556
120. Rajput VD, Singh A, Minkina TM, Verma KK, Singh AK (2023) Nanotechnology for sustainable agriculture. Apple Academic Press. https://doi.org/10.1201/9781003333128
121. R Raliya V Saharan C Dimkpa P Biswas 2018 Nanofertilizer for precision and sustainable agriculture: current state and future perspectives J Agric Food Chem 66 26 6487 6503 https://doi.org/10.1021/acs.jafc.7b02178
122. ME Ramadan MM El-Saber AE Adelhamid AA El-Sayed 2022 Effect of nano-chitosan encapsulated spermine on growth, productivity and bioactive compounds of chili pepper (*Capsicum annuum* L) under salinity stress Egypt J Chem 65 8 197 207
123. A Rasheed H Li MM Tahir A Mahmood M Nawaz AN Shah MT Aslam S Negm M Moustafa MU Hassan Z Wu 2022 The role of nanoparticles in plant biochemical, physiological, and molecular responses under drought stress: a review Front Plant Sci 13 976179https://doi.org/10.3389/fpls.2022.976179
124. MAS Raza B Zulfiqar R Iqbal MN Muzamil MU Aslam F Muhammad J Amin HM Aslam MA Ibrahim M Uzair M Habib-ur- Rahman 2023 Morpho-physiological and biochemical response of wheat to various treatments of silicon nano-particles under drought stress conditions Sci Rep 13 1 2700
125. Rienzie R, Adassooriya NM (2018) Toxicity of nanomaterials in agriculture and food. In: Nanomaterials: ecotoxicity, safety, and public perception. Springer International Publishing, pp 207–234. https://doi.org/10.1007/978-3-030-05144-0_11
126. Y Roiter M Ornatska AR Rammohan J Balakrishnan DR Heine S Minko 2008 Interaction of nanoparticles with lipid membrane Nano Lett 8 3 941 944 https://doi.org/10.1021/nl0800801
127. Sachdev S, Ansari SA, Ansari MI, Fujita M, Hasanuzzaman M (2021) Abiotic stress and reactive oxygen species: generation, signaling, and defense mechanisms. Antioxidants 10(2):1–37. https://doi.org/10.3390/antiox10020277
128. Salam U, Ullah S, Tang ZH, Elateeq AA, Khan Y, Khan J, Khan A, Ali S (2023) Plant metabolomics: an overview of the role of primary and secondary metabolites against different environmental stress factors. Life 13(3):706. https://doi.org/10.3390/life13030706
129. PB Santhosh S Penič J Genova A Iglič V Kralj-Iglič NP Ulrih 2012 A study on the interaction of nanoparticles with lipid membranes and their influence on membrane fluidity J Phys Conf Ser 398 1 012034https://doi.org/10.1088/1742-6596/398/1/012034
130. Sayed EG, Desoukey SF, Desouky AF, Farag MF, EL-kholy RI, Azoz SN (2024) Synergistic influence of arbuscular mycorrhizal fungi inoculation with nanoparticle foliar application enhances chili (*Capsicum annuum* L) antioxidant enzymes, anatomical characteristics, and productivity under cold-stress conditions. Plants 13(4):517
131. WM Semida A Abdelkhalik GF Mohamed ATA El-Mageed ASA El-Mageed MM Rady EF Ali 2021 Foliar application of zinc oxide nanoparticles promotes drought stress tolerance in eggplant (*Solanum melongena* L) Plants 10 2 421
132. Shang Y, Kamrul Hasan M, Ahammed GJ, Li M, Yin H, Zhou J (2019) Applications of nanotechnology in plant growth and crop protection: a review. Molecules 24(14):2558. https://doi.org/10.3390/molecules24142558
133. P Sharma V Pandey MMM Sharma A Patra B Singh S Mehta A Husen 2021 A review on biosensors and nanosensors application in agroecosystems Nanoscale Res Lett 16 136 1 24 https://doi.org/10.1186/s11671-021-03593-0
134. Sheng H, Chen S (2020) Plant silicon-cell wall complexes: Identification, model of covalent bond formation and biofunction. Plant Physiol Biochem 155:13–19. https://doi.org/10.1016/j.plaphy.2020.07.020

135. Shukla RK, Badiye A, Vajpayee K, Kapoor N (2021) Genotoxic potential of nanoparticles: structural and functional modifications in DNA. Front Genet 12:728250. https://doi.org/10.3389/fgene.2021.728250
136. KS Siddiqi A Husen 2016 Engineered gold nanoparticles and plant adaptation potential Nanoscale Res Lett 11 400 1 10
137. KS Siddiqi A Husen 2022 Plant response to silver nanoparticles: a critical review Crit Rev Biotechnol 42 7 773 990 https://doi.org/10.1080/07388551.2021.1975091
138. Sidhu AK, Sharma M, Bhickchand Agrawal S, Bhavsar P, Samota MK (2024) Nanomaterial strategies for enhancing plant resilience in the face of temperature stress. CABI Agric Biosci 5(1). https://doi.org/10.1186/s43170-024-00255-w
139. A Singh RS Sengar VD Rajput T Minkina RK Singh 2022 Zinc oxide nanoparticles improve salt tolerance in rice seedlings by improving physiological and biochemical indices Agriculture 12 7 1014
140. A Singh RS Sengar UP Shahi VD Rajput T Minkina KA Ghazaryan 2022 Prominent effects of zinc oxide nanoparticles on roots of rice (*Oryza sativa* L) grown under salinity stress Stresses 3 1 33 46
141. D Singh A Sharma SK Verma H Pandey M Pandey 2024 Impact of nanoparticles on plant physiology, nutrition, and toxicity: a short review Next Nanotechnol 6 100081https://doi.org/10.1016/j.nxnano.2024.100081
142. Singh M, Kumar J, Singh S, Singh VP, Prasad SM (2015) Roles of osmoprotectants in improving salinity and drought tolerance in plants: a review. Rev Environ Sci Biotechnol 14(3):407–426. https://doi.org/10.1007/s11157-015-9372-8
143. Singh S, Husen A (2019) Role of nanomaterials in the mitigation of abiotic stress in plants. Nanomater Plant Potent 441–471. https://doi.org/10.1007/978-3-030-05569-1_18
144. Singh S, Husen A (2019) Role of nanomaterials in the mitigation of abiotic stress in plants. In: Husen A, Iqbal M (eds) N*anomaterials* and plant potential. Springer International Publishing AG, Cham pp. 441–471. https://doi.org/10.1007/978-3-030-05569-1_18
145. Singh S, Husen A (2020) Behavior of agricultural crops in relation to nanomaterials under adverse environmental conditions. In: Husen, A, Jawaid M (eds) Nanomaterials for agriculture and forestry applications. Elsevier Inc., pp 219–256. https://doi.org/10.1016/B978-0-12-817852-2.00009-3
146. A Smertenko P Dráber V Viklický Z Opatrný 1997 Heat stress affects the organization of microtubules and cell division in *Nicotiana tabacum* cells Plant Cell Environ 20 12 1534 1542 https://doi.org/10.1046/j.1365-3040.1997.d01-44.x
147. Song WJ, Jeong MS, Choi DM, Kim KN, Wie MB (2019) Zinc oxide nanoparticles induce autophagy and apoptosis via oxidative injury and pro-inflammatory cytokines in primary astrocyte cultures. Nanomaterials 9(7):1043. https://doi.org/10.3390/nano9071043
148. Song Y, Jiang M, Zhang H, Li R (2021) Zinc oxide nanoparticles alleviate chilling stress in rice (*Oryza sativa* L.) by regulating antioxidative system and chilling response transcription factors. Molecules 26(8):2196
149. Subramanian KS, Manikandan A, Thirunavukkarasu M, Rahale CS (2015) Nano-fertilizers for balanced crop nutrition. Nanotechnol Food Agric 69–80. https://doi.org/10.1007/978-3-319-14024-7_3
150. Sutulien˙e R, Brazaityt˙e A,Małek S, Jasik M, Samuolien˙e G (2023) Response of oxidative stress and antioxidant system in pea plants exposed to drought and boron nanoparticles. Antioxidants 12(2):528
151. YY Syu JH Hung JC Chen HW Chuang 2014 Impacts of size and shape of silver nanoparticles on *Arabidopsis* plant growth and gene expression Plant Physiol Biochem 83 57 64
152. Szőllősi R, Molnár Á, Kondak S, Kolbert Z (2020) Dual effect of nanomaterials on germination and seedling growth: Stimulation vs. phytotoxicity. Plants 9(12):1–30. https://doi.org/10.3390/plants9121745
153. Tripathi D, Singh M, Pandey-Rai S (2022) Crosstalk of nanoparticles and phytohormones regulate plant growth and metabolism under abiotic and biotic stress. Plant Stress 6. https://doi.org/10.1016/j.stress.2022.100107

154. RS Varma 2012 Greener approach to nanomaterials and their sustainable applications Curr Opin Chem Eng 1 2 123 128 https://doi.org/10.1016/j.coche.2011.12.002
155. Vatankhah A, Aliniaeifard S, Moosavi-Nezhad M, Abdi S, Mokhtarpour Z, Reezi S, Tsaniklidis G, Fanourakis D (2023) Plants exposed to titanium dioxide nanoparticles acquired contrasting photosynthetic and morphological strategies depending on the growing light intensity: a case study in radish. Sci Rep 13(1). https://doi.org/10.1038/s41598-023-32466-y
156. Z Wang L Xu J Zhao X Wang JC White B Xing 2016 CuO nanoparticle interaction with *Arabidopsis thaliana*: toxicity, parent-progeny transfer, and gene expression Environ Sci Technol 50 11 6008 6016 https://doi.org/10.1021/acs.est.6b01017
157. Wei JL, Zou Y, Li P, Yuan XJ (2020) Titanium dioxide nanoparticles promote root growth by interfering with auxin pathways in *Arabidopsis thaliana*
158. H Wu N Tito JP Giraldo 2017 Anionic cerium oxide nanoparticles protect plant photosynthesis from abiotic stress by scavenging reactive oxygen species ACS Nano 11 11 11283 11297 https://doi.org/10.1021/acsnano.7b05723
159. Yadav S, Modi P, Dave A, Vijapura A, Patel D, Patel M (2020) Effect of abiotic stress on crops. In: Sustainable crop production. IntechOpen. https://doi.org/10.5772/intechopen.88434
160. Yan G, Fan X, Peng M, Yin C, Xiao Z, Liang Y (2020) Silicon improves rice salinity resistance by alleviating ionic toxicity and osmotic constraint in an organ-specific pattern. Front Plant Sci 11. https://doi.org/10.3389/fpls.2020.00260
161. KY Yang S Doxey JE McLean D Britt A Watson D Al Qassy A Jacobson AJ Anderson 2018 Remodeling of root morphology by CuO and ZnO nanoparticles: effects on drought tolerance for plants colonized by a beneficial pseudomonad Botany 96 3 175 186
162. L Yin S Wang K Tanaka S Fujihara A Itai X Den S Zhang 2016 Silicon-mediated changes in polyamines participate in silicon-induced salt tolerance in *Sorghum bicolor* L Plant Cell Environ 39 2 245 258 https://doi.org/10.1111/pce.12521
163. S Zafar Z Hasnain S Danish ML Battaglia S Fahad MJ Ansari SA Alharbi 2024 Modulations of wheat growth by selenium nanoparticles under salinity stress BMC Plant Biol 24 1 35
164. SM Zahedi M Abdelrahman MS Hosseini NF Hoveizeh LSP Tran 2019 Alleviation of the effect of salinity on growth and yield of strawberry by foliar spray of selenium-nanoparticles Environ Pollut 253 246 258
165. SM Zahedi F Moharrami S Sarikhani M Padervand 2020 Selenium and silica nanostructure-based recovery of strawberry plants subjected to drought stress Sci Rep 10 1 17672
166. Zahra N, Kausar A, Abdelghani HTM, Singh S, Vashishth DS, Bachheti A, Bachheti RK, Husen A (2024b) Serotonin improves plant growth, foliar functions and antioxidant defence system in Ethiopian mustard (*Brassica carinata* A. Br.) S Afr J Bot 170:1–9. https://doi.org/10.1016/j.sajb.2024.05.002
167. Zahra N, Kausar A, Sharma P, Wabaidur SM, Husen A (2024a) Serotonin application to Guizotia abyssinica improves growth, physiological functions, antioxidant system and alleviates the deleterious effects induced by salt stress. Plant Biosyst—Int J Deal All Asp Plant Biol 1-9. https://doi.org/10.1080/11263504.2024.2384972
168. Zhang C, Wang X, Du J, Gu Z, Zhao Y (2021) Reactive oxygen species-regulating strategies based on nanomaterials for disease treatment. Adv Sci 8(3). https://doi.org/10.1002/advs.202002797
169. Zhang L, Liu Z, Song Y, Sui J, Hua X (2024) Advances in the involvement of metals and metalloids in plant defense response to external stress. Plants 13(2):313. https://doi.org/10.3390/plants13020313
170. Zhao H, Lou Y, Sun H, Li L, Wang L, Dong L, Gao Z (2016) Transcriptome and comparative gene expression analysis of *Phyllostachys edulis* in response to high light. BMC Plant Biol 16(1). https://doi.org/10.1186/s12870-016-0720-9
171. H Zhou H Wu F Zhang Y Su W Guan Y Xie JP Giraldo W Shen 2021 Molecular basis of cerium oxide nanoparticle enhancement of rice salt tolerance and yield Environ Sci Nano 8 11 3294 3311

Role of Metal and Metal-Oxide Nanoparticles in Agricultural Crop Salinity Stress Adaptation and Mitigation

Ankita Ghosh, Pritom Biswas, and Megha Barot

Abstract Currently, more than 20% of global agricultural soils are impacted by salinity, with the extent of saline areas continuously expanding due to both natural and anthropogenic activities. To mitigate the adverse effects of salinity stress on crops, innovative approaches, including nanotechnology, have been developed. Among these, nanoparticles (NPs) offer a promising approach to lessen the harmful impacts of salinity stress on crops, presenting opportunities for sustainable agriculture. Due to their unique physicochemical characteristics and nanoscale dimensions, NPs have demonstrated considerable potential in reducing the negative consequences of salinity stress. This chapter delves into the function of metal and metal-oxide nanoparticles in enhancing salinity stress tolerance and adaptation in agricultural crops. It also examines the mechanisms through which metal and metal-oxide NPs contribute to improved salt tolerance. Additionally, the chapter provides a critical assessment of the harmful impacts of NPs on plants and discusses the challenges and concerns associated with their extensive use. A thorough grasp of the advantages and limitations of nanoparticle applications can foster the development of sustainable agricultural practices.

Keywords Salinity stress · Agricultural crops · Metal-oxide nanoparticles · Sustainable agriculture · Stress mitigation

---

A. Ghosh (✉)
Department of Environmental Sciences, Govind Ballabh Pant University of Agriculture and Technology, Pantnagar, Uttarakhand, India
e-mail: ankitaghoshdav@gmail.com

P. Biswas
Department of Biotechnology, Central University of South Bihar, Gaya, Bihar, India

M. Barot
Research and Development Cell, Parul University, Vadodara, Gujarat, India

Department of Environmental Science, Parul Institute of Applied Sciences, Parul University, Vadodara, Gujarat, India

A. Husen (ed.), *Nanobiotechnology for Abiotic Stress Adaptation and Mitigation in Agricultural Crops*, Smart Nanomaterials Technology,
https://doi.org/10.1007/978-981-96-9298-9_2

1 Introduction

The Green Revolution in the mid-twentieth century significantly boosted global food production, enabling us to feed a population now nearing 8 billion [15]. However, to meet the demands of a further expanding population by 2050, food production must increase substantially, likely by 60–80% [24]. Climate change has exacerbated several challenges, including land degradation, water shortages, and extreme heat, all of which hinder efficient crop production [31, 43–52, 70]. Among these challenges, soil salinization poses a particularly serious threat to crop yields [112]. Soil salinization is a persistent issue, and restoring saline soils to their original state is a time-consuming process. Salinity significantly impacts the growth of most crops (excluding halophytes), leading to reduced yields and compromised quality. Most crops are glycophytes, meaning they can tolerate only low levels of salt [102]. Plants subjected to salinity stress endure numerous adverse effects [37, 44, 43, 53, 56, 118, 119]. Initially, osmotic stress arises due to decreased water availability, caused by elevated salt concentrations, leading to stomatal closure to reduce water loss [115]. However, this action limits carbon dioxide absorption, impairing photosynthesis, and subsequently restraining plant growth [91]. Following osmotic stress, ion toxicity occurs, resulting from the increase in sodium (Na^+) and chloride (Cl^-), further hampering growth [75]. These primary stresses, both osmotic and ionic, disrupt essential biochemical and physiological processes [22, 69]. Consequently, secondary stresses such as oxidative stress, nutrient imbalances, and hormonal disruptions emerge, amplifying the effects of salinity on plant growth and development [69]. Salinity stress also affects mineral nutrition in plants, leading to imbalances. The excessive uptake of sodium (Na^+) interferes with cellular processes like reactive oxygen species (ROS) production, breakdown of proteins, disruptions in the production and function of plant hormones, and alterations in the structure and function of cellular membranes [3]. Additionally, salinity stress negatively impacts nitrogen (N) metabolism by inhibiting the activity of key enzymes [21]. Morphologically, salinity stress reduces leaf area, decreases photosynthetic pigment levels, impairs gas exchange, and lowers the quantum efficiency of photosystem II [12, 120]. Salinity stress impairs photosynthesis by causing the denaturation of critical membrane proteins, leading to chlorophyll degradation and disruptions in chloroplast structure, ultimately hindering photosynthetic efficiency [22, 38, 92]. Increased chlorophyllase activity disrupts chloroplast structure, resulting in chlorophyll breakdown and subsequently reduced photosynthesis and plant growth. Moreover, salt stress influences protein expression and gene regulation in plant organs, especially roots [57, 86]. It also impairs water absorption in roots by affecting aquaporins, membrane proteins essential for water transport [101]. Figure 1 illustrates the impact of salt stress on plants.

To survive in saline conditions, crops have developed intricate mechanisms, such as osmotic adjustment, ion balance maintenance, and the regulation of reactive oxygen species (ROS) [53, 102, 109]. These mechanisms directly address the primary stressors associated with salinity which are osmotic stress, sodium toxicity,

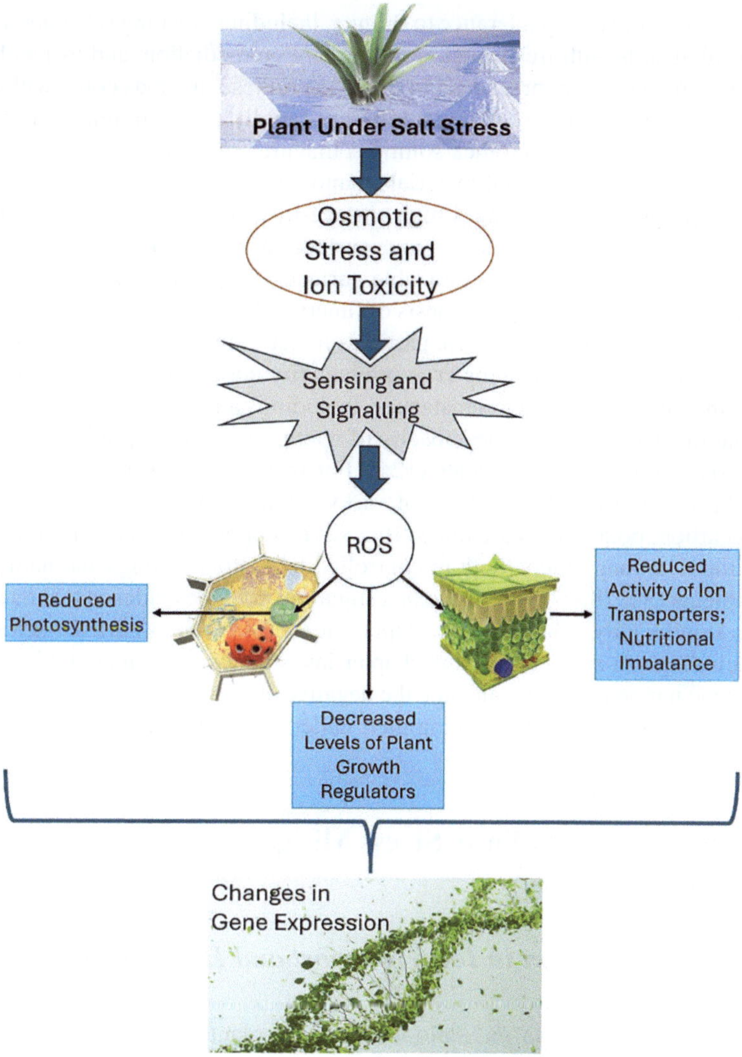

Fig. 1 Effects of salt stress on plants

and excessive ROS production [123]. Plants mitigate osmotic stress by synthesizing compatible solutes, a process that consumes significant energy [88]. Alternatively, they can store excess sodium in vacuoles to maintain osmotic balance [10]. To maintain ion homeostasis, plants prioritize a low sodium-to-potassium ratio within the cell, achieved through mechanisms such as sodium exclusion, vacuolar compartmentalization, and potassium retention. To manage ROS levels, plants employ both enzymatic and non-enzymatic antioxidant defense systems [103]. A range of strategies has been employed to improve crop tolerance to salinity, including breeding salt-tolerant cultivars, flushing soils with freshwater, utilizing phytoremediation, and using effective agricultural management practices [34]. However, these methods come with certain limitations. Nanotechnology in agriculture has rapidly grown in importance due to its potential for efficient and quick solutions that are crucial for sustainable farming [40, 124]. Unlike conventional materials, nanomaterials exhibit distinctive physicochemical properties, including a high surface-to-volume ratio, surface reactivity, stability in specific environments, unique surface structures, potential fluorescence, electrical conductivity, and customizable surfaces [58]. These inherent properties enable the tailored design and synthesis of functional nanomaterials to enhance crop yields [108]. The utilization of nanoparticles in agricultural settings is considered a promising strategy for enhancing crop productivity under both ideal conditions and challenging environmental circumstances, including salinity stress [116]. Extensive investigations have revealed the capacity of nanoparticles to improve salinity tolerance across a diverse plant species [124]. For instance, a considerable number of studies have investigated the effects of various nanoparticles comprising elements such as carbon, potassium, calcium, sulfur, silver, copper, iron, zinc, boron, silicon, and titanium, among others, both in controlled laboratory settings and natural field environments [32, 78, 97, 98]. These investigations provide evidence that nanoparticles can significantly enhance plant adaptability across various environmental stressors, including salt stress [80]. This chapter investigates the potential of metal and metal-oxide nanoparticles to alleviate the negative effects of salinity stress on plants, with a particular focus on elucidating their underlying mechanisms of action.

2 Mechanisms of Salinity Stress Mitigation by Nanoparticles

2.1 Gene Regulation in Ion Transport and Homeostasis

Maintaining a balanced Na^+/K^+ ratio is crucial for plant salt tolerance under saline conditions. To enhance this tolerance, plants employ various mechanisms to minimize Na^+ levels, maximize K^+ levels, and ultimately achieve a balanced Na^+/K^+ ratio, which is essential for normal cellular functions [14]. Nanomaterials have demonstrated potential in improving salt tolerance by helping plants regulate this balance under salt stress conditions. For example, cerium oxide nanoparticles (CeO_2 NPs) with an average size of 6.06 ± 0.48 nm, a zeta potential of -15.31 ± 0.12 mV, and applied at a concentration of 0.8 mM as a foliar spray have been observed to alter the gene expression associated with sodium and potassium transport in cotton plants subjected to salt stress (198 mM NaCl) [63]. Treatment with nanoceria in cotton plants resulted in increased expression of the *HKT1* gene, a key transporter for high-affinity potassium (K^+) uptake and sodium (Na^+) exclusion, leading to reduced leaf Na^+ accumulation. Conversely, the *KOR* gene, responsible for K^+ efflux, exhibited decreased expression [62]. Another study indicated that nanoceria priming improved K^+ retention and decreased Na^+ uptake, thereby enhancing the Na^+/K^+ balance in rapeseed during salt stress [82]. Liu et al. [67] stated that CeO_2 nanoparticles facilitated the development of root apical barriers in rapeseed plants exposed to salt stress. This resulted in reduced salt accumulation within the roots by limiting sodium (Na^+) retention and promoting its transport to the shoots. It is critical to manage Na^+ transported to shoots by storing it in vacuoles or leaf sheaths or recycling it back to roots and soil to prevent harm from excessive Na^+ accumulation. Overall, cerium oxide nanoparticles are effective in boosting crop salt tolerance. Despite being classified as a heavy metal, cerium has been shown to exert positive effects on plant growth at low concentrations [62]. In addition to cerium oxide nanoparticles, studies have shown that other nanomaterials can also enhance crop salt tolerance. For instance, foliar application of selenium nanoparticles (Se NPs) (10–44 nm, 0.4–1.54 mM) in kidney beans led to increased expression of the *ZmNHX1* gene, a Na^+/H^+ exchanger, facilitating sodium (Na^+) compartmentalization and improving salt tolerance [89]. Similarly, Cs–Se nanoparticles (22 mg/L) were found to elevate potassium (K^+) levels while reducing sodium (Na^+) levels in bitter melon under saline conditions [79]. Seed priming with silver nanoparticles (Ag NPs) at concentrations ranging from 0.04 to 2.4 mg/L enhanced the expression of the *SOS2* gene in salt-stressed tomato plants [77]. Likewise, seed priming with nano-silicon (nano-Si) (0.6–3.1 mg/L) resulted in increased expression of the *NHX6* gene in soybean seedlings, leading to enhanced K^+ retention and an improved K^+/Na^+ ratio in saline condition [8]. Notably, the *AKT, HAK, HKT*, and *SOS* gene families play crucial roles in maintaining Na^+/K^+ homeostasis under saline conditions. Wu [111] demonstrated that polyacrylic acid-coated nanoceria (PNC) significantly increased *HAK5* gene expression in *Arabidopsis thaliana*, enhancing K^+ retention in mesophyll cells. Additionally, CRISPR-Cas9 techniques showed that CsAKT1 is vital for PNC-improved salt tolerance in cucumbers, leading to increased K^+ content in leaves and roots exposed to salinity stress [27].

2.2 Reactive Oxygen Species (ROS) and Oxidative Stress Mitigation

Salinity stress disrupts the relationship between ROS production and removal, leading to ROS accumulation and subsequent oxidative stress. This stress has a deleterious effect on essential cellular components such as carbohydrates, DNA, lipids, and proteins, causing irreversible harm or even death of the plant [11]. Maintaining ROS homeostasis is crucial for plant salt tolerance. Catalase (CAT) is crucial in mitigating oxidative stress by catalyzing the rapid conversion of hydrogen peroxide into water and oxygen. This enzymatic activity contributes to increased photosynthetic capacity, evidenced by higher chlorophyll fluorescence, and elevated levels of photosynthetic pigments, while simultaneously preventing chloroplast aggregation, thereby alleviating the detrimental impact of salinity [42]. Ascorbate peroxidase (APX) is another critical enzyme involved in ROS detoxification within higher plants subjected to salt stress. For example, radish plants exposed to salinity exhibited enhanced APX activity in both foliar and root tissues [8]. Moreover, salt-tolerant rice cultivars demonstrate a more efficient ROS scavenging system compared to salt-sensitive cultivars [35]. The activity of superoxide dismutase (SOD), which plays a crucial function in scavenging ROS, is also linked to improved salt tolerance in species like tomato and chickpea [30]. For example, overexpression of Mn-SOD in *Arabidopsis* enhanced its resistance to salinity [110]. Kaur et al. [59] underscore the crucial role of antioxidant enzymes in conferring salt tolerance. Plants adapted to saline conditions often exhibit higher antioxidant enzyme activities, reinforcing the importance of ROS homeostasis for salt stress resilience. Additionally, treatments with antioxidants such as sodium nitroprusside, a nitric oxide donor, and exogenous hormones like melatonin have been shown to increase salt tolerance in plants [122]. Therefore, strategies that strengthen the plant's ability to eliminate excess ROS may hold remarkable prospects for enhancing salt resistance.

Nanozymes, which are nanomaterials exhibiting enzyme-like activities, such as those observed in catalase (CAT), peroxidase (POD), and superoxide dismutase (SOD) have shown promise in enhancing ROS homeostasis. Research indicates that various nanoparticles, such as nano-chitosan, silicon (Si) nanoparticles, selenium (Se) nanoparticles, manganese oxide (Mn_2O_3) nanoparticles, nano-silica (nano-SiO_2), and nanoceria, exhibit significant activities like antioxidant enzymes like ascorbate peroxidase (APX), POD, SOD, and CAT [87]. The utilization of nanozymes can therefore be instrumental in maintaining ROS homeostasis and enhancing plant salt stress tolerance. Research has demonstrated that nanoceria effectively improves salt tolerance across diverse plant species, such as *Arabidopsis*, cotton, rice, and rapeseed [82]. In addition to nanozymes, various nanoparticles, such as aluminum oxide (Al_2O_3), copper oxide (CuO), magnetite (Fe_3O_4), silicon dioxide (SiO_2), titanium oxide (TiO_2), and zinc oxide (ZnO), have been shown to augment plant stress endurance [2]. For instance, TiO_2 nanoparticles significantly increased antioxidant enzyme activity in *Dracocephalum moldavica* L., thereby promoting enhanced salt resistance [82]. Similarly, SiO_2 nanoparticles increase salt resistance in *Cucurbita pepo* L. by bolstering the plant's capacity to neutralize reactive oxygen species (ROS) [93]. The processes through which nanomaterials influence ROS homeostasis are intricate. Research indicates that nanomaterials may directly scavenge ROS or amplify the activities of the plant's antioxidant systems. For instance, foliar and root applications of polyacrylic acid-coated nanoceria (PNC) in cucumber significantly augmented the performance of CAT, POD, and SOD, while concurrently increasing glutathione (GSH) and ascorbate (ASA) levels, thereby improving salinity resistance [65]. RNA sequencing revealed that PNC treatment in cucumber leaves and roots upregulated the gene expression related to biosynthesis of these antioxidants, including CAT2, CSD3, FSD3, GME, GSH1, GSH2, MSD1, and POD [23]. In rapeseed, PNC treatment specifically enhanced SOD activity by upregulating the Cu–Zn

SOD gene in saline conditions, while the expression of Mn-SOD and Fe-SOD genes remained unaffected [23].

2.3 Phytohormone Regulation for Salinity Tolerance

Phytohormones like abscisic acid (ABA), auxin, cytokinin, ethylene, and gibberellins (GA) are crucial in managing plant responses to salt stress [125]. Salt-resistant maize hybrids have been observed to exhibit elevated leaf ABA levels under salt stress compared to their sensitive counterparts [84]. ABA significantly contributes to plant salt tolerance by upregulating the transcription of genes associated with salt stress responses. This regulatory mechanism subsequently influences ion selectivity in uptake and stimulates the levels of osmoprotectants, such as betaine, proline, polyamines, and soluble sugars [76]. In one study, silver nanoparticles (NPs) were shown to modulate wheat's ability to withstand salinity by reducing ABA levels by approximately 46% under saline conditions [107].

Nanomaterials exert a multifaceted influence on ABA signaling in plants, impacting both ABA levels and the gene expression associated with ABA. For instance, in salt-stressed tomato plants, silver nanoparticles (Ag NPs) were observed to downregulate the expression of DDF2, TAS14, and ZFHD1 genes known to be involved in ABA signaling pathways [8]. Similarly, seed priming with selenium and ZnO NPs in rapeseed resulted in decreased expression of ABA-related genes, including BnCYP707A1, BnCYP707A3, and BnCYP707A4 [85]. While ABA plays a pivotal factor in mediating plant responses to salinity, other phytohormones, such as auxin, brassinosteroids (BRs), cytokinin (CK), ethylene, gibberellins (GA), jasmonic acid (JA), and salicylic acid (SA), also contribute significantly to improving plant salt resistance [25]. For instance, priming rapeseed seeds with polyacrylic acid-coated nanoceria (PNC) elevated SA levels, thereby improving salt tolerance [61].

2.4 Regulating Gas Signals to Combat Salinity

Gas signaling molecules like carbon monoxide (CO), ethylene, hydrogen (H_2), hydrogen sulfide (H_2S), and nitric oxide (NO) are crucial in plant responses to salinity [96]. Among these gaseous signaling molecules, nitric oxide (NO) has been extensively studied for its role in alleviating salinity stress. NO, a crucial signaling molecule in plant physiology, significantly influences various developmental processes, including seed germination and root elongation [122]. NO production is intricately linked to hormonal signaling pathways, interacting with abscisic acid (ABA), auxin, cytokinin (CK), ethylene, and gibberellin (GA), while also regulating jasmonic acid (JA) and salicylic acid (SA) levels [121]. NO plays a pivotal role in adjusting antioxidant defense and safeguarding cellular redox homeostasis under saline condition [114]. For example, polyacrylic acid-coated nanoceria (PNCs) increased rice resistance to salt stress by promoting NO production through increased nitrate reductase (NIA2) activity and subsequent dephosphorylation, leading to improved antioxidant enzyme activity, NADPH oxidase function, and ion homeostasis [66]. Furthermore, encapsulating mercaptosuccinic acid (MSA), an NO precursor, within chitosan nanoparticles facilitated sustained NO release, thereby enhancing salt tolerance in maize [122].

3 Metal and Metal-Oxide Nanoparticles in the Management of Salinity Stress

Salinity stress poses a major threat to global agricultural productivity, prompting researchers to investigate the use of nanoparticles to alleviate its impact [96]. Research has shown that metal and metal-oxide nanoparticles can enhance plant tolerance to salinity stress [114]. Table 1 shows the application of various nanoparticles (NPs) in plants under salt stress conditions. Most utilized nanoparticles to mitigate salinity stress are given below.

3.1 Zinc Oxide Nanoparticles (ZnO NPs)

Zinc, a micronutrient crucial for plant growth and development, plays a vital role in regulating cellular processes, including signal transduction and gene expression [6]. Its importance is particularly evident under stress conditions, such as salinity. Considering the extensive utilization of zinc oxide nanoparticles (ZnO NPs) in diverse commercial sectors, their potential benefits for agriculture have garnered significant attention [99, 122]. ZnO NPs are recognized as a valuable metal oxide with the potential to enhance plant health [18]. Scientific investigations have unequivocally demonstrated that ZnO NPs significantly enhance plant growth and development under both optimal and environmental conditions, including salinity [72]. For instance, Farouk and Al-Amri [28] observed that ZnO NP application in okra (*Abelmoschus esculentus* L.) resulted in increased photosynthetic pigment and upregulated the activity of antioxidant enzymes (CAT and SOD), while concurrently reducing proline and total soluble sugar accumulation. Similarly, Mahmoud et al. [69] stated that Zn NPs mitigated the detrimental impact of osmotic stress in canola (*Brassica napus*) by bolstering antioxidant defense mechanisms, enhancing osmolyte biosynthesis, and optimizing ion regulation. Hezaveh et al. [39] reported that a combination of zeolite, Zn, silicon, and boron nanoparticles significantly enhanced potato growth, yield, and physiological performance in saline environments, likely due to improved water and nutrient uptake, enhanced photosynthesis, and increased antioxidant enzyme activity. While low concentrations of Zn NPs are generally beneficial, higher concentrations

Table 1 Potential roles of nanoparticles (NPs) in enhancing salt tolerance of plants

NPs	Salinity	Plant species	Effects	Key references
La_2O_3 nanoparticles	250 mM	*Medicago sativa*	Alfalfa seed vigor under salt stress was increased by seed priming with La_2O_3 NPs (0–500 mg/L). A priming dosage of 10 mg/L had considerably increased the root length, seedling length, germination potential, and seed germination rate	[29]
ZnO	50 mM NaCl	*Capsicum annum* L.	In a salt-stressed environment, foliar application of 100 ppm of the ZnO NPs promoted chili growth. Significant improvement in all growth parameters as well as reduced oxidative stress was observed	[5]
SeNPs (Selenium nanoparticles)	1% NaCl	Wheat (Ghazi variety)	0.1% SeNP application showed maximum wheat growth compared to 0.05 and 0.01% concentration. Increased wheat growth, chlorophyll content, and regulated antioxidant concentration	[117]
ZnO	200 mM NaCl	*Phaseolus vulgaris* L.	ZnO NPs found modulating the antioxidant defense mechanisms, and reducing oxidative stress resulted in improving plant growth as well as they regulate the ion uptake and their accumulation which influenced nutrient balance during salinity stress conditions	[36]
ZnO-Si	100 mM NaCl	*Pisum sativum* L.	Increased yield of the pea (*Pisum sativum* L.)	[26]

(continued)

Table 1 (continued)

NPs	Salinity	Plant species	Effects	Key references
Zn	Seawater EC 52 dS/m (20%)	*Gossypium barbadense* L.	Dry weight of roots, stems, leaves, bolls, shoot/root ratio, and N, K, Ca, Na content increased, while P and Zn content decreased	[55]
ZnO	150 mM NaCl	*Lycopersicon esculentum* Mill	Increased shoot and root length, biomass, leaf area, chlorophyll content, and photosynthetic parameters, along with higher protein content and enhanced POX, SOD, and CAT activities	[7]
Ag	20 dS/m	*Trigonella foenumgraceum*	Enhanced seed germination and seedling growth parameters	[41]
Ag	150 mM NaCl	*Triticum aestivum* L.	Fresh and dry weight of shoots increased, while MDA, H_2O_2 content, chlorophyll, soluble sugars, and proline levels decreased	[71]
TiO_2	200 mM NaCl	*L. esculentum* Mill	Enhanced activities of SOD, POX, carbonic anhydrase, and nitrate reductase, with increased proline and glycine betaine accumulation. Improved chlorophyll, lycopene, total phenolics, antioxidant capacity, plant growth, and yield, while H_2O_2 content and TBARS decreased	[60]
Fe_2O_3	150 mM NaCl	*Mentha piperita* L.	Increased leaf biomass and P, K, Zn, Fe, and Ca content, while proline, MDA, CAT, SOD, GPOX activity, and DPPH radical scavenging activity decreased	[13]

(continued)

Table 1 (continued)

NPs	Salinity	Plant species	Effects	Key references
Fe	100 mM NaCl	*Fragaria × ananassa*	Increased plant growth, chlorophyll, carotenoids, carbohydrates, membrane stability index, relative water content, K, and Fe, while Na content decreased	[74]
CeO_2	75 mM NaCl	*Cucumis sativus* L.	Increased biomass dry weight, chlorophyll, GSH, and ASA content, while H_2O_2, MDA, and activities of SOD, POD, and CAT decreased	[23]
CeO_2	50 mM NaCl	*Brassica napus* L.	Total biomass, leaf and root dry weight increased, with shortened apoplastic barriers in the root apex. Na content in roots and leaves decreased, while Ce content increased, and the F/Fm ratio rose	[83]

can exhibit toxicity even under optimal conditions [18]. Mahmoud et al. [69] demonstrated that ZnO NP application in rapeseed under salt stress mitigated ion leakage, improved photosynthetic efficiency, and modulated the expression of stress-related genes, including upregulation of ARP and downregulation of SKRD2, MYC, and MPK4.

3.2 *Silver Nanoparticles (Ag NPs)*

Silver nanoparticles (Ag NPs) are extensively utilized across diverse industries owing to their distinctive physicochemical properties [90]. Renowned for their potent antimicrobial activity, Ag NPs have also attracted considerable interest for their potential use in wastewater treatment [63]. The potential of Ag NPs to enhance crop growth and development is increasingly recognized within the agricultural sector. Studies have shown that Ag NPs can positively impact a range of plant growth parameters, such as germination and overall plant development, by modulating several physiological and biochemical traits [90]. Mohamed et al. [71] reported that Ag NP treatment of *Pennisetum glaucum* seeds alleviated oxidative stress under salinity stress conditions, primarily due to heightened antioxidant enzyme activity. Moreover, the treatment led to a decreased leaf Na^+/K^+ ratio and elevated levels of flavonoids and phenolics [100]. Khan et al. [63] showed that pre-sowing treatment of wheat seeds with Ag NPs significantly enhanced seedling growth, proline and soluble sugar content, and POD activity under saline conditions. Furthermore, Ag NPs have been shown to improve wheat germination and grain yield in the presence of salinity by enhancing photosynthetic efficiency and modulating phytohormone levels, including increased concentrations of 1-naphthalene acetic acid, indole-3-butyric acid, and 6-benzylaminopurine, while simultaneously reducing ABA levels [4]. However, research on the effectiveness of Ag NPs in mitigating salinity's negative impact on plants is still limited, highlighting the need for further studies at the physiological, biochemical, and molecular aspects. Additionally, researchers caution against the application of Ag NPs due to the potential release of toxic silver ions (Ag^+) into the environment, which can pose significant risks to various organisms [105, 113].

3.3 Titanium Dioxide Nanoparticles (TiO_2 NPs)

Titanium is the ninth most prevalent element in the Earth's crust, comprising 0.33% of its structure [17]. It is recognized for its beneficial effects on plant performance, including the modulation of enzyme activities, improvement of photosynthesis, and enhancement of chlorophyll content [19]. Furthermore, titanium significantly contributes to increasing plants' resilience to environmental stressors [68]. Gohari et al. [33] explored the consequences of TiO_2 on *Dracocephalum moldavica* under saline condition, showing that a 100 mg/L concentration of TiO_2 facilitated plant growth in saline conditions by boosting the activity of vital enzymes. Additionally, TiO_2 treatment enhanced the oil content, such as geraniol, geranial, geranyl acetate, and z-citral, under salt stress. Similarly, Abdel Latef et al. [2] evaluated the influence of various concentrations of TiO_2 nanoparticles on salt-stressed broad bean plants. It was observed that the lowest concentration (0.01%) exhibited the most significant enhancement of salt tolerance, characterized by increased activity of antioxidant enzymes and elevated levels of amino acids, proline, and soluble sugars.

3.4 Iron Oxide Nanoparticles (Fe_2O_3 NPs)

Iron is an indispensable inorganic component of all living systems, including plants, playing a crucial role in various biological processes [106]. It is fundamental to several essential cellular functions, such as chlorophyll synthesis, photosynthesis, and respiration [64]. Furthermore, iron is essential in synthesizing proteins involved in plant metabolism, cellular respiration, DNA repair, photosynthesis, and oxygen transport, hence affecting overall crop yield [106]. Iron is critical for plant growth and development, influencing various physiological processes. It significantly improves the ability of plants to tolerate salt stress by stimulating the activity of enzymes that protect the plant from oxidative damage [73, 94]. Chan-Rodriguez and Walker [20] investigated the effects of nano-Fe_2O_3 on *Dracocephalum moldavica* under saline conditions and observed that a concentration of 60 mg L^{-1} substantially increased plant growth and enzymatic activity compared to untreated controls. Abdoli et al. [3] investigated how iron oxide nanoparticles (at a concentration of 3 mM) affected the growth and health of a plant species called *Trachyspermum ammi* when it was exposed to saline conditions. Their findings revealed that a combined application of Fe_2O_3 NPs and salicylic acid (SA) significantly mitigated the negative impacts of salinity stress by enhancing the K^+/Na^+ ratio, iron content, and key antioxidant enzymes, including CAT, POD, polyphenol oxidase, and SOD. Moreover, this combined treatment resulted in increased levels of endogenous SA and crucial osmolytes. These enhancements resulted in improved plant vigor, as evidenced by increased root, and shoot growth, enhanced photosynthetic pigment content, improved membrane stability, and, ultimately, increased seed yield. Furthermore, Torabian et al. [104] observed that foliar application of Fe NPs in *Helianthus annuus* cultivated under salt stress significantly stimulated the activity of CAT, polyphenol oxidase, and POD.

3.5 Cerium Oxide Nanoparticles (CeO_2 NPs)

Cerium, a lanthanide series element, is extensively employed in sectors including semiconductors, cosmetics, optics, medicine, drug delivery, and fuel cells [54]. Cerium nanoparticles (Ce NPs) have been found to affect the physio-biochemical properties of plants under optimal growth conditions at low concentrations [16]. In wheat, for example, low concentrations of Ce NPs have been shown to stimulate growth and photosynthetic activity, whereas higher concentrations can have detrimental effects on these processes [95]. The deposition of Ce NPs within various plant tissues has been observed to exhibit a dose–response relationship [1]. Notably, it has been shown that Ce NPs can enhance plant resilience with different environmental challenges [54]. The utilization of Ce NPs to *Brassica napus* under salt stress conditions resulted in significant improvements in photosynthetic efficiency and biomass accumulation [81]. Furthermore, under saline conditions, Ce NP application resulted in anatomical changes that increased Na^+ flow to the shoots and lowered Na^+ levels in root cells, improving the plants' physiological state and salinity tolerance [83]. An et al. [9] explored the morphological, physiological, biochemical, and molecular mechanisms in *Gossypium hirsutum* under salinity stress induced by Ce NPs seed priming. Their findings indicated significant improvements in growth and biomass, alongside differential gene expression in root tissues, following Ce NP seed priming. Additionally, Ce NP-treated plants exhibited reduced ROS production. The researchers concluded that the salinity tolerance induced by Ce NPs seed priming was associated with Ca^{2+} signaling, ion homeostasis, and ROS pathways. Although promising results have been observed, our understanding of the molecular mechanisms underlying Ce NP-mediated enhancement of plant salinity tolerance remains limited [95].

4 Conclusion

Global agriculture grapples with the detrimental impact of soil salinity. Nanobiotechnology advancements have facilitated the synthesis and widespread application of nanoparticles (NPs). NPs like silver (Ag), cerium oxide (CeO_2), iron oxide (Fe_2O_3), titanium dioxide (TiO_2), and zinc oxide (ZnO) have shown promise in enhancing crop growth and productivity under saline conditions when applied at specific concentrations. However, exceeding optimal concentrations can lead to adverse effects on plant health. This dual nature of NPs necessitates careful consideration of their concentration and properties to maximize their beneficial role in mitigating salt stress. Understanding the molecular mechanisms underlying NP-mediated salt tolerance is paramount for successful implementation of this approach in crop production.

5 Future Perspectives

The future of NP-based salt stress mitigation in plants hinges on several critical factors. Firstly, field trials are essential to validating the efficacy of NPs observed in controlled laboratory settings. Secondly, a comprehensive assessment of the environmental impact of widespread NP use in agriculture is crucial to ensuring their sustainability and safety. Moving forward, research should prioritize the development of targeted, eco-friendly, and cost-effective NPs specifically designed for different crops, recognizing that NP responses can vary across species and salt concentrations. Expanding the scope of NP research to include their potential in mitigating multiple abiotic stresses, such as drought and extreme temperatures, is crucial for maximizing their application in enhancing crop resilience. Furthermore, a thorough and comprehensive investigation into the potential negative impacts of NPs on living organisms, encompassing humans, livestock, and the soil microbiome, is imperative. This necessitates a rigorous assessment of NP accumulation within edible plant parts to ensure the safe and responsible implementation of nanotechnology in agricultural practices.

References

1. Abbas Q, Liu G, Yousaf B, Ali MU, Ullah H, Munir MAM, Rehman A (2020) Biochar-assisted transformation of engineered-cerium oxide nanoparticles: effect on wheat growth, photosynthetic traits and cerium accumulation. Ecotoxicol Environ Saf 187:109845
2. Abdel Latef AAH, Abu Alhmad MF, Abdelfattah KE (2017) The possible roles of priming with ZnO nanoparticles in mitigation of salinity stress in lupine (*Lupinus termis*) plants. J Plant Growth Regul 36:60–70
3. Abdoli S, Ghassemi-Golezani K, Alizadeh-Salteh S (2020) Responses of ajowan (*Trachyspermum ammi* L.) to exogenous salicylic acid and iron oxide nanoparticles under salt stress. ESPR 27:36939–36953

4. Abou-Zeid H, Ismail G (2018) The role of priming with biosynthesized silver nanoparticles in the response of *Triticum aestivum* L to salt stress. Egypt J Bot 58(1):73–85
5. Adnan M, Mahmood F, Zhao Z, Khaliq H, Usman M, Muhammad T, Ashraf GA (2025) Effect of the foliar application of biogenic-ZnO nanoparticles on physio-chemical analysis of chilli (Capsicum annum L.) in a salt stress environment. Environ Sci: Adv 4(2):306–317
6. Alabdallah NM, Alzahrani HS (2020) The potential mitigation effect of ZnO nanoparticles on [*Abelmoschus esculentus* L. Moench] metabolism under salt stress conditions. Saudi J Biol Sci 27(11):3132–3137
7. Alharby HF, Metwali EM, Fuller MP, Aldhebiani AY (2017) Impact of application of zinc oxide nanoparticles on callus induction, plant regeneration, element content and antioxidant enzyme activity in tomato (*Solanum lycopersicum Mill*) under salt stress. Arch Biol Sci 68(4):723–735
8. Almutairi ZM (2016) Influence of silver nano-particles on the salt resistance of tomato (*Solanum lycopersicum*) during germination. Int J Agric Biol 18(2):449–457
9. An J, Hu P, Li F, Wu H, Shen Y, White JC, Giraldo JP (2020) Emerging investigator series: molecular mechanisms of plant salinity stress tolerance improvement by seed priming with cerium oxide nanoparticles. Environ Sci: Nano 7(8):2214–2228
10. Annunziata MG, Ciarmiello LF, Woodrow P, Maximova E, Fuggi A, Carillo P (2017) Durum wheat roots adapt to salinity remodeling the cellular content of nitrogen metabolites and sucrose. Front Plant Sci 7:2035
11. Arias-Moreno DM, Jiménez-Bremont JF, Maruri-López I, Delgado-Sánchez P (2017) Effects of catalase on chloroplast arrangement in *Opuntia streptacantha* chlorenchyma cells under salt stress. Sci Rep 7(1):8656
12. Ashraf M, Shahzad SM, Imtiaz M, Rizwan MS (2018) Salinity effects on nitrogen metabolism in plants-focusing on the activities of nitrogen metabolizing enzymes: a review. J Plant Nutr 41:1065–1081
13. Askary M, Talebi SM, Amini F, Bangan ADB (2017) Effects of iron nanoparticles on Mentha piperita L. under salinity stress. Biologija 63(1)
14. Assaha DV, Ueda A, Saneoka H, Al-Yahyai R, Yaish MW (2017) The role of Na^+ and K^+ transporters in salt stress adaptation in glycophytes. Front Physiol 8:509
15. Baillie J, Zhang YP (2018) Space for nature. Science 361(6407):1051–1051
16. Barrios AC, Rico CM, Trujillo-Reyes J, Medina-Velo IA, Peralta-Videa JR, Gardea-Torresdey JL (2016) Effects of uncoated and citric acid coated cerium oxide nanoparticles, bulk cerium oxide, cerium acetate, and citric acid on tomato plants. Sci Total Environ 563:956–964
17. Buettner KM, Valentine AM (2012) Bioinorganic chemistry of titanium. Chem Rev 112(3):1863–1881
18. Caldelas C, Weiss DJ (2017) Zinc homeostasis and isotopic fractionation in plants: a review. Plant Soil 411:17–46
19. Carbajal-Vazquez VH, Gomez-Merino FC, Herrera-Corredor JA, Contreras-Oliva A, Alcantar-Gonzalez G, Trejo-TéLlez LI (2020) Effect of titanium foliar applications on tomato fruits from plants grown under salt stress conditions. Notulae Botanicae Horti Agrobotanici Cluj-Napoca 48(2):924–937
20. Chan-Rodriguez D, Walker EL (2018) Analysis of yellow striped mutants of *Zea mays* reveals novel loci contributing to iron deficiency chlorosis. Front Plant Sci 9:157
21. Chang J, Cheong BE, Natera S, Roessner U (2019) Morphological and metabolic responses to salt stress of rice (*Oryza sativa* L.) cultivars which differ in salinity tolerance. Plant Physiol Biochem 144:427–435
22. Chaves MM, Flexas J, Pinheiro C (2009) Photosynthesis under drought and salt stress: regulation mechanisms from whole plant to cell. Ann Bot 103:551–560
23. Chen L, Peng Y, Zhu L, Huang Y, Bie Z, Wu H (2022) CeO_2 nanoparticles improved cucumber salt tolerance is associated with its induced early stimulation on antioxidant system. Chemosphere 299:134474
24. Devahastin S (2020) Food security: impact of climate change and technology. Dry Technol 38:1538

25. El-Badri AM, Batool M, Wang C, Hashem AM, Tabl KM, Nishawy E, Wang B (2021) Selenium and zinc oxide nanoparticles modulate the molecular and morpho-physiological processes during seed germination of *Brassica napus* under salt stress. Ecotoxicol Environ Saf 225:112695
26. Elshoky HA, Yotsova E, Farghali MA, Farroh KY, El-Sayed K, Elzorkany HE, Rashkov G, Dobrikova A, Borisova P, Stefanov M, Ali MA, Apostolova E (2021) Impact of foliar spray of zinc oxide nanoparticles on the photosynthesis of Pisum sativum L. under salt stress. Plant Physiol Biochem 167:607–618
27. Farhangi-Abriz S, Torabian S (2018) Nano-silicon alters antioxidant activities of soybean seedlings under salt toxicity. Protoplasma 255:953–962
28. Farouk S, Al-Amri SM (2019) Exogenous zinc forms counteract NaCl-induced damage by regulating the antioxidant system, osmotic adjustment substances, and ions in canola (*Brassica napus* L. cv. Pactol) plants. J Soil Sci Plant Nutr 19(4):887–899
29. Gao J, Ding Y, Liu Y, He Y, Zhao D, Su X, Gao L, Song K, He X (2025) Effects of seed priming with La_2O_3 nanoparticles on seed vigor of alfalfa (Medicago sativa) under salt stress. S Afr J Bot 176:1–11
30. Gapińska M, Skłodowska M, Gabara B (2008) Effect of short-and long-term salinity on the activities of antioxidative enzymes and lipid peroxidation in tomato roots. Acta Physiol Plant 30:11–18
31. Ghosh A, Sah D, Chakraborty M, Rai JPN (2024) Mechanism and application of bacterial exopolysaccharides: an advanced approach for sustainable heavy metal abolition from soil. Carbohydr Res 544:109247
32. Ghosh A, Sah D, Chakraborty M, Rai JPN (2023) Bio-mediated detoxification of heavy metal contaminated soil and phytotoxicity reduction using novel strain of *Brevundimonas vancanneytii* SMA3. Heliyon 9(11):e22344
33. Gohari G, Zareei E, Rostami H, Panahirad S, Kulak M, Farhadi H, Fotopoulos V (2021) Protective effects of cerium oxide nanoparticles in grapevine (*Vitis vinifera* L.) cv. Flame Seedless under salt stress conditions. Ecotoxicol Environ Saf 220:112402
34. Gorai M, Ennajeh M, Khemira H, Neffati M (2011) Influence of NaCl-salinity on growth, photosynthesis, water relations and solute accumulation in phragmites Australis. Acta Physiol Plant 33:963–971
35. Goswami K, Mittal D, Gautam B, Sopory SK, Sanan-Mishra N (2020) Mapping the salt stress-induced changes in the root miRNome in Pokkali rice. Biomolecules 10(4):498
36. Gupta A, Bharati R, Kubes J, Popelkova D, Praus L, Yang X, Severova L, Skalicky M, Brestic M (2024) Zinc oxide nanoparticles application alleviates salinity stress by modulating plant growth, biochemical attributes and nutrient homeostasis in Phaseolus vulgaris L. Front Plant Sci 15:1432258
37. Hafeez MB, Ghaffar A, Zahra N, Ahmad N, Al-Qahtani WH, Sharma P, Husen A, Li J (2024) Exploring the efficacy of plant growth promoters in nutrient acquisition and antioxidant defense responses of late sown wheat under saline conditions. S Afr J Bot 171:812–824. https://doi.org/10.1016/j.sajb.2024.06.042
38. Hassan MJ, Ali Raza M, Khan I, Ahmad Meraj T (2020) Selenium and salt interactions in black gram (*Vigna mungo* L): ion uptake, antioxidant defense system, and photochemistry efficiency. Plants 9(4):467
39. Hezaveh TA, Pourakbar L, Rahmani F, Alipour H (2019) Interactive effects of salinity and ZnO nanoparticles on physiological and molecular parameters of rapeseed (*Brassica napus* L.). Commun Soil Sci Plant Anal 50(6):698–715
40. Hofmann T, Lowry GV, Ghoshal S, Tufenkji N, Brambilla D, Dutcher JR, Wilkinson KJ (2020) Technology readiness and overcoming barriers to sustainably implement nanotechnology-enabled plant agriculture. Nat Food 1(7):416–425
41. Hojjat SS, Kamyab M (2017) The effect of silver nanoparticle on Fenugreek seed germination under salinity levels. Russ Agric Sci 43:61–65
42. Huang H, Ullah F, Zhou DX, Yi M, Zhao Y (2019) Mechanisms of ROS regulation of plant development and stress responses. Front Plant Sci 10:800

43. Husen A, Iqbal M, Aref MI (2014) Growth, water status and leaf characteristics of *Brassica carinata* under drought stress and rehydration conditions. Braz J Bot 37:217–227
44. Husen A, Iqbal M, Aref IM (2016) IAA-induced alteration in growth and photosynthesis of pea (*Pisum sativum* L.) plants grown under salt stress. J Environ Biol 37:421–429
45. Husen A (2021) The harsh environment and resilient plants: an overview. In: Husen A (ed) Harsh environment and plant resilience. Springer International Publishing AG, Cham, pp 1–23. https://doi.org/10.1007/978-3-030-65912-7_1
46. Husen A (2021) Morpho-anatomical, Physiological, biochemical and molecular responses of plants to air pollution. In: Husen A (ed) Harsh environment and plant resilience. Springer International Publishing AG, Cham, pp 203–234. https://doi.org/10.1007/978-3-030-659 12-7_9
47. Husen A (2021) Cross talk between autophagy and hormones for abiotic stress tolerance in plants. In: Husen A (ed) Plant performance under environmental stress. Springer International Publishing AG, Cham, pp 1–15. https://doi.org/10.1007/978-3-030-78521-5_1
48. Husen A (2021) Harsh environment and plant resilience (*Molecular and Functional Aspects*). Springer Nature Switzerland AG, Cham, Switzerland. https://doi.org/10.1007/978-3-030-659 12-7
49. Husen A (2021) Plant performance under environmental stress (*Hormones, Biostimulants and Sustainable Plant Growth Management*). Springer Nature Switzerland AG, Switzerland. https://doi.org/10.1007/978-3-030-78521-5
50. Husen A (2022) Environmental pollution and medicinal plants. CRC Press. https://doi.org/10.1201/9781003178866
51. Husen A (2023) Plants and their interaction to environmental pollution (Damage detection, adaptation, tolerance, physiological and molecular responses). Elsevier Inc. https://doi.org/10.1016/C2020-0-03319-0
52. Husen A, Iqbal M (2023) Medicinal plants—their response to abiotic stress. Springer Nature Singapore Pvt. Ltd., Singapore. https://doi.org/10.1007/978-981-19-5611-9
53. Husen A, Iqbal M, Sohrab SS, Ansari MKA (2018) Salicylic acid alleviates salinity-caused damage to foliar functions, plant growth and antioxidant system in Ethiopian mustard (*Brassica carinata* A. Br.). Agric Food Secur 7(44):1–14
54. Hussain I, Singh A, Singh NB, Singh P (2019) Plant-nanoceria interaction: toxicity, accumulation, translocation and biotransformation. S Afr J Bot 121:239–247
55. Hussein MM, Abou-Baker NH (2018) The contribution of nano-zinc to alleviate salinity stress on cotton plants. R Soc Open Sci 5(8):171809
56. Hussein M, Embiale A, Husen A, Aref IM, Iqbal M (2017) Salinity-induced modulation of plant growth and photosynthetic parameters in faba bean *(Vicia faba)* cultivars. Pak J Bot 49:867–877
57. Jia J, Liang Y, Gou T, Hu Y (2020) The expression response of plasma membrane aquaporins to salt stress in tomato plants. Environ Exp Bot 178:104190
58. Kang Z, Lee ST (2019) Carbon dots: advances in nanocarbon applications. Nanoscale 11(41):19214–19224
59. Kaur N, Dhawan M, Sharma I, Pati PK (2016) Interdependency of reactive oxygen species generating and scavenging system in salt sensitive and salt tolerant cultivars of rice. BMC Plant Biol 16:1–13
60. Khan MN (2016) Nano-titanium dioxide (nano-TiO_2) mitigates NaCl stress by enhancing antioxidative enzymes and accumulation of compatible solutes in tomato (*Lycopersicon esculentum Mill.*). J Plant Sci 11:1–11
61. Khan MN, Li Y, Fu C, Hu J, Chen L, Yan J, Li Z (2022) CeO_2 nanoparticles seed priming increases salicylic acid level and ROS scavenging ability to improve rapeseed salt tolerance. Glob Chall 6(7):2200025
62. Khan MN, Li Y, Khan Z, Chen L, Liu J, Hu J, Li Z (2021) Nanoceria seed priming enhanced salt tolerance in rapeseed through modulating ROS homeostasis and α-amylase activities. J Nanobiotechnol 19:1–19

63. Khan I, Raza MA, Awan SA, Shah GA, Rizwan M, Ali B, Huang L (2020) Amelioration of salt induced toxicity in pearl millet by seed priming with silver nanoparticles (AgNPs): the oxidative damage, antioxidant enzymes and ions uptake are major determinants of salt tolerant capacity. Plant Physiol Biochem 156:221–232
64. Kim SA, Guerinot ML (2007) Mining iron: iron uptake and transport in plants. FEBS Lett 581(12):2273–2280
65. Li Y, Liu M, Yang X, Zhang Y, Hui H, Zhang D, Shu J (2022) Multi-walled carbon nanotubes enhanced the antioxidative system and alleviated salt stress in grape seedlings. Sci Hortic 293:110698
66. Li ZG, Gou HQ, Li RQ (2019) Electrical stimulation boosts seed germination, seedling growth, and thermotolerance improvement in maize (*Zea mays* L.). Plant Signal Behav 14(12):1681101
67. Liu J, Fu C, Li G, Khan MN, Wu H (2021) ROS homeostasis and plant salt tolerance: plant nanobiotechnology updates. Sustainability 13(6):3552
68. Lyu S, Wei X, Chen J, Wang C, Wang X, Pan D (2017) Titanium as a beneficial element for crop production. Front Plant Sci 8:597
69. Mahmoud AWM, Abdelaziz SM, El-Mogy MM, Abdeldaym EA (2019) Effect of foliar ZnO and FeO nanoparticles application on growth and nutritional quality of red radish and assessment of their accumulation on human health. Agric (Pol'nohospodárstvo) 65(1):16–29
70. Maryum Z, Luqman T, Nadeem S, Khan SMUD, Wang B, Ditta A, Khan MKR (2022) An overview of salinity stress, mechanism of salinity tolerance and strategies for its management in cotton. Front Plant Sci 13:907937
71. Mohamed AKS, Qayyum MF, Abdel-Hadi AM, Rehman RA, Ali S, Rizwan M (2017) Interactive effect of salinity and silver nanoparticles on photosynthetic and biochemical parameters of wheat. Arch Agron Soil Sci 63(12):1736–1747
72. Molnár Á, Papp M, Kovács DZ, Bélteky P, Oláh D, Feigl G, Kolbert Z (2020) Nitro-oxidative signalling induced by chemically synthetized zinc oxide nanoparticles (ZnO NPs) in Brassica species. Chemosphere 251:126419
73. Moradbeygi H, Jamei R, Heidari R, Darvishzadeh R (2020) Investigating the enzymatic and non-enzymatic antioxidant defense by applying iron oxide nanoparticles in *Dracocephalum moldavica L.* plant under salinity stress. Sci Hortic 272:109537
74. Mozafari AA, Dedejani S, Ghaderi N (2018) Positive responses of strawberry (Fragaria× ananassa Duch.) explants to salicylic and iron nanoparticle application under salinity conditions. PCTOC 134:267–275
75. Munns R, Tester M (2008) Mechanisms of salinity tolerance. Annu Rev Plant Biol 59:651–681
76. Peleg Z, Blumwald E (2011) Hormone balance and abiotic stress tolerance in crop plants. Curr Opin Plant Biol 14(3):290–295
77. Peng Y, Chen L, Zhu L, Cui L, Yang L, Wu H, Bie Z (2022) CsAKT1 is a key gene for the CeO_2 nanoparticle's improved cucumber salt tolerance: a validation from CRISPR-Cas9 lines. Environ Sci: Nano 9(12):4367–4381
78. Pinedo-Guerrero ZH, Cadenas-Pliego G, Ortega-Ortiz H, González-Morales S, Benavides-Mendoza A, Valdés-Reyna J, Juárez-Maldonado A (2020) Form of silica improves yield, fruit quality and antioxidant defense system of tomato plants under salt stress. Agriculture 10(9):367
79. Rady MM, Desoky ESM, Ahmed SM, Majrashi A, Ali EF, Arnaout SM, Selem E (2021) Foliar nourishment with nano-selenium dioxide promotes physiology, biochemistry, antioxidant defenses, and salt tolerance in *Phaseolus vulgaris*. Plants 10(6):1189
80. Rastogi A, Tripathi DK, Yadav S, Chauhan DK, Živčák M, Ghorbanpour M, Brestic M (2019) Application of silicon nanoparticles in agriculture. 3 Biotech 9:1–11
81. Rossi L, Bagheri M, Zhang W, Chen Z, Burken JG, Ma X (2019) Using artificial neural network to investigate physiological changes and cerium oxide nanoparticles and cadmium uptake by *Brassica napus* plants. Environ Pollut 246:381–389
82. Rossi L, Zhang W, Lombardini L, Ma X (2016) The impact of cerium oxide nanoparticles on the salt stress responses of *Brassica napus* L. Environ Pollut 219:28–36

83. Rossi L, Zhang W, Ma X (2017) Cerium oxide nanoparticles alter the salt stress tolerance of *Brassica napus* L. by modifying the formation of root apoplastic barriers. Environ Pollut 229:132–138
84. Rowe JH, Topping JF, Liu J, Lindsey K (2016) Abscisic acid regulates root growth under osmotic stress conditions via an interacting hormonal network with cytokinin, ethylene and auxin. New Phytol 211(1):225–239
85. Ryu H, Cho YG (2015) Plant hormones in salt stress tolerance. J Plant Biol 58:147–155
86. Safdarian M, Askari H, Shariati V, Nematzadeh G (2019) Transcriptional responses of wheat roots inoculated with *Arthrobacter nitroguajacolicus* to salt stress. Sci Rep 9:1–12
87. Sen SK, Chouhan D, Das D, Ghosh R, Mandal P (2020) Improvisation of salinity stress response in mung bean through solid matrix priming with normal and nano-sized chitosan. Int J Biol Macromol 145:108–123
88. Shabala S (2013) Learning from halophytes: physiological basis and strategies to improve abiotic stress tolerance in crops. Ann Bot 112(7):1209–1221
89. Sheikhalipour M, Esmaielpour B, Gohari G, Haghighi M, Jafari H, Farhadi H, Kalisz A (2021) Salt stress mitigation via the foliar application of chitosan-functionalized selenium and anatase titanium dioxide nanoparticles in stevia (*Stevia rebaudiana Bertoni*). Molecules 26(13):4090
90. Sheng Z, Van Nostrand JD, Zhou J, Liu Y (2018) Contradictory effects of silver nanoparticles on activated sludge wastewater treatment. J Hazard Mater 341:448–456
91. Shu K, Qi Y, Chen F, Meng Y, Luo X, Shuai H, Zhou W (2017) Salt stress represses soybean seed germination by negatively regulating GA biosynthesis while positively mediating ABA biosynthesis. Front Plant Sci 8:1372
92. Shu S, Yuan LY, Guo SR, Sun J, Yuan YH (2013) Effects of exogenous spermine on chlorophyll fluorescence, antioxidant system and ultrastructure of chloroplasts in *Cucumis sativus* L. under salt stress. Plant Physiol Biochem 63:209–216
93. Siddiqui MH, Al-Whaibi MH, Faisal M, Al Sahli AA (2014) Nano-silicon dioxide mitigates the adverse effects of salt stress on *Cucurbita pepo* L. Environ Toxicol Chem 33(11):2429–2437
94. Singh N, Bhatla SC (2016) Nitric oxide and iron modulate heme oxygenase activity as a long distance signaling response to salt stress in sunflower seedling cotyledons. Nitric Oxide 53:54–64
95. Singh A, Hussain I, Singh NB, Singh H (2019) Uptake, translocation and impact of green synthesized nanoceria on growth and antioxidant enzymes activity of *Solanum lycopersicum* L. Ecotoxicol Environ Saf 182:109410
96. Singh S, Kumar V, Kapoor D, Kumar S, Singh S, Dhanjal DS, Singh J (2020) Revealing on hydrogen sulfide and nitric oxide signals co-ordination for plant growth under stress conditions. Physiol Plant 168(2):301–317
97. Singh S, Husen A (2019) Role of nanomaterials in the mitigation of abiotic stress in plants. In: Husen A, Iqbal M (eds) N*anomaterials* and plant potential. Springer International Publishing AG, Cham, pp 441–471. https://doi.org/10.1007/978-3-030-05569-1_18
98. Singh S, Husen A (2020) Behavior of agricultural crops in relation to nanomaterials under adverse environmental conditions. In: Husen, A, Jawaid M (eds) Nanomaterials for agriculture and forestry applications. Elsevier Inc. https://doi.org/10.1016/B978-0-12-817852-2.00009-3
99. Sofy MR, Elhindi KM, Farouk S, Alotaibi MA (2020) Zinc and paclobutrazol mediated regulation of growth, upregulating antioxidant aptitude and plant productivity of pea plants under salinity. Plants 9(9):1197
100. Soliman M, Qari SH, Abu-Elsaoud A, El-Esawi M, Alhaithloul H, Elkelish A (2020) Rapid green synthesis of silver nanoparticles from blue gum augment growth and performance of maize, fenugreek, and onion by modulating plants cellular antioxidant machinery and genes expression. Acta Physiol Plant 42:1–16
101. Sun L, Song G, Guo W, Wang W (2019) Dynamic changes in genome-wide histone3 lysine27 trimethylation and gene expression of soybean roots in response to salt stress. Front Plant Sci 10:1031

102. Tanveer M, Anjum SA, Hussain S, Cerdà A, Ashraf U (2017) Relay cropping as a sustainable approach: problems and opportunities for sustainable crop production. ESPR 24:6973–6988
103. Tewari RK, Yadav N, Gupta R, Kumar P (2021) Oxidative stress under macronutrient deficiency in plants. J Soil Sci Plant Nutr 21(1):832–859
104. Torabian S, Farhangi-Abriz S, Zahedi M (2018) Efficacy of $FeSO_4$ nano formulations on osmolytes and antioxidative enzymes of sunflower under salt stress. Indian J Plant Physiol 23:305–315
105. Tortella GR, Rubilar O, Durán N, Diez MC, Martínez M, Parada J, Seabra AB (2020) Silver nanoparticles: toxicity in model organisms as an overview of its hazard for human health and the environment. J Hazard Mater 390:121974
106. Tripathi DK, Singh S, Gaur S, Singh S, Yadav V, Liu S, Sahi S (2018) Acquisition and homeostasis of iron in higher plants and their probable role in abiotic stress tolerance. Front Environ Sci 5:86
107. Wahid I, Kumari S, Ahmad R, Hussain SJ, Alamri S, Siddiqui MH, Khan MIR (2020) Silver nanoparticle regulates salt tolerance in wheat through changes in ABA concentration, ion homeostasis, and defense systems. Biomolecules 10(11):1506
108. Wang P, Lombi E, Zhao FJ, Kopittke PM (2016) Nanotechnology: a new opportunity in plant sciences. Trends Plant Sci 21(8):699–712
109. Wang A, Lv J, Wang J, Shi K (2022) CO_2 enrichment in greenhouse production: towards a sustainable approach. Front Plant Sci 13:1029901
110. Wang Y, Ying Y, Chen J, Wang X (2004) Transgenic Arabidopsis overexpressing Mn-SOD enhanced salt-tolerance. Plant Sci 67(4):671–677
111. Wu H (2018) Plant salt tolerance and Na+ sensing and transport. Crop J 6(3):215–225
112. Xiong D, Flexas J, Huang J, Cui K, Wang F, Douthe C, Lin M (2022) Why high yield QTLs failed in preventing yield stagnation in rice? Crop and Environ 1(2):103–107
113. Yan A, Chen Z (2019) Impacts of silver nanoparticles on plants: a focus on the phytotoxicity and underlying mechanism. Int J Mol Sci 20(5):1003
114. Yang J, Duan G, Li C, Liu L, Han G, Zhang Y, Wang C (2019) The crosstalks between jasmonic acid and other plant hormone signaling highlight the involvement of jasmonic acid as a core component in plant response to biotic and abiotic stresses. Front Plant Sci 10:1349
115. Yang Y, Guo Y (2018) Elucidating the molecular mechanisms mediating plant salt stress responses. New Phytol 217:523–539
116. Ye Y, Cota-Ruiz K, Hernández-Viezcas JA, Valdes C, Medina-Velo IA, Turley RS, Gardea-Torresdey JL (2020) Manganese nanoparticles control salinity-modulated molecular responses in *Capsicum annuum* L. through priming: a sustainable approach for agriculture. ACS Sustain Chem Eng 8(3):1427–1436
117. Zafar S, Hasnain Z, Danish S, Battaglia ML, Fahad S, Ansari MJ, Alharbi SA (2024) Modulations of wheat growth by selenium nanoparticles under salinity stress. BMC Plant Biol 24(1):35
118. Zahra N, Kausar A, Abdelghani HTM, Singh S, Vashishth DS, Bachheti A, Bachheti RK, Husen A (2024) Serotonin improves plant growth, foliar functions and antioxidant defence system in Ethiopian mustard (*Brassica carinata* A. Br.). S Afr J Bot 170:1–9. https://doi.org/10.1016/j.sajb.2024.05.002
119. Zahra N, Kausar A, Sharma P, Wabaidur SM, Husen A (2024) Serotonin application to Guizotia abyssinica improves growth, physiological functions, antioxidant system and alleviates the deleterious effects induced by salt stress. Plant Biosystems—An Int J Deal All Asp Plant Biol 1–9. https://doi.org/10.1080/11263504.2024.2384972
120. Zhang Y, Kaiser E, Marcelis LF, Yang Q, Li T (2020) Salt stress and fluctuating light have separate effects on photosynthetic acclimation, but interactively affect biomass. Plant Cell Environ 43(9):2192–2206
121. Zhao G, Zhao Y, Lou W, Su J, Wei S, Yang X, Shen W (2019) Nitrate reductase-dependent nitric oxide is crucial for multi-walled carbon nanotube-induced plant tolerance against salinity. Nanoscale 11(21):10511–10523

122. Zhou H, Wu H, Zhang F, Su Y, Guan W, Xie Y, Shen W (2021) Molecular basis of cerium oxide nanoparticle enhancement of rice salt tolerance and yield. Environ Sci: Nano 8(11):3294–3311
123. Zhu JK (2016) Abiotic stress signaling and responses in plants. Cell 167(2):313–324
124. Zulfiqar F, Navarro M, Ashraf M, Akram NA, Munné-Bosch S (2019) Nanofertilizer use for sustainable agriculture: advantages and limitations. Plant Sci 289:110270
125. Zörb C, Geilfus CM, Mühling KH, Ludwig-Müller J (2013) The influence of salt stress on ABA and auxin concentrations in two maize cultivars differing in salt resistance. J Plant Physiol 170(2):220–224

Role of Metal and Metal-Oxide Nanoparticles in Agricultural Crop Drought Stress Adaptation and Mitigation

Pooja Singh, Himshikha Malviya, and Jay Singh

Abstract Drought is one of the abiotic stressors that crops encounter which reduces agricultural productivity around the world. It impacts all phases of plant development, including the vegetative and reproductive stages, which lowers yields and causes financial loss. Plants under stress will exhibit altered morphological, biochemical, and molecular responses due to oxidative stress, and hormonal, isotonic, and nutritional imbalances. The beneficial role of nanoparticles, which alter the cell wall and regulate stomatal closure during stresses to aid in water retention, has been shown in numerous studies. Furthermore, due to the enhanced enzymatic and non-enzymatic antioxidants, plants treated with nanoparticles have demonstrated increased pigment content and a faster rate of photosynthesis. Reactive oxygen species accumulation in plants under drought stress (DS) is also managed by nanoparticle therapy, which has decreased malondialdehyde from the lipid peroxidation process. To protect themselves against DS, these plants also control the expression of genes that respond to drought and the manufacture of phytohormones including cytokinin, auxin, gibberellin, and abscisic acid. These plants can preserve water content, improve nutrition, and water availability, and lessen oxidative and osmotic damage caused by DS in plants. The use of nanobiotechnology to reduce DS by controlling plant biochemical pathways is summarized in this chapter. Compared to traditional methods, this approach has demonstrated effective and rapid plant drought resistance. Therefore, the use of nanoparticles could result in innovations in controlling plant development and lessening the negative effects of DS in agriculture.

Keywords Abiotic stress · Drought · Nanoparticle · Enzymes · Photosynthesis · Mitigation

---

P. Singh
Department of Botany, Government Veer Surendra Say PG College, Gariaband, Chhattisgarh, India

H. Malviya · J. Singh (✉)
Department of Chemistry, Institute of Sciences, Banaras Hindu University, Varanasi, Uttar Pradesh, India
e-mail: jaysingh.chem@bhu.ac.in

A. Husen (ed.), *Nanobiotechnology for Abiotic Stress Adaptation and Mitigation in Agricultural Crops*, Smart Nanomaterials Technology,
https://doi.org/10.1007/978-981-96-9298-9_3

1 Introduction

Serious worries about sustainable development and food security have plagued the entire country. Consequently, one of the 17 Sustainable Development Goals set forth by the UN is to end hunger. Food production and distribution are still facing significant challenges in the modern day due to factors including population growth, climate change, environmental contamination, and rising water and energy demands. Furthermore, current agricultural practices use a variety of resources. For instance, yearly production in the United States uses more than two quadrillion British thermal units of energy, 2.7 trillion cubic meters of water (approximately 70% of all freshwater all over the world), 187 million tonnes of fertilisers, and 4 million tonnes of pesticides [1, 83, 116]. The FAO estimates that by 2050, there will be 10 billion people on the globe, which will lead to a 50% rise in food demand, particularly in developing nations like India. Agriculture is the backbone of the Indian economy, and 60% of the population depends on it for fuel, fibre, and folder. A decrease in food production results in food shortages and a fall in nutrition security while an increase in food production rates helps the country's gross domestic product, expand. Numerous factors, including soil contamination, soil nutrition, climate change, and others, may contribute to this decline in yield [18, 44–50, 83]. In this case, nanotechnology may be an effective tool for transforming various facets of agriculture, ranging from food packaging to soil remediation. In agriculture, nanomaterials can be employed extensively as growth regulators, pesticides, herbicides, insecticides, nanocarriers, nanosensors, and nanobarcodes [49–55]. Additionally, this nanotechnology can be used for soil remediation and water filtration [5, 6, 87]. Similarly, nanoparticles can transport genetic material and proteins by functioning as cargo. Aromatic and medicinal plants may be genetically modified as a result, improving their yield and nutrient intake while also strengthening their resilience to stress. Geospatial and nanosensor technologies can also be utilized to track crop yield, and nanobarcodes can be used to identify and manage crop pathogen infections by tagging protein-associated pathogenicity. Precision agriculture will benefit from this. Numerous abiotic elements are constantly present in plants and other sessile organisms [5, 44, 103, 109, 116]. Variations in the environment, such as dryness, salinity, alkalinity, flooding, air pollutants, and mineral toxicity or deficiencies, can significantly lower production [52–59]. While some plants stand in this unfavourable state, this is not always the case. Water is crucial to plant life since it facilitates the uptake and transportation of nutrients, and its shortage can stress plants by causing drought and physiological water inaccessibility, which are particularly prevalent in arid and semiarid areas [44, 70]. Plants under drought stress (DS) have poorer growth and productivity because of decreased water status, biomass, impaired photosynthesis, nutrient uptake, osmotic and antioxidant activities, and reactive oxygen species that are overproduced by photorespiration [29, 35, 46, 47]. These factors cause DNA damage, protein denaturation, and lipid peroxidation, which impedes cell growth and elongation. Numerous studies have demonstrated the beneficial effects of metal- and carbon-based nanoparticles in reducing DS by fostering tolerance [44, 69, 74, 105]. By improving the uptake of nutrients and water through stress tolerance and upregulating genes involved in cell growth, metal-based nanoparticles like zinc oxide, titanium oxide, iron, and copper nanoparticles, as well as carbon-based nanomaterials like fullerene, graphene, and fullerol, are highly useful in reducing the effects of DS. These nanomaterials are utilized to increase photosynthetic factors and germination characteristics including biomass, growth rate, and yield. These parameters control several physiological processes, including stomata conductance and transpiration rate. Additionally, they improve the assimilation of carbon dioxide in leaves and lessen the membrane's ion leakage [2, 20, 69, 101, 105]. Furthermore, the published research claims that these nanomaterials control the defence mechanism by boosting the activity of both enzymatic and non-enzymatic antioxidants, including glutathione, catalase, superoxide dismutase, ascorbate peroxidase, and peroxidase. By reducing oxidative damage brought on by the generation of reactive oxygen species like hydrogen peroxide and oxygen, as well as by raising the levels of osmolytes and osmoprotectants like prolines, glycine betaine, soluble sugars, and amino acids that assist in osmotic adjustment under DS conditions, this substance also lessens the effects of DS [25, 51, 52, 79, 83, 114]. Numerous factors, including the nanomaterial's content, surface characteristics, application method, and plant species, affect its impact. To enhance the plant yield characteristics, this work has summarized the processes of drought mitigation and tolerance mediated by nanomaterials. Furthermore, the current research has highlighted the potential of nanomaterials in creating resilient crops with sustained yield. Additionally, physio-biochemistry, phytohormone and gene expression, life cycle and nutritional quality, germination and vegetative traits, and secondary growth are all improved by nanoparticles. Nanoparticles aid plants in reducing their tolerance to drought, and this methodical graphic attempt to summarize all the work in a single glance, as shown in Fig. 1.

2 Effects of Drought Stress on Crop Plants

When plants regular growth and development deviate from the typical range of homeostatic control, stress is the cause. Stress can be classified as either biotic or abiotic. Moisture, drought, temperature, minerals, salinity, soil acidity, pH, and mineral shortage are some of the variables that affect abiotic stress. Drought is the most prevalent stressor on plants worldwide, which contributes to food scarcity. A small farmer's ability to produce crops is further hampered by irregular and little rainfall. Climate change is the primary cause of this kind of stress, which hurts crop yield. Increasing crop output is now crucial to meeting the growing population's demands. While long-term stress kills crop plants, short-term stress lowers crop output. The impact of stress is dependent on several factors, including the plant's developmental stages, its degree and duration, the species' genotypic ability, and its interactions with its surroundings. Climate change poses a challenge to global food security in the twenty-first century, and feeding the world's growing population is extremely difficult ("Effects of DS on Crop Production and Productivity", 2020).

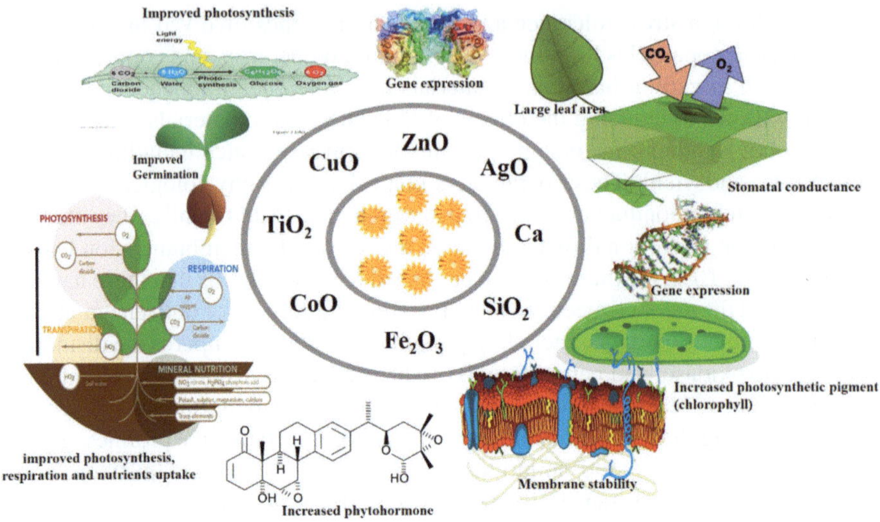

Fig. 1 This figure represents the schematic of the overall process

These elements have a detrimental effect on the dry and warm regions of the world as well as on plant growth and development during the favourable season. Additionally, it restricts crop output, particularly in arid and hot regions. This condition is caused by the quantity and distribution of water being insufficient during the agricultural plants life cycle, which limits the plants ability to express their full genetic potential. Through the reduction of leaf size, stem expansion, root multiplication, disruption of the plant-nutrient relationship, and inhibition of water efficiency, this stress has decreased crop plant yield. The world's food security is at risk due to this stress, which could make it difficult to meet the demand for food.

A plant experiences DS when it becomes difficult to deliver water to its roots, which occurs when transpiration rates are extremely high. Many parts of the world may experience drought because of climate change, which deprives crops of water and affects several physiological processes, including growth, membrane function, yield, pigment content, integrity, photosynthetic activity, and osmotic adjustment water relations [17, 86]. Plants respond to water scarcity through a variety of events, which alters physio-biochemical processes and plant growth. These changes include modifications to tissue osmotic potential, growth rate, antioxidant defences, and plant structure. It is now essential to understand how crops respond and adjust to water shortages, as well as to take action to make crop plants more drought-resistant and ensure greater crop yields when faced with adverse environmental stressors. This work provides an overview of the morphophysiological and biochemical responses of plants to DS [9, 27].

2.1 Morphological Response

When plants experience DS, their exterior form changes first, followed by changes to their internal structure. Loss of water causes the plant to grow very slowly or possibly die, which is one of the most important effects. Certain plant phenotypes can be adapted to changing conditions by plants under abiotic stress. For example, xerophytes have evolved morphological traits and have adapted to drought in their ontogenetic development. The morphological and structural traits of drought-resistant plants—such as their leaves, stems, roots, and so forth—were tailored to the arid climate [121].

(i). ***Growth and Yield:*** For the process of agricultural growth and development to necessitate agronomic approaches with the growing population, a thorough understanding is crucial. The extent to which plant vegetative and reproductive growth patterns can be controlled under varying environmental stress circumstances will determine plant performance and crop sustainability. The intricate interactions between the sources and sink constraints of a plant's two primary organs—the root system and the shoot—that create functional equilibrium determine how much a plant can develop. More than any other environmental component, a permanent or temporary water deficit significantly impairs plant growth and development. Impaired germination and poor stand establishment is the first and foremost side effect of DS [41]. Due to the decrease in turgor pressure, cell growth is the most sensitive physiological activity. This growth is the outcome of the merismatic cells' subsequent expansion and division, which produces daughter cells. The interruption of water passage from the xylem to the surrounding elongating cells during dry conditions can hinder the elongation of higher plants cells [84]. By lowering the soil water potential, this stress shortens the lifespan of leaves and diminishes their size and number per plant. Temperature, leaf turgor, and the assimilating supply for growth are some of the variables that affect the expansion of leaf area. The suppression of photosynthesis and the decrease in fresh and dry biomass production also inhibit leaf expansion [94, 126]. Six irrigation treatments were applied to maize in one study, which found that as water stress increased, the plant's height, stem diameter, and leaf area all reduced. Concurrently, there was a decrease in plant height, a decrease in cell size, a decrease in cob length, a decrease in shoot freshness and dry weight, and an increase in leaf senescence. Additionally, maize exhibits a 37% decrease in biomass accumulation at silking, a 34% decrease throughout the grain-filling period, and a 21% decrease during maturity [64, 73].

2.2 Physiological Response

Plants that experience DS will undergo several physical changes, which will result in several physiological and biochemical alterations. For instance, the varying degrees of effect of plants under DS are reflected in the variations in photosynthetic rate, osmotic regulating chemicals, drought-induced proteins, and antioxidant enzymes.

(a) ***Root Signalling:*** To promote plant growth in its early stages and to obtain water from the superficial soil layer—which is occasionally lost through evaporation—a vast root system is crucial. Numerous debatable pieces of research demonstrate how dryness affects root growth. Water stress was found to boost the rate of root growth in *Catharanthus roseus*, while the root of maize showed decreased root growth. The ratio of a plant's roots to shoots often rises when the water potential is low, but the shoots are more sensitive to growth restriction than the roots [61, 95]. Usually, when water availability is limited, the root: shoot ratio of plants rises because roots are less sensitive than shoots to growth inhibition by low water potentials [118]. When DS is present, roots send a signal cascade down the xylem to the shoots, triggering physiological changes that ultimately determine the degree of stress response. Plants root: shoot ratio rises when the water supply is low, because roots are less sensitive than shoots. When under stress, roots use the xylem to send a signal cascade to the shoots, which alters their physiology and ultimately determines how well the plant has adapted. Root-shoot signalling has been linked to ethylene, malate, cytokinins, abscisic acid, and other unknown substances. Stomatal closure, a crucial adaptation to the field's limited water supply, is brought on by root-to-leaf communication via the transpiration stream. ABA encourages the outflow of $K+$ ions from the guard cells by lowering turgor pressure and closing stomata. Under water stress, ABA levels can rise by up to 50 times because of cell membrane disruption or turgor loss [40]. Furthermore, studies conducted on plants under DS that used doses of exogenous ABA far lower than those needed to seal stomata in detached leaves cast doubt on the function of ABA. Cytokinin is a crucial signal that travels from the roots to the shoots, while ABA is a signal that regulates growth and transpiration. Cytokinins generated mostly in roots and implicated in responses to nutrient shortage may play a significant role in drought responses. It is currently unclear if all plant species react similarly to cytokinin at the amounts observed in the leaf and guard cells, despite recent studies showing lower cytokinin concentrations in the xylem under DS [26, 78, 98].

(b) ***Photosynthesis and Chlorophyll Content:*** Environmental stress disrupted several key aspects of photosynthesis, including the thylakoid electron transport, the carbon reduction cycle, the stomatal control of the carbon dioxide supply, an increase in the accumulation of carbohydrates, the peroxidative destruction of lipids, and the disturbance of water balance. Plants' capacity for photosynthesis, which affects physiological and biochemical processes and, eventually, the overall growth and production of the plant, is either directly or indirectly linked to their ability to adapt to different environments. Crop plants gas exchange parameters were significantly impeded by DS because of reduced leaf expansion, compromised photosynthetic machinery, early leaf senescence, oxidation of chloroplast lipids, and modifications to protein and pigment structures [8, 21, 75]. The entry and exit of carbon dioxide absorbability are controlled by stomata. The initial reaction to drought is the opening and closing of stomata, which lowers the rate of photosynthetic reaction. Closing stomata causes the leaves to lose carbon dioxide and reduces carbon uptake in favour of photorespiration. Numerous studies provide evidence that as DS increases, stomata gradually close. Even under DS, there is always a positive association between stomatal conductance and leaf water potential, and the two always interact. Drought-induced root-to-leaf signalling helped the soil by drying down the transpiration stream and causing stomatal closure. Changes in chlorophyll production, structural and functional alterations in chloroplasts, and disruptions in the assimilate accumulation, transport, and distribution processes are examples of "non-stomatal" mechanisms. In comparison to well water control, the DS in maize resulted in a significant decrease in net photosynthesis (33.22%), transpiration rate (37.84%), stomatal conductance (25.54%), water-use efficiency (50.87%), intrinsic water-use efficiency (11.58%), and intercellular CO_2 (5.86%), according to the report presented by Anjum et al. [9]. Additionally, several studies have documented a reduction in photosynthetic activity brought on by either stomatal or non-stomatal processes. Stomata are the initial reactions to DS, which lowers the rate of photosynthesis. They are also the entrance gate where water loss and CO_2 absorbability occur. When stomata close, the leaf is deprived of CO_2, and photorespiration replaces photosynthetic carbon absorption. It is clear from previous research and current data on drought-induced photosynthetic responses that stomata gradually close as DS increases. It is common knowledge among botanists that stomatal conductance and leaf water status are constantly correlated, even in drought-stressed environmental conditions. This discovery makes it evident that stomatal closure is caused by drought-induced root-to-leaf signalling that is aided by soil drying through the transpiration stream. Changes in chlorophyll synthesis, modifications to the structure and function of chloroplasts, and disruptions to the assimilates' accumulation, transport, and dispersion processes are examples of "non-stomatal" mechanisms [10, 19, 97]. Another crucial element is chlorophyll, which has a positive correlation with both the amount of chlorophyll and the rate of photosynthesis. On the other hand, the decrease in chlorophyll content, which results from pigment photooxidation and chlorophyll degradation, is an indication of oxidative stress. The decrease in chlorophyll concentration during DS may be due to pigment degradation; photosynthetic pigments are essential for light gathering and the production of reducing powers, but both chlorophyll a and b are susceptible to soil dryness. Depending on the length and intensity of the drought, numerous species have been shown to have decreased or unchanged chlorophyll levels during DS [31, 67, 125]. Another stressor that significantly reduced the amounts of chlorophyll, chlorophyll b, and total chlorophyll in many sunflower cultivars was drought. Additionally, two olive cultivars that were exposed to less irrigation had decreased Chl (a + b) levels; for Chemlali and Chetoui, these reductions were 29 and 42%, respectively. Furthermore, it is thought that the primary reason why photosynthesis becomes inactive under water stress is the decrease in chlorophyll concentrations. Moreover, the loss of chloroplast membranes, excessive swelling, distortion of the lamellae vesiculation, and the development of lipid droplets have all been implicated in the decrease in chlorophyll concentration caused by DS [39, 63, 72]. It is evident from the facts that low levels of photosynthetic pigments can directly impede primary generation and photosynthetic capacity. From a physiological point of view, the amount of chlorophyll in leaves is a highly significant characteristic. When plants experience a water deficit, the mesophyll cells lose the most chlorophyll, whereas the bundle sheath cells lose a less amount.

(c) ***Water Relations:*** Numerous factors, such as relative water content (RWC), leaf water potential, stomatal resistance, transpiration rate, leaf temperature, and canopy temperature, influence plant water relations. It is considered a measure of plant water status, represents the metabolic activity in tissues, and serves as an indicator of dehydration tolerance. RWC is higher in the early phases of leaf formation and falls as dry matter builds up and the leaf ages; it relates to both water absorption by the roots and water loss through transpiration. According to [81], when leaves are exposed to drought, they show significant decreases in RWC and water potential. RWC's response to DS decreases, which raises their leaf temperature while also drastically lowering their transpiration rate, relative water content, and leaf water potential. Additionally, when two poplar species were subjected to DS, the RWC in the water-stressed cuttings dropped by 16% in Populus kangdingensis and 23.3% in Populus cathayana. The interaction between species, drought event duration, and intensity had an impact on the data that was acquired [81, 108, 122]. In drought conditions, plant water relations are affected by decreased water availability as well as stomatal closure and opening. Changes in leaf temperature may assist control the water status of leaves and maintain water-use efficiency by reducing water loss. Even more harmful was the significant reduction in water-use efficiency in plant-growing areas.

(d) ***Osmolyte Accumulation:*** Numerous organic and inorganic solutes in the cytosol assist in lowering the osmotic potential to maintain turgor pressure. During drought, osmotic potential keeps leaf turgor, which improves water absorption from drying soil. Wheat is characterized by low concentrations of these solutes, and proline accumulation and mobilization enhance resilience to water stress [82, 111]. The accumulation of proline solute, which has been extensively researched in relation to stress tolerance, is the initial reaction of plants to water-deficit stress to lessen cell damage. It can build up in maize plants, increasing their stress levels and reaching a peak after 10 days, following which they start to decline after 15 days. Additionally, it operates as a signalling molecule to alter mitochondrial processes, affect cell division or death, and initiate gene expression, all of which can be crucial for plants to recover from stress. Stress tolerance and proline accumulation have been related, and plants that are more resilient to stress typically have higher proline concentrations than those that are more susceptible. Additionally, it affects protein solvation, protects complex proteins' quarternary structure, keeps membrane integrity under dehydration stress, and lessens photoinhibition or lipid membrane oxidation [11, 23, 113]. Additionally, it helps to buffer cellular redox potential under stress, scavenge free radicals, and stabilize subcellular structures [14].

2.3 Biochemical Response

(a) ***Reactive oxygen species (ROS):*** The oxidative burst, the earliest metabolic reaction of eukaryotic cells to biotic and abiotic stressors, is the generation of reactive oxygen species. It serves as a secondary messenger to initiate further defence reactions in plants and is an early event of the defence response to water stress. ROS, or free radicals, which comprise peroxides and oxygen ions, are a natural result of oxygen's regular metabolism and play a crucial part in cell communication. However, environmental stressors like drought cause a rapid increase in ROS levels in plants, which leads to oxidative damage to lipids, proteins, and DNA. Since ROS are extremely reactive, they can cause significant harm to plants by accelerating lipid peroxidation, protein deterioration, fragmentation, and eventually cell death [12, 31]. The lipid membrane is immediately attacked by ROS, and lipid peroxidation raises the amount of malondialdehyde (MDA), which is thought to be a good indicator of membrane lipid peroxidation. It reduces the membrane's stability, which is a sign of the frequency of free radical reactions in tissues and reflects the degree of lipid peroxidation brought on by ROS. The glycollate oxidase reaction in the peroxisomes during photorespiration and oxygen absorption loading on the tissues also result in the production of reactive oxygen species, particularly H_2O_2. The MDA and H_2O_2 contents in the water-stressed cuttings increased by 88.9 and 99.7% in *P. cathayana*, respectively, while they only increased by 44 and 63.6% in *P. kangdingensis*, according to [122]. Moreover, lipid peroxidation levels in the leaves of pea (*Pisum sativum*) plants rose two to four times as DS increased, and this was strongly connected with protein peroxidation [76–77, 122].

(b) ***Antioxidant enzymes:*** To prevent damage to active oxygen and ensure proper cellular function, plants have an internal defence mechanism called antioxidant enzymes that initiate a cleanup process. Whether oxidative signalling will cause harm depends on how well ROS generation and antioxidative enzyme activity are balanced. Plants have developed a sophisticated enzymatic and non-enzymatic antioxidant system to reduce oxidative stress. This system includes ROS scavenging enzymes like ascorbate peroxidase (APX), catalase (CAT), peroxidase (POD), and superoxide dismutase (SOD), as well as low-molecular-mass antioxidants like glutathione, ascorbate, and carotenoids [12, 45, 80]. Enzymatic components can either directly scavenge ROS or act by generating a non-enzymatic antioxidant, whereas non-enzymatic antioxidants work together to preserve the integrity of the photosynthetic membranes. According to [120], the elevated activities of CAT, SOD, POD, APX, and GR at 25% field capacity were 4.3, 103, 172, 208, and 56% in *P. cathayana*, respectively, compared to 8.1,

125, 326, 276, and 78% in *P. kangdingensis*. While SOD in the chloroplast, mitochondrion, cytoplasm, and peroxisome can convert O^{2-} into H_2O_2, antioxidant activity is necessary for the effective destruction of O2– and H2O2 in plant cells. Additionally, POD contributes to the scavenging of H_2O_2, which is created by the dismutation of O^{2^-}, which is catalysed by SOD. The primary enzyme that removes H_2O_2 from the microbody and mitochondria, CAT helps to lessen the negative consequences of oxidative stress. Peroxisomes contain it, although it is thought to be essential for breaking down H_2O_2 under stress and by boosting resistance to oxidative damage; maintaining a higher level of antioxidative enzyme activity may help induce drought. Plant tolerance to drought may be related to antioxidant enzymes' capacity to scavenge ROS and lessen their harmful effects [104, 106, 120].

3 Nanoparticle Uptake and Its Mechanism

Three processes are involved in the mechanism of nanoparticles in plants: deposition, access, and translocation. The nanoparticle first interacts with the plant's surface, which is controlled by hydrophobic, electrostatic, and Van der Waals forces. The application method, concentration, size, and shape of the nanomaterial particle are some of the variables that affect uptake, translocation, and distribution. Treatment/priming, foliage treatment, soil treatment, irrigation, and hydroponics treatment are some of the well-known techniques for treating nanoparticles for plant seeds [32, 112]. Through the process of endocytosis, foliar spray nanoparticles are absorbed by the plant's cell via the stomata or cuticles on the leaf surface. While very tiny nanoparticles, less than 5 nm, penetrate through a cuticular pathway, these stomatal spores are in the micrometre range, which permits the uptake of these nanomaterials. Numerous variables, including pore size, stomatal density, and leaf architecture, affect how well these nanomaterials are absorbed. Additionally, the hydrophobicity, size, and chemical coating of these nanomaterials affect their ability to penetrate the leaf's cuticle and mesophyll tissue. For example, compared to citrate-coated gold nanoparticles, PVP-coated gold nanoparticles, which are three nanometers in size, have a greater cuticular penetration rate. Due to variations in stomatal abundance in different leaf sections, the stomatal uptake of nanomaterial in *Allium porum* leaves treated with polystyrene (43 nm) NPs was found to exhibit fluctuation [15]. While the internalization of TiO2 NPs through stomata was validated by X-ray fluorescence microscopy in in vitro cultured *Arabidopsis thaliana* seedlings, carbon-based nanomaterial uptake by the surface of plant cells was demonstrated by pore size diameters in the nano range. The penetration rate is influenced by the development and deposition of nanoaggregates on the leaf surface. For instance, this nanoparticle was shown to be trapped in the epidermis of cucumber leaves, which resulted in a 30% decrease in penetration rate. Additionally, vascular bundles facilitate the transport of these nanomaterials from leaf tissue to other plant tissues [66, 42, 71]. The stomatal absorption and translocation of metal-oxide nanoparticles (NPs) with sizes ranging from 27.3 to 46.7 nm into watermelon plants stems and roots after foliar exposure were also documented by Wang et al. [117]. Nanoparticles are absorbed by the roots via the apoplastic or symplastic pathway after seedling and soil treatment. This process depends on several variables, such as osmotic pressure, capillary pressures, and the plant cell wall. Through the apoplastic pathway, nanoparticles penetrate the cell wall and membrane of the epidermis before moving on to the cortex, endodermis, and vascular bundles. Nevertheless, the symplastic pathway can be used to avoid the Casparian strips that impede entry into the xylem [32]. Through endocytosis, passive diffusion, and assisted transport—all of which are mediated by aquaporins and ion channels—nanoparticles enter root cells via the symplastic pathway. Additionally, the plasmodesmata facilitate intercellular transport via the symplastic route, which is carried out by endosomes inside cells. It has been proposed that Rab proteins control and guide intracellular nanoparticles towards the plasmodesmata. Rice, poplar, and Arabidopsis have all been studied for their plasmodesmata-mediated transport of carbon, Au, and silver (Ag) nanoparticles, respectively [22, 34, 127]. Similarly, another study shows that cell wall pores (5–20 nm) allow passive transport of gold nanoparticles smaller than 20 nm into the root cells. Additionally, [102] found that multi-walled carbon nanotubes directly permeate the cell membrane in *Catharanthus roseus*. Furthermore, endocytosis can also facilitate the entry of nanoparticles into root cells via clathrin-independent (NPs $\leq$ 1000 nm) and clathrin-dependent (NPs $\leq$ 80 nm) pathways. As an alternative, several metal ion transporters have been found in plants that aid in the transport of metal ions like Fe^{2+}/Fe^{3+}, Zn^{2+}, Mn^{2+}, Cd^{2+}, and Cu^{+}/Cu^{2+}. These include natural resistance-associated macrophage proteins (Nramps), iron-regulated transporter 1 (IRT1), and copper influx transporter 1 (COPT1), indicating their role in the transport of metal-based nanoparticles [68, 102, 117]. After adsorption, water, sap, and xylem and phloem sieve tube elements move the nanoparticles from the roots to the plant's aerial sections. In the same way, [117] documented the unidirectional and bidirectional transport of copper oxide nanoparticles in Zea mays through the xylem and phloem. The uptake and transport of carbon nanoparticles have been extensively studied using fluorescence microscopy and Fourier transform infrared research [7, 102, 110]. Flow rate, ionic strength, and sap composition all affect the dissolution, aggregation, and transformation of nanoparticles inside the phloem. For example, the sap may change into iron oxide nanoparticles if it contains inorganic materials like phosphate and sulphite. The species also affects the absorption and transport of nanoparticles. For instance, gold nanoparticle deposition was seen in *Oryza sativa* shoots but not in *Cucurbita pepo.* In a similar vein, Cerium oxide nanoparticle transport was seen in pumpkin but not in wheat [93, 99, 112, 127]. Numerous data indicate that there is variation in the transport of nanoparticles among species, this may be because distinct species have varied morphological traits and size exclusion limits. Nevertheless, different plants exhibit different reactions to the uptake and movement of nanoparticles, and the cause of this mechanism is still being studied. Consequently, we can propose that a great deal of research remains necessary to elucidate the mechanisms and transport of nanoparticles in plants [32]. The entire Fig. 2 is shown below,

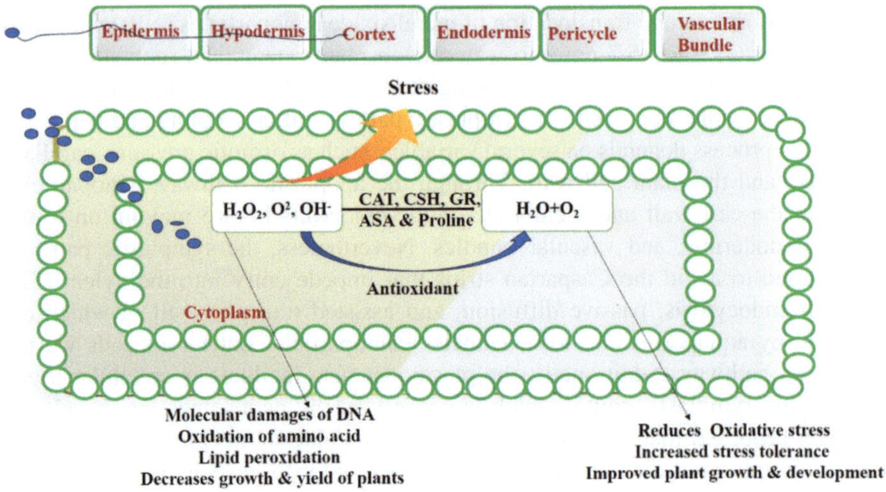

Fig. 2 This illustration showed the absorbance of nanoparticles, mechanism, and increase in stress tolerance in plant cells

showing how the nanoparticle enters or penetrates the plant cell through the cuticle, epidermis, and wound. Furthermore, it is stated that when plant cells are under DS, they begin to produce reactive oxygen species, which are further accountable for lipid peroxidation, DNA damage, and death. These nanoparticles participate in the synthesis of antioxidants and secondary metabolites that support the development of stress tolerance in plant cells.

4 Application of Nanoparticle in Plants Under Drought Stress

4.1 Nanoparticles Improve the Germination and Vegetative Traits

Seed germination is the earliest stage of plant life, and the establishment of seedlings is a very delicate stage that is extremely vulnerable to environmental stress. However, the reduced water uptake in seeds causes the hydrolytic enzymes, especially amylase, which is essential for embryo development, to become inactive, delaying seed germination in plants. Additionally, it has been claimed that the nanoparticle has improved crop plants' germination and growth rates [37, 88, 100]. Research is necessary because the exact mechanism is yet unclear. Several publications that demonstrate improved germination and growth metrics in crop plants mediated by nanoparticles under DS have been complied with and thoroughly examined in this chapter. Higher germination rates have been found for seeds primed with nanoparticles, this could be because the increased production of phytohormones, which act as the promotor for germination, is enhanced [37]. According to one study, soybean seeds treated with zinc oxide nanoparticles had a higher rate of germination than seeds in the control group. When plants are subjected to severe PEG-induced dehydration stress of − 1 MPa, 89.5% more germination occurs when 1 g of nanoparticle treatment per litre is applied. Additionally, it has been observed that nanoparticles function as a cofactor in the activation of hydrolytic enzymes, leading to enhanced embryonic development. According to published reports, soy-treated seeds had increased radicle lengths when treated with 1 g nanoparticle per litre [100]. Furthermore, another study found that increasing the amount of zinc oxide nanoparticles in soybean plants caused a significant change in their fresh and dry weight. Furthermore, [100] published similar data, indicating that zinc oxide nanoparticles also promoted cell growth and improved membrane stability. As the concentration of nanoparticles increased, the weight of fresh and dry samples decreased, indicating that the nanoparticle-treated seed stock was effectively used for the biosynthesis of phytohormones. Subsequently, an increase in the growth rate of plumule and radicle length was observed in soybean plants. From this data, it is inferred that nanoparticles stimulate the amino acid tryptophan [74, 100]. Similarly, when treated with a low concentration of carbon nanotube, *Alnus subcordate*, hop bush, chickpea, and wheat showed improved morphological traits like growth rate, radicle length, seed vigour index, fresh weight, dry weight, seedling growth, and mean germination time [62, 89, 115, 123]. Additionally, seeds treated with nanoparticles indicate that the water intake is controlled by the seed coat's aquaporins. Plants treated with silicon oxide and silicon dioxide nanoparticles have demonstrated improved production of phytohormones, including auxins and gibberellins, leading to increased marigold seedling germination and vigour index. Conversely, under extreme drought conditions, the control group displayed lower germination characteristics [88]. Another study by Faraji and Sepehri [30] found that when wheat seedlings were subjected to moderate and severe DS and treated with varying concentrations of titanium oxide nanoparticles, along with sodium nitroprusside as a nitrous oxide (NO) donor, the biomass of the seedlings increased. Wheat seedlings treated with nanoparticles under medium stress exhibited a notable increase in length; under severe stress, this was not the case. In contrast, during extreme DS, the control group had shorter seedlings, while the group treated with sodium nitroprusside nanoparticles displayed longer seedlings. This finding implies that NO from sodium nitroprusside enhances the protective impact of nanoparticles by promoting the development of wheat seedlings in conjunction with nanoparticle treatment. Similarly, when biosynthesized nano selenium was applied topically to wheat seedlings at a concentration of 30 mg/L, it increased several morphological characteristics, including plant height, root and shoot length, total number of leaves, leaf area, and fresh and dry weight under conditions of water deficit. However, a lesser dosage encouraged the growth of roots and shoots, and the buildup of nano selenium in the seedlings enhanced organogenesis [60]. Recent research on plants treated with metal and metal-oxide nanoparticles has demonstrated beneficial effects

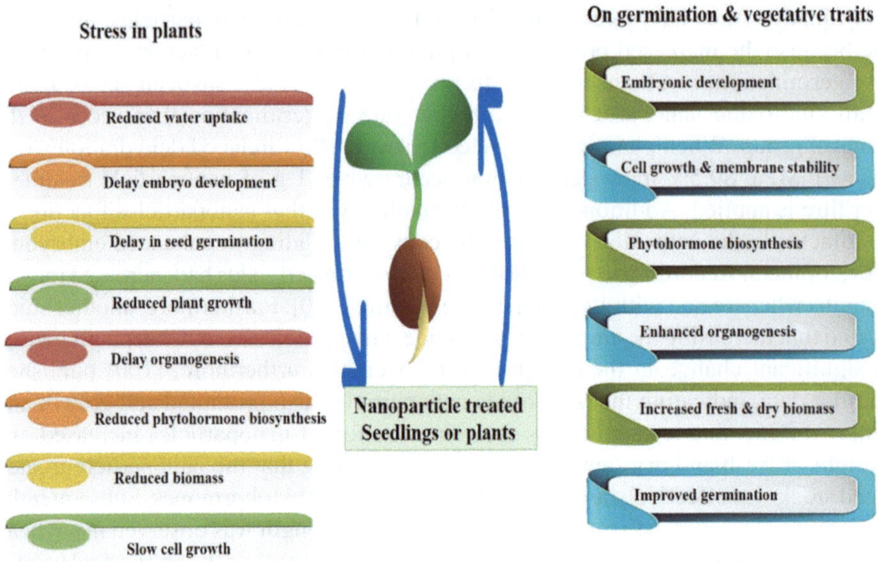

Fig. 3 The flow chart represents the effects of nanoparticles on plant stress and germination

on germination, growth, and agronomical parameters in crop plants. The application of nanoparticles in crop plants is accomplished through the appropriate dose-dependent uptake and accumulation of the nanoparticles as well as their interaction with the plant cells [3, 4]. This section contains a variety of statistics showing the application of nanoparticles, which has had several beneficial outcomes, such as increased crop plant germination and growth rates. Higher germination rates, growth metrics, increased radicle length, enhanced cell growth, enhanced membrane cell stability, increased fresh and dry biomass, enhanced morphological traits, and increased biosynthesis of phytohormones, and other improved vegetative traits are also mentioned. The work has arranged these results in a single glance in the provided Fig. 3. All these above characteristics are reduced or inhibited under DS conditions, but nanoparticles help the plants overcome this situation and increase their tolerance strength under unfavourable conditions.

4.2 *Nanoparticles Improve the Physio-Biochemistry*

By enhancing several physiological functions, including water relations, stomatal conductance, photosynthetic activity, and the uptake of micro and macronutrients, applying nanoparticles to plants can lessen DS, according to findings in physiochemistry. For example, Hawthorne seedling's xylem potential has been increased with the use of silicon dioxide nanoparticles. On the other hand, the increased xylary potential had less of a negative impact on Hawthorne seedlings, but no changes in the relative water content of water-stressed plants were noted as compared to the controls. These nanoparticle seed priming techniques improved the length of wheat seedlings at the crucial stages of water scarcity conditions. According to reports, the accumulation of nanoparticles on the cuticle also acts as a barrier, shielding the seedlings from stress and preserving water by reducing transpiration. It has also been demonstrated to enhance water and nutrient uptake, which may have led to increased biomass in the seedlings [13, 91, 107]. Additionally, it has been observed that silicon nanoparticles raise endogenous cytokinin, which helps in the restoration of photosynthetic pigments under stressful circumstances. Fullerenol nanoparticles were applied topically to *Beta vulgaris* plants to reduce oxidative stress and drought, according to a similar study. To help deal with water stress, this nanoparticle forms an extra water reserve by binding with the water molecule [20, 43]. The relative water content, drought tolerance index, and biomass loss rate of soybean seedlings treated with iron, copper, cobalt, and zinc oxide nanoparticles were examined. Although the relative water content was identical for all seedling groups, the iron and copper-treated groups had a greater relative water content (71%, and their accumulation improved their ability to retain water. When compared to other nanoparticle and control groups, iron-treated seedlings exhibit the best level of drought tolerance. When there was a water deficit, the biomass loss rate was lower in soybean seedlings treated with iron and cobalt, however, iron treatment has been effective in reducing DS [69]. Like this, eggplant was treated with foliar application of zinc oxide nanoparticles, which improve photosynthetic activity in both well-watered and drought-stressed conditions. In water-deficient conditions, the chlorophyll index increased in nanoparticle-treated seedling, while in the negative control group, the leaf chlorophyll index decreased. However, it has been seen that these nanoparticle treatments with 100 ppm improve physiological parameters, improve membrane stability, and play a major role in helping seedlings maintain their relative water content [101]. Additionally, another study by [96] Sallam et al. [96] and Ahmed et al. [3] discovered that treating wheat seedlings with copper and silver nanoparticles enhanced photosynthetic activity and greater levels of chlorophyll a and b pigments. Prior research has shown that copper is an essential macronutrient involved in photosynthesis, and wheat seedlings treated with copper nanoparticles showed regulated stomatal conductance at 0.3 mg/L under DS as well as increased photosynthetic activity that reduces DS. Abiotic stress results in the generation of O^2, OH^-, and H_2O_2 radicals, which profoundly damage plant tissues. These radicals can harm proteins, lipids, and carbohydrates, which lowers metabolism and damages membranes. All these damages are irreversible because they damage DNA and cause programmed cell death [3, 4, 65, 96]. The accumulation of hydrogen peroxide leads to lipid oxidation, which damages the membrane and is indicated by accumulation of MDA. In addition, linseed seedlings were treated with titanium dioxide nanoparticles, which prevented hydrogen peroxide accumulation in both drought and well-watered circumstances as compared to the control groups. The accumulation of MDA at a concentration of 4 mM and the reduction of reactive oxygen species generation improved cellular electron exchange pathways, hence reducing oxidative damage brought on by electron leakage and ROS production [2, 13]. Additionally, in another investigation, silicon dioxide nanoparticles were applied to Hawthorne seedlings, which lessens the formation of MDA under stressful and underwatered situations, where the seedling's MDA content decreases as the nanoparticle concentration rises. Treatment with nanoparticles has been shown to increase the antioxidant activities of superoxide dismutase, peroxidase, and catalase. It will also shield the seedlings from oxidative damage brought on by DS. In a related study, rapeseed seedlings treated with maghemite (iron oxide) nanoparticles under DS showed less hydrogen peroxide and MDA formation than the control, which decreased lipid peroxidation [13, 85]. Additionally, calcium nanoparticles were applied to rapeseed seedlings once more, improving photosynthetic parameters such as pigment levels, photosynthetic rate, and photosystem performance under 15% PEG-induced dehydration stress. The impact of nanoparticles on the cell organelle was examined using a transmission emission microscope. With enlarged thylakoids, the chloroplast seemed asymmetrical, but the chloroplast treated with nanoparticles was more regular than the stressed plant cells [16]. Furthermore, under DS conditions, zinc oxide nanoparticles treated with cucumber at a concentration of 100 mg/L decreased the production of 36% H_2O_2 and 40% $(O^2) -$, which dramatically decreased lipid peroxidation and electron leakage in comparison to the control group. SOD, CAT, POX, glutathione reductase (GR), and APX recycling enzymes, such as monodehydroascorbate reductase and dehydroascorbate reductase, are among the enzymatic and non-enzymatic antioxidant activities that are increased by this treatment. By reducing oxidative damage brought on by the generation of reactive oxygen species, all these enzymes shield seedlings from irreversible harm and even death. Since SOD is Zn-dependent, this therapy enhances Zn^{2+} absorption and utilization, which raises SOD activity noticeably. Based on this finding, we propose that this therapy strengthens the antioxidant enzymes that assist in eliminating reactive oxygen species (ROS) produced by DS, thereby shielding plants from oxidative damage. Additionally, it activates the antioxidant defence system, preventing oxidative damage brought on by DS [36]. In a related study, maize seedlings treated with copper nanoparticles showed an increase in carotenoid content and a decrease in chlorophyll content. The presence of carotenoid shields chlorophyll from drought-induced oxidative damage. By increasing scavenging enzymes that aid in eliminating ROS from the plant system, oxidative damage is lessened. By enhancing photosynthetic indices and enzymatic antioxidants, which strengthen and protect the seedlings from drought-induced oxidative stress, this treatment, according to the study, lowers oxidative stress in maize seedlings. Additionally, a variety of metabolites, such as osmoprotectants and osmolytes, such as polyol, proline, glycine betaine, total soluble sugars, and total proteins, are produced by plants. Since proteins, lipids, soluble sugar, and carbohydrates are essential cellular components that help stressed plants recover swiftly, these metabolites can effectively maintain osmotic equilibrium, activate the stress tolerance mechanism, and shield these substances from oxidative damage [28]. These new findings show that plants treated with zinc oxide nanoparticles during drought had higher amounts of soluble sugars, proline, glycine betaine, and total proteins. Additionally, this treatment raised the number of metabolites and stopped the decrease of soluble sugar. Proline is a crucial osmolyte that helps in reestablishing osmotic equilibrium, and this medication has increased its biosynthesis via regulating gene expression [36]. Stress causes the generation of reactive oxygen species, hydro anion, and hydrogen peroxides, which further deteriorate the cell membrane, raise lipid peroxidation, and reduce plant yield. According to the reports mentioned above, plants treated with nanoparticles are effective at reducing DS through several physiological processes, including stomatal conductance, photosynthetic activity, nutrient uptake, transpiration reduction, activation of the antioxidant defence system, and activation of the plant's stress tolerance mechanism. In the end, the author has attempted to provide a concise overview of all the ways that nanoparticles are used in plants to combat DS, as shown in Fig. 4.

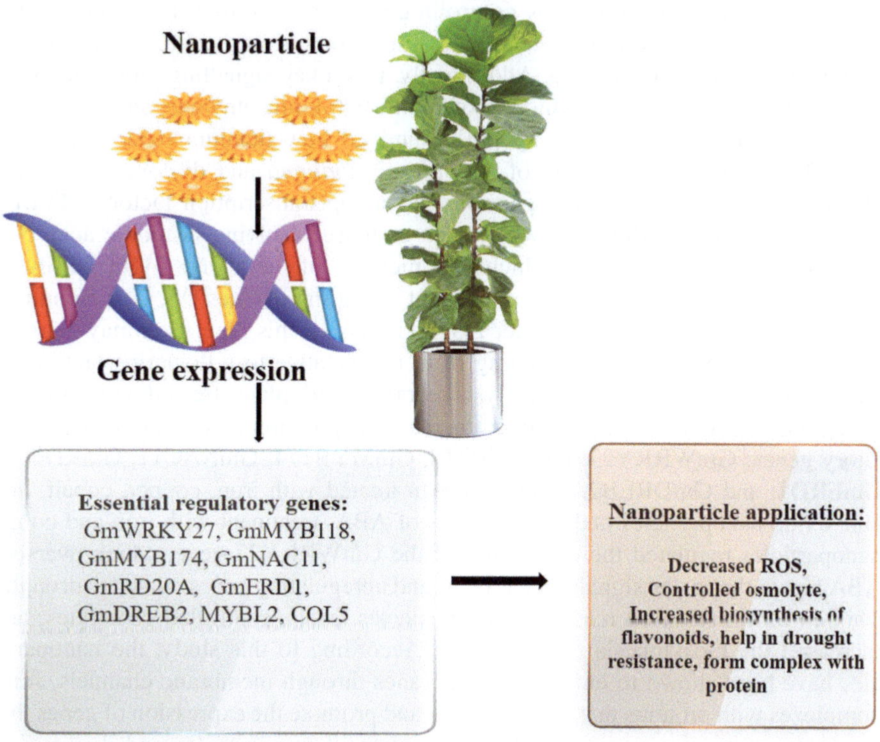

Fig. 4 The figure represents the impact of nanoparticles and their application on gene expression in plants

4.3 Nanoparticle Influence on Phytohormones and Gene Expression

DS has a significant effect on regulating the expression of genes that respond to drought and phytohormones, and it can initiate a variety of intricate signalling cascades. Metal/metal-oxide nanoparticles influence drought-induced gene expression through either the ABA-dependent or ABA-independent pathways, depending on the type of nanoparticle and the species of plant [69]. Phytohormones, particularly the abscisic acid (ABA) phytohormone, which is engaged in mechanisms that help plants overcome abiotic stress, including DS, have a significant impact on plant drought responses and defence mechanisms against tolerance. ABA participates in the development of tolerance, by controlling plant growth, root development, and leaf elongation. In addition, it modulates transpiration rate, cellular water retention, and stomatal conductance. Additionally, it is a key signalling molecule in the manufacture of other phytohormones, including cytokinin, ethylene, and gibberellic acid [96, 124]. Another study by [33] examined how titanium oxide nanoparticles affected the transcriptome of *Arabidopsis thaliana* and discovered that DS affected genes induced by nanoparticles, including transcription factors. MYBL2 controls several signalling pathways, such as those involving jasmonic acid, SA, and ABA, as well as COL5, which is a member of the CONSTANS family of flowering regulators. While ABA-dependent signalling reduces DS, COL5 controls the flowering period of plants under DS; additionally, this regulation may help the plant escape drought by transitioning from a vegetative to a flowering state [33]. According to a gene expression analysis report by [69], plants treated with nanoparticles in drought conditions exhibited increased expression of seven essential regulatory genes: GmWRKY27, GmMYB118, GmMYB174, GmNAC11, GmRD20A, GmERD1, and GmDREB2. While soybean treated with iron, copper, cobalt, and zinc oxide nanoparticles had higher levels of ABA, treatment with iron and cobalt nanoparticles regulated the expression of the GmWRKY27 gene, which oversees ABA biosynthesis, the signalling pathway, and upregulating genes related to drought. This, in turn, decreased reactive oxygen species levels, controlled osmolytes, and increased the biosynthesis of flavonoids. According to this study, the nanoparticles have been shown to enter cell membranes through membrane channels, form complexes with proteins that bind calcium, and promote the expression of genes that aid in drought resistance [69, 92]. The function of nanoparticles and their use in phytohormone and gene expression are summed up in this section. These nanoparticles aid in regulating the osmolytes and reducing the generation of reactive oxygen species. Treatment with nanoparticles also aids in maintaining the drought resistance gene in plants and boosting the manufacture of flavonoid phytohormone. At the very least, these nanoparticles help the proteins that bind calcium to form complexes. To illustrate the precise application of nanoparticles in crop plants, these identical results are grouped into a single Fig. 5.

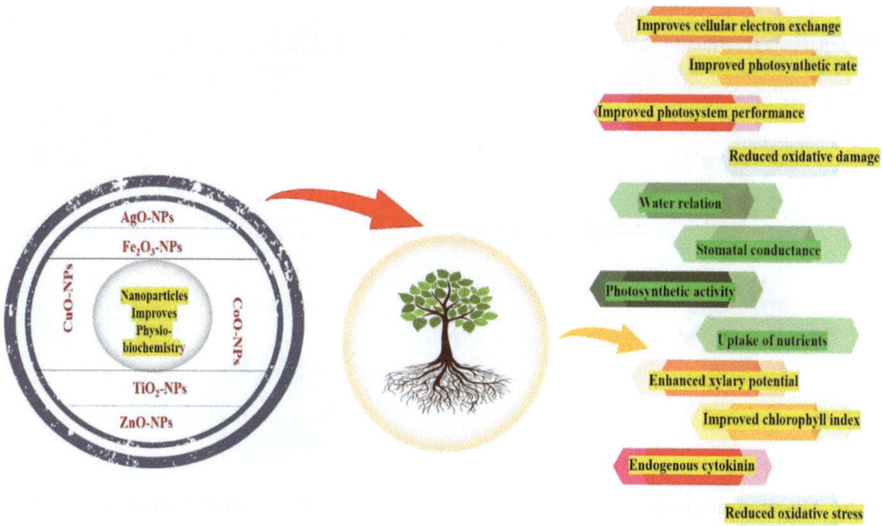

Fig. 5 The function of nanoparticles in enhancing different plant physio-biochemistry is depicted in this picture

4.4 Nanoparticle Influence on the Life Cycle, Yield, and Nutritional Quality of Crops

One unfavourable circumstance that has a major effect on crop physiology, yield, productivity, and quality is drought. It has negative impacts on entire farming communities and, in certain situations, entire nations in addition to individual farming homes [3]. During the vegetative and reproductive stages, soybeans that are exposed to DS show a decline in growth indices. Nonetheless, during vegetative maturation in non-drought settings, seedlings treated with micnobits—a mixture of micronutrients such ZnO, CuO, and B_2O_3 nanoparticles of varying sizes—improved the growth indices, such as plants grown, leaf area, number, and fresh shoot weight. Likewise, reproductive characteristics like pod fresh weight, number, fresh grain weight, grain count, and grain dry weight are adversely affected by stress. Micnobit treatments enhanced several parameters including the yield and helped the soybean plant recover from DS efficiently. The study also reported a significant spike in the translocation of nutrients, such as N and K, in the grain [24]. In a related study, wheat seedlings grown hydroponically and treated with silver nanoparticles have shown improved morphological and yield indices. Furthermore, in comparison to other treatments and control groups, plants treated with cupper nanoparticles (3 mg/L) and silver nanoparticles (30 mg/L) displayed increased spike length, spikelet density per spike, grain weight, and quantity in stressed seedlings. Additionally, applying titanium oxide nanoparticles to linseed increased its protein and oil content, improving yield quality in water-stressed environments. In both well-watered and stressed situations,

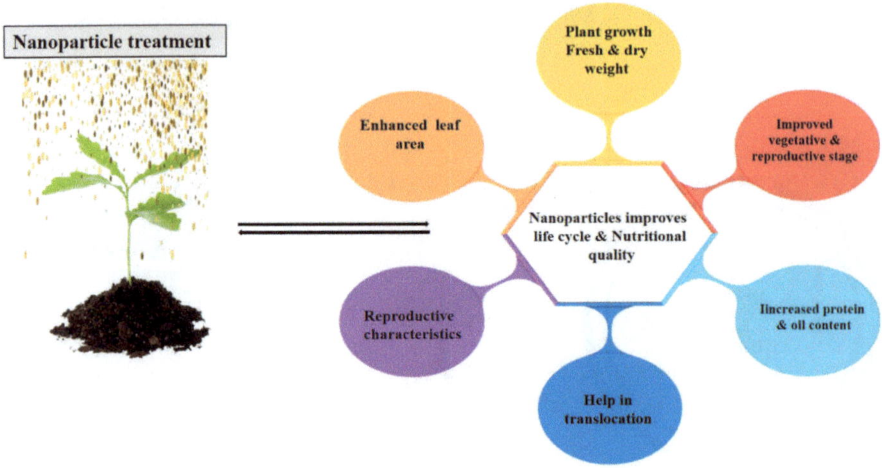

Fig. 6 The application of nanoparticles to enhance the nutritional value and life cycle of crop plants is depicted in this graphic

treatment at 100 mg/L increased the seed oil content to 37 and 35% and the seed protein content to 19 and 22%, respectively [2, 3]. Nanoparticles have been employed in this area to improve the life cycle and nutritional quality. Micnobits are used to increase fresh shoot weight, leaf area, and plant growth. Pod fresh weight, dry weight, and other reproductive traits are all improved by nanoparticles that enter plant cells. Following that, as seen in the provided Fig. 6, this nanoparticle treatment aids in the translocation of nitrogen and phosphorus nutrients. Nanoparticle treatments can help plants manage DS during both vegetative and reproductive stages. These treatments show potential for enhancing yield quantity and quality under challenging conditions. However, further research is urgently needed to confirm and better understand the role of nanoparticles in improving plant health throughout its life cycle under DS. Additionally, studies on the accumulation of nanoparticles in grains and their potential impact on human and livestock health upon consumption are essential.

4.5 Nanoparticles Enhance Secondary Metabolites

Secondary metabolites, like phenolics, flavonoids, and alkaloids, have therapeutic uses that attract research interest. Secondary metabolites are recognized as indicators of stressful conditions in plants; they are produced when high levels of reactive oxygen species are produced, which causes oxidative stress, and these secondary metabolites function as a plant defence system against stress [7, 119]. However, the mechanism by which nanoparticles are involved in the biosynthesis of secondary metabolites is still unknown. Nevertheless, it is still unclear how nanoparticles contribute to the creation of secondary metabolites. Here, water-stressed sorghum seedlings cultivated in soil treated with silicon nanoparticles exhibited a notable rise in flavonoid content and an improvement in total phenolic compounds. Mildly and moderately stressed plants had significantly higher levels of ferulic acid, a phenolic substance; severely stressed seedlings had lower levels. Compared to weakly and moderately stressed seedlings, highly stressed seedlings had significantly greater levels of vanillic acid. The fortification of the cell wall by Si NPs may be due to the production of phenolic chemicals [38]. Similarly, periwinkle seedlings treated with chitosan nanoparticles topically showed a substantial rise in alkaloid levels under stressful circumstances. Nonetheless, alkaloids accumulated more in root samples than in shoot samples, and improving water intake through osmotic adjustment for plant growth increased alkaloid concentration. According to this study, the high alkaloid content was responsible for photosynthate storage and transfer. Genes involved in the manufacture of terpenoid indole alkaloids, including strictosidine synthase, deacetylvin doline-4-O-acetyltransferase 15, peroxidase 1, and geissoschizine synthase, are increased in seedlings treated with nanoparticles. Research is currently ongoing to determine the precise mechanism underlying interspecies differences in drought amelioration and nanoparticle uptake. The types of nanoparticles and other parameters, such as pore size, density and distribution, stomatal size, epidermal thickness and composition are the primary determinants of these interactions, which vary by species [90]. Nanoparticles assist in increasing the amount of phenolics, flavonoids, ferulic acid, vanillic acid, terpenoids, indole alkaloids, and alkaloids in the metabolites section. As seen in Fig. 7, this nanoparticle promotes plant-nutrient uptake, antioxidant, and photosynthetic activities, and helps in transfer and photosynthetase storage.

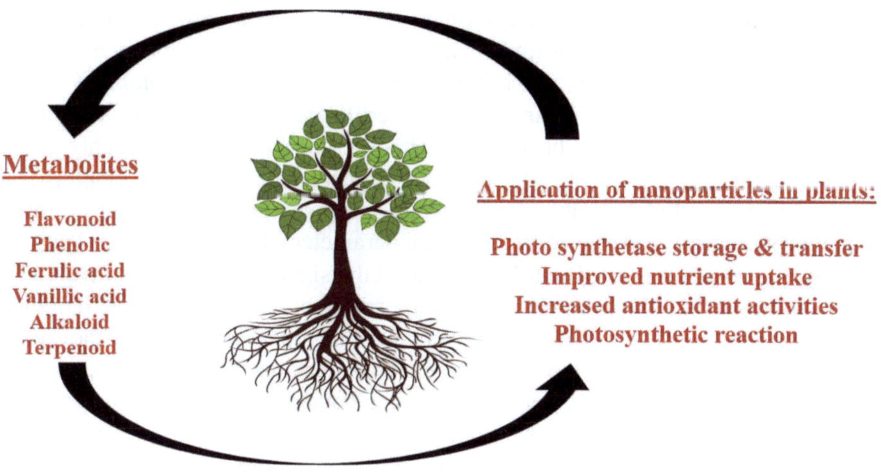

Fig. 7 The function of nanoparticles in plants to improve metabolites and their physiological processes is illustrated in this image

5 Conclusion and Prospects

Agriculture is currently dealing with several issues, such as the growing world population, hunger, and poverty because of a slowdown in the rate of rise in agricultural yields. Today's agricultural yield decreases are brought on by global stress. Numerous biotic and abiotic environmental factors influence plant development and production in agriculture; among abiotic elements, DS is one of the main abiotic stresses that negatively impacts crop growth and primarily restricts agricultural productivity globally. One of the main factors influencing the regulation of fitness and survival in healthy plants is their water capacity. DS significantly reduces agricultural productivity by altering several plant characteristics, including photosynthesis, respiration, water relations, and biomass production. Additionally, it significantly impacts plant development, phenology, respiration, assimilate partitioning, photosynthesis, water and nutrient relations, seed germination, and yield components. Plant growth is impacted at every phenological stage, demonstrating the morphological and molecular effects of DS. It is the primary abiotic factor that negatively impacts agricultural productivity, growth, and development. It is also one of the most major environmental stresses that significantly reduces yield, particularly in warm, arid regions of the world. The drought has a variability of effects on many crop plants' physical, physiological, and genetic traits, as shown in Fig. 8. Modern nanotechnology, which mostly uses nanoparticles and is permeating human life, is the most practical way to deal with all the circumstances. It is also the fastest growing technology. Nanotechnology's use in crop development and sustainable agriculture is still in its infancy. Therefore, developing a fundamental understanding of how nanoparticles interact with plants at the cellular and molecular level has become essential to maximizing the potential benefits of using their peculiar and unique qualities in the agriculture industry. By improving physiological parameters such as water relations, stomatal conductance, photosynthetic activity, and the uptake of micro- and macronutrients into nanoparticle-based fertilisers, nanoparticle treatment has been shown to reduce DS in host plants. This present work has thoroughly examined the methods of uptake and advantageous impacts of nanoparticles, emphasizing that using nanoparticle therapy might greatly improve agricultural plant germination, morphological, and physio-biochemical parameters. In addition to nanoparticle therapy, it was demonstrated that DS increased the signalling pathways involved in the manufacture of phytohormones and osmoprotectants. However, it is still unclear what exact mechanisms underlie the movement of nanoparticles and how they affect crops during drought at the molecular level. This work has combined recent developments in nanoparticles that help us better understand their positive impact in the host plants as well as their modes of uptake, compartmentalization, and interactions with plants. To comprehend the underlying principles, more research must be done on the use of nanoparticles to reduce DS in plants. It is necessary to investigate ionic channels and nutritional pathways because nanoparticles improve nutrient uptake and utilization. They raise antioxidant activities and enhance photosynthetic parameters which affects the intercellular pathways of photosynthetic processes and antioxidant

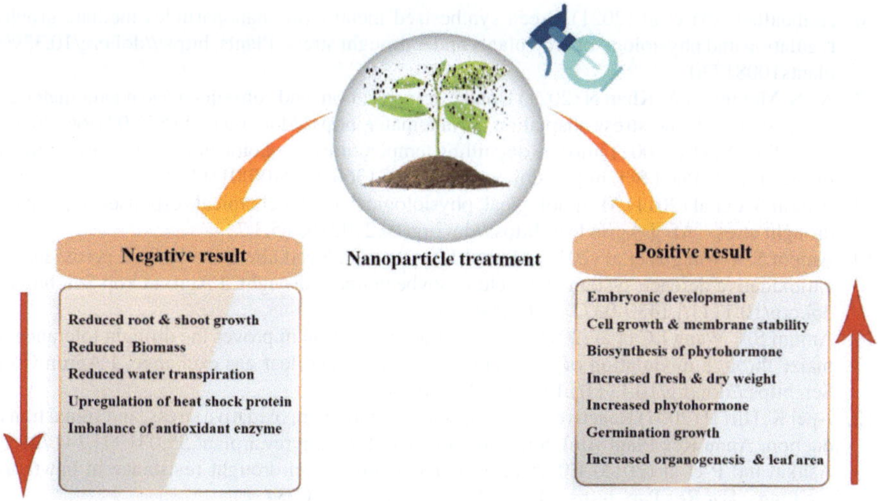

Fig. 8 The utilization of nanoparticles to lessen the adverse effects of drought stress is shown in this figure

activation, which is yet mostly unknown. Furthermore, the effects of nanoparticles on plants under DS and their stress tolerance mechanism may be revealed via an omics-based approach. It reduces the DS on crop plants, which may be a novel agricultural strategy. Overuse of nanoparticles, however, may result in excessive dose accumulation in food grains, which may then enter the food chain and pose a serious concern. Before launching nano-agri products, precautions should be taken to avoid harmful side effects from nanoparticles.

References

1. Adrees M et al (2020) Simultaneous mitigation of cadmium and drought stress in wheat by soil application of iron nanoparticles. Chemosphere. https://doi.org/10.1016/j.chemosphere.2019.124681
2. Aghdam MTB, Mohammadi H, Ghorbanpour M (2016) Effects of nanoparticulate anatase titanium dioxide on physiological and biochemical performance of Linum usitatissimum (Linaceae) under well-watered and drought stress conditions. Rev Bras de Bot. https://doi.org/10.1007/s40415-015-0227-x
3. Ahmed F et al (2021) Applications of copper and silver nanoparticles on wheat plants to induce drought tolerance and increase yield. IET Nanobiotechnol. https://doi.org/10.1049/nbt2.12002
4. Ahmed T et al (2021) Nanoparticle-based amelioration of drought stress and cadmium toxicity in rice via triggering the stress responsive genetic mechanisms and nutrient acquisition. Ecotoxicol Environ Saf. https://doi.org/10.1016/j.ecoenv.2020.111829
5. Al-Khayri JM et al (2023) The role of nanoparticles in response of plants to abiotic stress at physiological. Biochem Mol Levels Plants. https://doi.org/10.3390/plants12020292

6. Alabdallah NM et al (2021) Green synthesized metal oxide nanoparticles mediate growth regulation and physiology of crop plants under drought stress. Plants. https://doi.org/10.3390/plants10081730
7. Ali S, Mehmood A, Khan N (2021) Uptake, translocation, and consequences of nanomaterials on plant growth and stress adaptation. J Nanomater. https://doi.org/10.1155/2021/6677616
8. Allen DJ, Ort DR (2001) Impacts of chilling temperatures on photosynthesis in warm-climate plants. Trends Plant Sci. https://doi.org/10.1016/S1360-1385(00)01808-2
9. Anjum SA et al (2011) Morphological, physiological and biochemical responses of plants to drought stress. Afr J Agric Res. https://doi.org/10.21921/jas.5.3.7
10. Anjum SA, Wang L et al (2011) Methyl Jasmonate-induced alteration in lipid peroxidation, antioxidative defence system and yield in soybean under drought. J Agron Crop Sci. https://doi.org/10.1111/j.1439-037X.2011.00468.x
11. Anjum SA, Wang LC et al (2011) Brassinolide application improves the drought tolerance in maize through modulation of enzymatic antioxidants and leaf gas exchange. J Agron Crop Sci. https://doi.org/10.1111/j.1439-037X.2010.00459.x
12. Apel K, Hirt H (2004) Reactive oxygen species: metabolism, oxidative stress, and signal transduction. Annu Rev Plant Biol. https://doi.org/10.1146/annurev.arplant.55.031903.141701
13. Ashkavand P et al (2015) Effect of SiO_2 nanoparticles on drought resistance in hawthorn seedlings. For Res Pap. https://doi.org/10.1515/frp-2015-0034
14. Ashraf M, Foolad MR (2007) Roles of glycine betaine and proline in improving plant abiotic stress resistance. Environ Exp Bot. https://doi.org/10.1016/j.envexpbot.2005.12.006
15. Avellan A et al (2019) Nanoparticle size and coating chemistry control foliar uptake pathways. Trans Leaf-to-Rhizosphere Transp Wheat ACS Nano. https://doi.org/10.1021/acsnano.8b09781
16. Ayyaz A et al (2022) Calcium nanoparticles (Ca-NPs) improve drought stress tolerance in Brassica napus by modulating the photosystem II, nutrient acquisition and antioxidant performance. NanoImpact. https://doi.org/10.1016/j.impact.2022.100423
17. Benjamin JG, Nielsen DC (2006) Water deficit effects on root distribution of soybean, field pea and chickpea. Field Crop Res. https://doi.org/10.1016/j.fcr.2005.10.005
18. Bisht S, Sharma V, Kumari N (2022) Biosynthesized magnetite nanoparticles from Polyalthia longifolia leaves improve photosynthetic performance and yield of Trigonella foenum-graecum under drought stress. Plant Stress. https://doi.org/10.1016/j.stress.2022.100090
19. Del Blanco IA et al (2000) Physiological performance of synthetic hexaploid wheat-derived populations. Crop Sci. https://doi.org/10.2135/cropsci2000.4051257x
20. Borišev M et al (2016) Drought impact is alleviated in sugar beets (Beta vulgaris L.) by foliar application of fullerenol nanoparticles. PLoS ONE. https://doi.org/10.1371/journal.pone.0166248
21. Chandra S (2003) Effect of leaf age on transpiration and energy budget in Ficus glomerata Roxb., Physiol Mol Biol Plants
22. Cifuentes Z et al (2010) Absorption and translocation to the aerial part of magnetic carbon-coated nanoparticles through the root of different crop plants. J Nanobiotechnol. https://doi.org/10.1186/1477-3155-8-26
23. Demiral T, Türkan I (2004) Does exogenous glycinebetaine affect antioxidative system of rice seedlings under NaCl treatment? J Plant Physiol. https://doi.org/10.1016/j.jplph.2004.03.009
24. Dimkpa CO et al (2017) Composite micronutrient nanoparticles and salts decrease drought stress in soybean. Agron Sustain Dev. https://doi.org/10.1007/s13593-016-0412-8
25. Djanaguiraman M et al (2018) Cerium oxide nanoparticles decrease drought-induced oxidative damage in sorghum leading to higher photosynthesis and grain yield. ACS Omega. https://doi.org/10.1021/acsomega.8b01894
26. Dodd IC (2003) Hormonal interactions and stomatal responses. J Plant Growth Regul. https://doi.org/10.1007/s00344-003-0023-x
27. Duan B et al (2007) Interactions between water deficit, ABA, and provenances in Picea asperata. J Exp Bot. https://doi.org/10.1093/jxb/erm160

28. El-Bassiouny HMS et al (2022) Physiological and molecular response of wheat cultivars to titanium dioxide or zinc oxide nanoparticles under water stress conditions. Int J Agron. https://doi.org/10.1155/2022/3806574
29. Embiale A, Hussein M, Husen A, Sahile S, Mohammed K (2016) Differential Sensitivity of Pisum sativum L. cultivars to water-deficit stress: changes in growth, water status, chlorophyll fluorescence and gas exchange attributes. J Agron 15:45–57
30. Faraji J et al (2020) Exogenous nitric oxide improves the protective effects of TiO2 nanoparticles on growth, antioxidant system, and photosynthetic performance of wheat seedlings under drought stress. https://doi.org/10.1007/s42729-019-00158-0
31. Farooq M et al (2009) Plant drought stress: effects, mechanisms and management. Agron Sustain Dev. https://doi.org/10.1051/agro:2008021
32. Fiol DF et al (2021) Nanotechnology in plants: recent advances and challenges. J Chem Technol Biotechnol. https://doi.org/10.1002/jctb.6741
33. García-Sánchez S, Bernales I, Cristobal S (2015) Early response to nanoparticles in the Arabidopsis transcriptome compromises plant defence and root-hair development through salicylic acid signalling. BMC Genomics. https://doi.org/10.1186/s12864-015-1530-4
34. Geisler-Lee J et al (2012) Phytotoxicity, accumulation and transport of silver nanoparticles by Arabidopsis thaliana. Nanotoxicology. https://doi.org/10.3109/17435390.2012.658094
35. Getnet Z, Husen A, Fetene M, Yemata G (2015) Growth, water status, physiological, biochemical and yield response of stay green sorghum Sorghum bicolor (L.) Moench varieties-a field trial under drought-prone area in Amhara regional state, Ethiopia. J Agron 14:188–202. https://doi.org/10.3923/ja.2015.188.202
36. Ghani MI et al (2022) Foliar application of zinc oxide nanoparticles: An effective strategy to mitigate drought stress in cucumber seedling by modulating antioxidant defense system and osmolytes accumulation. Chemosphere. https://doi.org/10.1016/j.chemosphere.2021.133202
37. Gholami S et al (2022) Seed germination and physiological responses of quinoa to selenium priming under drought stress. Bragantia. https://doi.org/10.1590/1678-4499.20210183
38. Ghorbanpour M, Mohammadi H, Kariman K (2020) Nanosilicon-based recovery of barley (: Hordeum vulgare) plants subjected to drought stress. Environ Sci Nano. https://doi.org/10.1039/c9en00973f
39. Guerfel M et al (2009) Impacts of water stress on gas exchange, water relations, chlorophyll content and leaf structure in the two main Tunisian olive (Olea Europaea L.) cultivars. Sci Hortic. https://doi.org/10.1016/j.scienta.2008.08.006
40. Guerrero F, Mullet JE (1986) Increased abscisic acid biosynthesis during plant dehydration requires transcription. Plant Physiol. https://doi.org/10.1104/pp.80.2.588
41. Harris D, Tripathi RS, Joshi A (2002) On-farm seed priming to improve crop establishment and yield in dry direct-seeded rice. Dir Seed Res Strat Opportunities Int Res Inst Manila Philippines 1:231–240
42. Hasaneen MNA, Abdel-Aziz HMM, Omer AM (2016) Effect of foliar application of engineered nanomaterials: carbon nanotubes NPK and chitosan nanoparticles NPK fertilizer on the growth of French bean plant. Biochem Biotechnol Res
43. Hatami M et al (2019) Physiological and antioxidative responses to GO/PANI nanocomposite in intact and demucilaged seeds and young seedlings of Salvia Mirzayanii. Chemosphere. https://doi.org/10.1016/j.chemosphere.2019.05.268
44. Hayat F et al (2023) Nanoparticles and their potential role in plant adaptation to abiotic stress in horticultural crops: a review. Sci Hortic. https://doi.org/10.1016/j.scienta.2023.112285
45. Horváth E et al (2007) Exogenous 4-hydroxybenzoic acid and salicylic acid modulate the effect of short-term drought and freezing stress on wheat plants. Biol Plant. https://doi.org/10.1007/s10535-007-0101-1
46. Husen A (2010) Growth characteristics, physiological and metabolic responses of teak (Tectona grandis Linn. f.) clones differing in rejuvenation capacity subjected to drought stress. Silvae Genet 59:124–136. https://doi.org/10.1515/sg-2010-0015
47. Husen A, Iqbal M, Aref IM (2017) Plant growth and foliar characteristics of faba bean (Vicia faba L.) as affected by indole-acetic acid under water-sufficient and water-deficient conditions. J Environ Biol 38:179–186

48. Husen A (2024a) Biostimulants in plant protection and performance. Elsevier Inc., Cambridge. https://doi.org/10.1016/C2022-0-02207-8
49. Husen A (2024b) Essential minerals in plant-soil system (coordination, signalling, and interaction under adverse situations). Elsevier Inc., Cambridge. https://doi.org/10.1016/C2022-0-02437-5
50. Husen A (2025a) Agricultural crop improvement: plant and soil relationships. Taylor & Francis Group, LLC, 6000 Broken Sound Pkwy. NW. https://doi.org/10.1201/9781032630366
51. Husen A (2025b) Role of osmolytes in changing environment. Elsevier Inc., Cambridge
52. Husen A (2025c) Role of antioxidants in mitigating plant stress. Elsevier Inc., Cambridge
53. Husen A (2024c) Plant response to gold nanoparticles. Springer Nature Singapore Pte Ltd., Singapore. https://doi.org/10.1007/978-981-97-7174-5
54. Husen A (2025d) Plant response to silver nanoparticles. Springer Nature Singapore Pte Ltd., Singapore. https://doi.org/10.1007/978-981-97-7352-7
55. Husen A (2025e) Emerging carbon nanomaterials for sustainable agricultural practices (synthesis, plant growth, performance, production and protection). Springer Nature Singapore Pte Ltd., Singapore
56. Husen A (2021a) Harsh environment and plant resilience, Springer International Publishing AG, Gewerbestrasse. Cham, pp 1–23. https://doi.org/10.1007/978-3-030-65912-7_1
57. Husen A (2021b) Plant performance under environmental stress. Springer International Publishing AG, Gewerbestrasse. Cham, pp. 1–15. https://doi.org/10.1007/978-3-030-78521-5_1
58. Husen A (2022) Environmental pollution and medicinal plants. CRC Press 6000 Broken Sound Parkway NW, Suite 300, Boca Raton, FL, pp 33487–2742. https://doi.org/10.1201/9781003178866
59. Husen A (2023) Plants and their interaction to environmental pollution (damage detection, adaptation, tolerance, physiological and molecular responses). Elsevier Inc., Cambridge. https://doi.org/10.1016/C2020-0-03319-0
60. Ikram M et al (2020) Foliar applications of bio-fabricated selenium nanoparticles to improve the growth of wheat plants under drought stress. Green Process Synth. https://doi.org/10.1515/gps-2020-0067
61. Jaleel CA et al (2008) Differential responses in water use efficiency in two varieties of Catharanthus roseus under drought stress. Comptes Rendus Biol. https://doi.org/10.1016/j.crvi.2007.11.003
62. Joshi A et al (2018) Multi-walled carbon nanotubes applied through seed-priming influence early germination, root hair, growth and yield of bread wheat (Triticum aestivum L.). J Sci Food Agric. https://doi.org/10.1002/jsfa.8818
63. Kaiser WM et al (1981) Photosynthesis under osmotic stress—Differential recovery of photosynthetic activities of stroma enzymes, intact chloroplasts, protoplasts, and leaf slices after exposure to high solute concentrations. Planta. https://doi.org/10.1007/BF00394981
64. Kamara AY et al (2003) The influence of drought stress on growth, yield and yield components of selected maize genotypes. J Agric Sci. https://doi.org/10.1017/S0021859603003423
65. Khaleghi A et al (2019) Morphological, physiochemical and antioxidant responses of Maclura pomifera to drought stress. Sci Rep. https://doi.org/10.1038/s41598-019-55889-y
66. Kurepa J et al (2010) Uptake and distribution of ultrasmall anatase TiO2 alizarin red s nanoconjugates in arabidopsis thaliana. Nano Lett. https://doi.org/10.1021/nl903518f
67. Kyparissis A, Petropoulou Y, Manetas Y (1995) Summer survival of leaves in a soft-leaved shrub (Phlomis fruticosa L., labiatae) under mediterranean field conditions: avoidance of photoinhibitory damage through decreased chlorophyll contents. J Exp Bot. https://doi.org/10.1093/jxb/46.12.1825
68. Li H et al (2016) Effects of surface ligands on the uptake and transport of gold nanoparticles in rice and tomato. J Hazard Mater. https://doi.org/10.1016/j.jhazmat.2016.04.043
69. Linh TM et al (2020) Metal-based nanoparticles enhance drought tolerance in soybean. J Nanomater. https://doi.org/10.1155/2020/4056563

70. Luz LM et al (2023) Distinct physiological mechanisms underpin growth and rehydration of Hymenaea courbaril and Hymenaea stigonocarpa upon short-term exposure to drought stress. J For Res. https://doi.org/10.1007/s11676-022-01558-2
71. Ma X et al (2010) Interactions between engineered nanoparticles (ENPs) and plants: phytotoxicity, uptake and accumulation. Sci Total Environ. https://doi.org/10.1016/j.scitotenv.2010.03.031
72. Manivannan P, Abdul Jaleel C et al (2007) Changes in antioxidant metabolism of Vigna unguiculata (L.) Walp. by propiconazole under water deficit stress. Colloids Surf B: Biointerfaces. https://doi.org/10.1016/j.colsurfb.2007.01.004
73. Manivannan P, Jaleel CA et al (2007) Growth, biochemical modifications and proline metabolism in Helianthus annuus L. as induced by drought stress. Colloids Surf B: Biointerfaces. https://doi.org/10.1016/j.colsurfb.2007.05.002
74. Mazhar MW et al (2022) Seed nano-priming with zinc oxide nanoparticles in rice mitigates drought and enhances agronomic profile. PLoS ONE. https://doi.org/10.1371/journal.pone.0264967
75. Menconi M et al (1995) Activated oxygen production and detoxification in wheat plants subjected to a water deficit programme. J Exp Bot. https://doi.org/10.1093/jxb/46.9.1123
76. Mittler R (2002) Oxidative stress, antioxidants and stress tolerance. Trends Plant Sci. https://doi.org/10.1016/S1360-1385(02)02312-9
77. Moran JF et al (1994) Drought induces oxidative stress in pea plants. Planta. https://doi.org/10.1007/BF00197534
78. Munns R, King RW (1988) Abscisic acid is not the only stomatal inhibitor in the transpiration stream of wheat plants. Plant Physiol. https://doi.org/10.1104/pp.88.3.703
79. Mustafa H et al (2021) Biosynthesis and characterization of titanium dioxide nanoparticles and its effects along with calcium phosphate on physicochemical attributes of wheat under drought stress. Ecotoxicol Environ Saf. https://doi.org/10.1016/j.ecoenv.2021.112519
80. Møller IM, Jensen PE, Hansson A (2007) Oxidative modifications to cellular components in plants. Annu Rev Plant Biol. https://doi.org/10.1146/annurev.arplant.58.032806.103946
81. Nayyar H, Gupta D (2006) Differential sensitivity of C_3 and C_4 plants to water deficit stress: association with oxidative stress and antioxidants. Environ Exp Bot. https://doi.org/10.1016/j.envexpbot.2005.06.021
82. Nayyar H, Walia DP (2003) Water stress induced proline accumulation in contrasting wheat genotypes as affected by calcium and abscisic acid. Biol Plant. https://doi.org/10.1023/A:1022867030790
83. Van Nguyen D et al (2021) Copper nanoparticle application enhances plant growth and grain yield in maize under drought stress conditions. J Plant Growth Regul. https://doi.org/10.1007/s00344-021-10301-w
84. Nonami H (1998) Plant water relations and control of cell elongation at low water potentials. J Plant Res. https://doi.org/10.1007/bf02507801
85. Palmqvist NGM et al (2017) Maghemite nanoparticles acts as nanozymes, improving growth and abiotic stress tolerance in brassica napus. Nanoscale Res Lett. https://doi.org/10.1186/s11671-017-2404-2
86. Praba ML et al (2009) Identification of physiological traits underlying cultivar differences in drought tolerance in rice and wheat. J Agron Crop Sci. https://doi.org/10.1111/j.1439-037X.2008.00341.x
87. Prasad R, Bhattacharyya A, Nguyen QD (2017) Nanotechnology in sustainable agriculture: recent developments, challenges, and perspectives. Front Microbiol. https://doi.org/10.3389/fmicb.2017.01014
88. Rahimi M et al (2021) Antioxidant gene expression analysis and evaluation of total phenol content and oxygen-scavenging system in tea accessions under normal and drought stress conditions. BMC Plant Biol. https://doi.org/10.1186/s12870-021-03275-0
89. Rahimi D et al (2016) Increasing drought resistance of Alnus subcordata CA Mey. seeds using a nano priming technique with multi-walled carbon nanotubes. J For Sci. https://doi.org/10.17221/15/2016-JFS

90. Rasheed A et al (2022) The role of nanoparticles in plant biochemical, physiological, and molecular responses under drought stress: a review. Front Plant Sci. https://doi.org/10.3389/fpls.2022.976179
91. Raza MAS et al (2023) Morpho-physiological and biochemical response of wheat to various treatments of silicon nano-particles under drought stress conditions. Sci Rep. https://doi.org/10.1038/s41598-023-29784-6
92. Rezaei Cherati S et al (2021) Whole-transcriptome responses to environmental stresses in agricultural crops treated with carbon-based nanomaterials. ACS Appl Bio Mater. https://doi.org/10.1021/acsabm.1c00108
93. de la Rosa G et al (2021) Interactions of nanomaterials and plants at the cellular level: current knowledge and relevant gaps. Nanotechnol Environ Eng. https://doi.org/10.1007/s41204-020-00100-1
94. Rucker KS et al (1995) Identification of peanut genotypes with improved drought avoidance Traits1. Peanut Sci. https://doi.org/10.3146/pnut.22.1.0003
95. Sacks MM, Silk WK, Burman P (1997) Effect of water stress on cortical cell division rates within the apical meristem of primary roots of maize. Plant Physiol. https://doi.org/10.1104/pp.114.2.519
96. Sallam A et al (2019) Drought stress tolerance in wheat and barley: advances in physiology, breeding and genetics research. Int J Mol Sci. https://doi.org/10.3390/ijms20133137
97. Samarah NH et al (2009) The effect of late-terminal drought stress on yield components of four barley cultivars. J Agron Crop Sci. https://doi.org/10.1111/j.1439-037X.2009.00387.x
98. Schachtman DP, Shin R (2007) Nutrient sensing and signaling: NPKS. Annu Rev Plant Biol. https://doi.org/10.1146/annurev.arplant.58.032806.103750
99. Schwabe F et al (2013) Influence of two types of organic matter on interaction of CeO_2 nanoparticles with plants in hydroponic culture. Chemosphere. https://doi.org/10.1016/j.chemosphere.2012.12.025
100. Sedghi M, Hadi M, Toluie SG (2013) Effect of nano zinc oxide on the germination parameters of soybean seeds under drought stress, XVI(2), pp 2–7
101. Semida WM et al (2021) Foliar application of zinc oxide nanoparticles promotes drought stress tolerance in eggplant (Solanum melongena L.). Plants. https://doi.org/10.3390/plants10020421
102. Serag MF et al (2012) Spatiotemporal visualization of subcellular dynamics of carbon nanotubes. Nano Lett. https://doi.org/10.1021/nl3029625
103. Sharma P et al (2021) A review on biosensors and nanosensors application in agroecosystems. Nanoscale Res Lett. https://doi.org/10.1186/s11671-021-03593-0
104. Sharma P, Dubey RS (2005) Drought induces oxidative stress and enhances the activities of antioxidant enzymes in growing rice seedlings. Plant Growth Regul. https://doi.org/10.1007/s10725-005-0002-2
105. Shekhawat GS et al (2021) Role of engineered carbon nanoparticles (CNPs) in promoting growth and metabolism of Vigna radiata (L.) Wilczek: Insights into the biochemical and physiological responses. Plants. https://doi.org/10.3390/plants10071317
106. Shigeoka S (2002) Regulation and function of ascorbate peroxidase isoenzymes. J Exp Bot. https://doi.org/10.1093/jexbot/53.372.1305
107. Siddiqi KS, Husen A (2016) Engineered gold nanoparticles and plant adaptation potential. Nanoscale Res Lett 11(1):1. https://doi.org/10.1186/s11671-016-1607-2
108. Siddique MRB, Hamid A, Islam MS (2000) Drought stress effects on water relations of wheat, Bot Bull Acad Sin
109. Siddiqui MH et al (2015) Role of nanoparticles in plants. Nanotechnol Plant Sci Nanopart Impact Plants. https://doi.org/10.1007/978-3-319-14502-0_2
110. Smirnova EA et al (2011) Multi-walled carbon nanotubes penetrate into plant cells and affect the growth of onobrychis arenaria seedlings. Acta Naturae. https://doi.org/10.32607/20758251-2011-3-1-99-106
111. Strange K (2020) Cellular and molecular physiology of cell volume regulation. Cell Mol Physiol Cell Vol Regul. https://doi.org/10.1201/9780367812140

112. Su LX et al (2018) Carbon nanodots for enhancing the stress resistance of peanut plants. ACS Omega. https://doi.org/10.1021/acsomega.8b02604
113. Szabados L, Savouré A (2010) Proline: a multifunctional amino acid. Trends Plant Sci. https://doi.org/10.1016/j.tplants.2009.11.009
114. Taran N et al (2017) Effect of zinc and copper nanoparticles on drought resistance of wheat seedlings. Nanoscale Res Lett. https://doi.org/10.1186/s11671-017-1839-9
115. Tripathi S, Sonkar SK, Sarkar S (2011) Growth stimulation of gram (Cicer arietinum) plant by water soluble carbon nanotubes. Nanoscale. https://doi.org/10.1039/c0nr00722f
116. Usman M et al (2020) Nanotechnology in agriculture: current status, challenges and future opportunities. Sci Total Environ 721:137778. https://doi.org/10.1016/j.scitotenv.2020.137778
117. Wang Z et al (2012) Xylem- and phloem-based transport of CuO nanoparticles in maize (Zea mays L.). Environ Sci Technol. https://doi.org/10.1021/es204212z
118. Wu Y, Cosgrove DJ (2000) Adaptation of roots to low water potentials by changes in cell wall extensibility and cell wall proteins. J Exp Bot. https://doi.org/10.1093/jexbot/51.350.1543
119. Yadav B et al (2021) Secondary metabolites in the drought stress tolerance of crop plants: a review. Gene Rep. https://doi.org/10.1016/j.genrep.2021.101040
120. Yang F et al (2009) Responses to drought stress in two poplar species originating from different altitudes. Biol Plant. https://doi.org/10.1007/s10535-009-0092-1
121. Yang X et al (2021) Response mechanism of plants to drought stress. Horticulturae. https://doi.org/10.3390/horticulturae7030050
122. Yang F, Miao LF (2010) Adaptive responses to progressive drought stress in two poplar species originating from different altitudes. Silva Fennica. https://doi.org/10.14214/sf.160
123. Yousefi S, Kartoolinejad D, Naghdi R (2017) Effects of priming with multi-walled carbon nanotubes on seed physiological characteristics of Hopbush (Dodonaeaviscosa L.) under drought stress. Int J Environ Stud. https://doi.org/10.1080/00207233.2017.1325627
124. Zahedi SM et al (2021) Mitigation of the effect of drought on growth and yield of pomegranates by foliar spraying of different sizes of selenium nanoparticles. J Sci Food Agric. https://doi.org/10.1002/jsfa.11167
125. Zhang J, Kirkham MB (1996) Antioxidant responses to drought in sunflower and sorghum seedlings. New Phytol. https://doi.org/10.1111/j.1469-8137.1996.tb01856.x
126. Zhao TJ et al (2006) Regulating the drought-responsive element (DRE)-mediated signaling pathway by synergic functions of trans-active and trans-inactive DRE binding factors in Brassica napus. J Biol Chem. https://doi.org/10.1074/jbc.M510535200
127. Zhu ZJ et al (2012) Effect of surface charge on the uptake and distribution of gold nanoparticles in four plant species. Environ Sci Technol. https://doi.org/10.1021/es301977w

Role of Metal and Metal-Oxide Nanoparticles in Agricultural Crop Heavy Metal-Stress Adaptation and Mitigation

Pooja V. Nagime, Anil Kumar Singh, and Sudarshan Singh

Abstract Heavy metal (HM) stress poses a noteworthy confront in contemporary agronomy, leading to diminished crop yield, impaired quality, and risks to food security. The accumulation of hazardous metals, involving cadmium (Cd), lead (Pb), and arsenic (As), in agricultural soils disrupts plant physiological systems, such as nutrient absorption, photosynthesis, and enzyme functions. This requires the formulation of novel ways for HM stress adaption and alleviation. Metal and metal-oxide nanoparticles (NPs), including zinc oxide (ZnO), iron oxide (Fe_2O_3), and titanium dioxide (TiO_2), have emerged as pivotal instruments in sustainable agriculture, providing distinctive physicochemical characteristics that improve stress resilience in crops. NPs operate via several methods, such as the immobilization of HMs, the initiation of antioxidant resistance systems, and the intonation of gene expression associated with stress responses. Their capacity to improve photosynthetic efficiency, increase nutrient bioavailability, and diminish reactive oxygen species (ROS) levels establishes them as efficient agents for mitigating HM toxicity. The synergistic application of NPs, namely the combination of silicon (Si) and selenium (Se)-NPs, has shown improved effectiveness in alleviating stress, underscoring their potential for customized therapies. This chapter offers an in-depth analysis of the consequences of HM stress on crops and the function of metal and metal-oxide NPs in adaptation and mitigation strategies. It explores their methods of action, practical uses,

P. V. Nagime
Centre of Excellence in Innovative Biotechnology for Sustainable Utilization of Bioresources, Faculty of Agro-Industry, Prince of Songkla University, Hat Yai, Thailand

A. K. Singh
United Institute of Pharmacy, Naini Prayagraj, Uttar Pradesh, India

S. Singh (✉)
Office of Research Administration, Chaing Mai University, Chiang Mai, Thailand
e-mail: sudarshansingh83@hotmail.com; sudarshan.s@cmu.ac.th

Faculty of Pharmacy, Chaing Mai University, Chiang Mai, Thailand

A. Husen (ed.), *Nanobiotechnology for Abiotic Stress Adaptation and Mitigation in Agricultural Crops*, Smart Nanomaterials Technology,
https://doi.org/10.1007/978-981-96-9298-9_4

and the challenges related to their use, encompassing environmental issues and cost-effectiveness. Future perspectives highlight the enhancement of nanoparticle formulations and the incorporation of nano-enabled technologies into agricultural practices that ensure sustainability and food security.

Keywords Metal · Metal-oxide · Nanoparticles · Metallic nanoparticles · Agro-industry

1 Introduction

1.1 Overview of Heavy Metal Stress in Agriculture

Nanotechnology has integrated with other scientific fields, encouraging the production of novel goods and technologies designed to improve crop properties and tackle agricultural issues [29]. The emerging discipline of agro-nanotechnology aims to enhance crop yield by maximizing the effective use of inputs including fertilizers, herbicides, and agrochemicals, frequently supported by real-time monitoring systems. This strategy utilizes a wide range of nanotechnological applications, potentially enhancing crop output and quality throughout all phases of agricultural practices, from seed germination until post-harvest management [39].

NPs exhibit significant assurance in humanizing plant growth, production, and quality under adverse deficiency and salinity emphasis conditions, attributed to their distinctive physicochemical features and interactions with plant systems [32–36]. These interactions affect multiple physiological and metabolic processes, which improve plant resistance and overall performance. NPs are essential in facilitating plant development under stress by refining nutrient absorption, modulating hormonal equilibrium, and alleviating oxidative and cellular damage induced by stress-related ROS [8]. Silicon NPs (SiNPs) have been particularly beneficial in improving root architecture, augmenting shoot biomass, and promoting overall plant development under drought and salinity stress [79]. Zinc oxide NPs (ZnONPs) have demonstrated the ability to improve root and shoot development, resulting in increased plant functioning and stress resilience [105]. The impact of NPs transcends growth enhancement to improve crop production in unfavorable conditions markedly. NPs boost production by humanizing nutrient accessibility, maximizing photosynthetic efficacy, and increasing stress lenience, even in conditions of limited water and high salinity [4]. Substantial production enhancements in staple crops, including wheat, rice, and maize, when subjected to nanoparticle treatment under stress circumstances were documented. The enhancements are mostly due to the NPs' capacity to mitigate stress-induced growth limitations, increase physiological functions, and promote metabolic stability [87].

NPs show the potential to enhance the crop's nutritional eminence cultivated under drought as well as salinity stress. By affecting plant metabolism, they can increase the concentration of vital nutrients including vitamins and antioxidants, which are crucial for human health. Selenium NPs have been demonstrated to enhance selenium levels in plants, improving their nutritional value. Moreover, NPs enhance the production of bioactive substances and secondary metabolites, hence boosting the feature characteristics of crops and potentially helping both producers and consumers [127]. Moreover, NPs modulate numerous physiological and biochemical processes that enhance plants' resilience to abiotic stressors. This encompasses the modulation of osmotic equilibrium, the stabilization of membrane integrity, and the activation of antioxidant defense systems. Nonetheless, whereas NPs present an option for improving plant productivity and quality in adverse conditions, meticulous attention must be paid to issues like proper dosage, application techniques, and the possible environmental repercussions of NP utilization. Subsequent research must focus on overcoming these obstacles to facilitate the sustainable implementation of nanotechnology in agriculture [110].

1.2 Impacts of HM Stress on Crop Productivity and Quality

HM contagion has become a major environmental and agricultural issue owing to its persistence and toxicity. The amassing of HMs in agricultural environments negatively impacts crop productivity and quality, with significant consequences toward food security, soil health, and humans as well as animals' health.

Disruption of physiological functions in plants: HMs such as boron (B), cadmium (Cd), lead (Pb), mercury (Hg), and chromium (Cr) interfere with critical physiological functions in plants, encompassing gas exchange, photosynthesis, and nutrient uptake [139]. These disruptions hinder plant growth and yield, resulting in a substantial decline in agricultural productivity. HMs disrupt chlorophyll synthesis, leading to diminished photosynthetic efficiency, inhibited growth, and decreased biomass accumulation [115].

Effects on soil microbiota: HMs negatively impact soil microbial populations, which are essential for nutrient cycling and plant development. Beneficial soil microorganisms that improve plant health and production are especially susceptible to HM toxicity, leading to reduced microbial diversity and functionality. This deterioration undermines soil fertility and diminishes the capacity of soils to sustain sustainable agricultural operations [96].

Pollution of food chains: the enduring characteristics of HMs enable them to penetrate agricultural soils and bioaccumulate in crops. Ingestion of food polluted with HMs presents significant health hazards, encompassing both acute and chronic conditions, including respiratory ailments, cancer, and neurological diseases. This pollution compromises the safety and quality of agricultural products, resulting in economic losses and public health emergencies [115].

Impact of human activities: human activities, including smelting, mining, and industrial pollutants, substantially contribute to HM contamination. Tailings and waste produced during mineral processing and metallurgical extraction discharge significant extent of harmful metals into the environment, intensifying the adulteration of agricultural soils [13]. The escalating worldwide demand for industrial expansion exacerbates this dilemma, jeopardizing sustainable agriculture [50].

Prolonged soil and environmental deterioration: the continual buildup of HMs in soils results in prolonged environmental deterioration. HMs modify the physicochemical characteristics of soil, such as pH and nutrient accessibility, rendering soils less conducive to agricultural activities. The discharge of hazardous metals into surrounding ecosystems compromises water quality, impacts aquatic creatures, and endangers terrestrial wildlife, resulting in a chain of ecological harm [71].

Consequences for sustainable agriculture: the anticipated rise in the world population by 2030 necessitates sustainable agriculture and improved food security. The presence of HMs in soils constitutes a substantial impediment to attaining this objective, as it restricts crop yield and deteriorates agricultural land. Mitigating HM contamination is crucial for enhancing the resilience of agricultural systems and fulfilling future food requirements [133].

HM stress markedly affects crop yield and quality, resulting in serious implications for agricultural sustainability including food security. The enduring characteristics of HMs, along with their disruption of plant physiological systems and soil microbial activities, underscore the necessity for the formulation of efficient remediation solutions. Bioremediation, employing microorganisms and vegetation, is a viable method for rehabilitating HM-contaminated soils and stimulating sustainable agricultural practices. Additional research and legislative measures are required to alleviate the impacts of HMs and protect the future of agriculture.

1.3 Importance of NPs in Modern Agriculture

NPs, characterized by dimensions between 1 and 100 nm, have become essential instruments in contemporary agriculture owing to their distinctive physical, chemical, and biological attributes [81]. These characteristics arise from their elevated surface area-to-volume ratio, diverse morphologies and dimensions, and capacity to infiltrate biological systems at the molecular scale [84]. The utilization of NPs in agriculture is revolutionizing conventional methods, improving efficiency, sustainability, and productivity [109]. NPs can be classified into natural and engineered NPs. Natural NPs, including hemoglobin (5.5 nm) and cytochrome c (3.1 nm), are present in biological systems and possess intrinsic functional characteristics. Conversely, engineered NPs are produced in regulated laboratory environments to exhibit characteristics, including elevated reactivity, catalytic efficacy, and antibacterial capabilities. Examples encompass carbon nanotubes and metal-based NPs, which possess varied applications in agriculture and additional areas [75, 107]. The distinctive characteristics of NPs render them especially advantageous in agriculture. Engineered NPs can function as catalysts, improving the efficacy of agricultural chemicals like fertilizers, herbicides, and pesticides. Their antibacterial properties assist in managing plant infections, hence diminishing crop losses attributed to illnesses. Moreover, nanotechnology facilitates the creation of "magic bullet" systems, in which NPs transport bioactive chemicals like fertilizers, fungicides, herbicides, or genetic material directly to designated areas [114]. This precise distribution decreases waste, mitigates environmental pollution, and enhances the efficacy of agricultural inputs. NPs are utilized to improve soil fertility and promote water retention [56]. Carbon-based NPs enhance soil structure and nutrient accessibility, whereas metal-based NPs facilitate water purification and pollutant removal. Quantum dots and other nanocomposites are utilized in nano-sensors for the real-time assessment of soil health, plant development, and pest infestation. These sensors deliver essential data, facilitating precision agriculture methodologies [97].

Engineered NPs are categorized into many categories based on their composition, each with specific applications. Carbon-based NPs, including carbon nanotubes and fullerenes, improve the delivery of nutrients and agrochemicals. Metallic NPs, including silver, zinc oxide, and titanium dioxide, are recognized for their antibacterial and catalytic characteristics. Dendrimers, branching polymers, provide the controlled release of agrochemicals, while quantum dots, semiconductor NPs, are utilized in imaging and sensing applications. Nanocomposites, hybrid materials that integrate various nanoparticle kinds, provide improved functionality for agricultural applications [108]. The incorporation of nanotechnology in agriculture tackles significant issues such as resource inefficiency, environmental deterioration, and food security. Nanotechnology enhances sustainable agricultural operations by mitigating the excessive application of fertilizers and pesticides. Moreover, its function in waste management and pollution mitigation corresponds with international initiatives for sustainable development [82]. The transformational potential of NPs in agriculture, encompassing precision farming, tailored pesticide delivery, and environmental cleanup, highlights their importance in tackling modern agricultural concerns. As research progresses, the creation of safer and more efficacious NPs will reinforce their significance in the future of agriculture [101]. Figure 1 illustrates the multifaceted function of metallic NPs with an elevation of HM stress in plants.

1.4 HM Stress in Agricultural Crops

1.5 Sources of HM Pollution in Agricultural Soils

HM accumulation in agricultural soils originates from various natural and anthropogenic sources. The principal contributors comprise atmospheric deposition, livestock manure, irrigation of wastewater, wastewater sludge, phosphate-based fertilizers, and metallo-pesticides. These sources substantially impact soil quality and food safety [22]. Atmospheric deposition from industrial pollutants, vehicle particulate matter, and coal-fired power plants delivers metals such as Hg, Pb, and Cd into the soil, leading to their amassing in the food chain [30]. Prolonged ingestion of crops

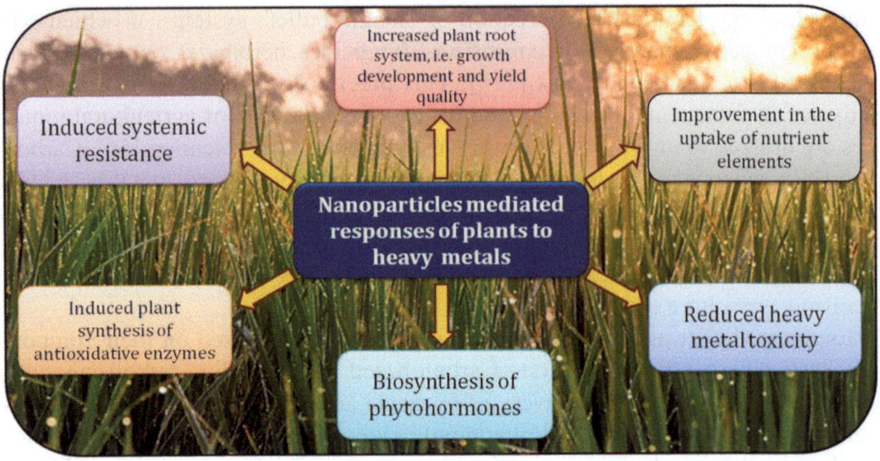

Fig. 1 Multifaceted function of metallic NPs with an elevation of heavy metal stress in plants

cultivated in contaminated soils presents significant health hazards [64]. The application of untreated sewage and wastewater sludge in agriculture intensifies pollution. These materials, frequently contaminated with HMs, are widely utilized due to water constraints and the necessity to sustain agricultural production. Nonetheless, they contribute Pb, Ni, Fe, Cu, Cr, and Mn to soils, resulting in detrimental environmental and public health consequences [17]. Likewise, sludge from electroplating and leather industries is characterized by preeminent concentrations of hazardous metals such as cadmium and zinc, negatively impacting crop development and soil health [129]. Phosphate fertilizers and phospho-gypsum waste augment soils with Cd, Cr, and Pb, resulting in bioaccumulation in crops such as tomatoes and green peppers. Although acute health hazards are minimal, the cumulative impacts from several exposure pathways are worrisome [2]. Indoor agriculture is susceptible to HM contamination, with greenhouse crops accumulating cadmium and other metals mostly due to anthropogenic activities [66]. Furthermore, soil serves as a conduit for environmental compartments, enabling the movement of HMs to groundwater and coastal areas [137]. The unchecked proliferation of nanotechnology has introduced nanoparticle-induced toxicity into agriculture. Toxic NPs, such as CuO-NPs, can impact crop physiology and provide considerable dangers to human health via dietary exposure [136]. Resolving these difficulties necessitates sophisticated contamination identification technologies and effective wastewater treatment systems to alleviate effects on soil and food safety [101].

1.6 Uptake and Toxicity Mechanisms in Plants of HMs

Pathways and transport mechanisms of metals in agricultural plants: HMs typically infiltrate crop plants via the roots, aided by the absorption of water as well as nutrients. The bio-speciation including amassing of HMs differs among plant species and their respective organs. Vegetables like lettuce, radishes, beans, and tomatoes demonstrate varying levels of HM absorption, with lettuce exhibiting greater lead buildup than other crops [145]. Root morphology and plasticity substantially affect metal absorption. Metals associate with cell wall constituents, including low-methyl-esterified pectins, which elevate under stress circumstances [53]. Lead binding to esterified pectins in moss protonemas exemplifies this mechanism. Polysaccharides in root cell walls enhance stress tolerance by chelating metals via functional groups (e.g., –COOH, –OH, –SH), while stress-induced remodeling may compromise cell membrane integrity [55]. Foliar absorption of HMs primarily transpires by adsorption to the cuticle or stomatal apertures, as evidenced in lead research on *Lactuca sativa.* Translocation encompasses xylem loading in roots and phloem transfer in leaves [112]. Figure 2 illustrate the HM toxicity occurrence that effects the development and productivity of crops with significant affect to human health.

Mechanisms of metal stress tolerance in crop plants: plants alleviate HM toxicity via avoidance and tolerance mechanisms. Avoidance methods encompass exocytosis and the reinforcement of root cell walls through lignin production. The plasma membrane is a physiological barrier, limiting metal ingress via lipid peroxidation

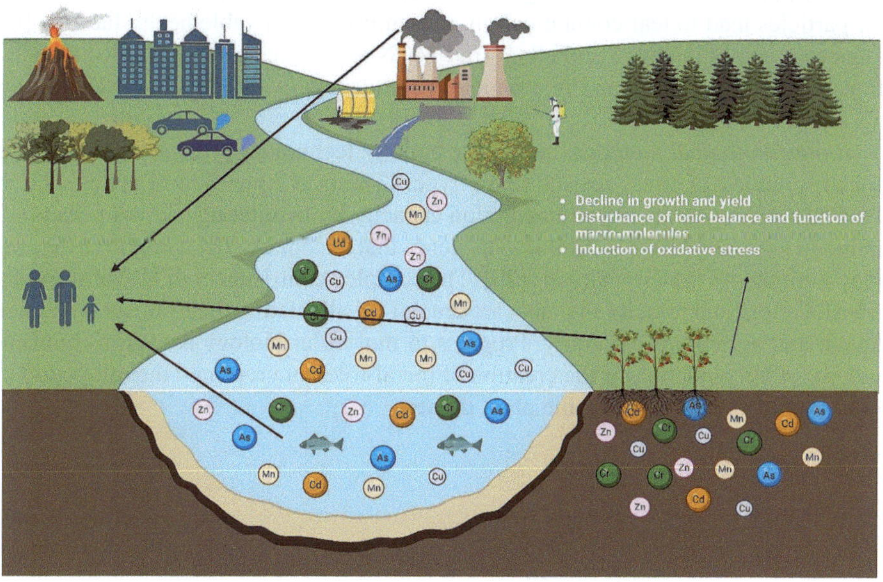

Fig. 2 The heavy metal toxicity occurrence that negatively affects the development and productivity of crops with a significant risk to human health due to heavy metal entrance into the food chain

[93]. Metallothioneins and phytochelatins are essential components in the scavenging of ROS during metal stress. Nonetheless, the overexpression of phytochelatin genes, such as TaPCS1 in wheat, can elevate Cd accumulation in shoots, increasing toxicity [93]. Molecular homeostasis strategies encompass intracellular compartmentalization and modifications in biochemical pathways. HMs cause DNA strand breakage, enzyme inactivation, and oxidative damage, which hinder plant growth and enhance phyto-availability [20].

Effects on plant physiology and biochemistry: HMs impede sprouting and seedling growth by blocking essential enzymes such as acid phosphatases, proteases, and α-amylases. The impacts are especially pronounced in rice and other crops affected by cadmium [104]. Multi-metal toxicity is inadequately studied, although cobalt demonstrates superior phytotoxicity relative to copper and chromium, adversely affecting biomass and physiological functions in cauliflower [111].

Molecular mechanisms and transporters: metal transporters, including ZIP, ATP-binding cassettes, and natural resistance-associated macrophages, govern HM absorption. OsHMA3 in rice sequesters cadmium in root vacuoles, hence diminishing its transfer to shoots and seeds [132]. Missense mutations in cellulose biosynthesis genes, such as bc13 in rice, markedly reduce Cd bioaccumulation [120]. Arsenic absorption in rice occurs via the Lsi1 and Lsi2 pathways, while phytochelatin complexes facilitate arsenic detoxification through vacuolar storage. Advanced imaging techniques, including Nano-SIMS, clarify subcellular compartmentalization of trace metals [20].

Genotoxicity and environmental influences: HMs impact soil–plant systems, with pH and pollution levels affecting bioaccumulation. In metropolitan environments, air particles lead to leaf contamination, presenting considerable health hazards [6]. Genotoxic investigations, including Trad-MCN assays, have demonstrated associations between pollution and micronucleus frequency, facilitating site selection for agriculture [121].

Biotechnological methods: molecular biology techniques provide understanding for the creation of stress-resistant crop types. The overexpression of genes such as PtoEXPA12 augments phytoremediation capacity by facilitating cadmium accumulation in tobacco plants [143]. Bacterial genes that encode As(III) methyltransferases may mitigate As toxicity in crops [20]. The complex contrivances of HM absorption, translocation, and toxicity in plants are crucial for alleviating their impact on human health and agricultural viability. Progress in molecular biology as well as biotechnology offers potential for the creation of metal-tolerant crops and the mitigation of bioaccumulation in consumable plant tissues.

2 Metal and Metal-Oxide NPs

2.1 *Types of Metal and Metal-Oxide NPs*

NPs have attained significant importance in numerous technological, medicinal, and agricultural domains, chiefly owing to their distinctive features and varied applications [83]. Recent breakthroughs in nanotechnology have broadened their scope, considerably enhancing human well-being. NPs are described as objects measuring between 1 and 100 nm, exhibiting unique physicochemical features [49]. Comprehensive studies validate the use of NPs in both in vitro and in vivo surroundings, illustrating their capacity to advance overall plant growth and development. The application of ZnONPs at a concentration of 1000 ppm markedly enhanced seed propagation and promoted shoot and root elaboration in peanut plants [100]. Likewise, the application of 1.5 ppm ZnONPs on chickpea leaves improved seed germination potential and augmented biomass accumulation [11]. The combined impacts of SiO_2 and TiO_2 NPs have demonstrated enhanced absorption of water and fertilizers [18]. SiO_2 NPs, when utilized on Zea mays L., improved biochemical production, efficacy of photosynthesis, and protein synthesis [123]. Similarly, TiO_2 NPs have demonstrated enhancements in photosynthetic efficiency and accumulation of biomass in spinach [144]. Treatment of Brassica napus seeds with TiO_2 NPs significantly enhanced seed germination efficiency [72]. Chemically produced apatite NPs (hydroxyapatite) exhibited a substantial enhancement in seed production and growth ratio in soybean cultures, surpassing traditional phosphorus fertilizers such as $Ca(H_2PO_4)_2$ [67]. In addition to serving as nutrients and fertilizers, NPs are utilized as transporters for plant growth regulators to enhance agricultural productivity. Polymeric NPs containing gibberellic acid, composed of chitosan and alginate, yielded an almost four-fold enhancement in fruit output [95]. These studies emphasize the adaptability of NPs in agriculture, demonstrating their capability to progress crop growth, yield, and biochemical efficiency while providing novel methods for sustainable farming techniques.

2.2 *Synthesis Methods of NPs for Agricultural Applications*

The fabrication of NPs generally adheres to two principal methodologies: top-down and bottom-up techniques. The top-down technique involves the derivation of NPs from bigger materials via breakdown. Methods such as electro-explosion, mechanical milling, sputtering, chemical etching, laser ablation, shows this approach, with the production of metal-oxide NPs frequently utilizing these techniques. The bottom-up technique entails the construction of NPs from more fundamental molecular or atomic antecedents. Common examples of this strategy include techniques such as spinning, sol–gel processes, condensation, and pyrolysis [80]. Furthermore, green synthesis where NPs are obtained from biological sources such as plants, fungi, and microbes also constitutes the bottom-up method [40]. In green synthesis, biological extracts from plants, fungus, and microorganisms interact with metal salt solutions via reduction and capping processes, leading to the creation of NPs enveloped in biomolecules. The diminutive dimensions of green-synthesized NPs promote efficient delivery to target locations, hence augmenting their efficacy and increasing results [113].

2.3 Characterization Methodologies to Produce Metallic NPs

The efficacy of nanoparticle synthesis depends on accurate characterization techniques as follows. *UV–Visible Spectroscopy:* It assesses light absorption in the UV–visible spectrum resulting from the stimulation of surface plasmon resonance in NPs of metal. Surface plasmon resonance is a collective oscillation on conduction electrons on the surface of NPs, stimulated by light at a particular wavelength [16]. The location and strength of the absorption peak yield insights into nanoparticle dimensions, morphology, and aggregation state [60]. *Dynamic Light Scattering:* It examines the scattering of light by NPs within a colloidal system. As a laser traverses the sample, NPs disperse the light, and their Brown's law of motion induces variations in the scattering intensity. The fluctuations are examined to ascertain particle size, size distribution, including polydispersity index, so facilitating the comprehension of colloidal stability [51, 54]. *Transmission Electron Microscopy:* It utilizes a high-energy electron beam that passes through a thin nanoparticle specimen. The electrons engage with the sample, producing contrast dependent on the material's density and thickness. This technique yields high-resolution images, facilitating the examination of nanoparticle shape, crystallinity, and size at the atomic scale [77, 119]. *Scanning Electron Microscopy*: It employs a concentrated electron beam to examine the nanoparticle surface. The interaction of electrons with the sample produces secondary and backscattered electrons that are detected to create detailed pictures. *Scanning Electron Microscopy* is optimal for analyzing surface morphology, particle dimension distribution, and structural uniformity [58, 59]. *X-ray diffraction:* It involves the scattering of X-rays by the crystalline lattice of NPs. Bragg's law examines the diffraction pattern to ascertain crystal structure, lattice parameters, and crystallite size. XRD is especially effective for distinguishing crystalline phases and verifying the quality of NPs [15, 42, 117]. *Fourier-Transform Infrared Spectroscopy:* It identifies chemical bond vibrations by quantifying infrared light absorption. Functional groups engaged in the synthesis and capping, along with stabilization of NPs display distinct absorption peaks [43, 89]. FTIR facilitates the identification of chemical bonds, stabilizing agents, and potentially functionalization of NPs [14, 42]. *Energy Dispersive X-Ray Spectroscopy:* It detects distinctive X-rays generated from a material when subjected to a high-energy electron beam, similar to Scanning Electron Microscopy. The emitted X-rays are examined to ascertain the chemical breakdown of NPs. EDS verifies the existence of metals and other elements in the sample [43, 125, 126]. *X-Ray Photoelectron Spectroscopy:* It uses X-rays to dislodge electrons from the nanoparticle surface, measuring the kinetic energy of the emitted electrons. The binding energy of the emitted electrons reveals details about the elemental composition, oxidation states, and chemical environment of the nanoparticle surface [7, 90, 92]. *Atomic Force Microscopy:* It uses a sharp probe to examine the surface of NPs. The probe's deflection caused by intermolecular forces is documented to provide topographical images. AFM delivers data on surface roughness, and morphological and particle distribution with nanoscale resolution [76]. *Zeta Potential Analysis:* It assesses the electrokinetic potential at the sliding plane of NPs within a colloidal suspension. A high zeta potential signifies robust electrostatic repulsion among particles, indicating favorable colloidal stability, whereas low values show a propensity for aggregation [57, 118, 124]. *Thermogravimetric Analysis:* It observes the variation in mass of NPs during controlled heating. The decomposition of organic stabilizers or adsorbed compounds on the nanoparticle surface, as well as thermal stability and oxidation behavior, can be investigated [45]. *Raman Spectroscopy:* It assesses the inelastic scattering of light due to molecular vibrations, yielding a molecular fingerprint of the specimen. Raman spectroscopy identifies functional groups, capping agents, and chemical interactions on nanoparticle surfaces [16]. *Nuclear Magnetic Resonance Spectroscopy:* It examines the interaction between atomic nuclei and a magnetic field to yield insights into molecular surroundings. NMR is valuable for examining the surface chemistry of NPs, their interactions with ligands, particularly the molecular structure of stabilizing agents [44].

3 Mechanisms of Nanoparticle-Mediated HM Stress Adaptation

HM contamination presents substantial environmental and agricultural concerns globally, impacting soil quality, plant health, and the general equilibrium of ecosystems. The capacity of NPs to alleviate HM stress in plants and remediate polluted environments has attracted significant interest in recent years owing to their distinctive physicochemical characteristics. NPs not only promote plant development and resilience to stress but also aid in the extraction of hazardous metals from contaminated environments, frequently as components of phytoremediation approaches [86, 134]. NPs can be classified as inorganic nonmetallic NPs, carbon-based NPs, metallic NPs, and organic polymeric materials, each engineered for different applications and functionalities. Phytoremediation, an environmentally sustainable plant-based technique, employs plant–nanoparticle interactions to augment the bioavailability of contaminants, facilitate plant development, and boost the extraction of HMs from polluted soils [116]. The simultaneous application of Lolium (ryegrass) along with nano hydroxyapatite has been shown to markedly improve the phytoextraction of Pb, achieving a removal of up to 44.39% of Pb from soil within three months [5].

NPs possess distinctive physical, chemical, and biological characteristics that render them effective in alleviating HM stress in plants. They interact with plants at molecular, cellular, and physiological levels to mitigate the deleterious effects of HMs. Comprehensive analyses of these mechanisms are presented below.

3.1 Reducing HM Uptake by Plants

NPs can diminish the absorption and movement of HMs in plants by altering the soil environment, sequestering metals, or creating stable complexes. For instance: SiNPs have demonstrated the ability to alleviate cadmium toxicity in rice plants by reducing cadmium absorption. This transpires as SiNPs trap cadmium in the rhizosphere by creating insoluble complexes, hence diminishing its bioavailability. The reduction of malondialdehyde (MDA) levels in rice further substantiates diminished oxidative stress resulting from decreased cadmium (Cd) accumulation [21].

Iron oxide NPs: These NPs diminish the accessibility of HMs such as lead, cadmium, zinc, and copper in soil, hence restricting their uptake by plants. Research on wheat seedlings has shown that Fe_3O_4 NPs form a protective barrier around the root surface, inhibiting HMs penetration [69].

3.2 Enhancing Antioxidant Enzyme Activity

HMs produce ROS in plants, resulting in oxidative stress and cellular damage. NPs can augment the activity of antioxidant enzymes that neutralize reactive oxygen species, thereby safeguarding plants from oxidative damage. We have the following examples.

Zinc Oxide NPs: ZnONPs augment the functionality of enzymes like superoxide dismutase, catalase, ands peroxidase, which mitigate ROS. In tomato plants subjected to cadmium, zinc oxide NPs enhanced antioxidant enzyme activity, mitigating oxidative damage and facilitating development [27, 130].

Iron oxide NPs: These NPs have been documented to mitigate oxidative stress in plants subjected to HM stress by enhancing antioxidant systems, particularly in crops such as wheat and rice [69].

3.3 Preventing Lipid Peroxidation and Membrane Injury

Lipid peroxidation, induced by reactive oxygen species, undermines cellular membrane integrity in plants subjected to HM stress. NPs can alleviate this damage by decreasing ROS levels and stabilizing membrane lipids:

Nano hydroxyapatite: This NP has demonstrated a reduction in MDA levels, an indicator of lipid peroxidation, in Brassica chinensis subjected to cadmium stress.

This decrease indicates that NPs safeguard cell membranes against oxidative damage [28].

Silicon NPs: The application of Si NP hydrosol on rice reduced MDA levels, stabilized cellular structures, and enhanced plant growth under cadmium stress [131].

3.4 Enhancing Plant Growth and Development

NPs promote growth and biomass accumulation even in the presence of HM stress by improving nutrient absorption, photosynthesis, and metabolic activities. Illustrations comprise the following.

Titanium Dioxide NPs: TiO_2NPs enhance the photosynthetic rate by augmenting chlorophyll concentration and light absorption. In soybean plants, TiO_2NPs mitigated cadmium toxicity, leading to enhanced growth and biomass accumulation [134].

Silicon NPs: SiNPs enhance nutrient absorption and water uptake, facilitating root and shoot development under HM stress. These effects correlate with elevated mineral content in plants, including heightened amounts of calcium and magnesium [131].

3.5 Adsorption and Retention of HMs

Certain NPs exhibit elevated adsorption capabilities, allowing them to bind and sequester HMs from soil and water, hence diminishing their bioavailability.

Reduced Graphene Oxide (rGO) hybrids: Hybrid NPs of rGO with magnetite or silver exhibit enhanced adsorption capabilities for HMs such as cadmium, lead, zinc, copper, and nickel. These nanohybrids surpass traditional adsorbents such as charcoal and resins, presenting a viable solution for soil remediation [94].

3.6 Alleviating Root Growth Suppression

HMs frequently impede root elongation and development, hence influencing plant nutrient and water absorption. Nutrient providers alleviate these effects by promoting root development via increased nutrient accessibility and hormonal modulation.

Iron oxide NPs: In tomato, cucumber, and lettuce, as well as carrot, Fe_3O_4 NPs mitigated root growth inhibition induced by Cd stress, promoting enhanced nutrient absorption and plant vitality [69].

ZnONPs: ZnONPs mitigated HM stress and enhanced nitrate reductase activity, crucial for root growth and nitrogen assimilation [27].

3.7 Modulating Physiological and Biochemical Pathways

NPs can affect essential biochemical and physiological processes, enhancing plant resilience to HM stress as follows.

Nano hydroxyapatite: This nanoparticle elevated chlorophyll concentrations and improved photosynthetic efficiency in Brassica chinensis, resulting in increased vitamin C content under cadmium stress [28].

Silicon NPs: Silicon NPs regulate the plant's stress-signaling pathways, mitigating HM-induced toxicity and improving resistance [131].

4 Mitigation of HM Stress Using NPs

Nanotechnology provides novel methods to alleviate the detrimental influences of abiotic stressors, like drought and salinity, which sternly hinder plant growth and yield. Drought stress, a substantial concern marked by water deficiency, impedes plant growth and metabolic processes. NPs significantly contribute to improving drought tolerance in multiple ways. Hydrogel NPs may absorb and release significant quantities of water slowly, thus establishing a moisture-rich microenvironment around plant roots, enhancing plant–water interactions in water-scarce circumstances [85]. Furthermore, NPs can augment the antioxidant defense systems in plants. Silver NPs enhance the activities of antioxidant enzymes, including catalase and superoxide dismutase, in wheat plants experiencing drought stress, thereby mitigating oxidative damage from ROS and fostering plant growth [23]. Zinc oxide NPs have proven effective in easing drought-induced stress by augmenting antioxidant activities and enhancing water use efficacy through the modulation of stomatal conductance and transpiration rates, allowing plants to sustain water balance in water-scarce environments [103]. Titanium dioxide NPs enhance drought tolerance by augmenting chlorophyll content, expanding photosynthetic efficiency, and activating antioxidant enzymes, thereby promoting the advance and throughput of plants under drought conditions [106].

Salinity stress, a significant abiotic issue, disturbs ionic equilibrium, water absorption, and nutrient accessibility, hindering vital physiological functions in plants. NPs mitigate these problems by regulating ion homeostasis and augmenting biological defense systems. Zinc oxide NPs have been proven to enhance maize development in saline environments by stimulating antioxidant enzymes such as catalase and peroxidase, along with superoxide dismutase, which mitigate damaging ROS and decrease lipid peroxidation [26]. These NPs modulate ion transport, mitigating sodium toxicity while enhancing the absorption of vital nutrients [138]. TiO_2NPs enhance nutrient absorption and utilization, mitigating nutritional imbalances induced by elevated salt levels [1]. SeNPs have been documented to promote barley development in saline soils by augmenting stress tolerance [127]. Furthermore, AgNPs alleviate salt stress by rectifying ionic imbalances, augmenting antioxidant activities, and enhancing nutrient absorption, hence promoting plant growth in saline environments [3].

NPs demonstrate potential in mitigating intricate stress interactions, including HM toxicity and nutritional shortages, which frequently occur alongside drought and salinity conditions. Biochar-derived NPs can adsorb HMs from soil, hence decreasing their absorption by plants and alleviating related toxicity [19]. Furthermore, certain NPs promote the development of metallothioneins, proteins that absorb HMs, hence enhancing the protection of plant cells against toxicity [102]. Due to their elevated surface area-to-volume ratio, nano-fertilizers improve nutrient solubility and availability, especially under stress circumstances that impede nutrient absorption, facilitating plant development and metabolic processes [8]. Although NPs hold significant promise, their use in agriculture requires careful consideration due to intricate interactions with plant systems and possible environmental consequences. An extended study is necessary to elucidate the molecular mechanisms governing NP–plant interactions and to guarantee sustainable practices. By tackling these issues, nanotechnology may facilitate groundbreaking progress in agriculture, especially in mitigating abiotic pressures [128].

4.1 Enhancement of Plant Growth and Yield Under HM Stress

The enhancement of plant development and production under HM stress via nanotechnology shows considerable potential in mitigating the complex issues associated with hazardous metal accumulation. A key mechanism pertains to the improvement of membrane stability. Abiotic stressors, such as HM exposure, lead to the excessive generation of ROS, which can harm cell membranes through lipid peroxidation. NPs, including AgNPs and TiO_2 NPs, alleviate oxidative damage by augmenting the activity of antioxidant enzymes like superoxide dismutase and catalase, thus maintaining membrane integrity [3]. Furthermore, silicon dioxide NPs have been documented to influence ion transporters and channels in the context of salinity stress, which frequently coincides with HM toxicity. This process reestablishes ionic equilibrium and safeguards cell membranes from the harmful impacts of surplus sodium ions [73]. Another essential function of NPs is the improvement of plant–water interactions, which is especially important under HM stress. The processes of water absorption, transport, and use are frequently impaired by changes in root membrane permeability resulting from the accumulation of hazardous metals. NPs can modify soil characteristics to improve its water retention capacity, thereby providing a consistent water supply to plants during periods of stress [49]. Nano-fertilizers enhance water use efficiency and reduce water loss by distributing nutrients in a controlled manner, hence facilitating plant growth [23]. At the molecular level, NPs affect plant stress response pathways by altering gene expression associated with osmotic equilibrium and water retention. Nanoparticle-mediated propagation of stress-responsive genes or hormones has been demonstrated to enhance cell turgor pressure, hence improving plant resilience to metal-induced water stress [3]. The utilization of nano-sensors enables accurate real-time monitoring of soil moisture and plant water stress, hence improving irrigation and averting both water scarcity and over-irrigation, which is essential for managing plants subjected to HMs [63].

Nanotechnology substantially enhances nutrition delivery in conditions of HM stress. This stress frequently induces ionic imbalances and diminishes nutrient availability by competing with vital ions for absorption. NPs, including SiO_2NPs, have shown the ability to rectify ionic imbalances and improve the uptake of essential nutrients such as nitrogen, phosphorus, and potassium [25]. Nano-fertilizers incorporate vital nutrients into NPs, enabling them regulated release and enhanced bioavailability during stress situations [88]. Moreover, certain NPs increase the expression of nutrient transporter genes, hence improving nutrient uptake in plants [47]. Biochar-based NPs enhance soil nutrient retention and water-holding capacity, fostering a more conducive environment for plant growth under HM stress [68]. NPs promote plant development and crop production under HM stress through stabilizing cellular membranes, regulating plant–water interactions, and enhancing nutrient delivery. These developments alleviate the direct effects of HMs and tackle the secondary repercussions of related conditions such as drought and salt. The potential of nanotechnology in agriculture is significant, yet, its extensive application necessitates additional research to assess long-term environmental effects and guarantee sustainable implementation [128].

4.2 Role in Detoxification and Tolerance Reduction of HMs

HM stress poses a continual challenge to plant systems, requiring the development of complex detoxification and tolerance mechanisms. These adaptive methods involve biochemical and physiological mechanisms that reduce HM toxicity while maintaining vital cellular functioning [9, 65]. Enzymatic detoxification serves as a fundamental defense mechanism, whereby plants employ specialized enzymes to mitigate HM-induced harm. Glutathione S-transferases (GSTs) are essential for conjugating hazardous metals with glutathione, hence promoting their sequestration and detoxification. Likewise, superoxide dismutases (SODs) alleviate oxidative stress induced by HMs, whereas phytochelatin synthases facilitate the production of phytochelatins that sequester metals, allowing their compartmentalization in vacuoles and diminishing cytotoxicity. Metallothioneins, diminutive cysteine-rich proteins, facilitate detoxification by sequestering HMs, therefore diminishing their reactivity and mitigating their detrimental effects [46].

The plant's reaction to HM exposure also includes the activation of intricate stress response pathways. These pathways activate stress-related genes and promote the production of stress proteins, essential for healing HM-induced damage and guaranteeing plant life. [78] elucidate the complex signaling networks and gene regulation mechanisms that plants utilize to detect and react to HM stress, providing significant insights into the molecular foundations of these adaptive responses. These reactions promote tolerance and establish a basis for advancing phytoremediation technology designed to alleviate HM contamination in soils [78].

Compartmentalization is a crucial detoxifying method in which plants sequester HMs in designated cellular compartments to avert disruption of essential metabolic processes. The interaction of enzymatic detoxification, stress pathway triggering, and metal sequestration highlights the ability of plants to adapt and flourish in metal-stress environments [142]. These mechanisms possess significant potential for enhancing environmentally sustainable solutions to address HM contamination. Understanding enzymatic detoxifying and stress-responsive pathways can boost phytoremediation strategies and improve the resilience of crops in polluted settings. Consequently, comprehending and utilizing these adaptive characteristics is essential for sustainable agricultural and environmental restoration efforts [99].

4.3 Synergistic Effects of NPs with Other Stress Mitigation Strategies

NPs demonstrate significant potential in alleviating abiotic stress, especially when utilized in conjunction with other NPs or methods. These combinations enhance the stress-relieving capabilities of individual NPs and create distinct biochemical and physiological interactions that bolster plant resistance in adverse environmental conditions. The simultaneous application of SiNPs as well as IONPs has demonstrated synergistic advantages in alleviating cadmium toxicity in *Phaseolus vulgaris*. Plants subjected to this mixture demonstrated augmented growth, enhanced gas exchange, and elevated net photosynthetic rates. The enhancements were associated with the synergistic augmentation of polyamine production and potassium (K^+) levels, which together alleviated Cd-induced stress [52]. Furthermore, the simultaneous application of ZnONPs and SeNPs mitigated arsenic (As) toxicity in soybean by enhancing ascorbic acid and glutathione levels, boosting antioxidant enzyme activity, and improving photosynthetic characteristics, while decreasing oxidative stress indicators such as MDA while ROS. The diminished activity of antioxidant enzymes in certain combined treatments underscores the intricate connections of NPs, indicating possible changes in signaling pathways, absorption processes, and interactions with biomolecules owing to their unique physicochemical characteristics [141].

The simultaneous application of FeNPs, SiNPs, and ZnONPs markedly enhanced growth parameters and yield in wheat seedlings cultivated in Cd-contaminated soil. The observed results were ascribed to the slow-release characteristics of NPs, which mitigated nutrient toxicity and improved the biofortification of important elements such as iron and zinc [37, 38, 105]. Likewise, the utilization of SiNPs in conjunction with TiO_2NPs enhanced shoot and root growth, dry root mass, and fresh shoot mass under cadmium stress. Improved nutrition absorption and regulation of water balance were essential processes driving these synergistic effects. At the molecular level, SiO_2 NPs increased the expression of chlorophyll production and defense against antioxidants genes in Brassica napus, thereby improving plant tolerance to HM stress [61]. The interaction between selenium and ZnONPs has shown advantageous effects under salt stress conditions. Their synergistic application enhanced seed germination, seedling vitality, and antioxidant enzyme activities, including phytohormonal control in rapeseed cultivars. The co-application increased the expression of gibberellin-related genes such as BnGA20ox, BnGA3ox, and BnCPS, resulting in improved seed germination and development in saline conditions [98]. A comparable synergistic effect was reported in Camelina sativa L. when SiNPs and CaNPs were co-administered, markedly enhancing biochemical and physiological characteristics [48]. The simultaneous application of ZnONPs and SeNPs in soybean plants improved arsenic tolerance by modulating WRKY transcription factors, decreasing electrolyte leakage, and facilitating the sequestration of arsenic into vacuoles. This interaction enhanced antioxidant levels and detoxification processes, including increased glutathione S-transferases (GSTs) and phytochelatins (PCs), hence promoting ROS scavenging. The elevated expression of stress-responsive transcription factors with phosphate transporter genes (GmPT1–8) underscored the molecular foundation of their synergistic interaction [12, 74, 140].

4.4 Challenges and Limitations

Effective management of HM pollution in agriculture is essential for sustained crop production along with environmental safety. Nonetheless, numerous techniques encounter obstacles that restrict their effectiveness, scalability, and ecological suitability. There are analysis of several remediation strategies and their limitations including the following.

Chlorination and thermal processing are efficacious techniques for diminishing concentrations of HMs such as lead, copper, zinc, and cadmium in polluted soil. These techniques utilize chemical agents like $CaCl_2$ and $MgCl_2$, along with elevated temperature processes, to immobilize metals and restore soil quality. Notwithstanding their efficacy, considerable limits emerge. For example, carcinogenic by-products such as trihalomethanes may be generated during the procedure [135]. Furthermore, excessive chlorine accumulation can harm soil microbiology, impair plant roots and foliage, and jeopardize aquatic ecosystems when leached into water bodies [122].

Adsorption is extensively employed to eliminate HMs such as lead, zinc, and copper from industrial effluents with materials including activated carbon, kaolinite, and zeolites. This approach is cost-effective and simple to run; nevertheless, it is hindered by low adsorption capabilities and separation difficulties. The low specificity of numerous adsorbents and their elevated costs further restrict large-scale usage [62].

Ion-exchange techniques, utilizing substances like zeolites and ion-exchange resins, efficiently diminish concentrations of HMs such as cadmium, copper, and lead in polluted soils. Nonetheless, the method presents limitations, such as the low thermal stability of resins, deactivation by cations like Na, and vulnerability to fouling due to organic matter. This approach may exhibit inefficiency under instances of elevated pollution [24].

Electro-kinetics remediation utilizes low-voltage direct current to mobilize HMs in soils with limited permeability. This technique attains substantial clearance rates and facilitates in situ remediation [10]. Nonetheless, it encounters obstacles like alterations in soil pH during extended treatment, inefficiency in very permeable soils, and interruptions in voltage gradients caused by buried metals. The laborious nature of this technique additionally hinders its wider adoption [41].

Chemical precipitation converts HMs into insoluble compounds using hydroxides or sulfides, offering a cost-efficient and uncomplicated cleanup method. Nonetheless, the limited solubility of metal sulfides and the emission of toxic hydrogen sulfide throughout the process provide considerable environmental hazards [98].

Bioleaching utilizes acidophilic bacteria, such as Thiobacillus ferrooxidans, to extract HMs from contaminated soils and sludge. This environmentally sustainable and economical method prevents secondary pollution. Nonetheless, bioleaching is a protracted process, and residual HMs may persist in the treated substrate, necessitating further treatments [31].

Recommendations for Improvements

Overcoming the constraints of HM cleanup solutions necessitates interdisciplinary research and creative methodologies. This encompasses the creation of high-capacity adsorbents, the optimization of ion-exchange resins for various conditions, and the improvement of bioleaching efficiency via the genetic engineering of microbes. Moreover, the amalgamation of techniques like adsorption with bioleaching or the incorporation of NPs into current methodologies may enhance remediation efficacy and sustainability. Consistent monitoring of HM contamination and the implementation of awareness campaigns are crucial for risk mitigation and the promotion of safe practices throughout agricultural and industrial environments [70, 91].

5 Future Perspectives

5.1 Innovations in Nanoparticle Synthesis and Functionalization

The progress in nanoparticle synthesis along with functionalization techniques has significant potential to transform agriculture and environmental remediation. Advanced techniques for synthesizing NPs, including eco-friendly synthesis methods, can enhance their stability, biocompatibility, and effectiveness. Functionalized NPs designed for specific applications such as focused nutrient delivery, controlled pesticide release, and HM adsorption could markedly improve the precision as well as the sustainability of agricultural operations and remediation efforts. In the realm of red mud cleanup, functionalized NPs may be designed to preferentially adhere to hazardous HMs while ensuring minimum environmental toxicity.

5.2 *Integration of Nanotechnology with Precision Agriculture*

The combination of nanotechnology with precision agriculture presents prospects for sustainable food production. NPs can function as intelligent carriers for nutrients and insecticides, enhancing their bioavailability and reducing environmental discharge. NPs could augment crop yields by enhancing plant physiological features, such as root system design and photosynthetic efficiency, even in inferior soils, including those contaminated by red mud. This connection can facilitate real-time soil monitoring systems through the utilization of nano-sensors, delivering accurate data on nutrient levels, moisture content, and metal toxicity.

5.3 *Potential for Sustainable and Eco-Friendly Agricultural Practices*

Nanotechnology provides a means to sustainable agricultural practices by diminishing dependence on the very traditional fertilizers and pesticides. Biochar–nanoparticle composites may improve soil health and immobilize pollutants in contaminated soils. This method may also be applied to the restoration of red mud, wherein NPs can be integrated with charcoal, gypsum, or organic fertilizers to mitigate soil salinity and alkalinity, hence enhancing plant growth in damaged ecosystems. Moreover, the integration of halophyte species with NPs may facilitate sustained phytoremediation strategies, utilizing their inherent resilience to adverse soil environments.

Developments into nano-phytoremediation for red mud remain still developing, yet the prospective advantages are encouraging. A primary focus for future research is the identification of plant species exhibiting advantageous characteristics, like rapid growth rates, broad root systems, and elevated resistance to HM deposition. These characteristics may be augmented by NPs that improve phytoremediation efficacy, facilitating a synergistic method for red mud remediation. A vital aspect is the investigation of nanoparticle transport pathways in plant systems, which would enhance their absorption and efficacy. Furthermore, thorough environmental risk evaluations are essential to analyze the interactions of NPs with distinct physicochemical characteristics of red mud, including its pH, electrical conductivity, that organic matter content, thereby assuring ecological safety. Additional innovation may emerge from the formulation of hybrid remediation systems that integrate NPs with various soil amendments, including gypsum, charcoal, and sewage sludge, to tackle the complex issues associated with red mud contamination. To effectively harness the promise of nano-phytoremediation, numerous problems must be addressed. This involves guaranteeing nanoparticle-based solutions' scalability and economic feasibility while addressing the environmental and ecological dangers linked to their prolonged use. Additionally, strong and progressive regulatory frameworks are necessary to facilitate the safe and successful implementation of nanotechnology in agricultural along with environmental remediation settings. Nanotechnology presents significant potential for enhancing sustainable agriculture as well as environmental remediation, especially in tackling the intricate issues of red mud contamination. By addressing existing challenges and promoting interdisciplinary cooperation, these ideas may be converted into meaningful and pragmatic solutions that substantially advance sustainability objectives.

6 Conclusion

HM stress presents a considerable challenge to worldwide agricultural output, jeopardizing food security along with ecosystem health. This review highlights the essential function of metal and metal-oxide NPs in mitigating HM stress in crops. Due to their distinctive physicochemical characteristics, NPs present a promising strategy for reducing HM toxicity by improving nutritional bioavailability, immobilizing harmful metals, and stimulating plant defense mechanisms. NPs enhance crop resistance under HM stress by modulating gene expression, promoting antioxidant enzyme activities, and reducing oxidative stress. The synergistic effects of combination nanoparticle applications, such as silicon and selenium NPs or zinc oxide and iron oxide NPs, boost their efficacy in stress relief. These methods not only reduce stress but also enhance crop growth, production, and nutritional quality, demonstrating their potential for sustainable agriculture techniques. Nonetheless, the utilization of NPs in agriculture presents several obstacles. Concerns about environmental safety, nanoparticle longevity, and cost-effectiveness require additional research to refine formulations and guarantee safe application. Future developments must concentrate on elucidating the molecular mechanisms governing NP–plant interactions, customizing NP formulations for crop and soil circumstances, and investigating environmentally sustainable synthesis methods. The amalgamation of nanotechnology with precision agriculture and biotechnology methods can facilitate creative and sustainable solutions to HM stress. Utilizing the capabilities of NPs, agriculture can advance toward improved productivity and environmental sustainability, responding to the pressing demand for resilient agricultural systems amid increasing HM contamination.

References

1. Abdel Latef AAH, Srivastava AK, El-sadek MSA, Kordrostami M, Tran LSP (2018) Titanium dioxide nanoparticles improve growth and enhance tolerance of broad bean plants under saline soil conditions. Land Degrad Dev 29:1065–1073
2. Al-Hwaiti M, Al-Khashman O (2015) Health risk assessment of heavy metals contamination in tomato and green pepper plants grown in soils amended with phosphogypsum waste materials. Environ Geochem Health 37:287–304
3. Al-Khayri JM, Rashmi R, Surya Ulhas R, Sudheer WN, Banadka A, Nagella P, Aldaej MI, Rezk AA-S, Shehata WF, Almaghasla MI (2023) The role of nanoparticles in response of plants to abiotic stress at physiological, biochemical, and molecular levels. Plants 12:292
4. Alharbi K, Hafez EM, Omara AE-D, Rashwan E, Alshaal T (2023) Zinc oxide nanoparticles and PGPR strengthen salinity tolerance and productivity of wheat irrigated with saline water in sodic-saline soil. Plant Soil 493:475–495
5. Ali H, Khan E, Sajad MA (2013) Phytoremediation of heavy metals—concepts and applications. Chemosphere 91:869–881
6. Amato-Lourenco LF, Lobo DJA, Guimarães ET, Moreira TCL, Carvalho-Oliveira R, Saiki M, Saldiva PHN, Mauad T (2017) Biomonitoring of genotoxic effects and elemental accumulation derived from air pollution in community urban gardens. Sci Total Environ 575:1438–1444
7. Attama AA, Momoh MA, Builders PF (2012) Lipid nanoparticulate drug delivery systems: a revolution in dosage form design and development. Recent Adv Novel Drug Carrier Syst 5:107–140
8. Azameti MK, Imoro AWM (2023) Nanotechnology: a promising field in enhancing abiotic stress tolerance in plants. Crop Des 2(2):100037
9. Borymski S, Cycoń M, Beckmann M, Mur LA, Piotrowska-Seget Z (2018) Plant species and heavy metals affect biodiversity of microbial communities associated with metal-tolerant plants in metalliferous soils. Front Microbiol 9:1425
10. Brusseau M (2019) Soil and groundwater remediation. Environmental and pollution science. Elsevier, pp 329–354
11. Burman U, Saini M, Kumar P (2013) Effect of zinc oxide nanoparticles on growth and antioxidant system of chickpea seedlings. Toxicol Environ Chem 95:605–612
12. Castrillo G, Sánchez-Bermejo E, de Lorenzo L, Crevillén P, Fraile-Escanciano A, Tc M, Mouriz A, Catarecha P, Sobrino-Plata J, Olsson S (2013) WRKY6 transcription factor restricts arsenate uptake and transposon activation in Arabidopsis. Plant Cell 25:2944–2957
13. Cenova A, Nestorovska M, Crcarevska MS, Mihailova L, Shalabalija D, Dodov MG (2022) Evaluation of antioxidant properties of herbal mixture and CBD oil for the treatment of chronic wound. Macedonian Pharm Bull 68:557–558
14. Chandak V, Kumbhar M, Kulal P (2024) Highly sensitive and selective acetone gas sensor-based La-doped ZnO nanostructured thin film. Mater Lett 357:135747
15. Chandak V, Kumbhar M, Kulal P (2024) Structural, optical and magnetic studies of Cr-doped ZnO nanoparticles prepared by co-precipitation method. Inorgan Chem Commun 172:113766
16. Chandak V, Kumbhar M, Talekar S, Gunjakar J, Kulal P (2024) Ultrasensitive and selective Cr-doped ZnO thin films synthesized via spray pyrolysis technique for detection of ammonia gas. Appl Phys A 130:285
17. Chandra R, Kumar V (2017) Phytoextraction of heavy metals by potential native plants and their microscopic observation of root growing on stabilised distillery sludge as a prospective tool for in situ phytoremediation of industrial waste. Environ Sci Pollut Res 24:2605–2619
18. Changmei L, Chaoying Z, Junqiang W, Guorong W, Mingxuan T (2002) Research of the effect of nanometer materials on germination and growth enhancement of Glycine max and its mechanism. Soybean Sci 21:168–171
19. Chausali N, Saxena J, Prasad R (2021) Nanobiochar and biochar based nanocomposites: advances and applications. J Agric Food Res 5:100191
20. Clemens S, Ma JF (2016) Toxic heavy metal and metalloid accumulation in crop plants and foods. Annu Rev Plant Biol 67:489–512

21. Cui J, Liu T, Li F, Yi J, Liu C, Yu H (2017) Silica nanoparticles alleviate cadmium toxicity in rice cells: mechanisms and size effects. Environ Pollut 228:363–369
22. El-Kady AA, Abdel-Wahhab MA (2018) Occurrence of trace metals in foodstuffs and their health impact. Trends Food Sci Technol 75:36–45
23. El-Saadony MT, Saad AM, Soliman SM, Salem HM, Desoky E-SM, Babalghith AO, El-Tahan AM, Ibrahim OM, Ebrahim AA, Abd El-Mageed TA (2022) Role of nanoparticles in enhancing crop tolerance to abiotic stress: a comprehensive review. Front Plant Sci 13:946717
24. Esmaeili H, Foroutan R (2015) Investigation into ion exchange and adsorption methods for removing heavy metals from aqueous solutions. Int J Biol Pharm Allied Sci 4:620–629
25. Etesami H, Fatemi H, Rizwan M (2021) Interactions of nanoparticles and salinity stress at physiological, biochemical and molecular levels in plants: a review. Ecotoxicol Environ Saf 225:112769
26. Faizan M, Bhat JA, Chen C, Alyemeni MN, Wijaya L, Ahmad P, Yu F (2021) Zinc oxide nanoparticles (ZnO-NPs) induce salt tolerance by improving the antioxidant system and photosynthetic machinery in tomato. Plant Physiol Biochem 161:122–130
27. Faizan M, Faraz A, Mir AR, Hayat S (2021) Role of zinc oxide nanoparticles in countering negative effects generated by cadmium in Lycopersicon esculentum. J Plant Growth Regul 40:101–115
28. Faraji J, Sepehri A, Salcedo-Reyes JC (2018) Titanium dioxide nanoparticles and sodium nitroprusside alleviate the adverse effects of cadmium stress on germination and seedling growth of wheat (Triticum aestivum L.). Univ Sci 23:61–87
29. Fraceto LF, Grillo R, de Medeiros GA, Scognamiglio V, Rea G, Bartolucci C (2016) Nanotechnology in agriculture: which innovation potential does it have? Front Environ Sci 4:186737
30. França FC, Albuuerque AM, Almeida AC, Silveira PB, Crescêncio Filho A, Hazin CA, Honorato EV (2017) Heavy metals deposited in the culture of lettuce (Lactuca sativa L.) by the influence of vehicular traffic in Pernambuco Brazil. Food Chem 215:171–176
31. Gao X, Jiang L, Mao Y, Yao B, Jiang P (2021) Progress, challenges, and perspectives of bioleaching for recovering heavy metals from mine tailings. Adsorpt Sci Technol 2021:9941979
32. Husen A (2022) Engineered nanomaterials for sustainable agricultural production, soil improvement and stress management. Elsevier Inc., Cambridge. https://doi.org/10.1016/C2021-0-00054-7
33. Husen A, Siddiqi KS (2023) Advances in smart nanomaterials and their applications. Elsevier Inc., Cambridge. https://doi.org/10.1016/C2021-0-02202-1
34. Husen A (2024) Plant Response to Gold Nanoparticles. Springer Nature Singapore Pte Ltd., Singapore. https://doi.org/10.1007/978-981-97-7174-5
35. Husen A (2025a) Plant response to silver nanoparticles. Springer Nature Singapore Pte Ltd., Singapore. https://doi.org/10.1007/978-981-97-7352-7
36. Husen A (2025b) Emerging carbon nanomaterials for sustainable agricultural practices (synthesis, plant growth, performance, production and protection). Springer Nature Singapore Pte Ltd., Singapore
37. Hussain A, Rizwan M, Ali Q, Ali S (2019) Seed priming with silicon nanoparticles improved the biomass and yield while reduced the oxidative stress and cadmium concentration in wheat grains. Environ Sci Pollut Res 26:7579–7588
38. Hussain A, Rizwan M, Ali S, Ur Rehman MZ, Qayyum MF, Nawaz R, Ahmad A, Asrar M, Ahmad SR, Alsahli AA (2021) Combined use of different nanoparticles effectively decreased cadmium (Cd) concentration in grains of wheat grown in a field contaminated with Cd. Ecotoxicol Environ Saf 215:112139
39. Ijaz M, Khan F, Ahmed T, Noman M, Zulfiqar F, Rizwan M, Chen J, Siddique KH, Li B (2023) Nanobiotechnology to advance stress resilience in plants: current opportunities and challenges. Mater Today Bio 22:100759
40. Jadoun S, Arif R, Jangid NK, Meena RK (2021) Green synthesis of nanoparticles using plant extracts: a review. Environ Chem Lett 19:355–374

41. Jamshidi-Zanjani A, Khodadadi Darban A (2017) A review on enhancement techniques of electrokinetic soil remediation. Pollution 3:157–166
42. Jayeoye TJ, Eze FN, Olatunde OO, Singh S, Zuo J, Olatunji OJ (2021) Multifarious biological applications and toxic Hg2+ sensing potentiality of biogenic silver nanoparticles based on securidaca inappendiculata hassk stem extract. Int J Nanomed 12:7557–74
43. Jayeoye TJ, Eze FN, Singh S, Olatunde OO, Benjakul S, Rujiralai T (2021) Synthesis of gold nanoparticles/polyaniline boronic acid/sodium alginate aqueous nanocomposite based on chemical oxidative polymerization for biological applications. Int J Biol Macromol 179:196–205
44. Jayeoye TJ, Nwude EF, Singh S, Prajapati BG, Kapoor DU, Muangsin N (2024) Sustainable synthesis of gold nanoparticles for drug delivery and cosmeceutical applications: a review. BioNanoSci 14(3):3355–84
45. Jayeoye TJ, Singh S, Eze FN, Olatunji O, Oguntimehin I, Tyopine AA, Odogiyon OB, Olatunji OJ (2024) Green synthesis of silver nanoparticles using cyto-compatible polymer derivative of tara gum for gold (iii) ion detection in water samples. J Polym Environ 32(12):6667–86
46. Jozefczak M, Remans T, Vangronsveld J, Cuypers A (2012) Glutathione is a key player in metal-induced oxidative stress defenses. Int J Mol Sci 13:3145–3175
47. Junedi MA, Mukhopadhyay R, Manjari KS (2023) Alleviating salinity stress in crop plants using new engineered nanoparticles (ENPs). Plant Stress 9:100184
48. Kazemi S, Rafati Alashti M, Hosseini SJ (2021) Response of biochemical and physiological properties of camellia (Camelina sativa L.) to foliar application of calcium and silicon nanoparticles. Silicon 14(12):6817–28
49. Khan I, Saeed K, Khan I (2019) Nanoparticles: properties, applications and toxicities. Arab J Chem 12:908–931
50. Kinuthia GK, Ngure V, Beti D, Lugalia R, Wangila A, Kamau L (2020) Levels of heavy metals in wastewater and soil samples from open drainage channels in Nairobi, Kenya: community health implication. Sci Rep 10:8434
51. Kiran NS, Vaishnavi G, Singh S, Yashaswini C, Parihar A, Pal S, Kaur J, Prajapati B, Mohite P, Puri A (2024) Biomaterials comprising implantable and dermal drug delivery targeting brain in management of Alzheimer's disease: a review. Regenerative Eng Trans Med 11(1):15–38
52. Koleva L, Umar A, Yasin NA, Shah AA, Siddiqui MH, Alamri S, Riaz L, Raza A, Javed T, Shabbir Z (2022) Iron oxide and silicon nanoparticles modulate mineral nutrient homeostasis and metabolism in cadmium-stressed Phaseolus vulgaris. Front Plant Sci 13:806781
53. Krzesłowska M (2011) The cell wall in plant cell response to trace metals: polysaccharide remodeling and its role in defense strategy. Acta Physiol Plant 33:35–51
54. Kumar A, Jayeoye TJ, Mohite P, Singh S, Rajput T, Munde S, Eze FN, Chidrawar VR, Puri A, Prajapati BG (2024) Sustainable and consumer-centric nanotechnology-based materials: an update on the multifaceted applications, risks and tremendous opportunities. Nano-Struct Nano-Objects 38:101148
55. Kumar A, Kumar A, Chaturvedi AK, Shabnam AA, Subrahmanyam G, Mondal R, Gupta DK, Malyan SK, Kumar SS (2020) Lead toxicity: health hazards, influence on food chain, and sustainable remediation approaches. Int J Environ Res Publ Health 17:2179
56. Kumar A, Mohite P, Munde S, Puri A, Edet HO, Ochoule DO, Chidrawar VR, Nagime PV, Uddin MJ, Singh S (2024) Revolutionizing the biomedical and environmental clean-up through polymeric nano-sponges: a review. Nano-Struct Nano-Objects 39:101193
57. Kumar A, Shah SR, Jayeoye TJ, Kumar A, Parihar A, Prajapati B, Singh S, Kapoor DU (2023) Biogenic metallic nanoparticles: Biomedical, analytical, food preservation, and applications in other consumable products. Front Nanotechnol 5:1175149
58. Kumbhar MB, Chandak VS, Kulal PM (2025) Enhanced ammonia gas sensing performance at room temperature of binder-free NiO, Cu and Co-doped NiO thin films synthesized via the SILAR method. Mater Chem Phys 329:130065
59. Kumbhar MB, Patil VV, Chandak VS, Gunjakar JL, Kulal PM (2025) Enhancing energy storage with binder-free nickel oxide cathodes in flexible hybrid asymmetric solid-state supercapacitors. J Alloy Compd 1010:177311

60. Kumbhar MB, Patil VV, Chandak VS, Shaikh SB, Chitare YM, Gunjakar JL, Kulal PM (2024) Exploring copper-doped nickel oxide as a superior cathode electrode material for flexible hybrid solid-state supercapacitor device. J Ind Eng Chem 147:422–35
61. Lai M, Ghouri F, Sarwar S, Alomrani SO, Riaz M, Haider FU, Liu J, Imran M, Ali S, Liu X (2023) Modulation of metal transporters, oxidative stress and cell abnormalities by synergistic application of silicon and titanium oxide nanoparticles: a strategy for cadmium tolerance in rice. Chemosphere 345:140439
62. Lakherwal D (2014) Adsorption of heavy metals: a review. Int J Environ Res Dev 4:41–48
63. Li D, Li G, Li J, Xu S (2022) Wearable crop sensor based on nano-graphene oxide for noninvasive real-time monitoring of plant water. Membranes 12:358
64. Li F-l, Shi W, Jin Z-f, Wu H-m, Sheng GD (2017) Excessive uptake of heavy metals by greenhouse vegetables. J Geochem Explor 173:76–84
65. Liu C, Lin H, Li B, Dong Y, Yin T (2020) Responses of microbial communities and metabolic activities in the rhizosphere during phytoremediation of Cd-contaminated soil. Ecotoxicol Environ Saf 202:110958
66. Liu J, Zhang X-H, Tran H, Wang D-Q, Zhu Y-N (2011) Heavy metal contamination and risk assessment in water, paddy soil, and rice around an electroplating plant. Environ Sci Pollut Res 18:1623–1632
67. Liu R, Lal R (2014) Synthetic apatite nanoparticles as a phosphorus fertilizer for soybean (Glycine max). Sci Rep 4:5686
68. Liu R, Lal R (2015) Potentials of engineered nanoparticles as fertilizers for increasing agronomic productions. Sci Total Environ 514:131–139
69. Lobzenko I, Burachevskaya M, Zamulina I, Barakhov A, Bauer T, Mandzhieva S, Sushkova S, Minkina T, Tereschenko A, Kalinichenko V (2022) Development of a unique technology for the pyrolysis of rice husk biochar for promising heavy metal remediation. Agriculture 12:1689
70. Lucchini RG, Kotelchuck D, Grandjean P (2015) Principles for prevention of the toxic effects of metals. Handbook on the Toxicology of Metals. Elsevier, pp 507–528
71. Madhav S, Mishra R, Kumari A, Srivastav A, Ahamad A, Singh P, Ahmed S, Mishra P, Sillanpää M (2024) A review on sources identification of heavy metals in soil and remediation measures by phytoremediation-induced methods. Int J Environ Sci Technol 21:1099–1120
72. Mahmoodzadeh H, Nabavi M, Kashefi H (2013) Effect of nanoscale titanium dioxide particles on the germination and growth of canola (Brassica napus)
73. Mahmoud LM, Dutt M, Shalan AM, El-Kady ME, El-Boray MS, Shabana YM, Grosser JW (2020) Silicon nanoparticles mitigate oxidative stress of in vitro-derived banana (Musa acuminata 'Grand Nain') under simulated water deficit or salinity stress. S Afr J Bot 132:155–163
74. Maity A, Natarajan N, Vijay D, Srinivasan R, Pastor M, Malaviya DR (2018) Influence of metal nanoparticles (NPs) on germination and yield of oat (Avena sativa) and berseem (Trifolium alexandrinum). Proceed Nat Acad Sci India Sect B: Biol Sci 88:595–607
75. Malhan A, Guleria M, Das U, Majee SB, Singh S, Nagime PV, Syukri DM, Chidrawar VR, Arania R, Shafi S (2024) Navigating the future of infectious diseases management through carbon nanotubes: a review. Nano LIFE 16(02):2530001
76. Mohite P, Puri A, Bharati D, Singh S (2024) Polyphenol-encapsulated nanoparticles for the treatment of chronic metabolic diseases. Role Flavonoids Chronic Metabol Dis From Bench Clin 18:375–416
77. Mohite P, Shah SR, Singh S, Rajput T, Munde S, Ade N, Prajapati BG, Paliwal H, Mori DD, Dudhrejiya AV (2023) Chitosan and chito-oligosaccharide: a versatile biopolymer with endless grafting possibilities for multifarious applications. Front Bioeng Biotechnol 11:1190879
78. Morkunas I, Woźniak A, Mai VC, Rucińska-Sobkowiak R, Jeandet P (2018) The role of heavy metals in plant response to biotic stress. Molecules 23:2320
79. Mukarram M, Petrik P, Mushtaq Z, Khan MMA, Gulfishan M, Lux A (2022) Silicon nanoparticles in higher plants: uptake, action, stress tolerance, and crosstalk with phytohormones, antioxidants, and other signalling molecules. Environ Pollut 310:119855

80. Nagime PV, Chandak VS (2024) A comprehensive review of nanomaterials synthesis: physical, chemical, and biological approaches and emerging challenges. Biocatal Agric Biotechnol 4:103458
81. Nagime PV, Shaikh NM, Shaikh SB, Lokhande CD, Patil VV, Shafi S, Syukri DM, Chidrawar VR, Kumar A, Singh S (2024) Facile synthesis of silver nanoparticles using Calotropis procera leaves: unraveling biological and electrochemical potentials. Discover Nano 19:139
82. Nagime PV, Singh S, Chidrawar VR, Rajput A, Syukri DM, Marwan NT, Shafi S (2024) Moringa oleifera: a plethora of bioactive reservoirs with tremendous opportunity for green synthesis of silver nanoparticles enabled with multifaceted applications. Nano-Struct Nano-Objects 40:101404
83. Nagime PV, Singh S, Shaikh NM, Gomare KS, Chitme H, Abdel-Wahab BA, Alqahtany YS, Khateeb MM, Habeeb MS, Bakir MB (2023) Biogenic fabrication of silver nanoparticles using Calotropis procera flower extract with enhanced biomimetics attributes. Materials 16:4058
84. Nagime PV, Syukri DM, Sjahriani T, Hermawan D, Shaikh NM, Shafi S, Chidrawar VR, Singh S, Kausar N, Elamin A (2024) Phyto-mediated biosynthesis of silver nanoparticles using Aloe barbadensis Miller leaves gel with improved antibacterial, anti-fungal, antioxidant, anti-inflammatory, anti-diabetic, and anti-cancer activities. Nano-Struct Nano-Objects 40:101368
85. Nair PMG, Chung IM (2014) Impact of copper oxide nanoparticles exposure on Arabidopsis thaliana growth, root system development, root lignificaion, and molecular level changes. Environ Sci Pollut Res 21:12709–12722
86. Naveed S, Oladoye PO, Alli YA (2023) Toxic heavy metals: a bibliographic review of risk assessment, toxicity, and phytoremediation technology. Sustain Chem Environ 2:100018
87. Nawaz S, Maqsood I, Batool F, Sandhu ZY, Hassan S, Akram F, Rashid B (2023) Improvement of abiotic stress tolerance in plants with the application of nanoparticles. Abiotic Stress Plants-Adapt Climate Change
88. Nongbet A, Mishra AK, Mohanta YK, Mahanta S, Ray MK, Khan M, Baek K-H, Chakrabartty I (2022) Nanofertilizers: a smart and sustainable attribute to modern agriculture. Plants 11:2587
89. Nwabor OF, Singh S, Ontong JC, Vongkamjan K, Voravuthikunchai SP (2021) Valorization of wastepaper through antimicrobial functionalization with biogenic silver nanoparticles, a sustainable packaging composite. Waste Biomass Valorization 12:3287–3301
90. Nwabor OF, Singh S, Wunnoo S, Lerwittayanon K, Voravuthikunchai SP (2021) Facile deposition of biogenic silver nanoparticles on porous alumina discs, an efficient antimicrobial, antibiofilm, and antifouling strategy for functional contact surfaces. Biofouling 37:538–554
91. Omotayo AO, Omotayo OP (2024) Potentials of microbe-plant assisted bioremediation in reclaiming heavy metal polluted soil environments for sustainable agriculture. Environ Sustain Indic 22:100396
92. Ontong JC, Singh S, Nwabor OF, Chusri S, Voravuthikunchai SP (2020) Potential of antimicrobial topical gel with synthesized biogenic silver nanoparticle using Rhodomyrtus tomentosa leaf extract and silk sericin. Biotech Lett 42:2653–2664
93. Ovečka M, Takáč T (2014) Managing heavy metal toxicity stress in plants: biological and biotechnological tools. Biotechnol Adv 32:73–86
94. Park CM, Wang D, Han J, Heo J, Su C (2019) Evaluation of the colloidal stability and adsorption performance of reduced graphene oxide–elemental silver/magnetite nanohybrids for selected toxic heavy metals in aqueous solutions. Appl Surf Sci 471:8–17
95. Pereira ADES, Oliveira HC, Fraceto LF (2019) Polymeric nanoparticles as an alternative for application of gibberellic acid in sustainable agriculture: a field study. Sci Rep 9:7135
96. Pham VHT, Kim J, Chang S, Chung W (2022) Bacterial biosorbents, an efficient heavy metals green clean-up strategy: prospects, challenges, and opportunities. Microorganisms 10:610
97. Phogat N, Khan SA, Shankar S, Ansary AA, Uddin I (2016) Fate of inorganic nanoparticles in agriculture. Adv Mater Lett 7:3–12
98. Pohl A (2020) Removal of heavy metal ions from water and wastewaters by sulfur-containing precipitation agents. Water Air Soil Pollut 231:503

99. Prasad M (2003) Phytoremediation of metal-polluted ecosystems: hype for commercialization. Russ J Plant Physiol 50:686–701
100. Prasad T, Sudhakar P, Sreenivasulu Y, Latha P, Munaswamy V, Reddy KR, Sreeprasad T, Sajanlal P, Pradeep T (2012) Effect of nanoscale zinc oxide particles on the germination, growth and yield of peanut. J Plant Nutr 35:905–927
101. Rai PK, Kumar V, Lee S, Raza N, Kim K-H, Ok YS, Tsang DC (2018) Nanoparticle-plant interaction: Implications in energy, environment, and agriculture. Environ Int 119:1–19
102. Rai PK, Song H, Kim K-H (2023) Nanoparticles modulate heavy-metal and arsenic stress in food crops: hormesis for food security, safety, and public health. Sci Total Environ 902:166064
103. Rehman A, Weng J, Li P, Shah IH, Ur Rahman S, Khalid M, Manzoor MA, Chang L, Niu Q (2023) Green synthesized zinc oxide nanoparticles confer drought tolerance in melon (Cucumis melo L.). Environ Exp Bot 212:105384
104. Rizwan M, Ali S, Adrees M, Ibrahim M, Tsang DC, Zia-ur-Rehman M, Zahir ZA, Rinklebe J, Tack FM, Ok YS (2017) A critical review on effects, tolerance mechanisms and management of cadmium in vegetables. Chemosphere 182:90–105
105. Rizwan M, Ali S, Ali B, Adrees M, Arshad M, Hussain A, ur Rehman MZ, Waris AA, (2019) Zinc and iron oxide nanoparticles improved the plant growth and reduced the oxidative stress and cadmium concentration in wheat. Chemosphere 214:269–277
106. Samanta S, Seth CS, Roychoudhury A (2023) The molecular paradigm of reactive oxygen species (ROS) and reactive nitrogen species (RNS) with different phytohormone signaling pathways during drought stress in plants. Plant Physiol Biochem PPB 206:108259–108259
107. Sangeetha J, Thangadurai D, Hospet R, Purushotham P, Karekalammanavar G, Mundaragi AC, David M, Shinge MR, Thimmappa SC, Prasad R (2017) Agricultural nanotechnology: concepts, benefits, and risks. Nanotechnology: an agricultural paradigm. Springer, Singapore, pp 1–17
108. Saritha GNG, Anju T, Kumar A (2022) Nanotechnology-big impact: how nanotechnology is changing the future of agriculture? J Agric Food Res 10:100457
109. Sekhon BS (2014) Nanotechnology in agri-food production: an overview. Nanotechnol Sci Appl 20:31–53
110. Selvakesavan RK, Kruszka D, Shakya P, Mondal D, Franklin G (2023) Impact of nanomaterials on plant secondary metabolism. Nanomater Interact Plant Cell Mech Macromol Agric Implications 2:133–170
111. Seneviratne M, Rajakaruna N, Rizwan M, Madawala H, Ok YS, Vithanage M (2019) Heavy metal-induced oxidative stress on seed germination and seedling development: a critical review. Environ Geochem Health 41:1813–1831
112. Shahid M, Dumat C, Khalid S, Schreck E, Xiong T, Niazi NK (2017) Foliar heavy metal uptake, toxicity and detoxification in plants: a comparison of foliar and root metal uptake. J Hazard Mater 325:36–58
113. Shang Y, Hasan MK, Ahammed GJ, Li M, Yin H, Zhou J (2019) Applications of nanotechnology in plant growth and crop protection: a review. Molecules 24:2558
114. Shweta SS, Sharma A, Chadha S, Guleria V (2021) Nanotechnology: a cutting-edge technology in vegetable production. J Hortic Sci Biotechnol 96:682–695
115. Singh A, Pal DB, Mohammad A, Alhazmi A, Haque S, Yoon T, Srivastava N, Gupta VK (2022) Biological remediation technologies for dyes and heavy metals in wastewater treatment: new insight. Biores Technol 343:126154
116. Singh J, Lee B-K (2016) Influence of nano-TiO2 particles on the bioaccumulation of Cd in soybean plants (Glycine max): a possible mechanism for the removal of Cd from the contaminated soil. J Environ Manag 170:88–96
117. Singh S, Chunglok W, Nwabor OF, Ushir YV, Singh S, Panpipat W (2022) Hydrophilic biopolymer matrix antibacterial peel-off facial mask functionalized with biogenic nanostructured material for cosmeceutical applications. J Polym Environ 30:938–953
118. Singh S, Dodiya TR, Dodiya R, Ushir YV, Widodo S (2022c) Lipid nanoparticulate drug delivery systems: approaches toward improvement in therapeutic efficacy of bioactive molecules. Drug carriers. IntechOpen

119. Singh S, Nwabor OF, Sukri DM, Wunnoo S, Dumjun K, Lethongkam S, Kusolphat P, Hemtanon N, Klinprathum K, Sunghan J (2022) Poly (vinyl alcohol) copolymerized with xanthan gum/hypromellose/sodium carboxymethyl cellulose dermal dressings functionalized with biogenic nanostructured materials for antibacterial and wound healing application. Int J Biol Macromol 216:235–250
120. Song X-Q, Liu L-F, Jiang Y-J, Zhang B-C, Gao Y-P, Liu X-L, Lin Q-S, Ling H-Q, Zhou Y-H (2013) Disruption of secondary wall cellulose biosynthesis alters cadmium translocation and tolerance in rice plants. Mol Plant 6:768–780
121. Spósito JCV, do Amaral Crispim B, Mussury RM, Grisolia AB (2015) Genetic instability in plants associated with vehicular traffic and climatic variables. Ecotoxicol Environ Saf 120: 445-448
122. Sun Y, Niu G, Dou H, Perez C, Alexander L (2022) Growth, gas exchange, and mineral nutrients of Hydrangea hybrids irrigated with saline water. HortScience 57:319–325
123. Suriyaprabha R, Karunakaran G, Yuvakkumar R, Prabu P, Rajendran V, Kannan N (2012) Growth and physiological responses of maize (Zea mays L.) to porous silica nanoparticles in soil. J Nanopart Res 14:1–14
124. Syukri D, Nwabor O, Singh S, Ontong J, Voravuthikunchai S (2024) Biogenic silver nanoparticles synthesized with eucalyptus camaldulensis aqueous leaf extract, a novel anti-virulence strategy for attenuating acinetobacter baumannii outer membrane protein A. Authorea Prepr
125. Syukri DM, Nwabor OF, Singh S, Ontong JC, Wunnoo S, Paosen S, Munah S, Voravuthikunchai SP (2020) Antibacterial-coated silk surgical sutures by ex situ deposition of silver nanoparticles synthesized with Eucalyptus camaldulensis eradicates infections. J Microbiol Methods 174:105955
126. Syukri DM, Nwabor OF, Singh S, Voravuthikunchai SP (2021) Antibacterial functionalization of nylon monofilament surgical sutures through in situ deposition of biogenic silver nanoparticles. Surf Coat Technol 413:127090
127. Thabet SG, Alqudah AM (2023) New genetic insights into improving barley cope with salt stress via regulating mineral accumulation, cellular ion homeostasis, and membrane trafficking. Environ Exp Bot 208:105252
128. Thabet SG, Alqudah AM (2024) Unraveling the role of nanoparticles in improving plant resilience under environmental stress condition. Plant Soil 503(1):313–30
129. Vareda JP, Valente AJ, Durães L (2019) Assessment of heavy metal pollution from anthropogenic activities and remediation strategies: a review. J Environ Manag 246:101–118
130. Venkatachalam P, Jayaraj M, Manikandan R, Geetha N, Rene ER, Sharma N, Sahi S (2017) Zinc oxide nanoparticles (ZnONPs) alleviate heavy metal-induced toxicity in Leucaena leucocephala seedlings: a physiochemical analysis. Plant Physiol Biochem 110:59–69
131. Wang F, Wang M, Liu Z, Shi Y, Han T, Ye Y, Gong N, Sun J, Zhu C (2015) Different responses of low grain-Cd-accumulating and high grain-Cd-accumulating rice cultivars to Cd stress. Plant Physiol Biochem 96:261–269
132. Wang M, Zhu Y, Cheng L, Andserson B, Zhao X, Wang D, Ding A (2018) Review on utilization of biochar for metal-contaminated soil and sediment remediation. J Environ Sci 63:156–173
133. Wennersten R, Qie S (2018) United Nations sustainable development goals for 2030 and resource use. Handbook of Sustainability Science and Research, pp 317–339
134. Wu S (1994) Effect of manganese excess on the soybean plant cultivated under various growth conditions. J Plant Nutr 17:991–1003
135. Xia Y, Tang Y, Shih K, Li B (2020) Enhanced phosphorus availability and heavy metal removal by chlorination during sewage sludge pyrolysis. J Hazard Mater 382:121110
136. Xiong T, Dumat C, Dappe V, Vezin H, Schreck E, Shahid M, Pierart A, Sobanska S (2017) Copper oxide nanoparticle foliar uptake, phytotoxicity, and consequences for sustainable urban agriculture. Environ Sci Technol 51:5242–5251
137. Yang Y, Chang AC, Wang M, Chen W, Peng C (2018) Assessing cadmium exposure risks of vegetables with plant uptake factor and soil property. Environ Pollut 238:263–269
138. Yasmin H, Mazher J, Azmat A, Nosheen A, Naz R, Hassan MN, Noureldeen A, Ahmad P (2021) Combined application of zinc oxide nanoparticles and biofertilizer to induce salt

resistance in safflower by regulating ion homeostasis and antioxidant defence responses. Ecotoxicol Environ Saf 218:112262

139. Zaynab M, Al-Yahyai R, Ameen A, Sharif Y, Ali L, Fatima M, Khan KA, Li S (2022) Health and environmental effects of heavy metals. J King Saud Univ Sci 34:101653
140. Zeeshan M, Hu YX, Afridi MS, Ahmad B, Ahmad S, Muhammad I, Hale B, Iqbal A, Farooq S, Wu HY (2022) Interplay of ZnONPs and/or SeNPs induces arsenic tolerance in soybean by regulation of antioxidants pool, WRKY genes, and expression of arsenic transporters. Environ Exp Bot 195:104783
141. Zeeshan M, Hu YX, Iqbal A, Salam A, Liu YX, Muhammad I, Ahmad S, Khan AH, Hale B, Wu HY (2021) Amelioration of AsV toxicity by concurrent application of ZnO-NPs and Se-NPs is associated with differential regulation of photosynthetic indexes, antioxidant pool and osmolytes content in soybean seedling. Ecotoxicol Environ Saf 225:112738
142. Zhakypbek Y, Kossalbayev BD, Belkozhayev AM, Murat T, Tursbekov S, Abdalimov E, Pashkovskiy P, Kreslavski V, Kuznetsov V, Allakhverdiev SI (2024) Reducing heavy metal contamination in soil and water using phytoremediation. Plants 13:1534
143. Zhang H, Ding Y, Zhi J, Li X, Liu H, Xu J (2018) Over-expression of the poplar expansin gene PtoEXPA12 in tobacco plants enhanced cadmium accumulation. Int J Biol Macromol 116:676–682
144. Zheng L, Hong F, Lu S, Liu C (2005) Effect of nano-TiO_2 on strength of naturally aged seeds and growth of spinach. Biol Trace Elem Res 104:83–91
145. Zhou H, Yang W-T, Zhou X, Liu L, Gu J-F, Wang W-L, Zou J-L, Tian T, Peng P-Q, Liao B-H (2016) Accumulation of heavy metals in vegetable species planted in contaminated soils and the health risk assessment. Int J Environ Res Public Health 13:289

Role of Metal and Metal-Oxide Nanoparticles in Crop Mineral Nutrient Deficiency Stress Adaptation and Mitigation

Izabela Jośko and Gianluca Brunetti

Abstract The increasing global population necessitates a heightened demand for food and energy, consequently prompting the intensification of agricultural production. This intensification often requires a greater use of fertilizers and pesticides, which may lead to progressive soil and groundwater contamination. How to stop this snowball effect? One potential solution to mitigate these challenges lies in nanotechnology, which could enhance plant fertilization efficiency through engineered nanoparticles (ENPs). Due to their nanoscale size, these particles exhibit unique and advantageous properties, leading to the exploration of new applications across various economic sectors, including agriculture. In the realm of fertilizers, ENPs present promising opportunities as their small size allows them to penetrate the cell wall barrier compared to larger particles more easily. Moreover, metal-based ENPs are characterized by gradual solubility, which enables them to release nutrients over an extended period, unlike traditional fertilizers that release their entire ion pool quickly. Despite the long-discussed fertilizing potential of ENPs since the advent of nanotechnology, there remain numerous uncertainties regarding the conditions of their application and their superiority over conventional fertilizers. This chapter conducts a comparative analysis of existing studies on the use of ENPs as micronutrient fertilizers for plants. The analysis of collected data aims to guide future research, potentially resulting in the optimized application of nano-agrochemicals.

Keywords Engineered nanoparticles · Nano-fertilizers · Micronutrients · Application conditions

---

I. Jośko (✉)
Institute of Plant Genetics, Breeding and Biotechnology, University of Life Sciences, Lublin, Poland
e-mail: izabela.josko@up.lublin.pl

G. Brunetti
Future Industries Institute, University of South Australia, Adelaide, Australia

A. Husen (ed.), *Nanobiotechnology for Abiotic Stress Adaptation and Mitigation in Agricultural Crops*, Smart Nanomaterials Technology,
https://doi.org/10.1007/978-981-96-9298-9_5

1 Introduction

Forecasts indicate that by 2050, the global population will reach nearly 10 billion, presenting new challenges related to the increased demand for food, energy, and water [124]. To address these needs, agricultural production must rise by 60% [5]. However, various factors such as land consumption, climate change, desertification, floods, erosion, and soil cover degradation are causing a reduction in arable land. By 2050, the arable land available per capita is projected to decrease by 18% [5]. Consequently, there is a pressing need to intensify plant production [56], which is crucial for both the food and bioenergy sectors. While fertilization has been the primary solution to boost crop yields, the use of fertilizers is often inefficient [68]. Less than 30% of applied fertilizers/pesticides actually reach their target within the plant, and over-application by farmers seeking higher yields leads to environmental contamination [68]. In response, the European Commission has implemented the "From Farm to Fork" strategy, mandating at least 50% and 20% reductions in nutrient loss and fertilizer use, respectively, and over 50% reduction in pesticide use and its risk by 2030 [35]. Moreover, recent disruptions like the COVID-19 pandemic and wars have highlighted the fragility of food supply chains, underscoring the need for crops to thrive in neighborhoods even in low-quality soils.

The ineffective fertilization of plants, coupled with the growing demand for plant crops, necessitates the exploration of new solutions. Nanotechnology has attracted significant interest and has been identified by the European Commission as one of the five "Key Enabling Technologies" [36], offering new hopes and opportunities for the advancement of various sectors, including agriculture [72]. Engineered nanoparticles (ENPs), due to their size (1–100 nm), possess specific and more effective properties, such as enhanced solubility and reactivity, compared to their bulk counterparts [72]. These unique properties have made nanomaterials a topic of interest within the fertilizer sector, as they hold potential both as components of fertilizers and as plant protection agents [67]. In plant fertilization, the application of ENPs is primarily aimed at two pathways: using nanomaterials as fertilizer components or as carriers of nutrients and ingredients capable of inducing transformations of fertilizer components [66]. Similar to the application of nanomaterials in medicine for targeted therapies, efforts are being made to develop precision nano-fertilizers that would gradually release active substances in accordance with a plant's specific nutrient requirements [72]. The on-demand release of active compounds would enable sustainable application of fertilizers and pesticides, thereby safeguarding the environment from an influx of contaminants [49].

Although numerous nano-agrochemicals are currently available in the market, significant knowledge gaps persist regarding ENP use in plant fertilization, starting with the generally accepted definition of nano-fertilizers [67]. To date, this term has encompassed both pure ENPs treated as sources of micro- and macronutrients, as well as various formulations containing ENPs that deliver active substances to plants [66]. Using macronutrients in nanomaterials is not a novel concept; certain substances, such as hydroxyapatite, have been utilized for an extended period [62].

In contrast, nano-formulations are distinguished by their diverse structure, form, and composition [60], and their nano-effect is challenging to verify due to the absence of larger equivalents. Consequently, this chapter aims to evaluate the fertilizing potential of pure metal-based ENPs, whose effects can be compared to both conventional fertilizer nutrients and the larger bulk counterparts of ENPs, to assess the "nano" effect accurately.

Metal-based ENPs, including metal and metal oxides, can serve as a source of micronutrients (such as Fe, Zn, Cu, and B) for plants and, subsequently, for humans [72]. Given that approximately two billion people suffer from micronutrient malnutrition [38], implementing strategies to enrich agricultural produce with these components is deemed essential. Plant nutrition using ENPs presents an effective method to combat "hidden hunger," which contributes to health issues such as anemia, immune deficiency, and intellectual development disorders, thus leading to significant and prolonged social and economic repercussions [2]. Due to their nanoscale size and unique properties, ENPs can deliver metal ions in a more controlled and plant-specific manner than conventional fertilizers [100]. Current knowledge regarding the potential application of ENPs as fertilizers often stems from ecotoxicological studies, where stimulating effects were observed at lower doses [72]. However, there is a notable lack of research specifically focused on the role of ENPs as fertilizers, which considers the conditions of their application and assesses the effects of such fertilization in comparison with conventional agents (Fig. 1). This knowledge is of contribution for developing a roadmap for the use of ENPs as a treatment in precision agriculture.

2 The Properties of ENPs Crucial for Plant Fertilization

The properties of ENPs, including size, surface charge, functionalization, and crystallinity, play a crucial role in determining the nature and progression of various processes that ENPs may undergo, such as dissolution, adsorption, reduction/oxidation, and aggregation, in different media [81, 87]. In the context of their potential application as nano-fertilizers, the characterization of ENPs must be accompanied by an assessment of their behavior, particularly in soils and solutions intended for application (Fig. 2). The strong tendency of ENPs to aggregate can impede their ability to penetrate the cell wall barrier [93]. To enhance stability, ENPs are often functionalized with coatings such as cysteine, citrate, and surfactants [87].

Surface modifications, particularly those affecting surface charge, can significantly alter the uptake and translocation possibilities of ENPs [11, 118]. Spielman-Sun et al. [118] observed that positively charged CeO_2 ENPs exhibited strong adhesion to the roots of *Triticum aestivum* L., resulting in a reduced uptake rate due to electrostatic interactions with negatively charged cell walls. Furthermore, the ability of ENPs to translocate within plants was also influenced by surface charge; neutral and negatively charged nano-CeO showed greater migration to the upper parts of

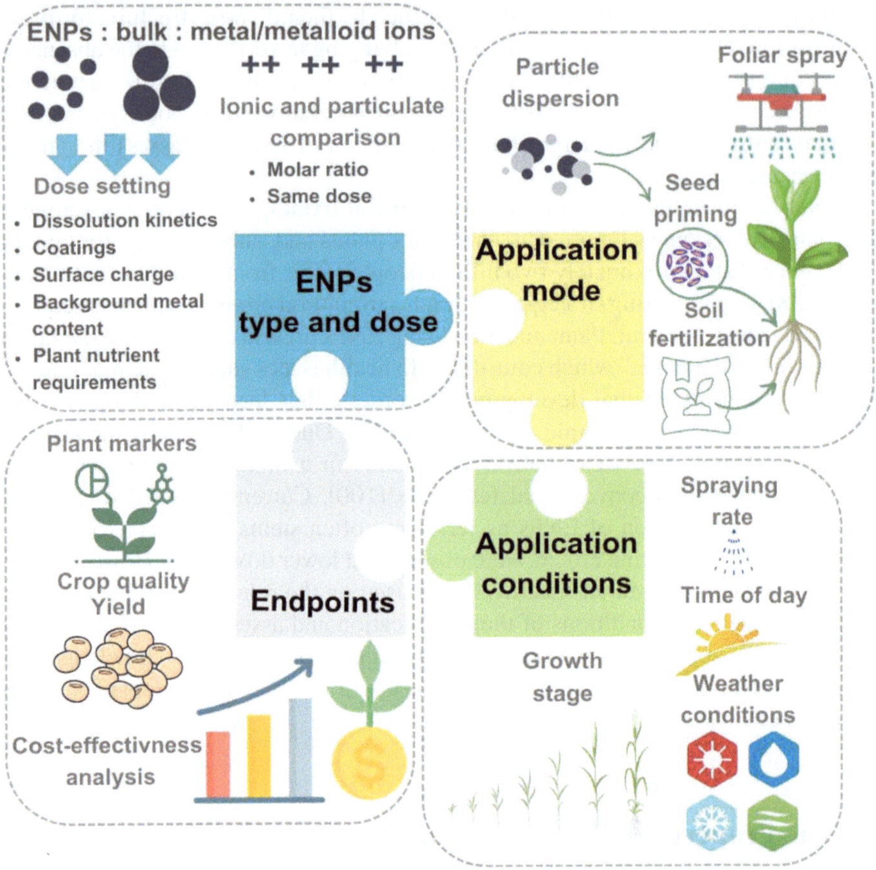

Fig. 1 Puzzle of efficiency optimization for the application of metal-based ENPs

plants [118]. The dissolution rate of ENPs is also critically important for plant fertilization. Theoretically, due to their smaller size and larger specific surface area, ENPs exhibit greater solubility compared to their bulk counterparts, ensuring faster delivery of microelements to plants [87]. Conversely, certain ENPs may release metal ions gradually, offering a valuable alternative to traditional fertilizers, such as soluble metal salts or complexing compounds, which deliver the complete pool of micronutrient ions immediately [67]. Analyzing dissolution kinetics should also consider the ENPs' chemical form; for instance, low-soluble CuO and CuS ENPs are more persistent than nano-$Cu(OH)_2$, allowing for prolonged Cu^{2+} supply to plants [117]. In addition, the processes and transformations affecting both ENPs and the ions released from them are influenced by various factors, including pH, ionic strength, and, in soils, natural organic matter (NOM) and organic ligands such as proteins. The probability and specific nature of ENP transformations depend on the method of application. For example, in foliar applications, ENPs are exposed to solar radiation,

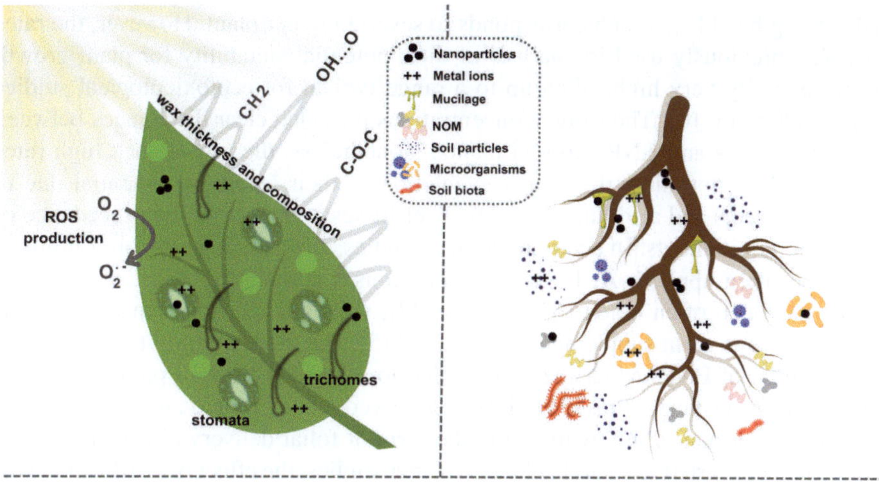

Fig. 2 Fate of ENPs on leaf surface and in soil

which may lead to reduction, oxidation, surface charge modification, and reactive oxygen species (ROS) generation [81]. In soil applications, the behavior of ENPs is influenced by interactions with soil components, such as microflora [22] or root exudates [84], determining the release and mobility of metals from ENPs. Given the multitude of factors affecting ENP action, analyzing their fate under application conditions is imperative to enhance the effectiveness of nano-fertilizers. This analysis should accompany evaluations of the quantitative and qualitative effects on final products.

3 The Application of Optimal ENP Rates

In both traditional agricultural practices and precision agriculture, the determination of fertilizer dosage is primarily based on the content of a specific element in the soil and the demand of a plant species for that particular microelement [71]. However, the total content of nutrients in the soil may not suffice as a criterion for fertilizer dosage [20]. For microelements such as Zn or Cu, only small percentages are considered potentially bioavailable fractions. Therefore, phyto-availability analysis using neutral salt extractants [86] can be instrumental in identifying nutrient deficiencies in specific plants. In existing studies on ENP fertilization, even when the total [15] or potentially bioavailable [58] content of microelements was analyzed (Table 2), these measurements primarily served to characterize the soil environment rather than to determine optimal ENP rates.

The micronutrient fertilization recommendations, depending on the type and soil content of a specific micronutrient and plant species, usually range from few to over a dozen kg/ha [116], which corresponds to several mg per plant. However, the rates of ENPs previously used in analysis of their potential suitability for plant growth were generally very high, often up to a range typical for ecotoxicological studies (e.g., 1000 ppm) [6]. Their high concentrations provided clear differences between control samples and ENP-exposed plants. Nonetheless, the use of such high rates of ENPs does not allow the proper evaluation of their fertilizing potential due to the toxic effects and also due to the fertilizer losses arising from the exceedance of the plant requirements. In ecotoxicological studies, a broad range of concentrations has been often applied. At lower concentrations a stimulating effect of ENPs on plants has been often noted [65, 77, 79, 133], which raised hopes regarding their potential use in plant supplementation [114, 115]. But there are only few studies on the potential of ENPs as fertilizers in which low doses (below 10 ppm) were used [96, 108]; for instance, Prasad et al. [97] observed a higher yield and content of Zn in peanut after nano-ZnO treatment in the form of foliar delivery (2 g/15 L–several ppm when converted to a single plant). In other studies, the effect of fertilization was observable only for individual parameters, e.g., biochemical, while no differences were noted in the amount of yield [108]. The use of low doses may be problematic, especially in the case of foliar application, due to the risk of random distribution. Rawat et al. [102] observed a significant increase in *Brassica juncea* L. yield only at a rate of 4 ppm, whereas the rates of 2, 6, 8, 10 ppm did not influence yield quantity. In the case of foliar application, apart from an appropriate dose, it is also necessary to adjust the proper volume of the nutrient solution applied. Application of a too large amount of solution will result in high losses of elements, while a too small volume of foliar fertilizer may cause non-uniform and random distribution of ENPs on plants.

4 Elucidation of the Effect of Nano-phenomenon

The potentially enhanced efficacy of ENPs in promoting plant growth requires rigorous comparison with conventional sources of micronutrients, which are most commonly available as salts or metal chelates, and less frequently as bulk metal oxides (Table 2). Major existing studies on the application of ENPs often utilized only water or soil as control samples [13, 16, 61, 123, 126–128], which cannot definitively establish an advantage of ENPs over conventional fertilizers. Only a few studies have compared the effects of ENPs with those of soluble salts [73, 97] or chelating compounds [106, 108]. However, these comparisons were frequently conducted during the early stages of plant growth, spanning from several days to a month. Despite observed differences between ENPs and salts during early growth stages, it remains uncertain whether these differences persist in subsequent growth phases. Moreover, few studies have considered the bulk counterparts of ENPs, such as metal micro-oxides [6, 73, 121], which would facilitate the assessment of particle size effects. This scarcity may be attributed to the larger particle size and consequently limited transport of microparticles into plants, as well as their lower solubility. A summary of existing studies performed by Kah et al. [67] indicates an 18% higher effectiveness of ENPs compared to their larger homologs based on indicators such as germination, plant growth, and yield observed in Petri dish tests, pot, and field experiments.

5 Mode of Application for ENPs

Similar to conventional fertilizers, the mode of application for ENPs plays a crucial role in determining the transport rate of nutrients to specific plant compartments. Research on the potential use of ENPs for the quantitative and qualitative enhancement of agricultural produce extends to the preparatory stage of sowing material. Seed nutripriming, which involves treating seeds with compounds including microelements, aims to improve both germination and plant growth during early developmental stages [40]. The potential application of ENPs for seed dressing has emerged from ecotoxicological studies demonstrating that seed germination and seedling growth are stimulated by ENPs [17, 43, 78]. Existing studies where seeds were soaked in ENP (nanopriming) solutions show greater germination vigor compared to untreated seeds or those treated with metal salts [82, 113, 120]. However, to substantiate the effects of nanopriming, further research is required to compare the development of plants derived from treated and untreated seeds at later growth stages, as seed priming with ENPs may fulfill the micronutrient requirement only during early growth stages. Consequently, seed dressing with ENPs should be considered an alternative solution in cases of minor micronutrient deficiencies in soil or as a supplementary fertilization treatment through foliar application or soil incorporation [40].

Given the distinct mechanisms regulating nutrient uptake by plant roots and aboveground parts, analyzing the fertilizer potential of ENPs necessitates consideration of their mode of application [112]. Numerous studies have examined the exposure of roots to ENPs applied to the soil, conducted in laboratory [59, 100, 126, 132, 133], greenhouse [6, 107, 108, 123], and field settings [63, 97, 121]. Under laboratory conditions, ENPs were applied to both nutrient solutions [9, 61, 96] and soils [58, 121, 132], which differently influenced their distribution and speciation. Studies on the effects of ENPs on hydroponically grown plants under specific deficiency conditions enable the estimation of contributions from elements originating from ENPs, minimizing or eliminating background effects. However, the distribution of ENPs in growing media remains a concern. ENP aggregates may sediment, rendering them unavailable or less available to plants, necessitating the refreshing of nutrient media in such experiments. In addition to hydroponic cultures, plant root exposure to ENPs has been conducted in nutrient media with agar addition [14, 74], stabilizing ENPs and preventing precipitation, though it does not address ENP mobility issues. Furthermore, such media preclude investigations spanning the entire plant life cycle. Soil, as the natural environment for plant growth, is a valid choice for assessing the suitability of ENPs in plant fertilization. Due to interactions with soil components, the fate of ENPs in soils markedly differs from that in nutrient media.

Wang et al. [125] observed the presence of nano-ZnO in solution culture, whereas in soil, nano-ZnO dissolved rapidly. Moreover, the speciation of Zn did not differ between nano-ZnO and $ZnCl_2$ applied to soil. Conversely, a study by Gao et al. [44] revealed higher extractability of Cu after 31 days in soil with nano-CuO compared to $(CuNO_3)_2$, indicating the gradual solubility of ENPs, potentially extending their availability to plants. It is essential to recognize that the justification for applying ENPs as soil-applied fertilizers necessitates prior research to investigate their fate in this matrix, which will depend on both the type of ENPs and soil properties.

Foliar application is often preferred over soil fertilization for micronutrient feeding of plants, primarily due to the shorter distance from the nutrient source to the target destination [37]. Unlike soil fertilization, ENPs interact directly with plant tissue, without the involvement of matrix components that may alter ENP properties and their potential translocation into plants. Although the absorption mechanism of elements is not fully understood, it generally occurs through the cuticle, trichomes, stomata, and lenticels, with varying intensities depending on plant species and leaf morphology [76]. Additionally, the uptake level of micronutrients is influenced by the form in which they are applied. Metal ions supplied as soluble salts or complex-forming compounds can penetrate plants more readily than ENPs. A study by Li et al. [76] demonstrated higher Zn absorption following $ZnSO_4$ treatment compared to nano-ZnO, indicating more efficient uptake of Zn ions versus nanoparticles [80]. This raises questions about the mobility and translocation potential of micronutrients within leaves, as many studies have reported slow movement of these nutrients in leaf tissue, irrespective of plant species. The absorption of elements is also contingent on the "technical" application conditions of ENPs, which influence solution distribution on plants. Despite numerous studies involving foliar application of ENPs [18, 64, 73, 101], many questions remain about the effectiveness of delivering nano-nutrients in this manner, particularly since most studies do not detail application conditions [63].

In addition to the quantity of nutrient solution applied, the degree of spraying is crucial to avoiding spotting [37]. Smaller droplets ensure more uniform coverage of the entire leaf area with the fertilizer, and they remain on the leaf surface for an extended period. In the context of foliar application, the time of day and weather conditions (such as temperature, wind, and humidity) are particularly significant, especially in field studies. To prevent leaf burning, fertilizers are typically applied in the morning or afternoon during calm, rain-free weather conditions to minimize drift and runoff [8]. Higher temperatures can reduce plant turgor, leading to decreased micronutrient absorption [83]. Foliar nutrition must also be conducted under appropriate humidity conditions to prevent rapid evaporation, allowing micronutrient solutions sufficient time (approximately 3–4 h) to penetrate the leaves [37]. However, only a limited number of studies provide detailed information on the conditions of ENP application, often focusing on individual parameters such as the time of day [6].

6 Timing of ENP Application

The timing of nano-fertilizer application is a critical factor influencing their effectiveness. The decision regarding the application time of ENPs depends on the mode of application and the plant's demand for a specific element at various growth stages. For soil applications, ENPs, like traditional fertilizers, are applied prior to sowing and before agro-technical practices to ensure uniform nutrient distribution [7]. It is important to note that due to their small doses, micronutrients can become immobilized by NOM and other soil particles [129]. Conversely, higher rates of ENP application are not recommended, as plant nutrient requirements may change (often increase) with growth stages, and excessive fertilization can lead to toxicity. Moreover, it has been observed that in studies on ENP soil applications, plant seedlings are predominantly used [29, 90, 108, 127, 128, 132], while seeds alone are less frequently involved [63, 85, 121].

As opposed to soil fertilization, foliar nutrition can be carried out throughout the entire growing season and therefore fertilizer application time is determined in such a way to meet the highest requirement of plant for a specific micronutrient (Fig. 3). In existing studies with foliar-applied-ENPs, plant nutrition was carried out at early growth stages, notably at the several-leaf stage [6, 30, 75, 99, 126, 132]. Although some plants display a strong demand for specific microelements already at an early growth stage, the choice of the optimal application time should be estimated by short exposure time and duration experiments. The most favorable application timing for traditional fertilizers is usually specified by recommendations which are based on years of both empirical and research experiences [4]. Moreover, as mentioned before, since the ions released from ENPs, rather than the ENPs themselves, would probably be the ones that would undergo the absorption processes, the fertilizer delivered as ENPs should be able to supply microelements for a longer period of time (only on the condition that they remain on the leaf surface). Thus, these considerations raise the question whether, in relation with the ion release from ENPs, the time of application should be shifted or the dose of ENPs should be increased so that the ENP concentration meets the actual plants demand. In addition, the possibility of emergency intervention, if micronutrient deficiency symptoms occur, is reported as one of the big advantages of foliar application. Nonetheless, nutrients should be supplied before deficiency symptoms become largely manifest since morphological and chemical changes in leaf surface can affect the uptake of nutrients. For example, Li et al. [76] found that under Zn deficiency, foliar absorption of Zn dropped by over 50% compared to plants with sufficient Zn. The lower uptake of Zn was related with a decrease of the leaf trichome density and also with changes in the chemical composition of the adaxial leaf surface.

In addition to meeting the plant's nutrient requirements for a specific element, the timing of fertilization should align with the purpose of the treatment. For instance, to enhance the microelement content or specific biochemical parameters in produce, fertilizers are typically applied during later plant development stages, once fruits or fructifications become visible. Applying microelements during anthesis or flowering

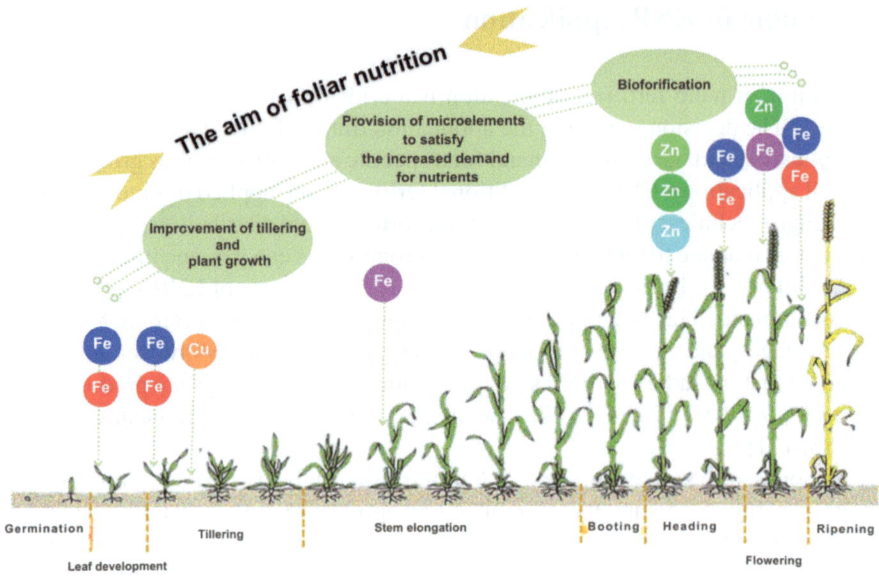

Fig. 3 The foliar application time of microelement-based nanoparticles to cereals

can increase protein content in kernels. To optimize nutritional effectiveness, fertilizer doses are often split and applied at different times to improve plant health and produce quality. In existing studies on foliar application, ENPs were administered in two or more doses [58, 123]. However, the impact of multiple applications is difficult to discern due to the lack of comparisons regarding application time and split dose effects, as time intervals were typically only in the order of days. Additionally, the plant growth stage influences morphology and physiology, which may affect ENP distribution and absorption. During phenological development, significant changes in plant habit may also modify the effectiveness of ENP treatments. For example, in cereal plants nearing the heading phase, flag leaves assume a vertical position, allowing fertilizer to cover both adaxial and abaxial leaf surfaces. However, this vertical orientation can make ENP retention on the leaf surface more challenging. Therefore, prior to application, the behavior of nano-fertilizers should be tested, and a surfactant may be necessary to enhance adhesion to leaves. To increase spray interception, applications should be performed at a suitable leaf area index (LAI), estimated at 3–4 [48] for wheat. Beyond leaf size and orientation, the thickness and wax composition of cuticles change with plant age [70], which can affect ENP penetration rates. Older plant leaves have a thicker wax layer, potentially hindering particle penetration into plant tissues.

7 Evaluation of ENPs Effectiveness in Enhancing Yield and Crop Quality

When assessing the effectiveness of fertilization with ENPs, a natural focus is on their impact on both yield quantity and quality. Many existing studies have evaluated the fertilizing potential at early plant growth stages and shortly after ENP application [6, 96, 106, 123, 126]. These studies primarily assessed key parameters indicating the physiological condition of plants during the growing season, such as plant pigment content, enzyme activity (mainly antioxidant enzymes) [3, 96, 106, 123], and the content of protein, phenols, and other metabolites [64, 75, 99, 121, 130]. Recently, there has been an increase in studies covering the entire plant life cycle, culminating in fruit production, where ENPs have generally been shown to contribute to increased yields [19, 24, 63]. Nevertheless, the effects of ENP fertilization were often compared to untreated samples, with fewer studies comparing ENPs to conventional fertilizer components (Table 1). Some studies demonstrated that Fe- and Zn-based ENPs resulted in higher yields in cereals such as maize [59], wheat [28, 46], and sorghum [25] compared to bulk compounds. Conversely, other studies found no significant differences in crop yields between the application of ENPs and bulk compounds [15]. Micronutrient fertilization is typically aimed at enhancing specific quality characteristics of yield, including seed biofortification. However, the application of ENPs has not consistently resulted in higher metal component loading in grains compared to larger counterparts [15, 103, 119]. Other basic nutritional parameters of seeds, such as elemental composition, total protein, carbohydrate, sugar, starch, gluten, vitamin C content, and amino acid composition, have also been evaluated. Some studies have assessed specific proteins related to the metal component of ENPs; for example, foliar application of nano-Fe_3O_4 resulted in higher ferritin content in maize kernels than Fe-EDDHA [15]. Many studies were conducted at high concentrations, often aiming to test worst-case scenarios for food security [27, 45, 46, 52, 90, 91], which sometimes led to a decrease in the nutritional quality of fruits. Moreover, despite the multiparametric evaluation of seeds, the applied rates of ENPs differed from those of bulk compounds, as ENPs were used at lower rates. This discrepancy further complicates the clear verification of the nano-effect (Table 2).

8 Use of ENPs in Mitigating Biotic and Abiotic Stress Responses

A significant purpose of using micronutrient-based ENPs in plant fertilization is to enhance plant resistance under biotic and abiotic stress conditions. Managing the nutritional status of plants is a key strategy for controlling plant diseases [111] (Fig. 4). By supplying micronutrients, ENPs activate plant defense mechanisms and indirectly contribute to disease management [12]. Their effectiveness in bolstering resistance to plant diseases has been particularly noted in ENPs with antifungal

Table 1 Comparison of the effects of ENPs and bulk compounds on quantitative and qualitative traits of yield

Type of ENPs/bulk compounds	Concentration	Plant	Effect of ENPs versus bulk compounds on yield			Key reference
			Quantity	Metal component content	Bioactivity	
Nano-Fe_3O_4 Fe-EDDHA	100 mg kg^{-1}	Maize	↑	↑	↑Ca content ↑ferritin content	[58]
Nano-Fe_xO_x Fe chelate Fe sulfate	2, 4 g L^{-1} 4, 8 g L^{-1}	Wheat	↑at 2 g L^{-1}	↑at 2 g L^{-}	↑at 2 g L^{-1} protein and carbohydrate content	[46]
Nano-ZnO-chitosan $ZnSO_4$	4, 40 mg Zn L^{-1} 40, 400 mg Zn L^{-1}	Wheat	↔	↔	↔protein content	[15]
Nano-ZnO-chitosan $ZnSO_4$	0.2%	Wheat	–	↓	↑enhancement of gamma gliadins ↓other gluten subunits decreased	[19]
Nano-Cu Micro-Cu	125, 250, and 500 mg kg^{-1}	Bell peppe	↔	↔	↔vitamin C content	[103]
Nano-ZnO $ZnSO_4$	10, 20, 50, 100, 200, 1000 mg kg^{-1}	Wheat	↑	↑	–	[28]
Nano-ZnO $ZnSO_4{\cdot}7H_2O$	6 mg kg^{-1}	Sorghum	↑	↓	↔N, K content ↓P content (high dose NPK) ↔P content (low dose NPK)	[25]

(continued)

Table 1 (continued)

Type of ENPs/bulk compounds	Concentration	Plant	Effect of ENPs versus bulk compounds on yield			Key reference
			Quantity	Metal component content	Bioactivity	
Nano-ZnO $ZnSO_4$ Micro-ZnO	50, 100, 200, 400, 600, 800, 1000, 1500, 2000 ppm 2000 ppm	Maize	↓	↔	–	[119]
Nano-CuO Micro-CuO	200, 400 mg kg^{-1}	Lettuce	↑	–	–	[127, 128]
Nano-ZnO Nano-Al_2O_3@ZnO Nano-KH550@ZnO Bulk-ZnO $ZnCl_2$	250, 1000 mg kg^{-1}	Green pea	↔	↑Al_2O_3@ZnO NPs at 1000 mg/kg	↔protein and carbohydrate profiles ↑doped ZnO NPs at 1000 mg kg^{-1} sucrose concentration	[91]

↑/↓—higher/lower effect of ENPs in comparison with bulk compounds; ↔—no difference between ENPs and bulk compounds effects; (–)—not determined

Table 2 Treatment conditions during foliar nutrition and soil amendment with ENPs

ENPs type and size	Control type			Dose	Matrix type	Background metal content		Plant	Plant growth stage during treatment	Duration of exposure	Endpoints	Key reference
	Matrix	Metal salts/ chelate compounds	Bulk compounds			Total	Potential bioavailable					
Foliar application												
Fe_3O_4 10–20 nm	✓	✓	✓	100 mg kg^{-1}	Loam	✓	✓	Maize *Zea mays* L.	Collar of fourth leaf/ visible collar of tenth leaf/ visible tassel fully extended/ milk	Till maturity	• Chlorophyll content • Antioxidant enzyme activity • Total protein, soluble sugar and ferritin content • Ca, Fe content in leaves and grains • 1000-grain weight	[58]
Fe_3O_4 15–20 nm	✓	✓		100 ppm	Soil	✓	✓	Maize *Zea mays* L.	Collar of fourth leaf/ visible collar of tenth leaf/ visible tassel fully extended/ milk	Till maturity	• Chlorophyll content • Antioxidant enzyme activity • Total protein, soluble sugar, and ferritin content • Microelement content in leaves and grains • Flavonoid contents	[59]

(continued)

Table 2 (continued)

ENPs type and size	Control type			Dose	Matrix type	Background metal content		Plant	Plant growth stage during treatment	Duration of exposure	Endpoints	Key reference
	Matrix	Metal salts/ chelate compounds	Bulk compounds			Total	Potential bioavailable					
Fe_xO_x	✓	✓	✓	ENPs: 2, 4 g L^- Sulfate/chelate: Fe: 4, 8 g L^{-1}	Silty clay loam			Wheat *Triticum aestivum* L.	Stem elongation/ flowering	Till maturity	• Harvest index • Chlorophyll and carotenoids content • Antioxidant enzyme activity seed weight, yield • Protein and carbohydrate content in grains	[46]
Nano-Fe Nano-Mg	✓	✓	✓	Fe: 0.25, 0.5 g L^{-1} Mg 0.05 g L^{-1}	n.i			Black eyed pea *Vigna unguiculata* L.	56 and 72 days after sowing	85 days from sowing	• Seed yield • Chlorophyll content • Seed protein • Fe, Mg content in stem and leaf	[18]
ZnO 20 nm	✓	✓	✓	2 g L^{-1}	Loam			Sunflower *Helianthus annus* L.	27 days after emergence at 4 fully expanded leaves/a week after first application	The beginning of flowering stage	• Shoot and root weight, plant height, leaf area • Chlorophylls content, gas exchange parameters • Na, K, Zn content in leaves	[123]

(continued)

Table 2 (continued)

ENPs type and size	Control type			Dose	Matrix type	Background metal content		Plant	Plant growth stage during treatment	Duration of exposure	Endpoints	Key reference
	Matrix	Metal salts/ chelate compounds	Bulk compounds			Total	Potential bioavailable					
ZnO 25 nm	✓	✓	✓	ENPs: 0.133 g L^{-1} $ZnSO_4$: 2 g L^{-1}	Red sandy loam	✓		Peanut *Archais hypogaea* L.	35 and 70 days after sowing	110 days from sowing	• Root and shoot length, dry weight • Initiation of flowering • Peanut yield • Zn content in leaf and kernels	[97]
ZnO	✓	✓	✓	10 mg L^{-1}	Sandy loam	✓		Mung bean *Vigna radiate* L.	2 weeks after sowing	14 days	• Plant growth and development (e.g., plant height, root length) • Soil enzymes activity • Protein and chlorophyll content in leaves • Biodistribution of Zn	[101]
ZnO 25 nm	✓	✓	✓	ENPs:50, 100, 200, 400, 600, 800, 1000, 1500, 2000 ppm Bulk ZnO/ $ZnSO_4$: 2000 ppm	Sandy clay loam	✓	✓	Maize *Zea mays* L.	Tasseling (48–58 days after sowing)/ milking stage (100–105 days after sowing	Till maturity	• Yield • Zn content in grains	[119]

(continued)

Table 2 (continued)

ENPs type and size	Control type			Dose	Matrix type	Background metal content		Plant	Plant growth stage during treatment	Duration of exposure	Endpoints	Key reference
	Matrix	Metal salts/ chelate compounds	Bulk compounds			Total	Potential bioavailable					
ZnO	✓	✓	✓	10 mg L^{-1}	Sungro® Sunshine LC1 medium			Coffee *Coffea Arabica* L.	One year old plants	45 days	• Fresh weight and dry weight • Zn content in leaves	[107]
ZnO-chitosan	✓	✓	✓	ENPs: 4, 40 mg Zn L^{-1} $ZnSO_4$: 40, 400 mg Zn L^{-1}	Clay loam		✓	Wheat *Triticum durum* Desf.	After anthesis (1 week for up to 5 weeks)	Till maturity	• Spike characteristics • Grain yield • Protein, Zn, Fe content in grains	[15]
ZnO-chitosan	✓	✓		0.2%	Soil			Wheat *Triticum durum* Desf	7 days after anthesis	Till maturity	• Zn content in grain • Gluten content in grain	[28]
ZnO	✓	✓		1000, 2000 mg L^{-1}	Steiner nutrient solution without Zn			Habanero pepper *Capsicum chinese* Jacq.	Vegetative growth, flowering, fruit development, maturity	Till maturity	• Plant growth and development • Chlorophyll content • Capsaicinoids and phenols content • Antioxidant capacity	[45]

(continued)

Table 2 (continued)

ENPs type and size	Control type			Dose	Matrix type	Background metal content		Plant	Plant growth stage during treatment	Duration of exposure	Endpoints	Key reference
	Matrix	Metal salts/ chelate compounds	Bulk compounds			Total	Potential bioavailable					
TiO_2 27 nm	✓	✓	✓	500, 2500, 4000 mg L^{-1}	Soil			Rape *Brassica napus* L.	44 days after sowing	Till maturity	• Tap root length, plant height, biomass of seedlings • Activities of antioxidant enzymes • Chlorophyll content • Photosynthetic gas exchange	(Li et al. 2015)
TiO_2	✓	✓	✓	0.25%	Hoagland solution			Spinach *Spinacia olearacea* L.	2 and 3 leaves plant	35 days	• Seed vigor, fresh and dry weight, chlorophyll and protein content • Contents of N and NH_4	[24]
Cu 50 nm	✓	✓	✓	50, 125, 250, 500 mg L^{-1}	Mixture of perlite–peat			Tomato *Solanum lycopersicum* L.	Flowering/ fruit setting	Till stage 6 (light red) of maturity	• K, Na, Cu content in fruits • Vitamin C, lycopene flavonoids, total phenols and protein content • Antioxidant enzyme activity	[80]

(continued)

Table 2 (continued)

ENPs type and size	Control type			Dose	Matrix type	Background metal content		Plant	Plant growth stage during treatment	Duration of exposure	Endpoints	Key reference
	Matrix	Metal salts/ chelate compounds	Bulk compounds			Total	Potential bioavailable					
Nano-Zn chelate Nano-B chelate 50 nm	✓	✓	✓	60, 120 mg Zn L^{-1} 3.25, 6,5 mg B L^{-1}	Coarse loamy			Pomegranate *Punica granatum* L.	8 years old trees before full bloom	Till general fruit appearance (for 2 seasons)	• Concentration of macro- and microelements in leaves • Fruit diameter, weight and length, fruit calyx diameter and peel thickness • Titratable acidity, total soluble solids, and juice pH • Total phenols and antioxidant activity • Total anthocyanin and total soluble sugars content in juice	[16]

(continued)

Table 2 (continued)

ENPs type and size	Control type			Dose	Matrix type	Background metal content		Plant	Plant growth stage during treatment	Duration of exposure	Endpoints	Key reference
	Matrix	Metal salts/chelate compounds	Bulk compounds			Total	Potential bioavailable					
Root exposure												
Fe_2O_3 20 nm	✓	✓	✓	ENPs: 2, 10, 50, 250, 1000 mg kg^{-1} Fe-EDTA: 45.87 mg kg^{-1}	Soil	✓		Peanut *Archais hypogea* L.	38-day seedlings	38 days	• Biomass • Chlorophyll content • Antioxidant enzymes and malondialdehyde • Phytohormone measurement • Total Fe content in roots and shoots	[108]
Fe_2O_3 ZnO, TiO_2	✓	✓	✓	1, 3, 6 g L^{-1}	Haplic Xerosol	✓		Common bean *Phaseolus vulgaris* L.	Before sowing	120 days	• Biomass • Chlorophyll content • Yield	[85]
Nano-Fe-chelate	✓	✓	✓	20 μM	Alkaline 1/4 Hoagland solution			Lettuce *Lactuca sativa* L.	Four-leaf-stage seedlings	6 weeks	• Shoot and root weight • Chlorophyll, carotenoids, soluble sugars content • Zn, P, Mn content in leaves	[106]
Fe_2O_3 50 nm	✓	✓	✓	100, 150, 200 mg L^{-1}	Hoagland medium Fe-deficiency			Spinach *Spinacea oleracea* L.	3-day seedlings	45 days	• Stem and root length • Biomass • Fe content	[13]

(continued)

Table 2 (continued)

ENPs type and size	Control type			Dose	Matrix type	Background metal content		Plant	Plant growth stage during treatment	Duration of exposure	Endpoints	Key reference
	Matrix	Metal salts/ chelate compounds	Bulk compounds			Total	Potential bioavailable					
Fe_3O_4	✓	✓	✓	27.54 μM	Hoagland solution			Maize *Zea mays* L.	12-day seedlings	60 days	• Biomass • Chlorophylls, proline, total soluble protein content • Antioxidant enzyme activity	[29]
γ-Fe_2O_3 17.7 nm	✓	✓	✓	ENPs: 20, 50, 100 mg L^{-1} $FeSO_4$: 50 μM –	1/2 Hoagland medium			Watermelon *Citrullus lantus* Thunb.	Three leaf stage	5 weeks	• Content of soluble sugar and protein, chlorophyll • Activity of antioxidant enzymes	[126]

(continued)

Table 2 (continued)

ENPs type and size	Control type			Dose	Matrix type	Background metal content		Plant	Plant growth stage during treatment	Duration of exposure	Endpoints	Key reference
	Matrix	Metal salts/ chelate compounds	Bulk compounds			Total	Potential bioavailable					
γ-Fe_2O_3 20 nm Fe_3O_4 20 nm	✓			100, 200, and 400 mg L^{-1}	Soil			Muskmelon *Cucumis melo* L.	Seven leaf stage (once a week for 4 weeks)	Ripening of fruit stage	• Height of melon plant • Chlorophyll content • Antioxidant enzymes activities and soluble protein content in leaves • Fe content in roots, shoots, stems, fruits • Yield • Vertical and transverse diameter, and peel thickness of fruit • Sugar, vitamin C, total amino acid content	[126]
TiO_2 15 nm Fe_3O_4 14.9 nm	✓	○	✓	50, 100, 150, 200, 250 mg kg^{-1}	Sandy loam			Lettuce *Lactuca sativa* L.	2-week seedlings	90 days	• Length of roots, shoots • Biomass • Concentration of P in roots, shoots • Primary metabolites • P speciation	[132]

(continued)

Table 2 (continued)

ENPs type and size	Control type			Dose	Matrix type	Background metal content		Plant	Plant growth stage during treatment	Duration of exposure	Endpoints	Key reference
	Matrix	Metal salts/ chelate compounds	Bulk compounds			Total	Potential bioavailable					
CuO 15 nm Fe_2O_3 20 nm TiO_2 20 nm	✓	○	✓	50, 500 mg kg^{-1}	Soil and sand (3:1)			Wheat *Triticum aestivum* L.	2-day seedlings	90 days	• Plant height, root length, and their respective fresh biomass • Grain yield • Content of Cu, Ti, Fe, and Zn in shoots, roots and grains • Amino acid and total protein analysis in grains	[127, 128]
SiO_2 20–40 nm	✓	○	✓	ENPs: 5, 10, 15, 20 kg ha^{-1} Bulk: 15, 20 kg ha^{-1}	Sandy loam			Maize *Zea mays* L.	Before sowing	38 days	• Number of shoot and roots, stem diameter, root length, leaf area • Protein, chlorophyll, and phenol determination in leaves • Elemental composition of root and leaf	[121]

(continued)

Table 2 (continued)

ENPs type and size	Control type			Dose	Matrix type	Background metal content		Plant	Plant growth stage during treatment	Duration of exposure	Endpoints	Key reference
	Matrix	Metal salts/chelate compounds	Bulk compounds			Total	Potential bioavailable					
Foliar application versus root exposure												
Fe_2O_3 6 nm	✓		✓	500, 1000 mg L^{-1}	Andosol			Soybean *Glycine max* L.	To soil: before sowing	To soil: before flowering	• Chlorophyll content • Parameters of photosynthesis (CO_2 assimilation rate, stomatal conductance, intercellular CO_2 concentration, transpiration rate)	[6]
									Foliar: 4 and 8-trifoliate leaf	Foliar: before pod filling		
CuO < 50 nm	✓	✓	✓	ENPs: 0.01, 0.02 ppm $CuSO_4$: 0.02 ppm	1 M Hoagland and Arnon solution			Maize *Zea mays* L.	To soil: 12-days seedlings	21 days	• Dry mass of roots and shoots • Accumulation of Cu in roots and shoots • Activity of antioxidant enzymes	[3]
	✓			ENPs foliar: 8 ppm					Foliar: on 3 weeks seedlings			
ZnO	✓	✓		6 mg L^{-1}	Sandy loam soil	✓		Sorghum *Sorghum bicolor* L.	At V-6 leaf stage	Till maturity	• Biomass • Yield • Zn content in grains • N, P, K content in grains	[21, 25]
	✓											

and antibacterial properties [12, 31, 32, 110]. For instance, Cu-based ENPs have been shown to suppress Fusarium wilt in watermelon [31] and tomatoes [32, 98], Verticillium wilt in eggplant [32], sudden death syndrome in soybean [94], and blast disease in finger millet [110]. The effects of ENPs on plant resistance have predominantly been studied in soil-grown plants infested with various pathogens, mostly fungi [12, 32, 34]. Fewer studies have examined aboveground plant tissue infections [50, 51, 110]. Regardless of the inoculation method, ENPs were typically applied foliarly [12, 31, 32, 47, 131]. While on plant leaves, ENPs can also exert antagonistic effects on pathogens, leading to reduced plant disease severity [51], some in vivo studies, however, did not observe a direct impact of ENPs on pathogens responsible for plant diseases, instead reinforcing the role of ENPs in enhancing plant resistance by altering the plant's nutritional status [32, 94]. Similar to nano-fertilizers, verifying the nano-effect in the context of enhanced plant resistance is necessary. However, a few studies have compared the effects of ENPs to their larger counterparts [31, 32], metal salts [32], or commercially used agrochemicals [31, 47]. Generally, these studies have found ENPs to be more effective in inhibiting disease development compared to bulk counterparts. In most cases, the reduced disease development due to ENPs resulted in increased yield quantity and quality. For example, the combined application of Cu-chitosan (as seed coating and foliar spray) led to a 75% inhibition of blast disease in finger millet, resulting in nearly a 90% yield increase compared to controls [110]. A study by Elmer and White [32] demonstrated that disease estimates for eggplants were more strongly reduced with ENPs of CuO, MnO, and ZnO compared to their bulk counterparts and salts. However, in terms of yield quantity, this trend was not observed for ZnO ENPs, as the application of $ZnSO_4$ resulted in more than double the yield compared to ENPs.

Apart from the risk associated with the occurrence of different pathogens, throughout the entire growing season plants can be exposed to abiotic stresses such as drought, salinity, and high- and low-temperatures [88]. Improvement in plant resistance to abiotic stresses through the application of ENPs has recently become the subject of many studies [1, 21, 41, 89, 122, 131] in response to built-up environmental pressures, frequently associated with climate change (Fig. 5). To date, most attention has been paid to the application of micronutrient-based ENPs for improving the condition of plants affected by salinity [1, 41, 55, 95, 123, 131]. Due to limited soil micronutrient availability, also under abiotic stress conditions, ENPs were predominantly foliar-applied [26, 41, 95, 105, 122]. Improvement in resistance to abiotic factors under the influence of ENPs was frequently evaluated at an early stage of plant growth, and stimulation of antioxidant mechanisms was primarily found [41, 42]. Research has also been conducted to investigate the entire plant life cycle and in these studies, enhanced yields were observed in crops grown under conditions of drought [21, 23], salinity [42, 55, 95], high temperature [26], flood [10], and heavy metals [53, 54, 104, 105]. Nevertheless, there is a lack of studies comparing the effects of ENPs with those of bulk compounds. Taking into account the fact that during the growing season plants are bound to experience the co-occurrence of different stresses, future research on the effectiveness of fertilization with ENPs should include a combination of different abiotic and biotic factors.

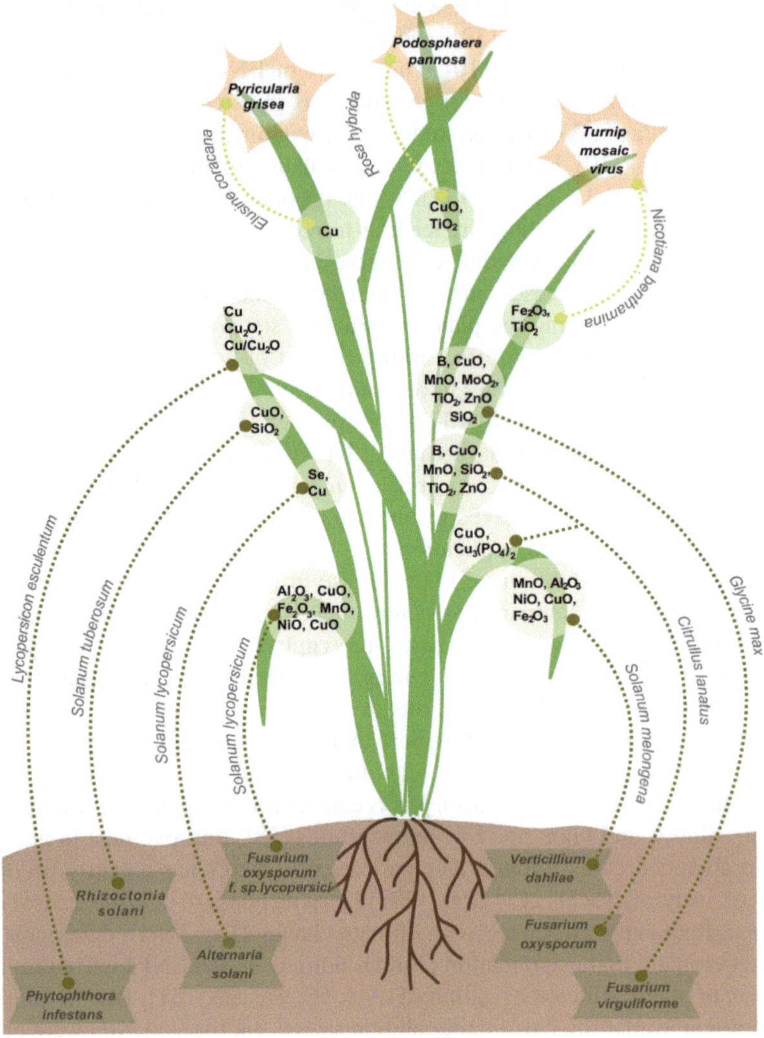

Fig. 4 Application of ENPs for suppression of plant pathogen severity

9 Challenges and Future Research Directions

ENPs have been touted as agrochemicals, smart delivery systems, nano-fertilizers, and on-demand compounds that could potentially revolutionize plant production by enabling more efficient nutrient use compared to traditional fertilizers. Despite the enthusiasm driving numerous studies since the advent of nanotechnology, the superiority of ENPs over conventional nutrient sources has not been fully verified. The

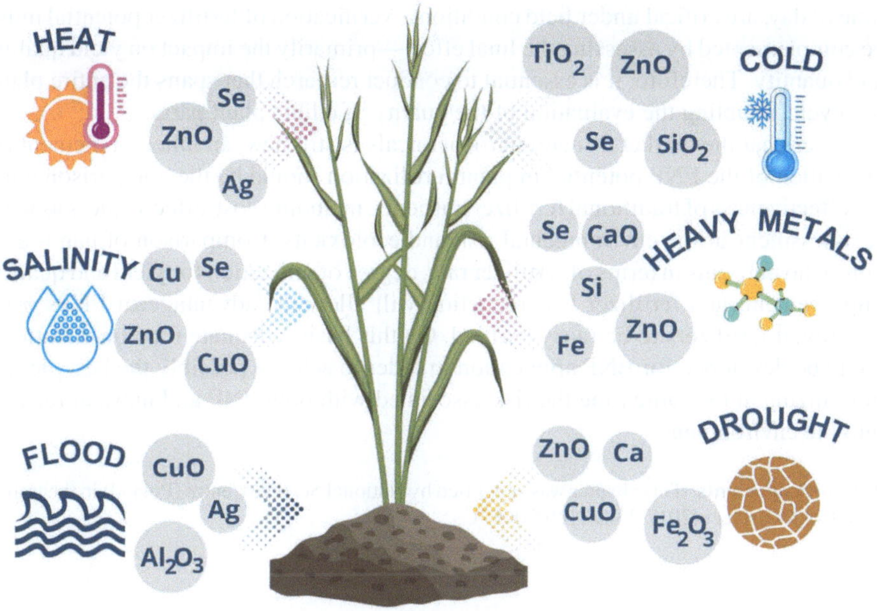

Fig. 5 Usage of metal-based ENPs in alleviating abiotic stress in plants. References [10, 33, 39, 57, 69, 92, 104, 109]

singular stimulating effects observed in a limited number of experiments have sometimes been overgeneralized, leading to imprudent advocacy for the use of ENPs as fertilizers. To avoid premature generalizations about the suitability of ENPs for plant fertilization, comprehensive research is essential, encompassing both laboratory and field studies. Existing studies aimed at evaluating the fertilizer potential of ENPs remain inadequate, often resembling ecotoxicological studies due to their use of very high concentrations and short exposure durations. Assessing the suitability of ENPs for plant nutrition should involve analyses tailored to the specific properties of ENPs, while also considering conventional fertilizer practices.

Given the distinct properties of ENPs that may offer advantages over conventional nutrients, it is crucial to studying their behavior, as well as the degree and rate of dissolution and absorption. Such analysis, combined with an investigation of plant requirements for specific nutrients and soil micronutrient availability, should inform the determination of ENP application rates for plant fertilization. Moreover, evaluating the effectiveness of ENP fertilization must consider the mode and conditions of application. This is particularly significant for foliar nutrition, the preferred method for delivering micronutrients to plants. Foliar application involves some randomness in fertilizer distribution; thus, to minimize the "spray and pray" approach, efforts should be made to control application conditions, including spraying degree, volume applied per leaf surface area, and the selection of an appropriate plant growth stage. Additionally, weather conditions (temperature, humidity, wind), which vary with the time of day, are critical under field conditions. Verification of fertilizer potential must be complemented by assessing the final effect—primarily the impact on yield quality and quantity. Therefore, it is essential to conduct research that spans the entire plant life cycle, enabling the evaluation of the quality of edible plant parts.

Given that the market of nano-agrochemicals is still new, an inherent element of evaluation of the ENP potential in plant fertilization should be the comparison with the effectiveness of traditional fertilizer nutrients, including cost-effectiveness as well as assessment of the environmental risk and ecotoxicity. Comparison of nano- and non-nano nutrients in terms of fertilizer rate, degree of utilization for yield and quality improvement, and fertilizer loss reduction will allow the advantage of ENPs over traditional fertilizers to be fairly verified. On this basis, adequate recommendations could be developed for ENP application in order to achieve possibly the best yields, minimizing at the same time the risk associated with both ENP and nutrient release into the environment.

Acknowledgements This chapter was supported by National Science Centre (Poland) in the frame of SONATA project (2017/26/D/NZ9/00067).

References

1. Abdel Latef AAH, Abu Alhmad MF, Abdelfattah KE (2017) The possible roles of priming with ZnO nanoparticles in mitigation of salinity stress in lupine (*Lupinus termis*) plants. J Plant Growth Regul 36:60–70. https://doi.org/10.1007/s00344-016-9618-x
2. ACC/SCN (2000) Ending malnutrition by 2020: an agenda for change in the millennium
3. Adhikari T, Sarkar D, Mashayekhi H, Xing B (2016) Growth and enzymatic activity of maize (*Zea mays* L.) plant: solution culture test for copper dioxide nano particles. J Plant Nutr 39:99–115. https://doi.org/10.1080/01904167.2015.1044012
4. Alexander A (1986) Optimum timing of foliar nutrient sprays. In: Alexander A (ed) Foliar fertilization: proceedings of the first international symposium on foliar fertilization. Organized by Schering Agrochemical Division, Special Fertilizer Group, Berlin (FRG) 14–16 Mar 1985. Developments in plant and soil sciences. Springer Netherlands, Dordrecht, pp 44–60. https://doi.org/10.1007/978-94-009-4386-5_4
5. Alexandratos N, Bruinsma J (2012) World agriculture towards 2030/2050: the 2012 revision 154
6. Alidoust D, Isoda A (2013) Effect of γFe_2O_3 nanoparticles on photosynthetic characteristic of soybean (*Glycine max* (L.) Merr.): foliar spray versus soil amendment. Acta Physiol Plant 35:3365–3375. https://doi.org/10.1007/s11738-013-1369-8
7. Alloway BJ (2008) Micronutrient deficiencies in global crop production. Brian J Alloway. https://www.springer.com/gp/book/9781402068591. Accessed 12 Mar 2018
8. Alshaal T, El-Ramady H (2017) Foliar application: from plant nutrition to biofortification. environ. Biodivers Soil Secur 1:71–83. https://doi.org/10.21608/jenvbs.2017.1089.1006
9. Arora S, Sharma P, Kumar S, Nayan R, Khanna PK, Zaidi MGH (2012) Gold-nanoparticle induced enhancement in growth and seed yield of Brassica juncea. Plant Growth Regul 66:303–310. https://doi.org/10.1007/s10725-011-9649-z
10. Ban Y, Tan J, Xiong Y, Mo X, Jiang Y, Xu Z (2023) Transcriptome analysis reveals the molecular mechanisms of *Phragmites australis* tolerance to CuO-nanoparticles and/or flood stress induced by *Arbuscular mycorrhizal* fungi. J Hazard Mater 442:130118. https://doi.org/10.1016/j.jhazmat.2022.130118

11. Barrios AC, Medina-Velo IA, Zuverza-Mena N, Dominguez OE, Peralta-Videa JR, Gardea-Torresdey JL (2017) Nutritional quality assessment of tomato fruits after exposure to uncoated and citric acid coated cerium oxide nanoparticles, bulk cerium oxide, cerium acetate and citric acid. Plant Physiol Biochem Effect Nanomater Plants 110:100–107. https://doi.org/10.1016/j.plaphy.2016.04.017
12. Borgatta J, Ma C, Hudson-Smith N, Elmer W, Plaza Pérez CD, De La Torre-Roche R, Zuverza-Mena N, Haynes CL, White JC, Hamers RJ (2018) Copper based nanomaterials suppress root fungal disease in watermelon (*Citrullus lanatus*): role of particle morphology, composition and dissolution behavior. ACS Sustain Chem Eng 6:14847–14856. https://doi.org/10.1021/acssuschemeng.8b03379
13. Chakravarty D, Erande MB, Late DJ (2015) Graphene quantum dots as enhanced plant growth regulators: effects on coriander and garlic plants. J Sci Food Agric 95:2772–2778. https://doi.org/10.1002/jsfa.7106
14. Cui D, Zhang P, Ma Y, He X, Li Y, Zhang J, Zhao Y, Zhang Z (2014) Effect of cerium oxide nanoparticles on asparagus lettuce cultured in an agar medium. Environ Sci Nano 1:459–465. https://doi.org/10.1039/C4EN00025K
15. Dapkekar A, Deshpande P, Oak MD, Paknikar KM, Rajwade JM (2018) Zinc use efficiency is enhanced in wheat through nanofertilization. Sci Rep 8:6832. https://doi.org/10.1038/s41598-018-25247-5
16. Davarpanah S, Tehranifar A, Davarynejad G, Abadía J, Khorasani R (2016) Effects of foliar applications of zinc and boron nano-fertilizers on pomegranate (*Punica granatum* cv. Ardestani) fruit yield and quality. Sci Hortic 210:57–64. https://doi.org/10.1016/j.scienta.2016.07.003
17. Dehkourdi EH, Mosavi M (2013) Effect of anatase nanoparticles (TiO_2) on parsley seed germination (*Petroselinum crispum*) in vitro. Biol Trace Elem Res 155:283–286. https://doi.org/10.1007/s12011-013-9788-3
18. Delfani M, Firouzabadi MB, Farrokhi N, Makarian H (2014) Some physiological responses of black-eyed pea to iron and magnesium nanofertilizers. Commun Soil Sci Plant Anal 45:530–540. https://doi.org/10.1080/00103624.2013.863911
19. Deshpande P, Dapkekar A, Oak M, Paknikar K, Rajwade J (2018) Nanocarrier-mediated foliar zinc fertilization influences expression of metal homeostasis related genes in flag leaves and enhances gluten content in durum wheat. PLoS ONE 13:e0191035. https://doi.org/10.1371/journal.pone.0191035
20. Dimkpa CO, Bindraban PS (2016) Fortification of micronutrients for efficient agronomic production: a review. Agron Sustain Dev 36:7. https://doi.org/10.1007/s13593-015-0346-6
21. Dimkpa CO, Bindraban PS, Fugice J, Agyin-Birikorang S, Singh U, Hellums D (2017) Composite micronutrient nanoparticles and salts decrease drought stress in soybean. Agron Sustain Dev 37:5. https://doi.org/10.1007/s13593-016-0412-8
22. Dimkpa CO, McLean JE, Britt DW, Anderson AJ (2015) Nano-CuO and interaction with nano-ZnO or soil bacterium provide evidence for the interference of nanoparticles in metal nutrition of plants. Ecotoxicology 24:119–129. https://doi.org/10.1007/s10646-014-1364-x
23. Dimkpa CO, Singh U, Bindraban PS, Elmer WH, Gardea-Torresdey JL, White JC (2019) Zinc oxide nanoparticles alleviate drought-induced alterations in sorghum performance, nutrient acquisition, and grain fortification. Sci Total Environ 688:926–934. https://doi.org/10.1016/j.scitotenv.2019.06.392
24. Dimkpa CO, Singh U, Bindraban PS, Elmer WH, Gardea-Torresdey JL, White JC (2018) Exposure to weathered and fresh nanoparticle and ionic Zn in soil promotes grain yield and modulates nutrient acquisition in wheat (*Triticum aestivum* L.). J Agric Food Chem 66:9645–9656. https://doi.org/10.1021/acs.jafc.8b03840
25. Dimkpa CO, White JC, Elmer WH, Gardea-Torresdey J (2017) Nanoparticle and Ionic Zn promote nutrient loading of sorghum grain under low NPK fertilization. J Agric Food Chem 65:8552–8559. https://doi.org/10.1021/acs.jafc.7b02961
26. Djanaguiraman M, Belliraj N, Bossmann SH, Prasad PVV (2018) High-temperature stress alleviation by selenium nanoparticle treatment in grain sorghum. ACS Omega 3:2479–2491. https://doi.org/10.1021/acsomega.7b01934

27. Du W, Tan W, Yin Y, Ji R, Peralta-Videa JR, Guo H, Gardea-Torresdey JL (2018) Differential effects of copper nanoparticles/microparticles in agronomic and physiological parameters of oregano (*Origanum vulgare*). Sci Total Environ 618:306–312. https://doi.org/10.1016/j.scitotenv.2017.11.042
28. Du W, Yang J, Peng Q, Liang X, Mao H (2019) Comparison study of zinc nanoparticles and zinc sulphate on wheat growth: from toxicity and zinc biofortification. Chemosphere 227:109–116. https://doi.org/10.1016/j.chemosphere.2019.03.168
29. Elanchezhian R, Kumar D, Ramesh K, Biswas AK, Guhey A, Patra AK (2017) Morpho-physiological and biochemical response of maize (*Zea mays* L.) plants fertilized with nano-iron (Fe_3O_4) micronutrient. J Plant Nutr 40:1969–1977. https://doi.org/10.1080/01904167.2016.1270320
30. El-Batal AI, Gharib FAE-L, Ghazi SM, Hegazi AZ, Hafz AGMAE (2016) Physiological responses of two varieties of common bean (*Phaseolus vulgaris* L.) to foliar application of silver nanoparticles. Nanomater Nanotechnol 6:13. https://doi.org/10.5772/62202
31. Elmer W, De La Torre-Roche R, Pagano L, Majumdar S, Zuverza-Mena N, Dimkpa C, Gardea-Torresdey J, White JC (2018) Effect of metalloid and metal oxide nanoparticles on Fusarium wilt of watermelon. Plant Dis 102:1394–1401. https://doi.org/10.1094/PDIS-10-17-1621-RE
32. Elmer WH, White JC (2016) The use of metallic oxide nanoparticles to enhance growth of tomatoes and eggplants in disease infested soil or soilless medium. Environ Sci Nano 3:1072–1079. https://doi.org/10.1039/C6EN00146G
33. Elsheery NI, Sunoj VSJ, Wen Y, Zhu JJ, Muralidharan G, Cao KF (2020) Foliar application of nanoparticles mitigates the chilling effect on photosynthesis and photoprotection in sugarcane. Plant Physiol Biochem 149:50–60. https://doi.org/10.1016/j.plaphy.2020.01.035
34. El-Shewy ES (2019) The efficacy of copper oxide, tri-calcium phosphate and silicon dioxide nanoparticles in controlling black scurf disease of potato. Ann Agric Sci Moshtohor 57:129–138. https://doi.org/10.21608/assjm.2019.42223
35. European Commission (2020) Farm to fork strategy
36. European Commission (2012) A European strategy for key enabling technologies—a bridge to growth and jobs
37. Fageria NK, Filho MPB, Moreira A, Guimarães CM (2009) Foliar fertilization of crop plants. J Plant Nutr 32:1044–1064. https://doi.org/10.1080/01904160902872826
38. FAO (2018) The future of food and agriculture: alternative pathways to 2050
39. Farajzadeh Memari-Tabrizi E, Yousefpour-Dokhanieh A, Babashpour-Asl M (2021) Foliar-applied silicon nanoparticles mitigate cadmium stress through physio-chemical changes to improve growth, antioxidant capacity, and essential oil profile of summer savory (*Satureja hortensis* L.). Plant Physiol Biochem 165:71–79. https://doi.org/10.1016/j.plaphy.2021.04.040
40. Farooq M, Wahid A, Siddique KHM (2012) Micronutrient application through seed treatments: a review. J Soil Sci Plant Nutr 12:125–142. https://doi.org/10.4067/S0718-95162012000100011
41. Fathi A, Zahedi M, Torabian S (2017) Effect of interaction between salinity and nanoparticles (Fe_2O_3 and ZnO) on physiological parameters of *Zea mays* L. J Plant Nutr 40:2745–2755. https://doi.org/10.1080/01904167.2017.1381731
42. Fathi A, Zahedi M, Torabian S, Khoshgoftar A (2017) Response of wheat genotypes to foliar spray of ZnO and Fe_2O_3 nanoparticles under salt stress. J Plant Nutr 40:1376–1385. https://doi.org/10.1080/01904167.2016.1262418
43. Feizi H, Rezvani Moghaddam P, Shahtahmassebi N, Fotovat A (2012) Impact of bulk and nanosized titanium dioxide (TiO_2) on wheat seed germination and seedling growth. Biol Trace Elem Res 146:101–106. https://doi.org/10.1007/s12011-011-9222-7
44. Gao X, Spielman-Sun E, Rodrigues SM, Casman EA, Lowry GV (2017) Time and nanoparticle concentration affect the extractability of Cu from CuO NP-amended soil. Environ Sci Technol 51:2226–2234. https://doi.org/10.1021/acs.est.6b04705
45. García-López JI, Zavala-García F, Olivares-Sáenz E, Lira-Saldívar RH, Díaz Barriga-Castro E, Ruiz-Torres NA, Ramos-Cortez E, Vázquez-Alvarado R, Niño-Medina G (2018) Zinc oxide

nanoparticles boosts phenolic compounds and antioxidant activity of *Capsicum annuum* L. during germination. Agronomy 8:215. https://doi.org/10.3390/agronomy8100215
46. Ghafari H, Razmjoo J (2013) Effect of foliar application of nano-iron oxidase, iron chelate and iron sulphate rates on yield and quality of wheat. Int J Agron Plant Prod 4:2997–3003
47. Giannousi K, Avramidis I, Dendrinou-Samara C (2013) Synthesis, characterization and evaluation of copper based nanoparticles as agrochemicals against *Phytophthora infestans*. RSC Adv 3:21743–21752. https://doi.org/10.1039/C3RA42118J
48. Gooding MJ, Davies WP (1992) Foliar urea fertilization of cereals: a review. Fertil Res 32:209–222. https://doi.org/10.1007/BF01048783
49. Guha T, Gopal G, Kundu R, Mukherjee A (2020) Nanocomposites for delivering agrochemicals: a comprehensive review. J Agric Food Chem. https://doi.org/10.1021/acs.jafc.9b06982
50. Hao Y, Fang P, Ma C, White JC, Xiang Z, Wang H, Zhang Z, Rui Y, Xing B (2019) Engineered nanomaterials inhibit *Podosphaera pannosa* infection on rose leaves by regulating phytohormones. Environ Res 170:1–6. https://doi.org/10.1016/j.envres.2018.12.008
51. Hao Y, Yuan W, Ma C, White JC, Zhang Z, Adeel M, Zhou T, Rui Y, Xing B (2018) Engineered nanomaterials suppress Turnip mosaic virus infection in tobacco (*Nicotiana benthamiana*). Environ Sci Nano 5:1685–1693. https://doi.org/10.1039/C8EN00014J
52. Hong J, Wang L, Sun Y, Zhao L, Niu G, Tan W, Rico CM, Peralta-Videa JR, Gardea-Torresdey JL (2016) Foliar applied nanoscale and microscale CeO_2 and CuO alter cucumber (*Cucumis sativus*) fruit quality. Sci Total Environ 563–564:904–911. https://doi.org/10.1016/j.scitotenv.2015.08.029
53. Hussain A, Ali S, Rizwan M, Rehman MZ, Qayyum MF, Wang H, Rinklebe J (2019) Responses of wheat (*Triticum aestivum*) plants grown in a Cd contaminated soil to the application of iron oxide nanoparticles. Ecotoxicol Environ Saf 173:156–164. https://doi.org/10.1016/j.ecoenv.2019.01.118
54. Hussain B, Lin Q, Hamid Y, Sanaullah M, Di L, Hashmi MLR, Khan MB, He Z, Yang X (2020) Foliage application of selenium and silicon nanoparticles alleviates Cd and Pb toxicity in rice (*Oryza sativa* L.). Sci Total Environ 712:136497. https://doi.org/10.1016/j.scitotenv.2020.136497
55. Hussein MM, Abou-Baker NH, (nd) The contribution of nano-zinc to alleviate salinity stress on cotton plants. R Soc Open Sci 5:171809. https://doi.org/10.1098/rsos.171809
56. IPCC (2019) Climate change and land: an IPCC special report on climate change, desertification, land degradation, sustainable land management, food security, and greenhouse gas fluxes in terrestrial ecosystems
57. Iqbal M, Raja NI, Mashwani Z-U-R, Wattoo FH, Hussain M, Ejaz M, Saira H (2019) Assessment of AgNPs exposure on physiological and biochemical changes and antioxidative defence system in wheat (*Triticum aestivum* L.) under heat stress. IET Nanobiotechnol 13:230–236. https://doi.org/10.1049/iet-nbt.2018.5041
58. Jalali M, Ghanati F, Modarres-Sanavi AM (2016) Effect of Fe_3O_4 nanoparticles and iron chelate on the antioxidant capacity and nutritional value of soil-cultivated maize (*Zea mays*) plants. Crop Pasture Sci 67:621–628. https://doi.org/10.1071/CP15271
59. Jalali M, Ghanati F, Modarres-Sanavi AM, Khoshgoftarmanesh AH (2017) Physiological effects of repeated foliar application of magnetite nanoparticles on maize plants. J Agron Crop Sci 203:593–602. https://doi.org/10.1111/jac.12208
60. Jampílek J, Kráľová K (2015) Applications of nanoformulations in agricultural production and their impact on food and human health. Proc ECOpole 9(2). https://doi.org/10.2429/proc.2015.9(2)054
61. Jeyasubramanian K, Thoppey UUG, Hikku GS, Selvakumar N, Subramania A, Krishnamoorthy K (2016) Enhancement in growth rate and productivity of spinach grown in hydroponics with iron oxide nanoparticles. RSC Adv 6:15451–15459. https://doi.org/10.1039/C5RA23425E
62. Jin Y, Liu W, Li X, Shen S, Liang S, Liu C, Shan L (2016) Nano-hydroxyapatite immobilized lead and enhanced plant growth of ryegrass in a contaminated soil. Ecol Eng 95:25–29. https://doi.org/10.1016/j.ecoleng.2016.06.071

63. Jośko I, Kusiak M, Różyło K, Baranowska-Wójcik E, Sierocka M, Sheteiwy M, Szwajgier D, Świeca M (2023) The life cycle study revealed distinct impact of foliar-applied nano-Cu on antioxidant traits of barley grain comparing with conventional agents. Food Res Int 164:112303. https://doi.org/10.1016/j.foodres.2022.112303
64. Jośko I, Kusiak M, Sozoniuk M, Feculak M, Wu KC-W, Fitzgerald M, Alyafei MS, Sheteiwy MS (2024) Analysis of multiple biomarkers revealed the size matters of Cu particles for barley response under foliar exposure. Sci Total Environ 918:170673
65. Jośko I, Oleszczuk P (2013) Influence of soil type and environmental conditions on ZnO, TiO_2 and Ni nanoparticles phytotoxicity. Chemosphere 92:91–99. https://doi.org/10.1016/j.chemosphere.2013.02.048
66. Kah M (2015) Nanopesticides and nanofertilizers: emerging contaminants or opportunities for risk mitigation? Front Chem 3. https://doi.org/10.3389/fchem.2015.00064
67. Kah M, Kookana RS, Gogos A, Bucheli TD (2018) A critical evaluation of nanopesticides and nanofertilizers against their conventional analogues. Nat Nanotechnol 13:677–684. https://doi.org/10.1038/s41565-018-0131-1
68. Kah M, Tufenkji N, White JC (2019) Nano-enabled strategies to enhance crop nutrition and protection. Nat Nanotechnol 14:532–540. https://doi.org/10.1038/s41565-019-0439-5
69. Kareem HA, Saleem MF, Saleem S, Rather SA, Wani SH, Siddiqui MH, Alamri S, Kumar R, Gaikwad NB, Guo Z, Niu J, Wang Q (2022) Zinc oxide nanoparticles interplay with physiological and biochemical attributes in terminal heat stress alleviation in Mungbean (*Vigna radiata* L.). Front Plant Sci 13. https://doi.org/10.3389/fpls.2022.842349
70. Kerstiens G (2016) Plant cuticle. In: eLS. American Cancer Society, pp 1–4. https://doi.org/10.1002/9780470015902.a0002088.pub3
71. Koch B, Khosla R (2003) The role of precision agriculture in cropping systems. J Crop Prod 9:361–381. https://doi.org/10.1300/J144v09n01_02
72. Kusiak M, Oleszczuk P, Jośko I (2022) Cross-examination of engineered nanomaterials in crop production: application and related implications. J Hazard Mater 424:127374. https://doi.org/10.1016/j.jhazmat.2021.127374
73. Kusiak M, Sierocka M, Świeca M, Pasieczna-Patkowska S, Sheteiwy M, Jośko I (2023) Unveiling of interactions between foliar-applied Cu nanoparticles and barley suffering from Cu deficiency. Environ Pollut 320:121044. https://doi.org/10.1016/j.envpol.2023.121044
74. Lee W-M, An Y-J, Yoon H, Kweon H-S (2008) Toxicity and bioavailability of copper nanoparticles to the terrestrial plants mung bean (*Phaseolus radiatus*) and wheat (*Triticum aestivum*): Plant agar test for water-insoluble nanoparticles. Environ Toxicol Chem 27:1915–1921. https://doi.org/10.1897/07-481.1
75. Lei Z, Mingyu S, Chao L, Liang C, Hao H, Xiao W, Xiaoqing L, Fan Y, Fengqing G, Fashui H (2007) Effects of nanoanatase TiO_2 on photosynthesis of spinach chloroplasts under different light illumination. Biol Trace Elem Res 119:68–76. https://doi.org/10.1007/s12011-007-0047-3
76. Li C, Wang P, Lombi E, Wu J, Blamey FPC, Fernández V, Howard DL, Menzies NW, Kopittke PM (2018) Absorption of foliar applied Zn is decreased in Zn deficient sunflower (*Helianthus annuus*) due to changes in leaf properties. Plant Soil 433:309–322. https://doi.org/10.1007/s11104-018-3841-0
77. Lin D, Xing B (2007) Phytotoxicity of nanoparticles: Inhibition of seed germination and root growth. Environ Pollut 150:243–250. https://doi.org/10.1016/j.envpol.2007.01.016
78. Liu R, Zhang H, Lal R (2016) Effects of stabilized nanoparticles of copper, zinc, manganese, and iron oxides in low concentrations on lettuce (*Lactuca sativa*) seed germination: nanotoxicants or nanonutrients? Water Air Soil Pollut 227:42. https://doi.org/10.1007/s11270-015-2738-2
79. López-Moreno ML, de la Rosa G, Hernández-Viezcas JÁ, Castillo-Michel H, Botez CE, Peralta-Videa JR, Gardea-Torresdey JL (2010) Evidence of the differential biotransformation and genotoxicity of ZnO and CeO_2 nanoparticles on soybean (*Glycine max*) Plants. Environ Sci Technol 44:7315–7320. https://doi.org/10.1021/es903891g

80. López-Vargas ER, Ortega-Ortíz H, Cadenas-Pliego G, De Alba Romenus K, Cabrera de la Fuente M, Benavides-Mendoza A, Juárez-Maldonado A (2018) Foliar application of copper nanoparticles increases the fruit quality and the content of bioactive compounds in tomatoes. Appl Sci 8:1020. https://doi.org/10.3390/app8071020
81. Lowry GV, Gregory KB, Apte SC, Lead JR (2012) Transformations of nanomaterials in the environment. Environ Sci Technol 46:6893–6899. https://doi.org/10.1021/es300839e
82. Mahakham W, Sarmah AK, Maensiri S, Theerakulpisut P (2017) Nanopriming technology for enhancing germination and starch metabolism of aged rice seeds using phytosynthesized silver nanoparticles. Sci Rep 7:8263. https://doi.org/10.1038/s41598-017-08669-5
83. Marschner H (1995) Mineral nutrition of higher plants. Elsevier. https://doi.org/10.1016/B978-0-12-473542-2.X5000-7
84. McManus P, Hortin J, Anderson AJ, Jacobson AR, Britt DW, Stewart J, McLean JE (2018) Rhizosphere interactions between copper oxide nanoparticles and wheat root exudates in a sand matrix: Influences on copper bioavailability and uptake. Environ Toxicol Chem 37:2619–2632. https://doi.org/10.1002/etc.4226
85. Medina-Pérez G (2018) Growth and development of common bean (*Phaseolus vulgaris* L.) var. pinto saltillo exposed to iron, titanium, and zinc oxide nanoparticles in an agricultural soil. Appl Ecol Environ Res 16:1883–1897. https://doi.org/10.15666/aeer/1602_18831897
86. Menzies NW, Donn MJ, Kopittke PM (2007) Evaluation of extractants for estimation of the phytoavailable trace metals in soils. Environ Pollut 145:121–130. https://doi.org/10.1016/j.envpol.2006.03.021
87. Misra SK, Dybowska A, Berhanu D, Luoma SN, Valsami-Jones E (2012) The complexity of nanoparticle dissolution and its importance in nanotoxicological studies. Sci Total Environ 438:225–232. https://doi.org/10.1016/j.scitotenv.2012.08.066
88. Mittler R (2006) Abiotic stress, the field environment and stress combination. Trends Plant Sci 11:15–19. https://doi.org/10.1016/j.tplants.2005.11.002
89. Mohammadi R, Maali-Amiri R, Abbasi A (2013) Effect of TiO_2 nanoparticles on chickpea response to cold stress. Biol Trace Elem Res 152:403–410. https://doi.org/10.1007/s12011-013-9631-x
90. Mukherjee A, Peralta-Videa JR, Bandyopadhyay S, Rico CM, Zhao L, Gardea-Torresdey JL (2014) Physiological effects of nanoparticulate ZnO in green peas (*Pisum sativum* L.) cultivated in soil. Met Integr Biometal Sci 6:132–138. https://doi.org/10.1039/c3mt00064h
91. Mukherjee A, Sun Y, Morelius E, Tamez C, Bandyopadhyay S, Niu G, White JC, Peralta-Videa JR, Gardea-Torresdey JL (2015) Differential toxicity of bare and hybrid ZnO nanoparticles in Green Pea (*Pisum sativum* L.): a life cycle study. Front Plant Sci 6:1242. https://doi.org/10.3389/fpls.2015.01242
92. Mustafa G, Sakata K, Hossain Z, Komatsu S (2015) Proteomic study on the effects of silver nanoparticles on soybean under flooding stress. J Proteomics 122:100–118. https://doi.org/10.1016/j.jprot.2015.03.030
93. Navarro E, Baun A, Behra R, Hartmann NB, Filser J, Miao A-J, Quigg A, Santschi PH, Sigg L (2008) Environmental behavior and ecotoxicity of engineered nanoparticles to algae, plants, and fungi. Ecotoxicology 17:372–386. https://doi.org/10.1007/s10646-008-0214-0
94. Peréz CDP, De La Torre Roche R, Zuverza-Mena N, Ma C, Shen Y, White JC, Pozza EA, Pozza AAA, Elmer WH (2020) Metalloid and metal oxide nanoparticles suppress sudden death syndrome of soybean. J Agric Food Chem 68:77–87. https://doi.org/10.1021/acs.jafc.9b06082
95. Pérez-Labrada F, López-Vargas ER, Ortega-Ortiz H, Cadenas-Pliego G, Benavides-Mendoza A, Juárez-Maldonado A (2019) Responses of tomato plants under saline stress to foliar application of copper nanoparticles. Plants 8:151. https://doi.org/10.3390/plants8060151
96. Pradhan S, Patra P, Das S, Chandra S, Mitra S, Dey KK, Akbar S, Palit P, Goswami A (2013) Photochemical modulation of biosafe manganese nanoparticles on *Vigna radiata*: a detailed molecular, biochemical, and biophysical study. Environ Sci Technol 47:13122–13131. https://doi.org/10.1021/es402659t

97. Prasad TNVKV, Sudhakar P, Sreenivasulu Y, Latha P, Munaswamy V, Reddy KR, Sreeprasad TS, Sajanlal PR, Pradeep T (2012) Effect of nanoscale zinc oxide particles on the germination, growth and yield of peanut. J Plant Nutr 35:905–927. https://doi.org/10.1080/01904167.2012.663443
98. Quiterio-Gutiérrez T, Ortega-Ortiz H, Cadenas-Pliego G, Hernández-Fuentes AD, Sandoval-Rangel A, Benavides-Mendoza A, Cabrera-de la Fuente M, Juárez-Maldonado A (2019) The application of selenium and copper nanoparticles modifies the biochemical responses of tomato plants under stress by *Alternaria solani*. Int J Mol Sci 20. https://doi.org/10.3390/ijms20081950
99. Raliya R, Nair R, Chavalmane S, Wang W-N, Biswas P (2015) Mechanistic evaluation of translocation and physiological impact of titanium dioxide and zinc oxide nanoparticles on the tomato (*Solanum lycopersicum* L.) plant. Metallomics 7:1584–1594. https://doi.org/10.1039/C5MT00168D
100. Raliya R, Saharan V, Dimkpa C, Biswas P (2018) Nanofertilizer for precision and sustainable agriculture: current state and future perspectives. J Agric Food Chem 66:6487–6503. https://doi.org/10.1021/acs.jafc.7b02178
101. Raliya R, Tarafdar JC, Biswas P (2016) Enhancing the mobilization of native phosphorus in the mung bean rhizosphere using ZnO nanoparticles synthesized by Soil Fungi. J Agric Food Chem 64:3111–3118. https://doi.org/10.1021/acs.jafc.5b05224
102. Rawat M, Nayan R, Negi B, Zaidi MGH, Arora S (2017) Physio-biochemical basis of iron-sulfide nanoparticle induced growth and seed yield enhancement in *B. juncea*. Plant Physiol Biochem 118:274–284. https://doi.org/10.1016/j.plaphy.2017.06.021
103. Rawat S, Adisa IO, Wang Y, Sun Y, Fadil AS, Niu G, Sharma N, Hernandez-Viezcas JA, Peralta-Videa JR, Gardea-Torresdey JL (2019) Differential physiological and biochemical impacts of nano vs micron Cu at two phenological growth stages in bell pepper (*Capsicum annuum*) plant. NanoImpact 14:100161. https://doi.org/10.1016/j.impact.2019.100161
104. Raza HZ, Shah AA, Noreen Z, Usman S, Zafar S, Yasin NA, Sayed SRM, Al-Mana FA, Elansary HO, Ahmad A, Habib F, Aslam M (2024) Calcium oxide nanoparticles mitigate lead stress in *Abelmoschus esculentus* though improving the key antioxidative enzymes, nutritional content and modulation of stress markers. Plant Physiol Biochem 206:108171. https://doi.org/10.1016/j.plaphy.2023.108171
105. Rizwan M, Ali S, Ali B, Adrees M, Arshad M, Hussain A, Zia ur Rehman M, Waris AA (2019) Zinc and iron oxide nanoparticles improved the plant growth and reduced the oxidative stress and cadmium concentration in wheat. Chemosphere 214:269–277. https://doi.org/10.1016/j.chemosphere.2018.09.120
106. Roosta HR, Jalali M, Shahrbabaki SMAV (2015) Effect of nano Fe-Chelate, Fe-Eddha and $FeSO_4$ on vegetative growth, physiological parameters and some nutrient elements concentrations of four varieties of lettuce (*Lactuca sativa* L.) In NFT system. J Plant Nutr 38:2176–2184. https://doi.org/10.1080/01904167.2015.1043378
107. Rossi L, Fedenia LN, Sharifan H, Ma X, Lombardini L (2019) Effects of foliar application of zinc sulfate and zinc nanoparticles in coffee (*Coffea arabica* L.) plants. Plant Physiol Biochem 135:160–166. https://doi.org/10.1016/j.plaphy.2018.12.005
108. Rui M, Ma C, Hao Y, Guo J, Rui Y, Tang X, Zhao Q, Fan X, Zhang Z, Hou T, Zhu S (2016) Iron oxide nanoparticles as a potential iron fertilizer for peanut (*Arachis hypogaea*) Front Plant Sci 7. https://doi.org/10.3389/fpls.2016.00815
109. Sardar R, Ahmed S, Yasin NA (2022) Titanium dioxide nanoparticles mitigate cadmium toxicity in *Coriandrum sativum* L. through modulating antioxidant system, stress markers and reducing cadmium uptake. Environ Pollut 292:118373. https://doi.org/10.1016/j.envpol.2021.118373
110. Sathiyabama M, Manikandan A (2018) Application of copper-chitosan nanoparticles stimulate growth and induce resistance in finger millet (*Eleusine coracana* Gaertn.) plants against blast disease. J Agric Food Chem 66:1784–1790. https://doi.org/10.1021/acs.jafc.7b05921
111. Servin A, Elmer W, Mukherjee A, De la Torre-Roche R, Hamdi H, White JC, Bindraban P, Dimkpa C (2015) A review of the use of engineered nanomaterials to suppress plant disease and enhance crop yield. J Nanopart Res 17:92. https://doi.org/10.1007/s11051-015-2907-7

112. Shahid M, Dumat C, Khalid S, Schreck E, Xiong T, Niazi NK (2017) Foliar heavy metal uptake, toxicity and detoxification in plants: a comparison of foliar and root metal uptake. J Hazard Mater 325:36–58. https://doi.org/10.1016/j.jhazmat.2016.11.063
113. Sheteiwy MS, Dong Q, An J, Song W, Guan Y, He F, Huang Y, Hu J (2017) Regulation of ZnO nanoparticles-induced physiological and molecular changes by seed priming with humic acid in *Oryza sativa* seedlings. Plant Growth Regul 83:27–41. https://doi.org/10.1007/s10725-017-0281-4
114. Siddiqi KS, Husen A (2022) Plant response to silver nanoparticles: a critical review. Crit Rev Biotechnol 42(7):773–990. https://doi.org/10.1080/07388551.2021.1975091
115. Siddiqi KS, Husen A (2016) Engineered gold nanoparticles and plant adaptation potential. Nanosc Res Lett 11(400):1–10. https://doi.org/10.1186/s11671-016-1607-2
116. Singh J, Singh M, Jain A, Bhardwaj S, Singh A, Singh DK, Bhushan B, Dubey SK (2013) An introduction of plant nutrients and foliar fertilization: a review. https://doi.org/10.13140/rg.2.1.1629.3844
117. Spielman-Sun E, Lombi E, Donner E, Avellan A, Etschmann B, Howard D, Lowry GV (2018) Temporal evolution of copper distribution and speciation in roots of *Triticum aestivum* exposed to CuO, $Cu(OH)_2$, and CuS nanoparticles. Environ Sci Technol 52:9777–9784. https://doi.org/10.1021/acs.est.8b02111
118. Spielman-Sun E, Lombi E, Donner E, Howard D, Unrine JM, Lowry GV (2017) Impact of surface charge on cerium oxide nanoparticle uptake and translocation by wheat (*Triticum aestivum*). Environ Sci Technol 51:7361–7368. https://doi.org/10.1021/acs.est.7b00813
119. Subbaiah LV, Prasad TNVKV, Krishna TG, Sudhakar P, Reddy BR, Pradeep T (2016) Novel effects of nanoparticulate delivery of zinc on growth, productivity, and zinc biofortification in Maize (*Zea mays* L.). J Agric Food Chem 64:3778–3788. https://doi.org/10.1021/acs.jafc.6b00838
120. Sundaria N, Singh M, Upreti P, Chauhan RP, Jaiswal JP, Kumar A (2018) Seed priming with iron oxide nanoparticles triggers iron acquisition and biofortification in wheat (*Triticum aestivum* L.) grains. J Plant Growth Regul. https://doi.org/10.1007/s00344-018-9818-7
121. Suriyaprabha R, Karunakaran G, Yuvakkumar R, Prabu P, Rajendran V, Kannan N (2012) Growth and physiological responses of maize (*Zea mays* L.) to porous silica nanoparticles in soil. J Nanopart Res 14:1294. https://doi.org/10.1007/s11051-012-1294-6
122. Taran N, Storozhenko V, Svietlova N, Batsmanova L, Shvartau V, Kovalenko M (2017) Effect of zinc and copper nanoparticles on drought resistance of wheat seedlings. Nanoscale Res Lett 12:60. https://doi.org/10.1186/s11671-017-1839-9
123. Torabian S, Zahedi M, Khoshgoftar AH (2016) Effects of foliar spray of two kinds of zinc oxide on the growth and ion concentration of sunflower cultivars under salt stress. J Plant Nutr 39:172–180. https://doi.org/10.1080/01904167.2015.1009107
124. United Nations. Department of Economic and Social Affairs (2017, n.d.) World population prospects: key findings and advance tables
125. Wang P, Menzies NW, Lombi E, McKenna BA, Johannessen B, Glover CJ, Kappen P, Kopittke PM (2013) Fate of ZnO nanoparticles in soils and cowpea (*Vigna unguiculata*). Environ Sci Technol 47:13822–13830. https://doi.org/10.1021/es403466p
126. Wang Y, Hu J, Dai Z, Li J, Huang J (2016) In vitro assessment of physiological changes of watermelon (*Citrullus lanatus*) upon iron oxide nanoparticles exposure. Plant Physiol Biochem PPB 108:353–360. https://doi.org/10.1016/j.plaphy.2016.08.003
127. Wang Y, Lin Y, Xu Y, Yin Y, Guo H, Du W (2019) Divergence in response of lettuce (var. ramosa Hort.) to copper oxide nanoparticles/microparticles as potential agricultural fertilizer. Environ Pollut Bioavailab 31:80–84. https://doi.org/10.1080/26395940.2019.1578187
128. Wang Y, Wang S, Xu M, Xiao L, Dai Z, Li J (2019) The impacts of γ-Fe_2O_3 and Fe_3O_4 nanoparticles on the physiology and fruit quality of muskmelon (*Cucumis melo*) plants. Environ Pollut 249:1011–1018. https://doi.org/10.1016/j.envpol.2019.03.119
129. Xing B, Senesi N, Vecitis CD (2016) Engineered nanoparticles and the environment: biophysicochemical processes and toxicity. Wiley

130. Yang F, Liu C, Gao F, Su M, Wu X, Zheng L, Hong F, Yang P (2007) The improvement of spinach growth by nano-anatase TiO_2 treatment is related to nitrogen photoreduction. Biol Trace Elem Res 119:77–88. https://doi.org/10.1007/s12011-007-0046-4
131. Zahedi SM, Abdelrahman M, Hosseini MS, Hoveizeh NF, Tran L-SP (2019) Alleviation of the effect of salinity on growth and yield of strawberry by foliar spray of selenium-nanoparticles. Environ Pollut 253:246–258. https://doi.org/10.1016/j.envpol.2019.04.078
132. Zahra Z, Arshad M, Rafique R, Mahmood A, Habib A, Qazi IA, Khan SA (2015) Metallic nanoparticle (TiO_2 and Fe_3O_4) application modifies rhizosphere phosphorus availability and uptake by *Lactuca sativa*. J Agric Food Chem 63:6876–6882. https://doi.org/10.1021/acs.jafc.5b01611
133. Zheng L, Hong F, Lu S, Liu C (2005) Effect of nano-TiO(2) on strength of naturally aged seeds and growth of spinach. Biol Trace Elem Res 104:83–92. https://doi.org/10.1385/BTER:104:1:083

Role of Metal and Metal-Oxide Nanoparticles in Agricultural Crop Extreme Temperature Stress Adaptation and Mitigation

Valentin Romanovski

Abstract The chapter considers the role of metal and metal-oxide nanoparticles in increasing crop resistance to extreme temperatures. It draws attention to the primary issues driven by climate change, such as frost and high temperatures, which have a negative impact on agricultural yields and product quality. The limitations of conventional strategies for resolving this issue, including the use of pesticides and breeding variants, are analyzed. Particular attention is paid to nanotechnology as a promising direction for enhancing plant defense mechanisms. Examples of using nanoparticles to stimulate photosynthesis and reduce temperature stress are given. The choosing of environmentally friendly approaches for synthesizing nanoparticles is emphasized to minimize the negative impact on the environment. The key mechanisms of the effect of nanoparticles on plant cells, and their effect on metabolism and antioxidant systems are described. The issues of long-term environmental consequences of the use of nanoparticles and the need for further research in the field of monitoring and control of their behavior in agroecosystems are raised. Promising research directions are formulated that aimed at developing sustainable agricultural technologies using engineered nanoparticles for plant adaptation to changing climate conditions.

Keywords Nanoparticles · Temperature stress · Agricultural crops · Metals · Metal oxides · Sustainable agriculture

1 Introduction

Modern agriculture faces a number of serious challenges related to climate change [21, 43]. One of the most critical factors affecting crop yields is extreme temperatures. The changes in the average temperature in different places on the planet lead to significant crop losses and deterioration in product quality [7]. In the context of

V. Romanovski (✉)
Department of Materials Science and Engineering, University of Virginia, Charlottesville, VA, USA
e-mail: rvd9ar@virginia.edu

A. Husen (ed.), *Nanobiotechnology for Abiotic Stress Adaptation and Mitigation in Agricultural Crops*, Smart Nanomaterials Technology,
https://doi.org/10.1007/978-981-96-9298-9_6

global climate change, there is a need to find innovative solutions that will help plants adapt to these extreme conditions. According to Pavlov et al. [37, 38, 54–56], traditional agricultural methods like controlling agricultural conditions or using resistant types that have been deliberately developed no longer always offer enough protection for plants. Nanotechnology is becoming more and more popular in this context [58, 69, 70].

The effects of extreme low and high temperatures on plant biochemical processes are numerous and complex. Plants get dehydrated at high temperatures due to accelerated water loss through stomata. Sustainable photosynthesis is a crucial mechanism that guarantees good plant growth. At low temperatures, metabolic processes and enzyme activity decrease what can lead to irreversible tissue damage. In addition, temperature extremes provoke the formation of reactive oxygen species (ROS) in plant cells [10, 50]. ROS have an oxidizing effect on cellular structures, damaging lipids, proteins, and DNA. To protect against ROS, plants activate antioxidant systems, but under conditions of severe temperature stress, natural protection may be insufficient.

Existing methods for increasing plant resistance to extreme temperatures include the use of growth stimulants, metabolic regulators, and anti-stress drugs [57]. But breeding techniques take a long time and don't always give promised results. In this regard, nanotechnology offers fundamentally new opportunities for increasing plant resistance to stress. Nanoparticles of metals and their oxides can act as growth stimulants, enhance plant defense mechanisms, and reduce the negative effects of temperature stress.

Nanoparticles of metals and their oxides have a number of unique properties, such as a high surface area, and the ability to penetrate into active cells and interact with biomolecules. These properties allow nanoparticles to have a multifunctional effect on plants, including (i) a substance for photosynthetic processes to improve electron transport; (ii) increasing resistance to oxidative stress; (iii) regulation of cell membrane permeability and metabolism; (iv) stimulation of the synthesis of protective proteins and enzymes. For example, zinc oxide (ZnO) nanoparticles can stimulate antioxidant activity and reduce the level of ROS in plant cells, which increases their resistance to low temperatures [14]. Iron oxide (Fe_2O_3) nanoparticles improve transport and technical energy metabolism, which is especially important at low temperatures [19].

One of the key issues in the use of nanoparticles in agriculture is their environmental safety. Although nanoparticles have a number of advantages, their impact on the environment and human health has not yet been sufficiently studied. Accumulation of nanoparticles in soils and marine ecosystems may lead to unpredictable consequences. Therefore, it is important to consider the following aspects: (i) choosing safe nanoparticle forms and concentrations; (ii) creating nanoparticles using environmentally friendly techniques; and (iii) tracking how the particles behave in soil and plants.

The green synthesis of nanoparticles, which produces nanoparticles without the use of toxic chemicals by using plant extracts or experiments, receives special attention. This makes the process more sustainable and reduces the possibility of negative environmental impacts. One of the promising developments in modern agrochemistry is the application of metal and metal-oxide nanoparticles in agriculture [58, 69–71, 73]. The processes by which nanoparticles interact with plants, their effects on soil, and the long-term environmental effects are still open. Further research should be aimed at (i) studying the penetration of nanoparticles into plant cells, (ii) assessing their production and manufacturing processes; (iii) developing safe and effective methods for using nanoparticles in agricultural technologies. Therefore, using metal and metal-oxide nanoparticles could be a key component of agricultural crops' adaptation strategy to high temperatures.

2 The Influence of Metals and Their Oxides on the Biochemical Processes of Plants and Adaptation to Extreme Temperatures

Metals are essential to plant biochemical processes [70] (enzymatic reactions, photosynthesis, and metabolism). Among the essential metals for plants are iron, copper, manganese, zinc, and magnesium. They provide resilience to stressors such as excessive salt, drought, and high temperatures by functioning as structural elements of cellular organelles and cofactors for enzymes. Metals can exist as complex compounds, oxides, or ions. Because of their many uses and environmental safety, metal oxides have recently attracted more attention in agrochemistry. To enhance plant growth and their ability to resist harsh climatic conditions, metal oxides such as magnesium oxide, zinc oxide, and titanium dioxide, and many others are utilized [2].

The impact of different metal concentrations on plant growth and development is depicted in Fig. 1. Depending on their impact, metals can be categorized into two classes. The first category consists of harmful metals that harm plants at any quantity, including lead, cadmium, arsenic, and a few others. Even at low concentrations, these metals have toxic effects on plants. Iron, copper, cobalt, manganese, and some others are required for plant well growth. At the right dosage, they support plant growth and development and take part in significant metabolic events. It should be mentioned that plants suffer from a shortage of vital metals which has a negative impact on their biochemical functions. Essential metals have a beneficial influence on plant growth and development when taken in the right amounts. But when allowable levels are exceeded, these metals are poisonous and start to harm plants, inhibiting their growth and resulting in poisoning. Because both an excess and a shortage of these elements can have negative impacts on plants, it is crucial to maintain a balance for critical metals.

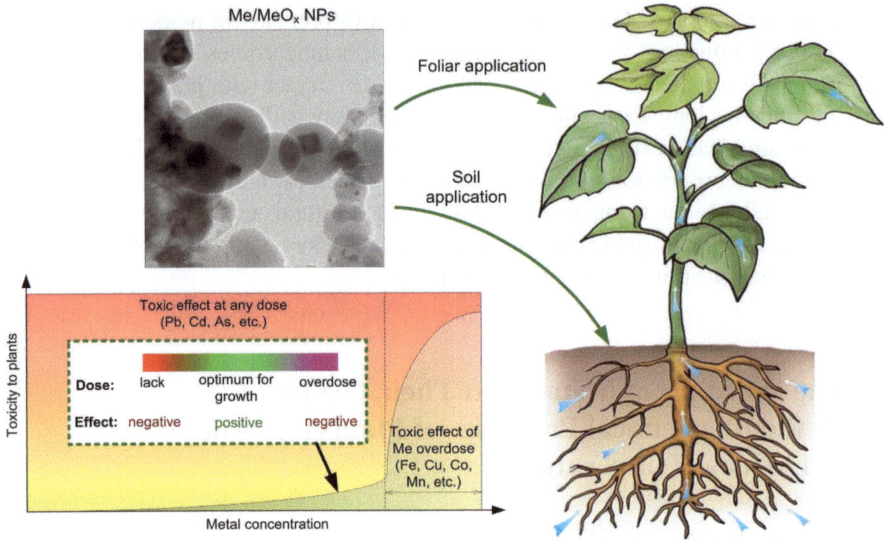

Fig. 1 Effect of metal dose on plant growth efficiency, and effect of metal dose on plant toxicity

2.1 Biochemical Functions of Metal Nanoparticles and Their Oxides in Plants

Plant biochemical processes depend highly on metals (such as iron, copper, zinc, magnesium, manganese, and calcium) and their oxides. They take part in the main biochemical processes in plants, provide enzymes, chlorophyll, etc., and help to control crucial metabolic processes. This section discusses the functions of metals and their oxides in plants, as well as the effects of their deficiency or excess on physiological processes. They perform structural, catalytic, and regulatory functions, maintaining normal plant metabolism. Iron is one of the most important micronutrients required by plants [60]. It participates in the synthesis of chlorophyll and is the main component of enzymes that catalyze important processes such as photosynthesis and respiration. Iron oxides, particularly Fe_2O_3 and Fe_3O_4, affect the absorption of iron by plant roots [86]. For plants, excessive levels of iron can be harmful because it can lead to oxidative stress, which damages cells. Iron is necessary for the production of enzymes and chlorophyll, but too much of it can harm cells, induce oxidative stress, and reduce photosynthetic activity. A lack of iron slows down the growth of plants and results in chlorosis.

Magnesium is a vital part of the chlorophyll molecule [83]. It plays a role in the activation of enzymes that catalyze the production of amino acids and carbohydrates. Magnesium deficiency leads to the deterioration of the photosynthetic activity of plants since magnesium is the activation center of the enzyme that helps capture solar energy to form organic matter. Manganese is an important component of the photosynthetic enzyme system, including the enzyme involved in the splitting of water molecules during photolysis. Zinc is involved in over 300 enzymatic reactions, including protein and carbohydrate metabolism [51]. Zinc deficiency causes damage to leaf tissue, yellowing, and reduced root and stem growth. Copper plays a key role in the formation of redox enzymes that participate in the exchange of oxygen and carbon dioxide during photosynthesis [39]. They are also necessary for the synthesis of some vitamins. Copper oxides are frequently used to treat fungal-induced plant diseases like powdery mildew.

On the other hand, some metals are inherently toxic to plants at any concentration. These include arsenic, cadmium, mercury, and lead [59]. These elements do not perform biological functions in plants and can accumulate in tissues, causing intoxication. For example, cadmium causes root damage, decreased photosynthetic activity, and growth inhibition, while arsenic affects metabolism and can disrupt cellular structure. High concentrations of these toxic metals also lead to damage to cell membranes, disruption of ionic balance, and inhibition of vital metabolic processes. Thus, metals can be both beneficial and toxic to plants depending on their concentration, which makes monitoring their levels in soils an important aspect of agriculture and ecology.

2.2 *The Influence of Metals and Their Oxides on Plant Adaptation to Extreme Temperatures*

Both high and low temperatures are among the main factors limiting plant growth [9]. Temperature stress can cause damage to cellular structures, disrupt metabolism, and reduce photosynthetic activity leading to reduced crop yields. Many published studies stated that some metals, like iron, copper, zinc, and magnesium, play a key role in strengthening plant defense mechanisms under temperature stresses. Enzymes that synthesize antioxidant molecules, such as glutathione and superoxide dismutase, which protect cells from oxidative stress brought on by high temperatures, can use metals and their oxides as cofactors (Table 1). They may also have an impact on physiological functions such as water exchange, membrane stabilization, and improved photosynthetic activity, which contributes to the resistance of plants to heat or cold. However, for each metal, there are optimal concentrations at which they have a positive effect, while excess can cause toxicity and negatively affect plants.

Thus, the study of the influence of metals and their oxides on the adaptation of plants to extreme temperatures opens new horizons for the development of agricultural technologies aimed at increasing the resistance of agricultural crops in the face of global climate change.

Table 1 Role of metal and metal-oxide nanoparticles in agricultural crop extreme temperature stress adaptation and mitigation

Me/MeO_x NPs	Type of temperature stress	Feeding method	Key reference	Comments
Se NPs	High temperature	Folliary	Sidhu et al. [81]	The application of selenium nanoparticles at low concentrations can reduce the impact of heat stress, improving plant development
TiO_2 NPs	Low temperature	Folliary	Al-Khayri et al. [3] and Thakur et al. [82]	The application of titanium nanoparticles can promote the development of plant tolerance to cold stress by controlling the damage pressure from low temperature and changing the metabolism for plant growth Alleviated membrane damage indexes, improved redox status Reduced H_2O_2 content, increased photosynthetic activity
TiO_2 NPs	High temperature	Folliary	Ghabel and Karamian [16] and Hasanpour et al. [22]	Improved plant morphology and antioxidant defense system (SOD, GPX), reduced H_2O_2 content
Fe NPs	High temperature	Folliary	Shoukat et al. [79] and Hassan et al. [23]	Iron nanoparticles have demonstrated the ability to reduce oxidative stress in plants grown under drought conditions Improved yield and antioxidant enzymes activity
ZnO NPs	High temperature	Folliary	Al-Khayri et al. [3] and Hassan et al. [82]	The application of zinc oxide nanoparticles can be effective in enhancing the physiological and metabolic activities of plants under drought Improved plant morphology and antioxidant defense system (SOD, GPX), reduced H_2O_2 content
Zn NPs	High temperature	Folliary	Hassan et al. [23]	Improved yield and antioxidant enzymes activity
Si NPs	High temperature	Folliary	Shoukat et al. [79]	Silicon nanoparticles have demonstrated the ability to reduce oxidative stress in plants grown under drought conditions

(continued)

Table 1 (continued)

Me/MeO$_x$ NPs	Type of temperature stress	Feeding method	Key reference	Comments
Mo NPs	High temperature	Folliary	Khalid et al. [40]	The application of molybdenum nanoparticles can improve plant growth at high temperatures by increasing their resistance to heat stress
Mg NPs	Low temperature	Folliary	Aguirre-Becerra et al. [1]	Magnesium nanoparticles can help increase plant resistance to cold stress by improving their physiological parameters
Ni NPs	High temperature	Folliary	Kumari et al. [42]	Nickel nanoparticles can reduce the negative impact of heat stress on plants, improving their growth and development
Cr NPs	Low temperature	Folliary	Aguirre-Becerra et al. [1]	Chromium nanoparticles can enhance plant resistance to cold stress by improving their physiological parameters
Co NPs	High temperature	Folliary	Khalid et al. [40]	Cobalt nanoparticles can improve plant growth at high temperatures by increasing their resistance to heat stress
MoO_3 NPs	High temperature	Folliary	Khalid et al. [40]	Molybdate nanoparticles can improve plant growth at high temperatures by increasing their resistance to heat stress
AgNP	High temperature		Iqbal et al. [28]	Improved plant morphological characteristics

2.2.1 Adaptation to High Temperatures

High temperatures are among the major problems for agriculture. High temperatures reduce photosynthetic activity, leading to cell membrane damage, and metabolic problems [75]. Nonetheless, plants possess adaption processes that enable them to endure in hot environments. Metals and their oxides are among the key elements that support temperature adaptation. Many mentioned metals serve as cofactors for antioxidant enzymes, which protect cells from oxidative stress-induced damage at high temperatures. Numerous enzymes involved in the metabolism of proteins, fats, and carbohydrates require it as a cofactor. Cell death can result from free radical damage of DNA, lipids, and cell membranes [6].

Zinc enhances photosynthetic activity and encourages the production of antioxidant molecules, including glutathione and vitamins C and E, which shield cells from oxidative damage at high temperatures [76, 91]. Zinc improved the plant's resistance to heat by lowering cell damage and boosting photosynthetic capacity, according to a study conducted on rice. Zinc oxide, for instance, has been demonstrated in experiments on sunflowers to improve photosynthetic efficiency and decrease cell damage, hence increasing plant resilience to high temperatures [13, 77].

Copper oxide significantly raises plant tolerance to heat, lowering oxidative stress and minimizing cell damage [93]. Superoxide dismutase (SOD) production depends on copper [8]. This enzyme is essential for protecting plants from heat-induced free radical formation, which strengthens plants' resilience to heat stress [36]. CuO raises the number of substances like glutathione.

In hot conditions, magnesium is essential for preserving normal chloroplast activity because increasing temperatures can reduce photosynthesis. Magnesium also helps plants maintain their energy balance at high temperatures by activating enzymes involved in the metabolism of lipids and carbohydrates. Magnesium helps plants avoid dehydration and maintain water balance during periods of intense heat, and helps plants under drought stress. Magnesium, as part of the oxide, activates the antioxidant systems of plants and helps maintain water balance in hot weather. In studies on crop plants such as soybeans, it was found that the use of MgO in soils helps improve their resistance to temperature stress by reducing water loss and maintaining normal photosynthetic activity.

The way by which plants adapt to high temperatures is complex and involves a number of physiological and biochemical processes. Metals and their oxides are essential for improving defensive systems in plants, including water balance and antioxidant molecule synthesis. However, it is important to remember that the effective use of these elements should consider their optimal concentrations since their excess can be toxic to plants.

2.2.2 Adaptation to Low Temperatures

Plant growth is seriously impacted by low temperatures as well as high temperatures. Water freezing in cells leads to membrane damage, a reduction in photosynthetic activity, and other cold-related disruptions. The presence of several metals and their oxides, which are important for protecting cell structure, is one of the fundamental reasons why plants can survive at low temperatures.

Microelements such as iron, copper, zinc, and magnesium, as well as their oxides, perform also important functions in the biochemical processes of plants that ensure resistance to low temperatures. Under low-temperature conditions, iron actively participates in maintaining the normal functioning of the photosynthetic system [30]. The enhanced synthesis of antioxidants, such as the enzymes catalase and superoxide dismutase, which shield cells from harm brought on by cold stress, is one characteristic of plants that have adapted to cold. These enzymes require iron as a cofactor, which lowers oxidative stress and preserves cellular integrity. Zinc is an important microelement that participates in the regulation of metabolism, and activates the synthesis of proteins and carbohydrates, which is especially important under cold stress. This element affects the water balance of plants, preventing excessive freezing of cells. In addition, zinc helps improve the stabilization of membrane structures that can be damaged at low temperatures. Experiments with rice and barley have found that adding zinc to the soil increases the plants' cold tolerance by improving their ability to photosynthesize and reduce frost damage. Another crucial element that is necessary for plants to survive in cold temperatures is copper. It plays a role in the production of antioxidant enzymes and initiates mechanisms that stabilize cell membranes that are sensitive to damage from cold exposure. Additionally, copper enhances plants' water metabolism and photosynthetic activity, which strengthens their defenses against cold stress. By enhancing the synthesis of antioxidant molecules and reducing cell damage, copper addition to the soil has been demonstrated to improve plant resilience to high cold in trials using winter crops like barley.

Metals play a key role in protecting plants from temperature stress, performing similar functions at both high and low temperatures, including activation of the antioxidant system, stabilization of cell membranes, and participation in osmoregulation processes. Oxides of copper, magnesium, zinc, and other elements affect membrane stabilization, water balance, and the activity of antioxidant systems, which helps plants cope with cold stress. One significant substance that can be used to improve a plant's resilience to cold is copper oxide. Copper oxide's antioxidant abilities assist in the activation of defense mechanisms in plants, such as the production of superoxide dismutase, which shields cells from oxidative stress-induced cell damage. CuO additionally helps to maintain cell membranes, improving a plant's ability resist to cold. As an example, research on sunflowers [12] can be mentioned. Zinc oxide also helps protect plants from the damaging effects of cold and impacts their photosynthetic activity. It induces the activation of antioxidant enzymes like catalase and superoxide dismutase, which are crucial to shielding cells from free radicals produced during low-temperature stress. Additionally, zinc oxide helps stabilize cell membranes and keeps them from degrading when frozen [15]. According to studies conducted on potatoes, using ZnO improves the water balance of the plants and lowers the degree of damage by increasing their resistance to cold [4]. Magnesium oxide, due to its properties to improve soil structure and regulate water balance, can also play a role in increasing plant resistance to cold conditions. Magnesium activates the antioxidant system and helps maintain normal metabolism at low temperatures. Magnesium oxide improves the functioning of enzymes involved in carbohydrate and lipid metabolism, which is especially important for maintaining the energy balance of plants in cold conditions. The complex process of plants adapting to low temperatures involves a number of physiological and biochemical processes meant to reduce the harm that cold can do. Enhancing plant defensive systems, such as the production of antioxidant molecules, stabilizing membrane structures, and preserving the regular functioning of the photosynthetic system at low temperatures are all made possible by metals and their oxides. The activation of enzymes by iron, copper, zinc, magnesium, and their oxides helps plants adapt to cold stress by enhancing their water balance and minimizing damage. Utilizing these components in agriculture might greatly boost crops' ability to withstand extreme cold, which is crucial given climate change and the rise in cold waves.

3 Green Synthesis of Nanoparticles for Agricultural Applications

There are many different methods for nanoparticle synthesis. They can be classified into physical, chemical, biological, and combinations of these mentioned. Most of the existing approaches require high energy, time, and reagent consumption, as well as advanced equipment. Chemical approaches, such as redox reactions and hydrothermal synthesis, are more common because they can provide the ability to control the size and shape of nanoparticles [5, 17, 18, 48, 49, 85]. In the last decades, more and more attention was paid to promising methods, such as exothermic combustion synthesis in solutions, which provides rapid and efficient crystallization of particles due to the release of a large amount of energy during the reaction [66, 73]. This method is easy to implement and does not require complex conditions, which makes it especially attractive for industrial applications. In recent years, green synthesis of nanoparticles has been actively developing, based on the use of plant extracts, bacteria, or enzymes as reducing agents and stabilizers [26, 27, 29, 53, 80]. Green synthesis approaches help to minimize or exclude the use of high toxic reagents and reduce negative impact on the environment. Solution combustion synthesis and green synthesis are considered promising due to their simplicity, cost-effectiveness, and environmental safety.

4 Future Directions and Research Prospects

The use of metal and metal-oxide nanoparticles (NPs) in agriculture is a promising direction for improving crop resistance to temperature stress, especially in the context of climate change. Given the increasing frequency and intensity of extreme temperatures, the study of the use of nanoparticles in combating these stresses is becoming increasingly important. Although the results of numerous studies have already demonstrated the positive effect of nanoparticles, further study of their mechanism of action and optimization of their use in real agricultural conditions remain critical tasks. Future research should be aimed at a deeper understanding of the mechanisms through which metal and metal-oxide nanoparticles affect plant cells under temperature stress. At present, it has been proven that nanoparticles of metals such as TiO_2, CuO, and ZnO can affect cell membranes, activate protective enzymes, and improve photosynthetic activity. However, it is not fully understood how exactly NPs interact with various molecular structures of plants, and how they affect stress processes, including oxidation, photosynthesis, water exchange, and synthesis of protective proteins. Further research can be focused on better understanding these mechanisms and optimize the use of nanoparticles in agriculture.

As was mentioned many times, metal and metal-oxide nanoparticles play many roles in plants' biochemical reactions, which is an important aspect of plants' sustainability under temperature stress. However, the mechanisms of activation of antioxidant activity, such as stimulation of superoxide dismutase (SOD), catalase, and peroxidase enzymes, require further investigation. It is necessary to clarify how nanoparticles can affect the antioxidant system at the molecular level, as well as to determine which types of NPs are most effective in activating these processes in different plants.

An important aspect of future research is the optimization of nanoparticle characteristics, such as size, shape, and concentration. These parameters can significantly affect the effectiveness of their application. For example, smaller nanoparticles can penetrate cells faster and have a more pronounced effect, but their excessive amount can cause toxicity. Future research can be focused on establishing the optimal characteristics of nanoparticles for the temperature stress stability of plants.

Another promising direction for further research is the long-term effects of nanoparticles on plants. Studies show that nanoparticles can significantly improve plant resistance to temperature stress when applied for a short period, but it is necessary to find out how long these effects last. It is necessary to study how nanoparticles affect plants under repeated stress conditions, as well as their ability to recover from extreme temperatures. This will help to assess whether nanoparticles can be used in the long term to protect crops from climate change.

One of the key issues in the use of nanoparticles in agriculture is their environmental safety. It is necessary to study in more detail how nanoparticles affect soil, microorganisms, and other components of the ecosystem. For example, nanoparticles can be toxic to some types of soil bacteria and fungi, which can disrupt the ecosystem and lead to negative consequences for soil health. It is important to conduct studies that will assess the impact of NPs on ecosystems in the long term and develop safe methods for their use, minimizing negative consequences.

Another important area of research is the transferring of controlled laboratory results into field conditions [78], to determine how nanoparticles behave in different climate zones, soil types, and agricultural practices. Problems may arise due to changes in soil composition, pH, and other factors that may affect the effectiveness of nanoparticles.

Another interesting direction is the development of multifunctional nanoparticles, like HEA nanoparticles, that can not only increase plant resistance to temperature stress but also address other agricultural issues, such as disinfection of seeds at germination stage or before the planting [65]. Different metals in the HEA alloy composition play their unique exact roles. For example, it is possible to create nanoparticles that simultaneously activate plant antioxidant systems and improve plant nutrition by increasing photosynthetic activity and water uptake. Such multifunctional nanoparticles may be especially useful in conditions where several stress factors act simultaneously.

Waste is a valuable source of various useful substances, which makes its processing promising for the extraction of these components and their further use in various technological and production processes [24]. A promising direction is the use of various metal-containing waste as an alternative raw material for the synthesis of nanoparticles, which contributes to the efficient disposal of waste, reducing the burden on the environment and the creation of materials with unique functional properties [41, 45–47, 89, 90]. Another promising direction is the use of various metal-containing waste as sources of bioavailable metals for plants, which allows for the simultaneous utilization of waste and enrichment of soils with essential microelements to increase the yield and sustainability of agricultural crops [31–35, 63, 64, 66–68, 72–74].

Increasing crop yields is part of a comprehensive approach in modern agriculture, covering everything from water treatment [25, 61, 62, 87, 88] and the implementation of technical solutions to optimize crop growth [11, 44, 72, 92] to preventing pollution of natural waters [20], as well as the use of automation, including digital twins, online data processing and forecasting using machine learning technologies [52].

5 Conclusion

Nanoparticles of metals and their oxides are a promising direction for increasing the resistance of agricultural crops to various types of temperature stress, including both high and low temperatures. In the context of global climate change and increasing instability of weather conditions, this problem is becoming one of the key ones for the agricultural sector. Traditional methods of plant adaptation to extreme temperatures, such as breeding and the use of chemicals, have limitations and often do not provide the proper level of plant resistance in extreme temperatures. The use of nanoparticles of metals such as zinc, copper, iron, magnesium, and their oxides (ZnO, Fe_2O_3, MgO) allows activating various physiological processes in plants, increasing their resistance to temperature stress. The main mechanisms of action of nanoparticles include improved photosynthesis, increased antioxidant activity, stimulation of root growth, and increased levels of protective compounds such as proline and antioxidants, which help plants withstand stressful conditions. Nanoparticles also help activate genes responsible for stress resistance and improve the water balance of plants, which is especially important at high temperatures and drought. One of the key aspects of introducing nanoparticles into agricultural technology is the method of their synthesis. Modern methods of nanoparticle synthesis include physical, chemical, and biological approaches. Among them, special attention is paid to exothermic combustion synthesis in solutions and green synthesis. Solution combustion synthesis and green synthesis are among the environmentally friendly and sustainable methods. The right dosage of nanoparticles helps increase their tolerance to temperature stress without negative side effects. It is crucial to keep in mind that too many nanoparticles might have harmful consequences and hinder plant growth. To achieve the maximum effect, it is necessary to carefully select dosages depending on the type of nanoparticles and growing conditions. The application methods of nanoparticles also play an important role in increasing their effectiveness. Research shows that both root and foliar applications can be effective, but their effectiveness depends on the type of nanoparticles and the type of stress.

From the other side, the use of nanoparticles in agriculture has a potential environmental risk. The accumulation of nanoparticles in soils and plants can affect microbial communities and biochemical processes in agroecosystems. So, it is very important to follow the right dosages for their application to the soil or foliar spraying.

Promising areas of research include the creation of complex nanomaterials with improved properties and multifunctional effects. For example, combined nanoparticles of metals and their oxides can simultaneously protect against temperature stress and improve the nutrient composition of the soil. An important direction is also the development of nanoparticles that will be slowly released into the soil or plants, providing a long-term effect and reducing the need for frequent application. The conclusion leads to the understanding that the use of nanoparticles of metals and their oxides can become an important tool for the adaptation of agricultural crops to changing climatic conditions. However, for the successful implementation of these technologies in agricultural practice, further research is needed to optimize dosages, develop environmentally friendly synthesis methods, and minimize potential risks. Only an integrated approach will ensure sustainable agriculture and food security in the context of global climate change.

References

1. Aguirre-Becerra H, Feregrino-Perez AA, Esquivel K, Perez-Garcia CE, Vazquez-Hernandez MC, Mariana-Alvarado A (2022) Nanomaterials as an alternative to increase plant resistance to abiotic stresses. Front Plant Sci 13:1023636
2. Ali S, Mehmood A, Khan N (2021) Uptake, translocation, and consequences of nanomaterials on plant growth and stress adaptation. J Nanomater 2021(1):6677616
3. Al-Khayri JM, Rashmi R, Surya Ulhas R, Sudheer WN, Banadka A, Nagella P, Almaghasla MI et al (2023) The role of nanoparticles in response of plants to abiotic stress at physiological, biochemical, and molecular levels. Plants 12(2):292
4. Al-Zaidi WJ, Ali AM, Muhsen TA (2023) Efficacy of nanoparticle zinc oxide in the resistance of fungus *Rhizoctonia solani* causing black scurf disease in local potatoes. Caspian J Environ Sci 21(1):95–103
5. Bakhsh EM, Khan ShB, Akhtar K, Danish EY, Fagieh TM, Qiu C, Sun Y, Romanovski V, Su X (2022) Simultaneous preparation of humic acid and mesoporous silica from municipal sludge and their adsorption properties for U(VI). Colloids Surf A 647:129060. https://doi.org/10.1016/j.colsurfa.2022.129060
6. Bano A, Gupta A, Rai S, Fatima T, Sharma S, Pathak N (2021) Mechanistic role of reactive oxygen species and its regulation via the antioxidant system under environmental stress. In Plant stress physiology—perspectives in agriculture, pp 1–18
7. Brás TA, Seixas J, Carvalhais N, Jägermeyr J (2021) Severity of drought and heatwave crop losses tripled over the last five decades in Europe. Environ Res Lett 16(6):065012
8. Cakmak I, Brown P, Colmenero-Flores JM, Husted S, Kutman BY, Nikolic M, Zhao FJ et al (2023) Micronutrients. In Marschner's mineral nutrition of plants. Academic Press, pp 283–385
9. Chaudhry S, Sidhu GPS (2022) Climate change regulated abiotic stress mechanisms in plants: a comprehensive review. Plant Cell Rep 41(1):1–31

10. Devireddy AR, Tschaplinski TJ, Tuskan GA, Muchero W, Chen JG (2021) Role of reactive oxygen species and hormones in plant responses to temperature changes. Int J Mol Sci 22(16):8843
11. Duan H, Zhao Y, Hu X, Liang M, Yang X, Yu L, Oranj BT, Romanovski V, Li P, Zhang Z (2024) Rolling circle amplification-enabled ultrasensitive point-of-care test method for aflatoxin B1 in the environment and food. Foods 13(19):3188. https://doi.org/10.3390/foods13193188
12. El-Saadony MT, Saad AM, Soliman SM, Salem HM, Desoky ESM, Babalghith AO, AbuQamar SF et al (2022) Role of nanoparticles in enhancing crop tolerance to abiotic stress: a comprehensive review. Front Plant Sci 13:946717
13. Ernst D, Kolenčík M, Šebesta M, Ďurišová L, Ďúranová H, Kšiňan S, Aydın E et al (2023) Agronomic investigation of spray dispersion of metal-based nanoparticles on sunflowers in real-world environments. Plants 12(9):1789
14. Faizan M, Bhat JA, Chen C, Alyemeni MN, Wijaya L, Ahmad P, Yu F (2021) Zinc oxide nanoparticles (ZnO–NPs) induce salt tolerance by improving the antioxidant system and photosynthetic machinery in tomato. Plant Physiol Biochem 161:122–130
15. Farhadi F, Towhidi A, Shakeri M, Seifi-Jamadi A (2022) Zinc oxide nanoparticles have beneficial effect on frozen-thawed spermatozoa of Holstein bulls. Iran J Appl Animal Sci 12(1):49–55
16. Ghabel VK, Karamian R (2020) Effects of TiO_2 nanoparticles and spermine on antioxidant responses of *Glycyrrhiza glabra* L. to cold stress. Acta Bot Croat 79:137–147
17. Glinskaya A, Petrov G, Romanovski V (2021) Crystal structure, physicochemical and sensory properties of solid solutions $Bi_{1-x}La_xFe_{1-x}Co_xO_3$ ($x = 0, 0.05, 0.1$). J Mater Sci Mater Electron 32(17):22579–22587. https://doi.org/10.1007/s10854-021-06743-3
18. Glinskaya A, Petrov G, Vialikanava I, Romanovski V (2023) Crystal structure, magnetic and photocatalytic properties of solid solutions $Bi_{2-x}La_xFe_4O_9$ ($x = 0.05, 0.1$). Chem Select 8(8):e202204285. https://doi.org/10.1002/slct.202204285
19. Guo WB, Wu C, Pan K, Yang L, Miao AJ (2023) Pre-exposure to Fe_2O_3 or TiO_2 nanoparticles inhibits subsequent biological uptake of 55Fe-labeled Fe_2O_3 nanoparticles. Environ Sci Technol 57(12):4831–4840
20. Gurgenidze D, Romanovski V (2023) The pharmaceutical pollution of water resources using the example of the Kura River (Tbilisi, Georgia). Water 15(14):2574. https://doi.org/10.3390/w15142574
21. Habib-ur-Rahman M, Ahmad A, Raza A, Hasnain MU, Alharby HF, Alzahrani YM, El Sabagh A et al (2022) Impact of climate change on agricultural production; Issues, challenges, and opportunities in Asia. Front Plant Sci 13:925548
22. Hasanpour H, Maali-Amir R, Zeinali H (2015) Effect of TiO_2 nanoparticles on metabolic limitations to photosynthesis under cold in chickpea. Russ J Plant Physiol 62:779–787
23. Hassan NS, El Din TAS, Hendawey MH, Borai IH, Mahdi AA (2018) Magnetite and zinc oxide nanoparticles alleviated heat stress in wheat plants. Curr Nanomater 3:32–43
24. Hurynovich A, Kwietniewski M, Romanovski V (2021) Evaluation of the possibility of utilization of sewage sludge from wastewater treatment plant—case study. Desalin Water Treat 227:16–25. https://doi.org/10.5004/dwt.2021.27199
25. Hurynovich A, Romanovski V (2018) Artificial recharge of deep aquifers. VI Międzynarodowa Konferencja Naukowo-Techniczna nt. „Nowoczesne miasta. Infrastruktura i środowisko" INFRAEKO 2018, 07–08 June 2018, Kraków/E3S Web Conf 45:00025. https://doi.org/10.1051/e3sconf/20184500025
26. Husen A (2023) Secondary metabolites based green synthesis of nanomaterials and their applications. Springer Nature Singapore Pte Ltd., Singapore. https://doi.org/10.1007/978-981-99-0927-8
27. Husen A (2023) Nanomaterials from agricultural and horticultural products. Springer Nature Singapore Pte Ltd., Singapore. https://doi.org/10.1007/978-981-99-3435-5
28. Iqbal MS, Singh AK, Singh SP, Ansari MI (2020) Nanoparticles and plant interaction with respect to stress response. In: Nanomaterials and environmental biotechnology. Springer, Cham, pp 1–15

29. Jayalekshmi C, Periakaruppan R, Romanovski V, Selvaraj KSV, Al-Dayan N (2024) Calotropis gigantea latex-derived zinc oxide nanoparticles: biosynthesis, characterization, and biofunctional applications. Eng 5(3):1399–1406. https://doi.org/10.3390/eng5030073
30. Ji W, Hu X, Kang M, Qiu X, Liu B, Tang L, Liu L et al (2023) Effects of pre-anthesis low-temperature stress on the mineral components in wheat grains. Front Plant Sci 14:1221466
31. Kamarou M, Korob N, Hil A, Moskovskikh D, Romanovski V (2021) Low-energy technology for producing anhydrite in the $CaCO_3$–H_2SO_4–H_2O system derived from industrial wastes. J Chem Technol Biotechnol 96(7):2065–2071. https://doi.org/10.1002/jctb.6740
32. Kamarou M, Korob N, Romanovski V (2021) Structurally controlled synthesis of synthetic gypsum derived from industrial wastes: sustainable approach. J Chem Technol Biotechnol 96(11):3134–3141. https://doi.org/10.1002/jctb.6865
33. Kamarou M, Moskovskikh D, Chan HL, Wang H, Li T, Akinwande AA, Romanovski V (2023) Low energy synthesis of anhydrite cement from waste lime mud. J Chem Technol Biotechnol 98:789–796. https://doi.org/10.1002/JCTB.7284
34. Kamarou M, Moskovskikh D, Kuskov K, Yudin S, Akinwande AA, Smorokov A, Özkılıç YO, Abdulwahid MY, Bhowmik A, Romanovskaia E, Korob N, Paspelau A, Romanovski V (2024) High-strength gypsum binder with improved water resistance coefficient derived from industrial wastes. Waste Manag Res. https://doi.org/10.1177/0734242X241240042
35. Kamarou M, Tan H, Moskovskikh D, Kulinich N, Ma X, Yang F, Beletskii E, Alekseeva E, Yudin S, Romanovski V (2024) High strength anhydrite cement based on lime mud from water treatment process: one step synthesis in water environment, characterization and technological parameters. Eng Rep. https://doi.org/10.1002/eng2.13054
36. Kamel SM, Elgobashy SF, Omara RI, Derbalah AS, Abdelfatah M, El-Shaer A, Elsharkawy MM et al (2022) Antifungal activity of copper oxide nanoparticles against root rot disease in cucumber. J Fungi 8(9):911
37. Karpov DF, Vafaeva KM, Pavlov MV, Dixit S, Ravikanth P, Nautiyal RD, Punia A, Madan P (2024) Parametric analysis of a radiant gas heating system for controlled-environment agriculture with preheated ventilation. In: E3S Web Conf 511:01010. https://doi.org/10.1051/e3sconf/202451101010
38. Karpov DF, Vafaeva KM, Pavlov MV, Verma R, Satish S, Gupta M, Shah SK, Gudainiyan J (2024) Indoor air temperature in the cultivation facility following an emergency heating system shutdown. E3S Web Conf 581:01045. https://doi.org/10.1051/e3sconf/202458101045
39. Kaur H, Kaur H, Kaur H, Srivastava S (2023) The beneficial roles of trace and ultratrace elements in plants. Plant Growth Regul 100(2):219–236
40. Khalid MF, Iqbal Khan R, Jawaid MZ, Shafqat W, Hussain S, Ahmed T, Alina Marc R et al (2022) Nanoparticles: the plant saviour under abiotic stresses. Nanomaterials 12(21):3915
41. Kozlovskaya I, Martsul V, Romanovski V (2024) Value-added lanthanum-containing products recovered from spent cracking catalyst. ACS Sustain Resourc Manag. https://doi.org/10.1021/acssusresmgt.4c00364
42. Kumari A, Gupta AK, Sharma S, Jadon VS, Sharma V, Chun SC, Sivanesan I (2024) Nanoparticles as a tool for alleviating plant stress: mechanisms, implications, and challenges. Plants 13(11):1528
43. Malhi GS, Kaur M, Kaushik P (2021) Impact of climate change on agriculture and its mitigation strategies: a review. Sustainability 13(3):1318
44. Malyukova LS, Martyushev NV, Tynchenko VV, Kondratiev VV, Bukhtoyarov VV, Konyukhov VY, Bashmur KA, Panfilova TA, Brigida V (2023) Circular mining wastes management for sustainable production of *Camellia sinensis* (L.) O. Kuntze. Sustainability 15:11671. https://doi.org/10.3390/su151511671
45. Martsul VN (2012) Inventory of galvanic sludge and deposits of the treatment facilities formed at the enterprises of the Republic of Belarus. In: Martsul VN, Likhacheva AV, Shibeka LA, Zalygina OS, Ramanouski VI, Khodin VV (eds) Proceedings of BSTU. Chemistry and technology of inorganic substances, vol 3, pp 71–77
46. Martsul VN (2012) Some applications of galvanic manufacture waste. In: Martsul VN, Zalygina OS, Shibeka LA, Likhacheva AV, Ramanouski VI (2012) Proceedings of BSTU. Chemistry and technology of inorganic substances, vol 3, pp 66–70

47. Martsul VN (2013) Treatment of electroplating shop wastewater at enterprises of the Republic of Belarus. In: Martsul VN, Zalygina OS, Likhacheva AV, Romanovski VI (eds) Proceedings of BSTU. Chemistry and technology of inorganic substances, vol 3, pp 57–62
48. Matsukevich I, Kulinich N, Kulbitskaya L, Kuznetsova T, Popkov V, Chebanenko M, Moskovskikh D, Kuskov K, Romanovski V (2023) Mesoporous nanocomposites based on CeO_2 and MgO: preparation, structure and photocatalytic activity. J Chem Technol Biotechnol 98(10):2497–2505. https://doi.org/10.1002/jctb.7476
49. Matsukevich I, Lipai Y, Romanovski V (2021) Cu/MgO and Ni/MgO composite nanoparticles for fast, high-efficiency adsorption of aqueous lead (II) and chromium (III) ions. J Mater Sci 56(8):5031–5040. https://doi.org/10.1007/s10853-020-05593-4
50. Mittler R, Zandalinas SI, Fichman Y, Van Breusegem F (2022) Reactive oxygen species signalling in plant stress responses. Nat Rev Mol Cell Biol 23(10):663–679
51. Nandal V, Solanki M (2021) The Zn as a vital micronutrient in plants. J Microbiol Biotechnol Food Sci 11(3):e4026
52. Nuriev M, Zaripova R, Yanova O, Koshkina I, Chupaev A (2024) Enhancing MongoDB query performance through index optimization. E3S Web Conf 531:03022
53. Palanimuthu V, Periakaruppan R, Romanovski V, Bharathi A, Selvaraj KSV, Anukeerthana S, Nishanthi R, Vanajadevi G (2024) Synthesis and structural characterization of SiO_2 nanoparticles using extract of *Gracilaria crassa* via green chemistry approach. ChemistryOpen. https://doi.org/10.1002/open.202400356
54. Pavlov MV, Vafaeva KM, Karpov DF, Dixit S, Kumar P, Joshi A, Mishra R, Brar M (2024) Impact of environmental factors on indoor air temperature in gas-fired radiant heated cultivated structures. E3S Web Conf 511:01036. https://doi.org/10.1051/e3sconf/202451101036
55. Pavlov MV, Vafaeva KM, Karpov DF, Pinjari M, Gandhi A, Kalele G, Ghalwan M, Abhilash P, Islam R (2024) Investigation of the thermal regime of a cultivation structure following an emergency shutdown of the heating system. E3S Web Conf 581:01043. https://doi.org/10.1051/e3sconf/202458101043
56. Pavlov MV, Karpov DF, Sinitsyn AA, Gudkov AG (2020) Winter greenhouse combined heating system. Mag Civ Eng 95(3):131–139. https://doi.org/10.18720/MCE.95.12
57. Pawar VS, Shivakumar H (2012) A current status of adaptogens: natural remedy to stress. Asian Pac J Trop Dis 2:S480–S490
58. Periakaruppan R, Romanovski V, Thirumalaisamy SK, Palanimuthu V, Sampath MP, Anilkumar A, Sivaraj DK, Ahamed NAN, Murugesan S, Chandrasekar D, Selvaraj KSV (2023) Innovations in modern nanotechnology for the sustainable production of agriculture. ChemEngineering 7(4):61. https://doi.org/10.3390/chemengineering7040061
59. Podar D, Maathuis FJ (2022) The role of roots and rhizosphere in providing tolerance to toxic metals and metalloids. Plant Cell Environ 45(3):719–736
60. Rai S, Singh PK, Mankotia S, Swain J, Satbhai SB (2021) Iron homeostasis in plants and its crosstalk with copper, zinc, and manganese. Plant Stress 1:100008
61. Ramanouski VI, Andreeva NA (2012) Purification of washing waters of iron removal stations. In: Proceedings of BSTU. Chemistry and technology of inorganic substances, vol 3, pp 62–65
62. Romanovski V (2020) New approach for inert filtering media modification by using precipitates of deironing filters for underground water treatment. Environ Sci Polluti Res 27:31706–31714. https://doi.org/10.1007/s11356-020-09514-5
63. Romanovski V, Klyndyuk A, Kamarou M (2021) Green approach for low-energy direct synthesis of anhydrite from industrial wastes of lime mud and spent sulfuric acid. J Environ Chem Eng 9(6):106711. https://doi.org/10.1016/j.jece.2021.106711
64. Romanovski V, Moskovskikh D, Tan H, Kuskov K, Volodko S, Akinwande AA, Periakaruppan R, Kong F, Ma X, Yang F, Kamarou M (2024) Gypsum binder with increased water resistance derived from membrane water desalination waste. Eng Rep. https://doi.org/10.1002/eng2.13028
65. Romanovski V, Roslyakov S, Trusov G, Periakaruppan R, Romanovskaia E, Chan HL, Moskovskikh D (2023) Synthesis and effect of CoCuFeNi high entropy alloy nanoparticles on seed germination, plant growth, and microorganisms inactivation activity. Environ Sci Pollut Res 30(2):23363–23371. https://doi.org/10.1007/s11356-022-23918-5

66. Romanovski V, Sdobnyakov N, Kolosov A, Savina K, Nepsha N, Moskovskikh D, Dobryden I, Zhang Z, Beletskii E, Romanovskaia E (2024) Structure patterns of one-step synthesis of CuNi nanopowders in air environment: experiment and atomistic simulations. Nano Struct Nano Obj 40:101377. https://doi.org/10.1016/j.nanoso.2024.101377
67. Romanovski V, Su X, Zhang L, Paspelau A, Smorokov A, Sehat AA, Akinwande AA, Korob N, Kamarou M (2023) Approaches for filtrate utilization from synthetic gypsum production. Environ Sci Pollut Res. https://doi.org/10.1007/s11356-022-24584-3
68. Romanovski V, Zhang L, Su X, Smorokov A, Kamarou M (2022) Gypsum and high quality binders derived from water treatment sediments and spent sulfuric acid: chemical engineering and environmental aspects. Chem Eng Res Des 184:224–232. https://doi.org/10.1016/j.cherd.2022.06.008
69. Romanovski V, Zhang Z, Akbarisehat A (2025) Nano-biosensors for precision farming and sustainable agriculture. In: Radhakrishnan EK (ed) Agriculture sustainability through nanotechnology. CRC Press, pp 156–174. https://doi.org/10.1201/9781003354253-10
70. Romanovski V (2024) Positive and negative effects of different metal-based nanoparticles on seed germination and plant growth. In: Nanofertilizer delivery, effects and application methods. Elsevier, pp 251–270. https://doi.org/10.1016/B978-0-443-13332-9.00006-X
71. Romanovski V, Periakaruppan R (2023) Why metal oxide nanoparticles are superior to other nanomaterials for agricultural application? In Nanometal oxides in horticulture and agronomy. Academic Press, pp 7–18. ISBN: 9780323918091. https://doi.org/10.1016/B978-0-323-91809-1.0014-7
72. Romanovski V, Dubina A, Sehat AA, Su X, Moskovskikh D (2024) Green and bio-waste-based materials for energy production, conversion, storage, and hybrid technologies. In: Materials for energy production, conversion, and storage. CRC Press, pp 118–135
73. Romanovski V, Matsukevich I, Romanovskaia E, Periakaruppan R (2023) Nano metal oxide as nanosensors in agriculture and environment. In: Nanometal oxides in horticulture and agronomy. Academic Press, pp 321–352. ISBN: 9780323918091. https://doi.org/10.1016/B978-0-323-91809-1.00016-0
74. Romanovski V, Sdobnyakov N, Roslyakov S, Kolosov A, Podbolotov K, Savina K, Khort A et al (2024) Bimetallic CuNi nanoparticle formation: solution combustion synthesis and molecular dynamic approaches. Inorg Chem. https://doi.org/10.1021/acs.inorgchem.4c04260
75. Sadura I, Janeczko A (2021) Brassinosteroids and the tolerance of cereals to low and high temperature stress: photosynthesis and the physicochemical properties of cell membranes. Int J Mol Sci 23(1):342
76. Saeed F Kausar A, Ali Q, Mahmood A, Javed T, Shabbir R, Ali MF (2024) Glutathione and zinc application upregulate the antioxidant system increasing photosynthetic pigments and growth under water stress in maize (*Zea mays* L.). Brazilian J Bot 47(3):795–808
77. Šebesta M, Kurtinová S, Kolenčík M, Illa R (2022) Enhancement of stress tolerance of crop plants by ZnO nanoparticles. Sustainable agriculture reviews 53: nanoparticles: a new tool to enhance stress tolerance. Springer International Publishing, Cham, pp 287–325
78. Selvaraj KSV, Periakaruppan R, Palanimuthu V, Romanovski V, Bharathi A, Mohan M (2024) Analysis of heavy metal accumulation in *Ulva rigida* and its effects on seed germination of *Pennisetum glaucum*. ChemistryOpen. https://doi.org/10.1002/open.202400386
79. Shoukat A, Pitann B, Zafar MM, Farooq MA, Haroon M, Nawaz A, Saqib ZA et al (2024) Nanotechnology for climate change mitigation: enhancing plant resilience under stress environments. J Plant Nutr Soil Sci 187(5):604–620
80. Siddiqi KS, Husen A, Rao RAK (2018) A review on biosynthesis of silver nanoparticles and their biocidal properties. J Nanobiotechnol 16(14):1–28
81. Sidhu AK, Sharma M, Bhickchand Agrawal S, Pradip Bhavsar P, Samota MK (2024) Nanomaterial strategies for enhancing plant resilience in the face of temperature stress. CABI Agric Biosci 5(1):60
82. Thakur S, Asthir B, Kaur G, Kalia A, Sharma A (2021) Zinc oxide and titanium dioxide nanoparticles influence heat stress tolerance mediated by antioxidant defense system in wheat. Cereal Res Commun 1–12

83. Tian XY, He DD, Bai S, Zeng WZ, Wang Z, Wang M, Chen ZC et al (2021) Physiological and molecular advances in magnesium nutrition of plants. Plant Soil 468:1–17
84. Viktorov I, Gibadullin R (2023) The principles of building a machine-learning-based service for converting sequential code into parallel code. E3S Web Conf 431:05012
85. Wang H, Wang Y, Liu Z, Luo S, Romanovski V, Huang X, Czech B, Sun H, Li T (2022) Rational construction of micron-sized zero-valent iron/graphene composite for enhanced Cr(VI) removal from aqueous solution. J Environ Chem Eng 10(6):109004. https://doi.org/10.1016/j.jece.2022.109004
86. Yousaf N, Ishfaq M, Qureshi HA, Saleem A, Yang H, Sardar MF, Zou C (2023) Characterization of root and foliar-applied iron oxide nanoparticles (α-Fe_2O_3, γ-Fe_2O_3, Fe_3O_4, and bulk Fe_3O_4) in improving Maize (*Zea mays* L.) performance. Nanomaterials 13(23):3036
87. Yushchenko V, Velyugo E, Romanovski V (2023) Influence of ammonium nitrogen on the treatment efficiency of underground water at iron removal stations. Groundw Sustain Dev 22:100943. https://doi.org/10.1016/j.gsd.2023.100943
88. Yushchenko V, Velyugo E, Romanovski V (2023) Development of a new design of deironing granulated filter for joint removal of iron and ammonium nitrogen from underground water. Environ Technol 45(14):2735–2742. https://doi.org/10.1080/09593330.2023.2185820
89. Zalyhina V, Cheprasova V, Belyaeva V, Romanovski V (2021) Pigments from spent Zn, Ni, Cu, and Cd electrolytes from electroplating industry. Environ Sci Pollut Res 28:32660–32668. https://doi.org/10.1007/s11356-021-13007-4
90. Zalyhina V, Cheprasova V, Romanovski V (2021) Pigments from spent chloride-ammonium zinc plating electrolytes. J Chem Technol Biotechnol 96(10):2767–2774. https://doi.org/10.1002/jctb.6822
91. Zandi P, Schnug E (2022) Reactive oxygen species, antioxidant responses and implications from a microbial modulation perspective. Biology 11(2):155
92. Zeng J, Xu R, El-Kady AA, Oranj BT, Ahmed R, Valentin R, Hu X, Wu W, Wang D, Mao J, Wu H, Gu X, Li P, Xu W, Zhang Z (2023) Nanomaterials enabled photoelectrocatalysis for removing pollutants in the environment and food. TrAC Trends Anal Chem 117187. https://doi.org/10.1016/j.trac.2023.117187
93. Zhang T, Wang Y, Ma X, Ouyang Z, Deng L, Shen S, Sun K et al (2022) Melatonin alleviates copper toxicity via improving ROS metabolism and antioxidant defense response in tomato seedlings. Antioxidants 11(4):758

Role of Metal and Metal-Oxide Nanoparticles in Agricultural Crop Flooding Stress Adaptation and Mitigation

Rabia Nazir

Abstract Crops are exposed to several abiotic stresses among which floods are the second largest disaster causing high economic losses in crop and livestock production. Flooding, submerging, and waterlogging cause hypoxia in plants leading to many physiological processes that activate plant defense mechanism to fight the induced stress. Prolonged exposure can lead to death of plants causing an alarming situation. Climatic change can worsen this situation and hence, there is a need to design such practices that can help the plants to tolerate stress so that the effects of these abiotic factors can be tackled. Nanotechnology have played a revolutionary role in agriculture, and the nanoparticles particularly of metal and metal oxides owing to their easy synthetic route and availability of raw materials are a deemed useful and cost-effective solution to multiple agro-problems. In this regard, various nanoparticles have been used that have impacted the plant biomass, photosynthetic activity, antioxidant activity, generation of metabolites, osmolytes, enzymes, etc., that can enhance the plant tolerance to fight against stress. The response of plant to these nanoparticles varies with particle size, dosage, mode of application, type of nanoparticle, plant varieties, etc. The chapter presents a bird's eye-view related to different stresses the plants are exposed to as a result of over-watering and the role of metal-based NPs in mitigating the stress in plants thereby increasing their yield.

Keywords Flooding · Waterlogging · Submergence · Plant stress · Silica · Alumina · Silver · Nanoparticles

R. Nazir (✉)
Pakistan Council of Scientific and Industrial Research Laboratories Complex, Lahore, Pakistan
e-mail: rabiapcsir@yahoo.com

A. Husen (ed.), *Nanobiotechnology for Abiotic Stress Adaptation and Mitigation in Agricultural Crops*, Smart Nanomaterials Technology,
https://doi.org/10.1007/978-981-96-9298-9_7

1 Introduction

Flooding has been the major disaster that causes immense damage to the crop. There are several incidences reported over the years, and most of the crops are intolerant to high levels of water. Flooding is a compound stress that inculcates several impacts to the plants depending upon the flooding levels and time plants are exposed to it. The stress hence imposed on plants as a result of flooding or waterlogging hence results in poor yield of crops. Hence there is a need to design practices that can help the flood-prone plants to recover from the stress and avoid plant damage. This chapter highlights the major role of nanotechnology in inducing stress-tolerance in plants that are prone to flooding stress.

Nanotechnology has brought revolution to the agriculture sector by replacing the traditional agricultural practices with new innovative ones in almost every domain. In recent years due to excessive use of agrochemicals and poor farming practices, crop productivity and quality have suffered a lot. By use of nanotechnology, the agriculture sector way can be paved toward sustainability achieving high-quality food by undergoing proper crop management practices [2, 13, 14, 16, 17, 41]. This will promote reduction in inputs (agrochemicals, etc.) while promoting maximization of agricultural productivity which consequently leads to environmental sustainability (Fig. 1). This multifaceted field has impacted agriculture by upheaving soil and water quality by designing nanomaterials that can help in the removal of contaminants, immobilization of heavy metals, improving soil health, using safe agro-chemicals, etc. [38]. The designing of fertilizers with controlled release of nutrients enables high nutrient uptake efficacy while reducing nutrient wastage and resulting in environmental pollution [21]. Similarly, the use of nano-pesticides offers targeted delivery, high efficacy, less dosage requirements, and lesser damage to non-target organisms promoting higher crop yields and sustainable agriculture [46]. Furthermore, nanotechnology has helped in giving a new direction to crop monitoring and sensing technologies by designing sensors for the early detection of diseases, residual pesticides, heavy metals and soil parameters, and irrigation water quality [3, 15, 25, 28, 40]. Crop productivity is also hampered by different abiotic and biotic stresses, and use of nanomaterials is found highly effective in stress management of affected plants [6, 35, 39].

2 Abiotic and Biotic Stresses

Agriculture is one of the highly susceptible sectors affected by the environmental impacts and disaster risks owing to its strong dependence on climatic conditions and natural resources. Crops are subjected to various stresses, i.e., abiotic and biotic that affect the crop productivity (Fig. 2). The abiotic stresses refer to environmental effects caused by non-living factors including light (UV radiation), flooding, waterlogging, drought, heavy metals, nutrients, and extreme temperatures, while biotic stresses are

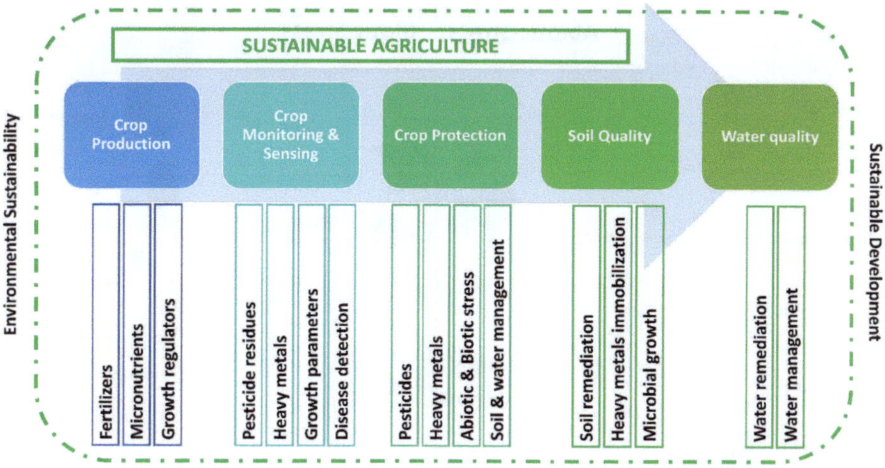

Fig. 1 Role of nanoparticles in sustainable agriculture

caused by living organisms, i.e., insects and microbes [45]. The biotic stresses are dealt with the use of different pesticides, herbicides, etc. The abiotic stresses hamper the plant growth and health depending upon the duration of stress and extent of deviations from the ideal conditions required by the plants for growth. The impacts caused by these abiotic stresses can fall in two categories, i.e., reversible (elastic) and irreversible (plastic) depending upon the part of plant, cell and tissue affected, and degree of extent [5, 24]. Drought and flooding are the two major disasters that have affected the crop productivity of the world [23]. The current chapter is focused on the flood-induced crop losses and the role of nanoparticles in mitigating the negative impacts induced by over-watering in the crops.

3 Flooding and Waterlogging

According to the FAO, floods are the second largest disaster that caused US 21 billion loss in crop and livestock production during 2008–18 [9] owing to on average 149 floods/year [23] accounting for approximately $123 billion loss per year and total of $3.8 trillion over a period of the last 3 decades. The crops highly affected in this tenure are cereals (69 million tons/year) and fruits and vegetables (40 million tons/year) [9]. In coming years owing to climatic changes, there is an increasing probability of rise in frequency as well as severity of these floods and hence the economic losses suffered by the farmers [7].

The crops are exposed to different levels of water that can be categorized as waterlogging and flooding (Fig. 3). These can be caused by storms, heavy rainfalls, flood, over-irrigation, and poor soil permeability or drainage [27]. Waterlogging refers to the situation where underground water comes to the surface leading to

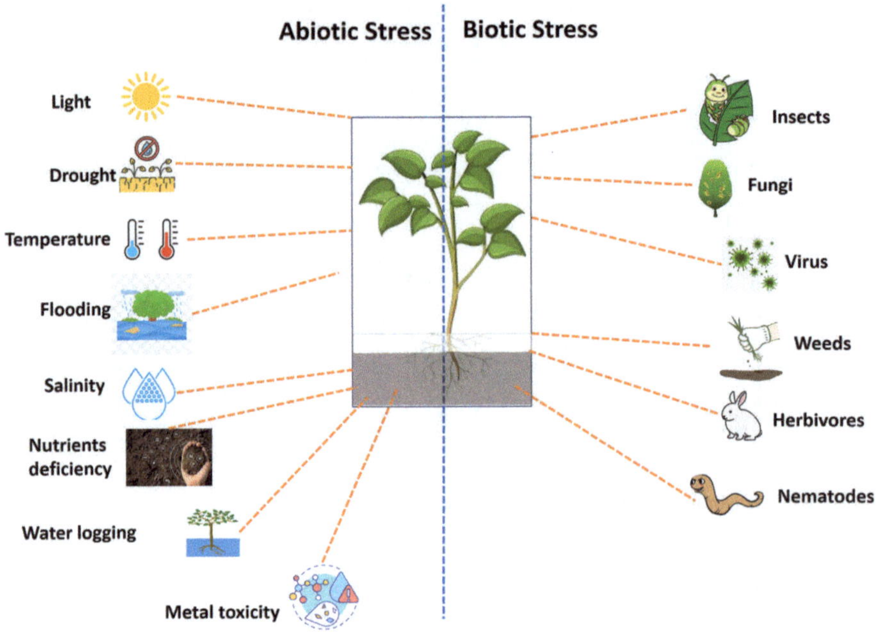

Fig. 2 Abiotic and biotic stresses in plants

excess water in the root zone which adversely affects plant growth. This leads to depletion of oxygen from the soil as there are not ample air pockets available for air diffusion leading to a natural condition known as hypoxia (< 21% oxygen) or anoxia (complete lack of oxygen) [10, 11, 22]. Conferring to the FAO report, almost 4.6 M ha arable land is affected in Pakistan and India owing to waterlogging [8]. Waterlogging is altogether a different scenario than flooding and poorly drained soil [8]. Flooding can be classified into two subclasses, i.e., submerged and completely merged. The former refers to partial dipping of stem and leaves of the plants in flood water while in the latter case the whole plant is sunk in water [27]. In some cases, these three situations are named waterlogging, submergence, and flooding, respectively [27]. In such conditions, the plant deals with low light exposure and poor gas exchange and becomes susceptible to more diseases. Hence, flooding is referred to as compound abiotic stress as it prone plant to further abiotic stresses [26].

The main reasons of flooding include enhancement in prevalence of exorbitant precipitation and wet days. In addition to that, water table level depth, soil structure, land topography, etc. can lead to waterlogging [22].

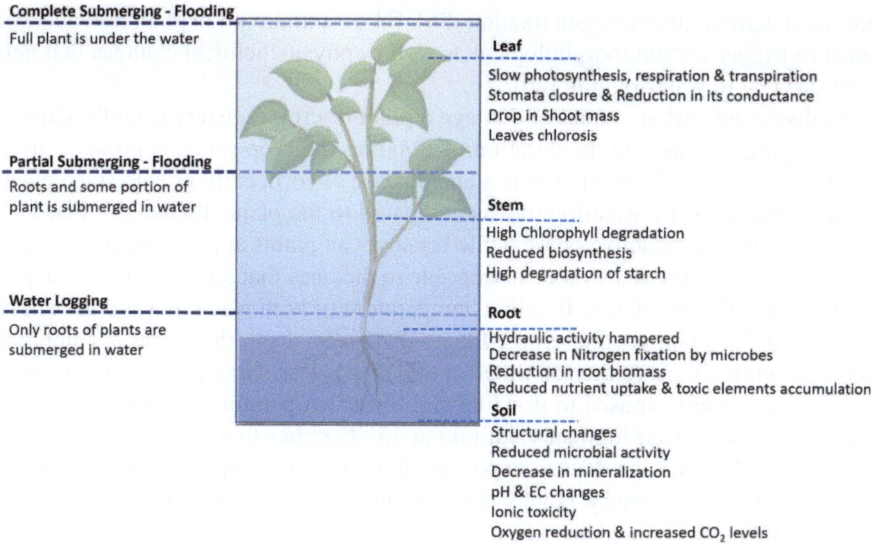

Fig. 3 Classification of flooding based on water level height (left side) and impacts of over-watering on soil and different plant parts (right)

4 Plant Responses to Flooding Stress

Mostly all crops are intolerant to excessive water levels and result in oxygen shortage resulting in highly negative impacts on crop growth and yield (Fig. 3). The foremost and detrimental effect is on the roots' respiration such that under submerged conditions the roots are not able to get enough oxygen necessary to carry out their normal aerobic processes [43]. Under prolonged stress conditions, plants as a hostility measure shift to anaerobic metabolism sustaining basal metabolic rate (BMR). Further prevalence of stress that likely occurs in case of stagnant water left after flooding or complete submerged conditions can critically hamper BMR and eventually plant death. Therefore, duration of waterlogging and genotype have critical relevance in stress defense [43]. Oxygen deficiency also leads to reduction in adenosine triphosphate (ATP) generation, energy crises, and excessive ROS generation. This in addition to generation of other toxic compounds (EtOH, lactic acid, aldehydes, nitrous oxide, etc.) under anaerobic conditions causes plant tissue damage and consequently death. Furthermore, prevalence of hypoxia hampers root metabolism, growth, and expansion. It also leads to abscisic acid and Ethylene (ET) buildup followed by stomata closure, CO_2 accumulation, hence inhibiting leaf growth. Photosynthetic activity is also affected by reduced light transmission to the plants which consequently results in reduced chlorophyll levels and nitrogen accumulation. Failure to carry out root processes impacts aerial parts affecting root-to-shoot functions, microbiomes, physiochemical, and biochemical traits [26, 27, 44]. In addition to negative affects on plants, flooding also effects soil pH, redox potential, nutrient availability, microbial density, and nitrogen fixation [26]. ET accumulation in plant also acts as a signal to trigger certain morphological as well as physiochemical changes that help plant survival to waterlogging [26].

As discussed earlier, extent of damage a plant or crop registers from flooding or waterlogging depends on the duration of flooding the crop remains prone to in its life cycle as well as the plant growth stage. In case of corn, early growth stage plants are more impacted by waterlogging as compared to the plants that are in their later vegetative and reproductive phase while for soybean plants at early vegetative stage and reproductive stages are more susceptible to the ones that are at their later stages. In addition to that the during flooding, temperature of both water and soil is deemed as the main factor impacting crops. Plants that suffer from dual stress of flooding and high temperature (either of water or soil) have less survival rate as compared to the ones that are exposed to flooding and lower temperatures. Corn plants under flooding stress showed high survival rate at 15 °C rather than at > 25 °C. In these situations, enhancing the days to exposure (flooding and temperature) also impacts the extent of flooding injury. High soil temperature, i.e., 10–18 °C also enhanced the degree of damage to the plants [22].

As a result of flooding stress, plants undergo adaptive changes to survive. One such change is development of air spaces, aerenchyma, to accommodate more oxygen supply to the plant roots as well as remove gases produced like CO_2, ET, and CH_4. This also helps in absorption of toxic ions like nitrites, sulfides. Development of adventitious roots from stem submerged parts close to the water surface is also recorded to make up for the oxygen supply. In addition to that generation of ROS (superoxide (O_2^-), singlet oxygen (1/2 O_2), H_2O_2, •OH) that is recorded in stress is counter-balanced by excessive activity of antioxidant system leading to enhanced generation of antioxidant enzymes (peroxidase (POD), catalase (CAT), superoxide dismutase (SOD), ascorbate peroxidase (APX), dehydroascorbate reductase glutathione (DHAR), guaiacol peroxidase (GPX), monodehydroascorbic acid reductase (MDAR), and glutathione reductase (GR)) and non-enzymatic antioxidants (ascorbic acid (AA), glutathione (GSH), tocopherols (lipid soluble), and carotenoids) [22, 27]. Harmonal changes also occur in stressed plants resulting in biosynthesis of jasmonic acid (JA), salicylic acid (SA), brassinosteroids (BR), gibberellin (GA), abscisic acid (ABA), auxin (IAA), and melatonin. Under waterlogging stress conditions, JA facilitates generation of sugars, phenolic compounds, and flavonoids while SA reduces electrolyte leakage and levels of proline, malondialdehyde (MDA), and H_2O_2 and enhances production of antioxidant enzymes. Shallow root formation facilitated by formation of adventitious roots and retardation of roots' elongation was also seen to be handled by SA. Furthermore, BR led to enhancement in antioxidant and enzymes activities, seedling growth, and growth attributes. Submergence stress in plants is registered with triggering of GA that limits stem elongation, and promotes internode and leaf sheath elongation regulating carbohydrate metabolism. ABA regulates stomatal opening and closing, formation of root aerenchyma, and accumulation of ROS under waterlogging stress. ET generation in waterlogged stressed plant facilitates auxin accumulation in the flooded plant parts leading to triggers of adventitious root formation. Melatonin helps plants in bearing waterlogging stress by facilitating protection of root tissues, maintaining aerobic respiration and photosynthesis, decreasing levels of ET, H_2O_2, MDA, and chlorosis, and enhancing antioxidant activity, growth attributes, and chlorophyll levels [27].

5 Nanoparticles and Flooding Stress Management

There is not much work done for exploring the role of metal and metal-oxide NPs in plants that are prone to stress induced by sub-mergence, waterlogging, and flooding. A few of the NPs that have been explored are silver, alumina, and silica (Table 1) detailed as follows.

5.1 Silver Nanoparticles

Silver NPs have played a favorable role in mitigating abiotic stress induced in plants by improving defense system of the plants by promoting enzymatic reactions, antioxidant activity, nutrient availability, photosynthetic activity, osmolytes regulations, germination rates, etc. [1]. The major impact of Ag NPs is through blocking of ethylene signals by down-regulating the associated genes while genes related to oxidative stress are up-regulated [31, 32].

The flooding stress induced in Saffron plants has been effectively tackled by the introduction of Ag NPs. The negative impacts of flooding stress in Saffron have been traced to ET accumulation, and Ag NPs are found helpful in inhibiting ET-induced damage to the plants. Soaking of saffron corns in Ag NPs (40, 80, 120 ppm) when compared with control (water treated) was found effective for overcoming stress-induced effects in the plants, like root number and length, root dry and fresh weights, and bud dry and fresh weights improved especially when NPs are employed at 40–80 ppm. Leaf, bud length and number are not impacted by flooding but application of Ag NPs improved these attributes [36, 37]. Foliar application of Ag NPs on saffron was also able to moderate some of the flood-affected parameters like plant height and corn number when used in dosages of 50–100 ppm. Plant weight, plant length, and root number though impacted by flood stress didn't register any significant influence after Ag NP foliar application [42].

Application of Ag NPs alone as well as in combination with nicotinic acid and KNO_3 was found helpful in mitigating the effects of flooding on soyabean seedlings. The application not only enhanced growth (e.g., root and hypocotyl length/weight) as compared to the control but also played a role in managing proteomes (proteins) related to stress, signaling, cell metabolism, protein degradation, protein synthesis, protein metabolism, stress, transport, RNA, and amino acid metabolism proteins [12,

Table 1 Role of nanoparticles in mitigating flooding or waterlogging effects

Plants	NPs	Variables	Additives	Application protocol	Parameters studied	Key references
Saffron	Ag	NP concentration	–	Saffron corms soaking prior to planting	Root number, length, fresh and dry weight; leaf bud, number, length, fresh and dry weight	Rezvani et al. [37]
Saffron	Ag	NP concentration	–	Saffron corms soaking prior to planting	Root number, length, dry weight, wet weight; Bud number, length, dry weight, fresh weight	Rezvani et al. [36]
Saffron	Ag	NP concentration	–	Foliar application	Root number, length; Leaf bud number, length; Plant weight	Sorooshzadeh et al. [42]
Soybean	Ag	Mode of NP synthesis (chemical & biogenic); NP concentration	–	Seedling treatment	Root length, weight; Hypocotyl length, weight; Proteomic analysis	Mustafa et al. [29]
Soybean	Ag	Mode of NP synthesis (organic & inorganic); Nicotinic acid	Nicotinic acid, KNO_3	Seedling treatment	Root length, weight; Hypocotyl length, weight; Proteomic analysis; Enzymatic analysis; Biochemical assays	Hashimoto et al. [12]
Soybean	Ag	NP concentration; $AgNO_3$ treatment	–	Seedling treatment	Seedling weight; Proteomic analysis; mRNA expression	Mustafa et al. [31, 32]

(continued)

Table 1 (continued)

Plants	NPs	Variables	Additives	Application protocol	Parameters studied	Key references
Soybean	Ag	NPs size	–	Seedling treatment	Root length; Hypocotyl length; Proteomic analysis; Bioinformatic analysis; Enzymatic analysis; mRNA expression	Mustafa et al. [33]
Soybean	Al_2O_3	NPs size; NP concentration; ZnO NPs; Ag NPs	–	Seedling treatment	Root length; Hypocotyl length; Proteomic analysis; Bioinformatic analysis; Enzymatic analysis; mRNA expression; in silico protein–protein interaction	Mustafa et al. [31, 32]
Soybean	Al_2O_3	NP size	–	Seedling treatment	Root length; Hypocotyl length; Proteomic analysis; Bioinformatic analysis; Enzymatic analysis; mRNA expression	Mustafa and Komatsu [30]

(continued)

Table 1 (continued)

Plants	NPs	Variables	Additives	Application protocol	Parameters studied	Key references
Blue berry	SiO_2	Bulk silica and NPs; Aerated and non-aerate system; Bulk and NP concentration	–	Foliar, root + foliar	Shoot & root dry weight; Leaf greenness; Photosynthetic activity; Enzymatic & non-enzymatic antioxidant activities; Osmolytes concentration; Lipid peroxidation; ROS concentration: Nutrients uptake	Iqbal et al. [19, 20]
Citrus fruits (Carrizo Citrange, Rubidoux, Rich 16-6)	SiO_2	Plant varieties; Aerated & non-aerated system; Mode of application	–	Foliar, root	Dry matter of dry leaf, stem, root, plants; Photosynthetic activity; Leaf greenness; Osmolytes concentration; Lipid peroxidation; Enzymatic antioxidant activity; Sugar, starch, enzyme activity (SPS, SS)	Iqbal et al. [18]

(continued)

Table 1 (continued)

Plants	NPs	Variables	Additives	Application protocol	Parameters studied	Key references
Peach	SiO_2	Bulk silica and NPs; aerated & non-aerate system; Bulk and NP concentration	–	Foliar	Plant biomass; Photosynthetic activity; Leaf greenness; Antioxidant activity; Osmolytes concentration; Lipid peroxidation; ROS; Nutrient uptake	Clavijo-Herrera et al. [4]
Muscadine grapes	SiO_2	Aerated and non-aerated system; Bulk and NPs	–	Root	Plant biomass; Photosynthetic activity; Stomatal conductance; CO_2 levels; Antioxidant enzymatic and non-enzymatic activities; Osmolytes concentration; Nutrients level	Iqbal et al. [19, 20]

31–33]. Under stress, ubiquitin proteasome system plays an important part in developing resistance against stress by preventing accumulation of damaged proteins, promoting their degradation and facilitating generation of new ones [12]. Changes in 919 root proteins in Ag NP-treated plants were detected as compared to flooded (733) and untreated (1274) plants. The down-regulation of genes, i.e., alcohol dehydrogenase 1 and pyruvate decarboxylase 2 was observed in Ag NP-treated plants that otherwise were up-regulated under flooding stress. Ag NPs also regulated levels of glyoxalase II 3 protein that otherwise were escalated in stressed plants indicating the recovery of plant [31, 32].

Soybean plants prone to flooding stress were exposed to different sizes (2, 15, 50–80 nm) and concentrations (0.2, 2, 20 ppm) of Ag NPs, and results were compared with positive (water) and negative control (2 ppm silver nitrate). The size-dependent interaction of NPs with proteins was noticed, and it was observed that 15 nm size was effective for mitigating flooding stress. The particle sizes less than (2 nm) and above (50–80 nm) have lesser impact on growth of plants [33]. In addition to that, the synthetic route as well as applied concentration of NPs impact growth attributes of soybean whereby biosynthesized seeds (16 nm) were found effective than chemically synthesized Ag NPs (15 nm). Among selected doses (1, 10, and 50 ppm), 10 ppm gave the best results, enhanced dose resulted in reduced root weight, lesions on root surface, and hypocotyl length [29].

5.2 *Aluminum Oxide Nanoparticles*

Aluminum NPs also have the ability to alter plant growth attributes, nutrient uptake, and oxidative stress and gene expression [34]. Flood-exposed soybean plants showed decreases in plant fresh weight and root length but on exposure to Al_2O_3 NPs (30–60 nm) at dosage of 50 ppm, these attributes showed enhancement. The other two dosages (5 and 500 ppm) inculcate negatively leading to reduction in weights even more than the flooded plants. In comparison, ZnO (< 50 nm) when employed at similar concentrations, a downward trend was observed in plant fresh weight and root length which was even less than the untreated flooded plants. Similar comparison with Ag NPs (15 nm; 50 ppm) portrayed better performance than the ZnO and flood-stressed plants but still not better than Al_2O_3 NPs. Functional analysis of root proteins in this case depicted changes in 586, 782, and 453 proteins in case of control, flood-stressed, and Al_2O_3 NP-treated plants. These proteins are related to proteins, cells, and stress in case of control; protein, cell, and amino acid in case of affected plants and protein; cell and glycolysis in case of treated ones. Proteins that predominantly showed improved response to Al_2O_3 NPs include the ones related to proteins, glycolysis, and lipid metabolism with sub-categorization pointing to proteins belonging to protein synthesis, degradation, and post-translational modification. In this case as well, reduction in glyceraldehyde-3-phosphate dehydrogenase with application of NPs ensures shifting to aerobic system. In addition to that the cell death rate in roots considerably reduced with application of NPs that was much enhanced under stressed conditions [31, 32]. Size-dependent influence of Al_2O_3 NPs (5, 135, 30–60 nm) was also witnessed on Soybean registering best results for NPs falling in the range of 30–60 nm following the order for root length: 30–60 nm > 135 nm > 5 nm > flood stressed. The number of altered proteins (protein metabolism, cell, and amino acid metabolism) also varies with size of NPs following the sequence: 30–60 nm (519) > 135 nm (478) > 5 nm (412) for proteins involved in protein metabolism; 30–60 nm (109) > 135 nm (107) > 5 nm (99) for cell proteins; and 30–60 nm (100) > 135 nm (99) > 5 nm (81) for amino acids [30].

The introduction of Al_2O_3 NPs facilitates changes in protein that occur majorly in mitochondria, an important part for aerobic respiration and generation of ATP (via tricarboxylic acid cycle, TCA cycle) as well as many other important compounds actively involved in photorespiration. The mitochondrial membrane permeability, activity of TCA cycle, and mitochondrial protein regulation are highly dependent on the NP size [30].

5.3 Silicon Dioxide Nanoparticles

Silicon, an important micronutrient, is found effective in overcoming abiotic stress in plants. Introduction of silicon can enhance antioxidant activity, leaf water potential, and SOD levels thereby promoting tolerance to stress. Moreover, it is also known to promote thickening of epidermis, halting toxic elements uptake, and enhance stability of chlorophyll, DNA, and RNA hence facilitating plant recovery [19, 20].

Plants under flooding stress present diverse response to different fertigation styles that vary from one specie to another. Citrus plants (Carrizo Citrange, Rubidoux, Rich 16-6) subjected to different treatments under flooding stress, i.e., flooding with aeration (FA), flooding without aeration (FWA), flooding with aeration and root (FAR)/foliar (FAF) NPs application, and flooding without aeration and root (FR)/foliar (FF) NP application. FWA caused hypoxia condition in all the three types of plants leading to reduction in leaf, stem, and root dry weights, photosynthesis, stomatal conductance, and leaf greenness to different extents. Root application (FR) of NPs facilitates improvement in weights as well as photosynthetic parameters, but foliar application leads to commendable results in both aerated (FAF) and non-aerated (FF) systems. The best results in case of leaf, root, and stem weights were shown by Rich 16-6 followed by Rubidoux and Carrizo Citrange while for photosynthetic parameters, the order follows Rubidoux, Carrizo Citrange, and Rich 16-6. In case of antioxidant enzymatic activity (SOD, POD, CAT, APX, and GPX) as well as osmoprotectants (proline and glycine betaine), the maximum levels were achieved with application of NPs. Rich 16-6 registered high levels with root application (FR), while foliar application (FF) was suitable for enhancing enzymes, proline, and glycine levels in other two species. MDA content that enhanced significantly in flooded plants (FWA) was reduced by root (FR) as well as foliar application (FF) especially under aerated conditions (FAR and FAF). These osmoprotectants play an important role in compensating oxidative damage induced in flood affected plants by promoting antioxidant activity in addition to cell membrane maintenance. Metabolic activity of plants is indicated by levels of acid invertase, neural invertase, sucrose synthase, sucrose phosphate synthase, fructokinase, hexokinase, phosphofructokinase, and pyruvate kinase that if enhanced are helpful in mitigating oxidative damage to plant. In flood-aerated plants, the reduced activities of neural invertase as well as of sucrose and sucrose phosphate synthase were observed as compared to the flood-stressed non-aerated plants while that of acid invertase is enhanced. On the other hand, the activities of fructokinase, phosphofructokinase, hexokinase, and pyruvate kinase were lowest for flood-stressed plants with no aeration. Among all NP treatments, root induction of NPs under non-aerated conditions in case of Rich 16-6 resulted in the highest increase of these enzyme activities while for Rubidoux these activities are enhanced by NP foliar application in non-aerated setup. For Carrizo Citrange neural invertase, acid invertase, sucrose synthase, and sucrose phosphate synthase, activities are promoted with foliar application with no air and those of fructokinase, phosphofructokinase, hexokinase, and pyruvate kinase escalate when NPs are applied through foliar spray with aeration. Root dosing of Si NPs in non-aerated system was also found effective for enhancing levels of glucose, fructose, and total soluble sugars in case of Rich 16-6 plants, and foliar one with no air was effective for increase in these attributes for other two varieties [18].

As discussed earlier, in addition to mode of application, SiO_2 NP size and dosage also affect the plant's ability to tolerate stress. Peach plants grown hydroponically under artificially induced hypoxia conditions (aerated and non-aerated) were cultivated by application of bulk sodium silicate solution (SiSol) and SiO_2 NP suspension at same dosage rates, i.e., 1000 and 1500 ppm to compare the affect of particle size. 7% enhancement in photosynthetic rates of plants treated with SiO_2 NPs (1000 ppm) was registered as compared to the ones treated with similar concentration of SiSol. Leaf greenness didn't present any significant difference among control plants (stressed plants grown under aerated conditions), plants exposed to hypoxia condition, and plants treated with Silica NPs and SiSol under hypoxia condition. The aerated plants treated with SiO_2 NPs showed 17% increase in lead greenness as compared to the ones exposed to SiSol. SOD, POD, CAT, glycine betaine, and proline levels improved in stressed plants treated with silicon NPs as compared to the SiSol. Hypoxia plants' leaves treated with 1000 ppm SiO_2 NPs recorded 39%, 45, and 41% decrease while roots registered 37, 52, and 53% reduction in MDA, O_2^-, and H_2O_2 contents, respectively when compared with SiSol treated ones. The Silica NPs when used at 1500 ppm concentration facilitate the accumulation of nutrients (N, P, K, Zn) in leaves both in aerated and non-aerated conditions while Mn and Fe levels enhance in SiSol-treated plants. As compared to SiSol-treated plants, Si content itself rose approximately 18% in non-hypoxic seedlings and 41% in hypoxic seeds. SiO_2 NPs are more helpful in mitigating the negative consequences of flooding as compared to bulk SiO_2 solution owing to their enhanced uptake by the plant [4]. Positive impact of Silicon NPs as compared to the bulk Silica in mitigating hypoxia stress was also observed in Muscadine grapes. The shoot dry weight, root dry weight, and total dry weight that registered 50, 45, and 50% decrease in stressed plants when compared to control were observed to improve by 15, 30, and 20% by introduction of bulk silica and 30, 46, and 20% by treatment with Silica NPs. Similarly, leaf greenness, photosynthetic activity, stomatal conductance, and CO2 levels that falls 60, 60, 65, and 60% in hypoxia plants recovered in silica NP-treated stressed plants by 120, 145, 133, and 160% respectively while the bulk silica ones only showed improvement of 90, 97, 87, and 125%, respectively. Similarly, the enhancement in enzymatic (SOD, POD, and CAT) and non-enzymatic (AsA and GSH) antioxidants of leaves and roots was also observed in case of Silica NPs as compared to the bulk silica. In comparison to control, proline and GB contents that rise in leaves (120 and 300%) and roots (220 and 180%) in the stressed plants only accede 60 and 40% in leaves and 30 and 40% in roots when treated with conventional silica. In case of silica NP-treated stressed plants' leaves and roots, proline escalates to 90 and 50% and GB grows to 110 and 100%. Lipid peroxidase (MDA), O_2^-, and H_2O_2 that escalate in hypoxia plants were considerably suppressed in roots and leaf nano silica when compared with conventional silica. Likewise, enhancement in N (90, 80%), P (140, 115%), K (45, 45%), and Zn (120, 35%) by bulk and nano silica, respectively that otherwise have dropped to 50, 40, 30, and 50% in hypoxia plants, respectively. There is also relative increase in the Si contents in plants treated with silica NPs [19, 20] as was also supported by previous studies carried out on peach [4].

Among different dosages (250, 500 ppm), particle types (nano-SiO_2 (13–30, 30–40, 40–50, and 80 nm), and bulk-SiO_2) and application (simple foliar; root + foliar) induced to high bush blueberry through hydroponic cultivation, it was observed that root + foliar application of 20–30 nm NPs at 250 ppm dosage was quite effective in relieving flooding stress symptoms. This led to improved growth traits, leaf gas exchange attributes, enzymatic and non-enzymatic antioxidant activities, osmolyte formation, and nutrient uptake (N, P, K, Zn) along with reduction in Fe and Mn toxicity levels [19, 20].

6 Conclusion

The plants exposed to flood or waterlogging stress register decrease in biomass along with reduction in photosynthetic attributes. Introduction of NPs can help in mitigating stress in plants and promoting better plant growth and hence yield. The response of stressed plants is determined by nanoparticles' size, dosage, application mode, and prevailing hypoxia conditions but each plant and related species respond to these factors differently owing to their own physiochemical nature. Generally, the introduction of NPs has resulted in improving enzymatic and non-enzymatic antioxidant activities and concentrations of osmolytes while effectively reducing the ROS as well as lipid peroxidase levels that are helpful in managing oxidative stress. Moreover, metabolic rates have also been positively impacted by application of NPs in the stressed system. There is lot more avenues that still need to be explored by use of different metals and metal oxides especially by introduction of bi and trimetallic oxides either in pure form or with amelioration by use of organic materials. The use of these nanomaterials can help in reducing the crop loses and promoting the cultivation of crops under this main abiotic factor.

References

1. Abasi F, Raja NI, Mashwani ZUR, Amjad MS, Ehsan M, Mustafa N, Haroon M, Proćków J (2022) Biogenic silver nanoparticles as a stress alleviator in plants: a mechanistic overview. Molecules 27(11):3378
2. Balusamy SR, Joshi AS, Perumalsamy H, Mijakovic I, Singh P (2023) Advancing sustainable agriculture: a critical review of smart and eco-friendly nanomaterial applications. J Nanobiotechnol 21(1):372
3. Beegum S, Das S (2022) 17 - Nanosensors in agriculture. In: Ghosh S, Thongmee S, Kumar A (eds) Agricultural nanobiotechnology. Woodhead Publishing, pp 465–478
4. Clavijo-Herrera J, Shahid MA, Rossi L, Sarkhosh A (2022) Exogenous silicon applications enhance peach a seedling response to flooding-induced hypoxia stress
5. Cramer GR, Urano K, Delrot S, Pezzotti M, Shinozaki K (2011) Effects of abiotic stress on plants: a systems biology perspective. BMC Plant Biol 11:1–14

6. Choudhary A, Sharma S, Kaur H, Sharma N, Gadewar MM, Mehta S, Husen A (2024) Plant system, abiotic stress resilience, reactive oxygen species, and coordination of engineered nanomaterials: a review. S Afr J Botany 171:45–59. https://doi.org/10.1016/j.sajb.2024.05.053
7. Djanibekov U, Polyakov M, Craig H, Paulik R (2024) Flood impacts on agriculture under climate change: the case of the Awanui Catchment, New Zealand. Econ Disasters Climate Change, 1–34
8. FAO (1994) Land degradation in south Asia: its severity, causes and effects upon the people
9. Food and Agriculture Organization (2023) The impact of disasters on agriculture and food security. Food and Agriculture Organization
10. Fukao T, Barrera-Figueroa BE, Juntawong P, Peña-Castro JM (2019) Submergence and waterlogging stress in plants: a review highlighting research opportunities and understudied aspects. Front Plant Sci 10:340
11. George E, Horst WJ, Neumann E (2012) Chapter 17—Adaptation of plants to adverse chemical soil conditions. In: Marschner P (ed) Marschner's mineral nutrition of higher plants, 3rd edn. Academic Press, San Diego, pp 409–472
12. Hashimoto T, Mustafa G, Nishiuchi T, Komatsu S (2020) Comparative analysis of the effect of inorganic and organic chemicals with silver nanoparticles on soybean under flooding stress. Int J Mol Sci 21(4):1300
13. Husen A (2025a) Plant response to silver nanoparticles. Springer Nature Singapore Pte Ltd., Singapore. https://doi.org/10.1007/978-981-97-7352-7
14. Husen A (2025b) Emerging carbon nanomaterials for sustainable agricultural practices (synthesis, plant growth, performance, production and protection). Springer Nature Singapore Pte Ltd., Singapore. https://doi.org/10.1007/978-981-97-5104-4
15. Husen A (2025c) Nanosensors for agricultural and other related sectors (towards sustainable use and future prospects). Springer Nature Singapore Pte Ltd., Singapore
16. Husen A (2023) Nanomaterials from agricultural and horticultural products. Springer Nature Singapore Pte Ltd., Singapore. https://doi.org/10.1007/978-981-99-3435-5
17. Husen A (2024) Plant response to gold nanoparticles. Springer Nature Singapore Pte Ltd., Singapore. https://doi.org/10.1007/978-981-97-7174-5
18. Iqbal S, Hussain M, Sadiq S, Seleiman MF, Sarkhosh A, Chater JM, Shahid MA (2024) Silicon nanoparticles confer hypoxia tolerance in citrus rootstocks by modulating antioxidant activities and carbohydrate metabolism. Heliyon 10(1)
19. Iqbal Z, Sarkhosh A, Balal RM, Gómez C, Zubair M, Ilyas N, Khan N, Shahid MA (2021) Silicon alleviate hypoxia stress by improving enzymatic and non-enzymatic antioxidants and regulating nutrient uptake in muscadine grape (*Muscadinia rotundifolia* Michx.). Front Plant Sci 11:618873
20. Iqbal Z, Sarkhosh A, Balal RM, Rauf S, Khan N, Altaf MA, Camara-Zapata JM, Garcia-Sanchez F, Shahid MA (2021) Silicon nanoparticles mitigate hypoxia-induced oxidative damage by improving antioxidants activities and concentration of osmolytes in southern highbush blueberry plants. Agronomy 11(11):2143
21. Jakhar AM, Aziz I, Kaleri AR, Hasnain M, Haider G, Ma J, Abideen Z (2022) Nano-fertilizers: a sustainable technology for improving crop nutrition and food security. NanoImpact 27:100411
22. Kaur G, Singh G, Motavalli PP, Nelson KA, Orlowski JM, Golden BR (2020) Impacts and management strategies for crop production in waterlogged or flooded soils: a review. Agron J 112(3):1475–1501
23. Kim W, Iizumi T, Hosokawa N, Tanoue M, Hirabayashi Y (2023) Flood impacts on global crop production: advances and limitations. Environ Res Lett 18(5):054007
24. Koyro H-W, Ahmad P, Geissler N (2012) Abiotic stress responses in plants: an overview. In: Environmental adaptations and stress tolerance of plants in the era of climate change, pp 1–28
25. Kumar A, Choudhary A, Kaur H, Mehta S, Husen A (2021) Metal-based nanoparticles, sensors and their multifaceted application in food packaging. J Nanobiotechnol 19(256):1–25. https://doi.org/10.1186/s12951-021-00996-0
26. Leeggangers HA, Rodriguez-Granados NY, Macias-Honti MG, Sasidharan R (2023) A helping hand when drowning: the versatile role of ethylene in root flooding resilience. Environ Exp Bot 213:105422

27. Manghwar H, Hussain A, Alam I, Khoso MA, Ali Q, Liu F (2024) Waterlogging stress in plants: unraveling the mechanisms and impacts on growth, development, and productivity. Environ Exp Botany, 105824
28. Mondal R, Dam P, Chakraborty J, Paret ML, Katı A, Altuntas S, Sarkar R, Ghorai S, Mandal AK, Husen A (2022) Potential of nanobiosensor in sustainable agriculture: the state-of-art. Heliyon 8(12):e12207. https://doi.org/10.2139/ssrn.4134873
29. Mustafa G, Hasan M, Yamaguchi H, Hitachi K, Tsuchida K, Komatsu S (2020) A comparative proteomic analysis of engineered and bio synthesized silver nanoparticles on soybean seedlings. J Proteomics 224:103833. https://doi.org/10.1016/j.jprot.2020.103833
30. Mustafa G, Komatsu S (2016) Insights into the response of soybean mitochondrial proteins to various sizes of aluminum oxide nanoparticles under flooding stress. J Proteome Res 15(12):4464–4475. https://doi.org/10.1021/acs.jproteome.6b00572
31. Mustafa G, Sakata K, Hossain Z, Komatsu S (2015) Proteomic study on the effects of silver nanoparticles on soybean under flooding stress. J Proteomics 122:100–118. https://doi.org/10.1016/j.jprot.2015.03.030
32. Mustafa G, Sakata K, Komatsu S (2015) Proteomic analysis of flooded soybean root exposed to aluminum oxide nanoparticles. J Proteomics 128:280–297
33. Mustafa G, Sakata K, Komatsu S (2016) Proteomic analysis of soybean root exposed to varying sizes of silver nanoparticles under flooding stress. J Proteomics 148:113–125. https://doi.org/10.1016/j.jprot.2016.07.027
34. Pokharel BR, Sheri V, Kumar M, Zhang Z, Zhang B (2024) The update and transport of aluminum nanoparticles in plants and their biochemical and molecular phototoxicity on plant growth and development: a systematic review. Environ Pollut 340:122875. https://doi.org/10.1016/j.envpol.2023.122875
35. Rajpal VR, Dhingra Y, Khungar L, Mehta S, Minkina T, Rajput VD, Husen A (2024) Metal and metal-oxide nanoparticles in nanosensing as well as managing biotic stress in plant system. Curr Res Biotechnol 7:100219. https://doi.org/10.1016/j.crbiot.2024.100219
36. Rezvani N, Sorooshzadeh A (2014) Effect of nano-silver on root and bud growth of saffron in flooding stress condition. Saffron Agron Technol 2(1):91–104
37. Rezvani N, Sorooshzadeh A, Farhadi N (2012) Effect of nano-silver on growth of saffron in flooding stress. Int J Agric Biosyst Eng 6(1):11–16
38. Sarkar A, Sengupta S, Sen S (2019) Nanoparticles for soil remediation. In: Gothandam KM, Ranjan S, Dasgupta N, Lichtfouse E (eds) Nanoscience and biotechnology for environmental applications. Springer International Publishing, Cham, pp 249–262
39. Shoala T (2023) The dual role of nanotechnology in the management of biotic and abiotic stresses in plants. In: Nanotechnology in agriculture and agroecosystems. Elsevier, pp 51–70
40. Sharma P, Pandey V, Sharma MMM, Patra A, Singh B, Mehta S, Husen A (2021) A review on biosensors and nanosensors application in agroecosystems. Nanoscale Res Lett 16(136):1–24. https://doi.org/10.1186/s11671-021-03593-0
41. Somkuwar SR, Sawant RC, Ingale PP, Masram DT, Chaudhary RR (2022) 6—Nanoparticles for sustainable agriculture: innovative potential with current and future perspectives. In: Pratap Singh R, Rai AR, Abdala A, Chaudhary RG (eds) Biogenic sustainable nanotechnology. Elsevier, pp 131–148
42. Sorooshzadeh A, Hazrati S, Oraki H, Govahi M, Ramazani A (2012) Foliar application of nano-silver influence growth of saffron under flooding stress. Brno, Czech Republic, EU 10:23–25
43. Striker GG (2012) Flooding stress on plants: anatomical, morphological and physiological responses. Botany 1(1):3–28
44. Tyagi A, Ali S, Park S, Bae H (2023) Exploring the potential of multiomics and other integrative approaches for improving waterlogging tolerance in plants. Plants 12:1544
45. Umar OB, Ranti LA, Abdulbaki AS, Bola A, Abdulhamid AK, Biola M, Victor K (2021) Stresses in plants: biotic and abiotic. Current Trends Wheat Res, 1–8
46. Wang D, Saleh NB, Byro A, Zepp R, Sahle-Demessie E, Luxton TP, Ho KT, Burgess RM, Flury M, White JC, White JC (2022) Nano-enabled pesticides for sustainable agriculture and global food security. Nat Nanotechnol 17(4):347–360

Role of Metal and Metal-Oxide Nanoparticles in Agricultural Crops Under UV Radiation, Stress Adaptation, and Mitigation

Aisha Kamal, Nida Sultan, Sazia Siddiqui, and Ayeesha Khatoon

Abstract Plants, as primary producers, are vulnerable to ultraviolet (UV) radiation, particularly UV-B (280–315 nm), which disrupts cellular processes, impairs photosynthesis, and reduces crop productivity. The ozone layer's breakdown has exacerbated UV-B exposure, intensifying oxidative stress in plants. To counter these challenges, nanotechnology has emerged as a transformative solution. Metal and metal-oxide nanoparticles, such as Si, Ag, TiO_2, and CeO_2, exhibit unique properties, including UV absorption, antioxidant mimicry, and the ability to enhance photosynthesis. These nanoparticles mitigate UV-B-induced damage by acting as UV shields, scavenging reactive oxygen species, and promoting the synthesis of protective metabolites like flavonoids. This chapter explores recent advancements in nanoparticle applications to alleviate UV stress, emphasizing their mechanisms of action and agricultural implications. By bridging nanotechnology and plant science, these innovations hold potential to enhance crop resilience, support sustainable agriculture, and address the challenges posed by climate change.

Keywords UV radiation stress · Nanoparticles · Antioxidant defense · Metal and metal oxide · ROS

1 Introduction

Plants, as primary producers in the ecosystem, are frequently exposed to various abiotic stresses that can significantly impact their growth, development, and productivity. One such stress is ultraviolet (UV) radiation, particularly UV-B (280–315 nm), which has intensified due to the depletion of the ozone layer. Increased UV radiation

A. Kamal (✉)
Department of Bioengineering, Integral University, Lucknow, India
e-mail: aishakamal04@gmail.com

N. Sultan · S. Siddiqui · A. Khatoon
Department of Bioscience, Integral University, Lucknow, India

A. Husen (ed.), *Nanobiotechnology for Abiotic Stress Adaptation and Mitigation in Agricultural Crops*, Smart Nanomaterials Technology,
https://doi.org/10.1007/978-981-96-9298-9_8

causes a cascade of detrimental effects, including DNA damage, protein oxidation, disruption of photosynthetic machinery, and the generation of reactive oxygen species (ROS). These oxidative stresses impair cellular functions and compromise the overall health and yield of plants, posing a serious threat to agricultural sustainability and food security [7, 8, 14, 15].

The advancement of nanotechnology in recent years has opened new avenues for mitigating the harmful effects of UV stress on plants. Nanoparticles (NPs), with their unique physicochemical properties such as a high surface area-to-volume ratio, catalytic activity, and adjustable size, have shown promising potential in enhancing plant resilience to UV-induced damage [46]. Engineered nanoparticles, including metal-based (e.g., ZnO, TiO_2, and CeO_2) and carbon-based nanomaterials, can serve as UV filters, ROS scavengers, and protective agents for cellular components.

The interaction of nanoparticles with plant systems is a complex phenomenon influenced by their physicochemical characteristics and the plant's metabolic responses. While certain nanoparticles can act as shields by absorbing or scattering UV radiation, others can boost the plant's antioxidant defense mechanisms, facilitating the maintenance of UV-induced cellular damage [30, 40]. Additionally, some nanoparticles enhance the biosynthesis of secondary metabolites, such as flavonoids and phenolic compounds, they are essential for protecting plants from several abiotic stresses including UV radiation [23].

This chapter explores the multifaceted role of nanoparticles in mitigating UV stress in plants, with a focus on their mechanisms of action, applications in agriculture, and potential challenges. It delves into recent advancements in nanoparticle-based strategies for UV stress management, highlighting their implications for crop protection, yield improvement, and sustainable agricultural practices. By bridging nanotechnology with plant science, this emerging field holds promise for addressing the adverse impacts of environmental stresses in a rapidly changing climate.

2 Types of Ultraviolet (UV) Radiation

Ultraviolet (UV) radiation constitutes approximately 7–9% of the solar spectrum and is separated into UV-A (315–400 nm), UV-B (280–315 nm), and UV-C (100–280 nm) subtypes according to wavelength. The majority of UV-B and almost all UV-C radiation are absorbed by the ozone layer, protecting terrestrial organisms from their damaging effects. However, human activities, particularly the production of chlorofluorocarbons (CFCs) and other ozone-depleting compounds, have weakened the outermost layer of ozone, forcing more harmful ultraviolet (UV) rays to reach Earth's surface [4, 44].

UV radiation impacts plants differently based on wavelength and exposure. UV-A, constituting 95% of UV radiation reaching Earth's surface, promotes plant growth and stress resistance by stimulating defense compounds like flavonoids. However, excessive UV-A can damage DNA, affecting plant development [9]. Despite its prevalence, the effects of UV-A on plants remain underexplored, even though it penetrates deeper into leaf tissues.

UV-B, making up 5% of surface UV radiation, is highly biologically active and can harm cellular components, causing growth retardation, oxidative stress, and reduced photosynthetic efficiency. While long-term exposure is damaging, controlled short-term UV-B exposure can induce defense-related compounds like flavonoids, enhancing plant resilience against pests and diseases [10].

UV-C, the most harmful type, hardly makes it to the surface after being mostly consumed by the ozone layer. It can cause severe cellular damage, including DNA destruction, leading to rapid cell death and tissue necrosis. UV-C is occasionally used in controlled environments for sterilization purposes but is not a natural concern due to ozone protection [5].

3 Impact of UV Radiation on Crops

Ultraviolet (UV) radiation, a component of solar radiation, plays a significant role in influencing plant growth, development, and productivity. Increased UV radiation due to stratospheric ozone depletion poses a potential threat to agricultural systems by affecting photosynthesis, nutrient uptake, and overall crop yield. Although plants have evolved adaptive mechanisms to cope with UV exposure, excessive UV radiation can induce physiological, biochemical, and morphological changes (Fig. 1). Its impact on crops varies depending on factors such as plant species, UV intensity, duration of exposure, and environmental conditions.

UV radiation, particularly UV-A and UV-B, alters shoot and leaf morphology, which affects light absorption and photosynthesis [29]. UV-A can induce elongation of leaves and alter petiole length, though some plant accessions (e.g., *A. thaliana* Di-1) exhibit reduced leaf elongation under UV-A exposure [45]. UV-B radiation has profound impacts on the growth of crops, development, and yield. Important physiological functions including photosynthesis, pigment synthesis, and nutrient cycling are disrupted under increased UV-B levels [17]. In crops like maize and rice, UV-B decreases chloroplast density, rubisco activity, and stomatal function, leading to reduced photosynthetic efficiency and growth [23, 35]. Similarly, rice, a staple food crop, exhibits reduced biomass, altered coleoptile growth, and increased oxidative stress from exposure to UV-B [16]. UV-B stress is more severe, damaging chloroplast structures like grana, thylakoids, and chlorophyll components [22, 39]. UV-B exposure impacts photosynthesis through reduced Rubisco or PEP carboxylase activity and increased reactive oxygen species (ROS). Stomatal conductance also significantly decreased under UV-B exposure [17, 49]. It is also reported that UV-B radiation directly impacts biomolecules, including proteins, lipids, and nucleic acids, resulting in metabolic, biochemical, and morphological changes. It induces the formation of pyrimidine dimers in DNA, causing mutations and impairing replication processes [3].

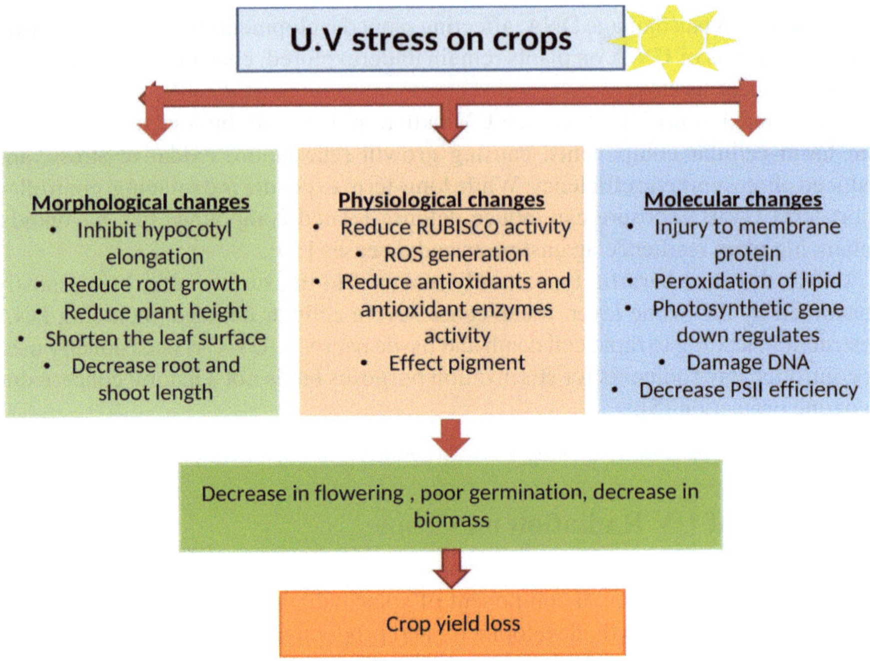

Fig. 1 Impact of UV radiation on morphological, physiological, and molecular traits of plants

4 Plant Defense Strategies Against UV-B Damage

Plants have evolved sophisticated mechanisms to protect against UV-B stress, including DNA repair pathways, antioxidant defenses, and specialized metabolites like flavonoids. UVR8 photoreceptor-mediated signaling activates protective responses, while ROS serve as both stress indicators and signaling molecules. Melatonin further enhances UV-B tolerance by mitigating oxidative damage and regulating protective gene expression (Fig. 2).

4.1 DNA Damage Repair and Antioxidant Defense

UV-B can cause DNA damage, including pyrimidine dimer formation, and ROS production [11]. Plants mitigate this through DNA repair pathways and ROS-scavenging systems. Enzymatic antioxidants such as glutathione reductase (GR), catalase (CAT), and superoxide dismutase (SOD), as well as non-enzymatic antioxidants including ascorbic acid, glutathione, and flavonoids, play a crucial role in DNA repair [6]. UV-B-specific photoreceptors, like UVR8, also reported to regulate

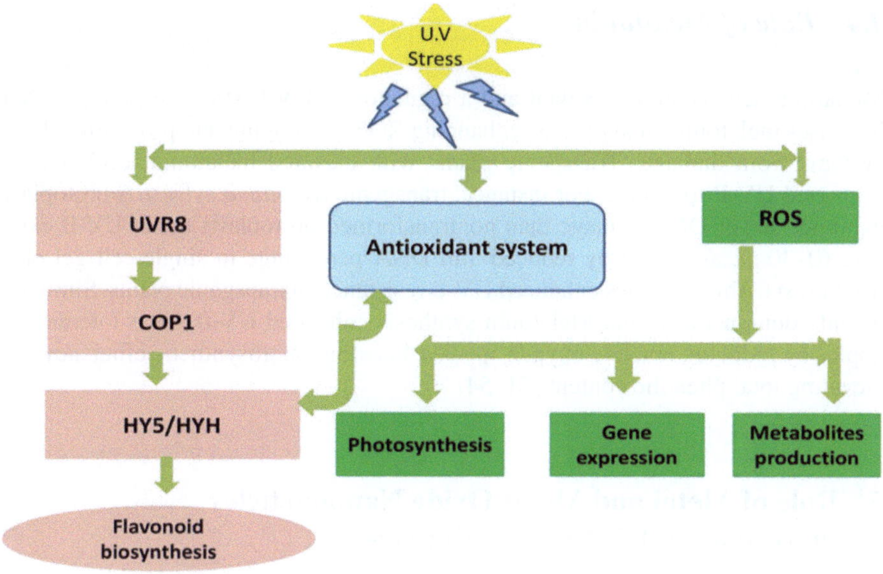

Fig. 2 Adaptive strategies of plants to UV-B radiation

these defenses by activating downstream signaling pathways to promote protective measures [12].

4.2 *Flavonoid Biosynthesis as a Sunscreen*

Flavonoids, synthesized in exposure to UV-B stress, absorb harmful radiation and reduce cellular damage. Key regulators like UVR8, HY5, and MYB transcription factors enhance flavonoid accumulation. The UVR8-COP1-HY5 signaling axis stabilizes HY5, promoting transcription of flavonoid biosynthesis genes [6].

4.3 *Gene Regulation and Signal Integration*

Under UV-B stress, plants use both UVR8-dependent and independent pathways to activate protective genes. UVR8 interacts with transcription factors like MYB13 to induce flavonoid biosynthesis [20, 28]. Other pathways, such as MAPK cascades and ATR-dependent responses, also contribute to stress tolerance, highlighting the complexity of UV signaling networks [18].

4.4 *Role of Melatonin*

Melatonin acts as an antioxidant and a regulator of UV-B stress responses. UV-B increases melatonin biosynthesis, enhancing ROS scavenging and protecting photosystems from damage. Transgenic plants with elevated melatonin levels exhibit improved UV-B tolerance. For instance, transgenic *Nicotiana sylvestris* protoplasts exhibited lower DNA damage than nontransformed protoplasts under UV-B exposure (0–30 s), as shown by reduced Tail DNA percentage in single-cell gel electrophoresis. This indicates enhanced UV-B resistance in transgenic plants. Similarly, a study demonstrated that Melatonin synthesis enhanced UV-B stress tolerance in apples by reducing reactive oxygen species, boosting photosynthetic efficiency, and elevating total phenolic content [21, 54].

5 Role of Metal and Metal-Oxide Nanoparticles in Mitigating UV-Induced Damage

The detrimental effects of UV radiation threaten agricultural productivity and necessitate innovative strategies to mitigate UV-B-induced stress in plants. To reduce the harmful effects of UV-B radiation on agriculture, several strategies have been proposed. Breeding and genetic engineering efforts aim to develop UV-B-tolerant crop varieties with enhanced antioxidant and DNA repair capabilities [13]. Additionally, protective agricultural practices, such as shading and mulching, can minimize UV-B exposure in sensitive crops.

In recent years, metal and metal-oxide nanoparticles (NPs) have gained a lot of attention because of their special qualities and numerous scientific and technological uses. Because of their special optical, catalytic, and photoprotective qualities, metal and metal-oxide nanoparticles (NPs) have become powerful tools for reducing UV stress [1]. However, certain metal and metal-oxide nanoparticles, including titanium dioxide (TiO_2) and silver oxide (AgNPs), have shown potential in alleviating UV-B-induced damage.

5.1 *Titanium Dioxide Nanoparticles (TiO_2-NPs)*

TiO_2-NPs, or titanium dioxide nanoparticles, are essential for reducing UV-B induced stress in plants by acting as an effective "sunscreen." These nanoparticles can absorb and scatter UV-B radiation, reducing its penetration into plant tissues and minimizing cellular damage. They reduce UV-B-induced mortality in *Daphnia magna* and influence photosynthesis in *Chlorella pyrenoidosa,* highlighting their potential in protecting aquatic organisms and mitigating UV-B stress [19, 25]. For instance, a study on *Arabidopsis thaliana* demonstrated that pretreatment with 10 mg/L TiO_2-NPs significantly alleviated UV-B toxicity by enhancing antioxidant enzyme activity, for instance superoxide dismutase (SOD), and promoting the synthesis of flavonoids, which serve as natural UV shields [47]. TiO_2-NPs also preserved the structural integrity of microtubules and reduced DNA damage caused by UV-B in saffron, as evidenced by decreased cyclobutane pyrimidine dimer (CPD) content and fewer mitotic abnormalities. It was also found that 50 mg/L foliar spray of TiO_2-NPs improved plant biomass under UV-B exposure, increasing shoot and root weights significantly [26]. While TiO_2-NPs improve plant adaptability to UV-B stress, careful evaluation of their biological safety is essential. Research has confirmed that concentrations of 10 mg/L are both safe and effective, showing no significant phytotoxic effects on plants like *Arabidopsis*, making TiO_2-NPs a promising nanotool for agricultural applications under enhanced UV-B radiation stress [47].

5.2 Silicon Nanoparticles (SiNP)

Silicon (Si) and silicon nanoparticles (SiNPs) perform an important part in reducing UV radiation-induced stress in plants by enhancing physiological and biochemical processes. When exposed to UV-B light, plants experience increased impact of oxidative stress, lipid peroxidation, and disruption to cellular structures like chloroplasts. Si supplementation has been shown to alleviate these effects by boosting antioxidant defenses, including increased activity of enzymes like ascorbate peroxidase and reduced malondialdehyde levels and superoxide radical (O_2^-), which indicate lower lipid peroxidation. Si improves chlorophyll content, photosynthetic efficiency, and the accumulation of flavonoids, anthocyanins, and UV-absorbing compounds, which act as natural sunscreens to shield plant tissues from UV-B damage [24, 37, 51].

When wheat seedlings exposed to UV-B radiation were supplemented with silicon, either as potassium silicate or silicon nanoparticles (SiNPs), plant biomass, chlorophyll content, and soluble sugar levels increased, while oxidative damage and electrolyte leakage decreased [51]. SiNPs further amplify these protective effects due to their higher surface area and bioactivity, offering improved stress resistance and nutrient uptake [43]. These findings highlight the potential of Si and SiNPs as eco-friendly strategies to improve plant resilience under increasing UV-B radiation caused by climate change.

5.3 Silver Nanoparticles (AgNPs)

Silver nanoparticles (AgNPs) have emerged as an effective tool for alleviating UV-B-induced stress in plants by improving their morphological, physiological, and biochemical traits. AgNPs mitigate UV stress-induced effects through multiple mechanisms, including their ability to enhance antioxidant enzyme activity, such as the scavengers of reactive oxygen species (ROS), peroxidase (POD), catalase (CAT), and superoxide dismutase (SOD) generated during UV-B exposure. Studies have shown that AgNPs increase the uptake of essential nutrients like nitrogen, magnesium, and iron, crucial for chlorophyll biosynthesis, thereby improving photosynthetic efficiency and chlorophyll content in garden thyme (*Thymus vulgaris* L.) [2]. Furthermore, biosynthesized AgNPs have been reported to increase root length, dissolved carbohydrate content, and secondary metabolites, such as flavonoids and essential oils, which act as natural UV protectants in many crops like like corn (*Zea mays* L.), common beans (*Phaseolus vulgaris* L.), and Indian mustard (*Brassica juncea* L.) [34, 36]. AgNPs also reduce the detrimental impact of UV-B on photosynthetic pigments highlighting their potential to improve plant resilience under UV stress [2].

5.4 Cerium Oxide Nanoparticles (CeNPs)

Cerium oxide nanoparticles (CeNPs) play a pivotal role in safeguarding plants against UV radiation-induced toxicity. These nanoparticles function as potent reactive oxygen species (ROS) scavengers due to their unique ability to switch between Ce^{3+} and Ce^{4+} oxidation states, simulating antioxidant enzyme activity [33]. By mitigating oxidative stress, CeNPs protect cellular components, including lipids, proteins, and DNA, from UV-B radiation-induced damage. Research on mung bean seedlings (*Vigna radiata*) revealed that CeNPs alleviate membrane damage and oxidative stress caused by enhanced UV-B exposure. They achieve this by modulating antioxidant enzyme activities, such as superoxide dismutase (SOD) and catalase (CAT), while also reducing malondialdehyde (MDA) levels, a marker of oxidative damage [41]. Furthermore, CeNPs improve root and shoot growth under UV stress by regulating hormonal balance, particularly auxin and cytokinin levels, and restoring starch deposition in root tips, which is critical for gravity sensing. These actions collectively enhance plant resilience by minimizing structural damage, improving physiological responses, and restoring normal growth patterns [50]. Consequently, CeNPs offer a promising nanotechnology-driven approach for agricultural sustainability under increasing UV radiation stress, fostering plant growth and productivity even in challenging environmental conditions.

6 Mechanisms of UV Stress Mitigation Using Metal Nanoparticles and Metal Oxides

Metal and metal-oxide nanoparticles (NPs) offer innovative strategies to mitigate UV-induced stress in plants by combining their unique physicochemical properties with biological interactions (Fig. 3).

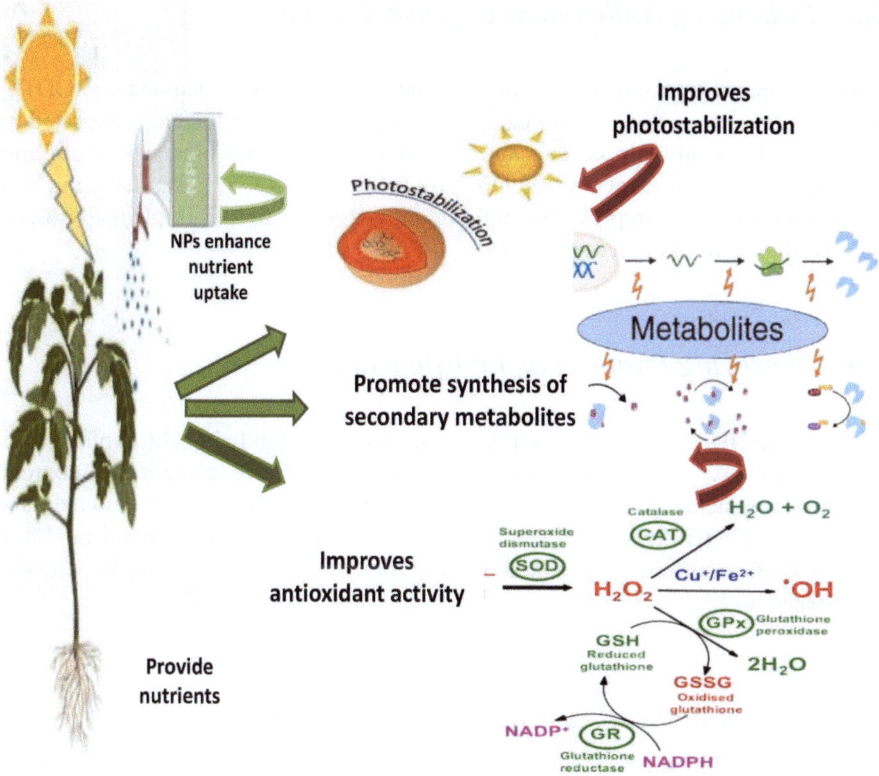

Fig. 3 Adopted strategies by the nanoparticles (NPs) in mitigating UV-induced stress in plants

6.1 UV Absorption and Scattering

Nanoparticles like titanium dioxide (TiO_2), cerium oxide (CeO_2), and zinc oxide (ZnO) effectively absorb and scatter ultraviolet (UV) radiation, reducing its penetration into plant tissues [38]. Their nanoscale size ensures efficient interaction with UV light, minimizing cellular harm brought on by reactive oxygen species (ROS) produced by UV and photo-degradation of plant biomolecules [2, 7]. For example, TiO_2 and CeO_2 NPs act as UV shields due to their wide band gaps, which provide high UV-blocking efficiency. These properties make NPs vital in preventing UV-induced photo-damage and maintaining plant health [27, 53].

6.2 Enhancing Antioxidant Defense Systems

CeO_2 nanoparticles mimic natural antioxidants like superoxide dismutase (SOD) by redox cycling between Ce^{3+} and Ce^{4+} states, scavenging free radicals generated by UV-induced oxidative stress. This antioxidant activity protects plant cells from lipid peroxidation, protein damage, and DNA fragmentation [42, 52]. Similarly, AgNPs, SiNPs and TiO_2-NPs improve the antioxidant enzyme activities in plants, boosting their resilience to oxidative stress [48].

6.3 Improving Photosynthetic Efficiency

Metal-oxide NPs enhance photosynthetic efficiency by stabilizing chloroplast structures under UV stress. TiO_2 NPs, for instance, improve chlorophyll content, stomatal conductance, and water-use efficiency in plants exposed to water and UV stress. These AgNPs, SiNPs and TiO_2-NPs maintain mitochondrial ultrastructure, ensuring energy production and efficient photosynthesis under stressful circumstances [2, 24, 32].

6.4 Promoting Synthesis of Protective Metabolites

Nanoparticles stimulate the production of secondary metabolites such as flavonoids and phenolics, which act as natural sunscreens, absorbing UV light and protecting plant tissues. Additionally, nano fertilizers deliver essential nutrients like nitrogen, phosphorus, and potassium in a controlled manner, promoting plant growth even under adverse UV conditions [2, 31].

6.5 Photo-Stabilization and Material Longevity

The photostability of nanoparticles like TiO_2 and Si ensures energy dissipation from absorbed UV radiation, preventing damage to cellular components. These properties not only protect plants but also enhance material longevity in industrial and agricultural applications [40].

7 Conclusion

Nanoparticles offer a promising avenue for mitigating the harmful impact of UV-B radiation on plants. By enhancing antioxidant defenses, promoting photosynthetic efficiency, and improving nutrient delivery, NPs can help plants adapt to UV-B-induced stress and maintain productivity. However, it is necessary to conduct more research to elucidate the precise mechanisms of NP-plant interactions and to develop guidelines for their safe and effective use in agriculture. As nanotechnology continues to evolve, it holds immense potential to address the challenges of environmentally friendly farming and ensure food security in the face of climate change and environmental stressors.

References

1. Aigbe UO, Osibote OA (2024) Green synthesis of metal oxide nanoparticles, and their various applications. J Hazard Mater Adv 13:100401. https://doi.org/10.1016/j.hazadv.2024.100401
2. Azadi M, Siavash Moghaddam S, Rahimi A, Pourakbar L, Popović-Djordjević J (2021) Biosynthesized silver nanoparticles ameliorate yield, leaf photosynthetic pigments, and essential oil composition of garden thyme (*Thymus vulgaris* L.) exposed to UV-B stress. J Environ Chem Eng 9(5):105919. https://doi.org/10.1016/j.jece.2021.105919
3. Bastian S, Poon GMK, Wyrick JJ (2022) Molecular mechanism of UV damage modulation in nucleosomes. Comput Struct Biotechnol J 20:5393–5400. https://doi.org/10.1016/j.csbj.2022.08.071
4. Bernhard GH, Neale RE, Barnes PW, Neale PJ, Zepp RG, Wilson SR, Andrady AL, Bais AF, McKenzie RL, Aucamp PJ, Young PJ, Liley JB, Lucas RM, Yazar S, Rhodes LE, Byrne SN, Hollestein LM, Olsen CM, Young AR et al (2020) Environmental effects of stratospheric ozone depletion, UV radiation and interactions with climate change: UNEP environmental effects assessment panel, update 2019. Photochem Photobiol Sci 19(5):542–584. https://doi.org/10.1039/d0pp90011g
5. Çavuşoğlu K, Kalefetoğlu Macar T, Macar O, Çavuşoğlu D, Yalçın E (2022) Comparative investigation of toxicity induced by UV-A and UV-C radiation using Allium test. Environ Sci Pollut Res 29(23):33988–33998. https://doi.org/10.1007/s11356-021-18147-1
6. Chen Z, Dong Y, Huang X (2022) Plant responses to UV-B radiation: signaling, acclimation and stress tolerance. Stress Biology 2(1):51. https://doi.org/10.1007/s44154-022-00076-9
7. De Jager TL, Cockrell AE, Du Plessis SS (2017) Ultraviolet light induced generation of reactive oxygen species. In: Ahmad SI (ed) Ultraviolet light in human health, diseases and environment, vol 996. Springer International Publishing, pp 15–23. https://doi.org/10.1007/978-3-319-56017-5_2
8. Deepti, Bachheti AJ, Chauhan K, Bachheti RK, Husen H (2022) Impact of UV radiation on the growth and pharmaceutical properties of medicinal plants. In: Husen A (ed) Environmental pollution and medicinal plants. CRC Press, Boca Raton, pp 47–64. https://doi.org/10.1201/9781003178866-3
9. Ferreyra MLF, Serra P, Casati P (2021) Recent advances on the roles of flavonoids as plant protective molecules after UV and high light exposure. Physiol Plant 173(3):736–749. https://doi.org/10.1111/ppl.13543
10. Golovatskaya IF, Laptev NI (2023) Effect of UV-B radiation on plants growth, active constituents, and productivity. In: Plants and their interaction to environmental pollution. Elsevier, pp 25–60. https://doi.org/10.1016/B978-0-323-99978-6.00024-8

11. He YY, Häder DP (2002) Reactive oxygen species and UV-B: effect on cyanobacteria. Photochem Photobiol Sci 1:729–736. https://doi.org/10.1039/b110365m
12. Heijde M, Ulm R (2012) UV-B photoreceptor-mediated signalling in plants. Trends Plant Sci 17(4):230–237. https://doi.org/10.1016/j.tplants.2012.01.007
13. Hu S, Chen Y, Qian C, Ren H, Liang X, Tao W, Chen Y, Wang J, Dong Y, Han J, Ouyang X, Huang X (2024) Nuclear accumulation of rice UV-B photoreceptors is UV-B- and OsCOP1-independent for UV-B responses. Nat Commun 15(1):6396. https://doi.org/10.1038/s41467-024-50755-6
14. Husen A (2022) Environmental pollution and medicinal plants. CRC Press, Boca Raton. https://doi.org/10.1201/9781003178866
15. Husen A (2023) Plants and their interaction to environmental pollution (damage detection, adaptation, tolerance, physiological and molecular responses). Elsevier, Cambridge, MA, USA. https://doi.org/10.1016/C2020-0-03319-0
16. Idris M, Seo N, Jiang L, Kiyota S, Hidema J, Iino M (2021) UV-B signalling in rice: response identification, gene expression profiling and mutant isolation. Plant, Cell Environ 44(5):1468–1485. https://doi.org/10.1111/pce.13988
17. Kataria S, Jajoo A, Guruprasad KN (2014) Impact of increasing Ultraviolet-B (UV-B) radiation on photosynthetic processes. J Photochem Photobiol, B 137:55–66. https://doi.org/10.1016/j.jphotobiol.2014.02.004
18. Lin L, Wu J, Jiang M, Wang Y (2021) Plant mitogen-activated protein kinase cascades in environmental stresses. Int J Mol Sci 22(4):1543. https://doi.org/10.3390/ijms22041543
19. Liu J, Wang W-X (2017) The protective roles of TiO_2 nanoparticles against UV-B toxicity in Daphnia magna. Sci Total Environ 593–594:47–53. https://doi.org/10.1016/j.scitotenv.2017.03.155
20. Liu W, Jenkins GI (2024) Recent advances in UV-B signalling: interaction of proteins with the UVR8 photoreceptor. J Exp Botany, erae132. https://doi.org/10.1093/jxb/erae132
21. Liu X, Wei Z, Gao T, Zhang Z, Tan K, Li C, Ma F (2021) Ectopic expression of *AANAT* or *HIOMT* improves melatonin production and enhances UV-B tolerance in transgenic apple plants. Fruit Res 1(1):1–13. https://doi.org/10.48130/FruRes-2021-0004
22. Mahdavian K (2024) Effects of ultraviolet radiation on plants and their protective mechanisms. Russ J Plant Physiol 71(6):184. https://doi.org/10.1134/S1021443724607481
23. Mathur S, Bheemanahalli R, Jumaa SH, Kakar N, Reddy VR, Gao W, Reddy KR (2024) Impact of ultraviolet-B radiation on early-season morpho-physiological traits of indica and japonica rice genotypes. Front Plant Sci 15:1369397. https://doi.org/10.3389/fpls.2024.1369397
24. Mavrič Čermelj A, Golob A, Vogel-Mikuš K, Germ M (2021) Silicon mitigates negative impacts of drought and UV-B radiation in plants. Plants 11(1):91. https://doi.org/10.3390/plants11010091
25. Middepogu A, Hou J, Gao X, Lin D (2018) Effect and mechanism of TiO_2 nanoparticles on the photosynthesis of *Chlorella pyrenoidosa*. Ecotoxicol Environ Saf 161:497–506. https://doi.org/10.1016/j.ecoenv.2018.06.027
26. Moradi Rikabad M, Pourakbar L, Siavash Moghaddam S, Popović-Djordjević J (2019) Agro-biological, chemical and antioxidant properties of saffron (*Crocus sativus* L.) exposed to TiO_2 nanoparticles and ultraviolet-B stress. Ind Crops Prod 137:137–143. https://doi.org/10.1016/j.indcrop.2019.05.017
27. Parwaiz S, Khan MM, Pradhan D (2019) CeO_2-based nanocomposites: an advanced alternative to TiO_2 and ZnO in sunscreens. Mater Express 9(3):185–202. https://doi.org/10.1166/mex.2019.1495
28. Qian C, Chen Z, Liu Q, Mao W, Chen Y, Tian W, Liu Y, Han J, Ouyang X, Huang X (2020) Coordinated transcriptional regulation by the UV-B photoreceptor and multiple transcription factors for plant UV-B responses. Mol Plant 13(5):777–792. https://doi.org/10.1016/j.molp.2020.02.015
29. Qian M, Rosenqvist E, Prinsen E, Pescheck F, Flygare A-M, Kalbina I, Jansen MAK, Strid Å (2021) Downsizing in plants—UV light induces pronounced morphological changes in the absence of stress. Plant Physiol 187(1):378–395. https://doi.org/10.1093/plphys/kiab262

30. Rajasekar M, Mary J, Sivakumar M, Selvam M (2024) Recent developments in sunscreens based on chromophore compounds and nanoparticles. RSC Adv 14(4):2529–2563. https://doi.org/10.1039/D3RA08178H
31. Rajput VD, Minkina T, Kumari A, Harish, Singh VK, Verma KK, Mandzhieva S, Sushkova S, Srivastava S, Keswani C (2021) Coping with the challenges of abiotic stress in plants: new dimensions in the field application of nanoparticles. Plants 10(6):1221. https://doi.org/10.3390/plants10061221
32. Rehman A, Khan S, Sun F, Peng Z, Feng K, Wang N, Jia Y, Pan Z, He S, Wang L, Qayyum A, Du X, Li H (2024) Exploring the nano-wonders: unveiling the role of nanoparticles in enhancing salinity and drought tolerance in plants. Front Plant Sci 14:1324176. https://doi.org/10.3389/fpls.2023.1324176
33. Rico CM, Hong J, Morales MI, Zhao L, Barrios AC, Zhang J-Y, Peralta-Videa JR, Gardea-Torresdey JL (2013) Effect of cerium oxide nanoparticles on rice: a study involving the antioxidant defense system and in vivo fluorescence imaging. Environ Sci Technol 47(11):5635–5642. https://doi.org/10.1021/es401032m
34. Salama HMH (2012) Effects of silver nanoparticles in some crop plants, common bean (*Phaseolus vulgaris* L.) and corn (*Zea mays* L.). Int Res J Biotechnol 3(10):190–197
35. Santos I, Almeida JM, Salema R (1993) Plants of *Zea mays* L. developed under enhanced UV-B radiation. I. Some ultrastructural and biochemical aspects. J Plant Physiol 141(4):450–456. https://doi.org/10.1016/S0176-1617(11)80193-8
36. Sharma P, Bhatt D, Zaidi MGH, Saradhi PP, Khanna PK, Arora S (2012) Silver nanoparticle-mediated enhancement in growth and antioxidant status of *Brassica juncea*. Appl Biochem Biotechnol 167(8):2225–2233. https://doi.org/10.1007/s12010-012-9759-8
37. Shen X, Zhou Y, Duan L, Li Z, Eneji AE, Li J (2010) Silicon effects on photosynthesis and antioxidant parameters of soybean seedlings under drought and ultraviolet-B radiation. J Plant Physiol 167(15):1248–1252. https://doi.org/10.1016/j.jplph.2010.04.011
38. Silva MRF, Alves MFRP, Cunha JPGQ, Costa JL, Silva CA, Fernandes MHV, Vilarinho PM, Ferreira P (2023) Nanostructured transparent solutions for UV-shielding: recent developments and future challenges. Mater Today Phys 35:101131. https://doi.org/10.1016/j.mtphys.2023.101131
39. Singh R, Upadhyay AK, Singh DV, Singh JS, Singh DP (2019) Photosynthetic performance, nutrient status and lipid yield of microalgae Chlorella vulgaris and *Chlorococcum humicola* under UV-B exposure. Curr Res Biotechnol 1:65–77. https://doi.org/10.1016/j.crbiot.2019.10.001
40. Soni S, Jha AB, Dubey RS, Sharma P (2022) Application of nanoparticles for enhanced UV-B stress tolerance in plants. Plant Nano Biology 2:100014. https://doi.org/10.1016/j.plana.2022.100014
41. Sun C, Zhao C, Han R (2025) The repair effects of cerium oxide nanoparticles on enhanced UV-B radiation in mung bean (*Vigna radiata*) seedling roots. Pak J Botany 57(4). https://doi.org/10.30848/PJB2025-4(16)
42. Thangavelu RM, Da Silva WL, Zuverza-Mena N, Dimkpa CO, White JC (2024) Nano-sized metal oxide fertilizers for sustainable agriculture: balancing benefits, risks, and risk management strategies. Nanoscale 16(43):19998–20026. https://doi.org/10.1039/D4NR01354A
43. Tripathi DK, Singh S, Singh VP, Prasad SM, Dubey NK, Chauhan DK (2017) Silicon nanoparticles more effectively alleviated UV-B stress than silicon in wheat (*Triticum aestivum*) seedlings. Plant Physiol Biochem 110:70–81. https://doi.org/10.1016/j.plaphy.2016.06.026
44. UNEP (United Nations Environment Programme) (2014) Environmental effects of ozone depletion and its interactions with climate change. ISBN 978-9966-076-04-5
45. Verdaguer D, Jansen MAK, Llorens L, Morales LO, Neugart S (2017) UV-A radiation effects on higher plants: exploring the known unknown. Plant Sci 255:72–81. https://doi.org/10.1016/j.plantsci.2016.11.014
46. Wahab A, Muhammad M, Ullah S, Abdi G, Shah GM, Zaman W, Ayaz A (2024) Agriculture and environmental management through nanotechnology: eco-friendly nanomaterial synthesis

for soil-plant systems, food safety, and sustainability. Sci Total Environ 926:171862. https://doi.org/10.1016/j.scitotenv.2024.171862

47. Wang J, Li M, Feng J, Yan X, Chen H, Han R (2021) Effects of TiO_2-NPs pretreatment on UV-B stress tolerance in *Arabidopsis thaliana*. Chemosphere 281:130809. https://doi.org/10.1016/j.chemosphere.2021.130809
48. Wang Z, Wang S, Ma T, Liang Y, Huo Z, Yang F (2023) Synthesis of zinc oxide nanoparticles and their applications in enhancing plant stress resistance: a review. Agronomy 13(12):3060. https://doi.org/10.3390/agronomy13123060
49. Williams TB, Dodd IC, Sobeih WY, Paul ND (2022) Ultraviolet radiation causes leaf warming due to partial stomatal closure. Hortic Res 9:uhab066. https://doi.org/10.1093/hr/uhab066
50. Wu H, Tito N, Giraldo JP (2017) Anionic cerium oxide nanoparticles protect plant photosynthesis from abiotic stress by scavenging reactive oxygen species. ACS Nano 11(11):11283–11297. https://doi.org/10.1021/acsnano.7b05723
51. Yao X, Chu J, Cai K, Liu L, Shi J, Geng W (2011) Silicon improves the tolerance of wheat seedlings to ultraviolet-B stress. Biol Trace Elem Res 143(1):507–517. https://doi.org/10.1007/s12011-010-8859-y
52. Yi L, Yu L, Chen S, Huang D, Yang C, Deng H, Hu Y, Wang H, Wen Z, Wang Y, Tu Y (2024) The regulatory mechanisms of cerium oxide nanoparticles in oxidative stress and emerging applications in refractory wound care. Front Pharmacol 15:1439960. https://doi.org/10.3389/fphar.2024.1439960
53. Yousif E, Ahmed D, Zainulabdeen K, Jawad A (2023) Photo-physical and morphological study of polymers: a review. Phys Chem Res 11(2):409–424. https://doi.org/10.22036/pcr.2022.342751.2105
54. Zhang L, Jia J, Xu Y, Wang Y, Hao J, Li T (2012) Production of transgenic Nicotiana sylvestris plants expressing melatonin synthetase genes and their effect on UV-B-induced DNA damage. In Vitro Cellular Dev Biol Plant 48(3):275–282. https://doi.org/10.1007/s11627-011-9413-0

Nanoparticle Cross-Talk or Interaction with Phytohormones for Agricultural Crop Abiotic Stress Adaptation and Mitigation

Shifa Shaffique, Muhammad Aaqil khan, Muhammad Imran, Sang-Mo kang, and In-Jung Lee

Abstract Climate change and the rapid increase in the human population are the biggest challenges for agriculture production. Abiotic stresses adversely affect the agriculture production globally. They interfere with the developmental and metabolic processes of plants. In response to exposure to stress, plants have evolved several defensive mechanisms such as biochemical and metabolic reprogramming of plant cells to minimize the stress but up to a certain limit. In the present era, nanotechnology is gaining greater attention due to its promising approach to mitigating stress. Nanoparticles work as regulatory molecules to enhance stress tolerance in plants via several complex mechanisms such as modulation in antioxidant defense systems, phytohormones, and secondary metabolites. Moreover, the phytohormones are also termed signaling molecules that try to mitigate stress by several regulatory mechanisms. Recently, a few studies have been published that show either synergistic or antagonistic phytohormones cross-talk with the NP that regulates protective response to tolerate the abiotic stress.

Keywords Cross-talk · Phytohormones · Plants · Protective · Potential

---

Shifa Shaffique, Muhammad Aaqil khan these authors contributed equally.

S. Shaffique · M. Imran · S.-M. kang · I.-J. Lee (✉)
College of agriculture and life sciences, Department of Applied Biosciences, Kyungpook National University, Daegu, South Korea
e-mail: Ijlee@knu.ac.kr

M. A. khan
Department of Chemical and Life Sciences, Qurtuba University Peshawar, Peshawar, Pakistan

M. Imran
Biosafety Division, National Institute of Agriculture Science, Rural Development Administration, Jeonju, Republic of Korea

A. Husen (ed.), *Nanobiotechnology for Abiotic Stress Adaptation and Mitigation in Agricultural Crops*, Smart Nanomaterials Technology,
https://doi.org/10.1007/978-981-96-9298-9_9

Abbreviations

ROS	Reactive oxygen species
HMs	Heavy metals
IAA	Auxins
CK	Cytokinin
ABA	Abscisic acids
GA	Gibberellic acids
NPs	Nanoparticles
Zn	Zinc
Cu	Copper
Ag	Silver

1 Introduction

This chapter explores the cross-talk between the phytohormones and nanoparticles (NPs) in enhancing plant resilience to several abiotic stresses. Abiotic stresses such as heavy metals (HMs), salt, drought, temperature, and cold significantly disrupt various morphological, anatomical, and metabolic processes which ultimately lead to adversely affecting the growth and productivity of plants [9, 14, 16–20]. Abiotic stress reduces crop production globally [21, 22, 63, 64]. NPs are ultrafine particles ranging in size 1–100 nm, having unique properties that help in mitigation of the adverse effects caused by reducing oxidative stress [40, 43]. The most important phytohormones are auxins (IAA), cytokinins (CKs), abscisic acids (ABA), gibberellic acids (GA), jasmonic acid (JA), and salicylic acid (SA) that act as a regulatory molecule in the mitigation of stress tolerance [23-27, 38, 48–50]. NPs act as regulatory molecules that enhance stress tolerance via a complex signaling network [6, 52, 53, 58], they interact with each other and form the complex signaling network that ultimately leads to enhanced stress tolerance. The detailed mechanism is discussed below.

2 Abiotic Stress and Nanoparticles

Different types of abiotic stresses such as HMs, salinity, heat, cold, and drought significantly impact plants' growth, development, and productivity [32]. Stress conditions adversely affect various morphological, metabolic, biochemical, and physiological processes, ultimately negatively affecting photosynthesis, growth, and yields [63]. It is essential to understanding the detrimental effects of these stress factors on plant physiology and growth to develop effective sustainable agriculture practices and to improve plants' resilience against various stress factors [2]. Upon exposure to stress reactive oxygen species (ROS) rapidly generating that increase the oxidative stress in plants are enough sophisticated that try to regulate the stress by modulation of secondary metabolites expression, activation of phytohormones and defensive mechanisms, stress-related genes, etc. After a certain limit, plant cells show signs of stress such as chlorosis, impaired photosynthesis, and stunted growth and production [44, 45]. There are a lot of technologies that are being practiced to optimize the growth of plants under stress such as biostimulants, agrochemicals, breeding techniques, growth regulators, and nanotechnology [12, 35]. Among all, nanotechnology is an innovative technique that is being used. Applications of NPs in sustainable agriculture recently gained significant attention for research worldwide due to their microscopic size (1–100 nm) and unique characteristics like larger surface area and high reactivity [1, 4, 46]. Currently, different types of NPs such as Zinc (Zn), copper (Cu), and silver (Ag) as well as carbon-based NPs have been widely used for increased plant growth and development improvement of soil quality, and increased nutrient uptake from the soil. The main purpose of using NPs in agriculture is to mitigate the adverse effects of abiotic stresses on different crops [47]. Figure 1 represents the production of ROS via several abiotic factors and the application of NPs inhibiting ROS production.

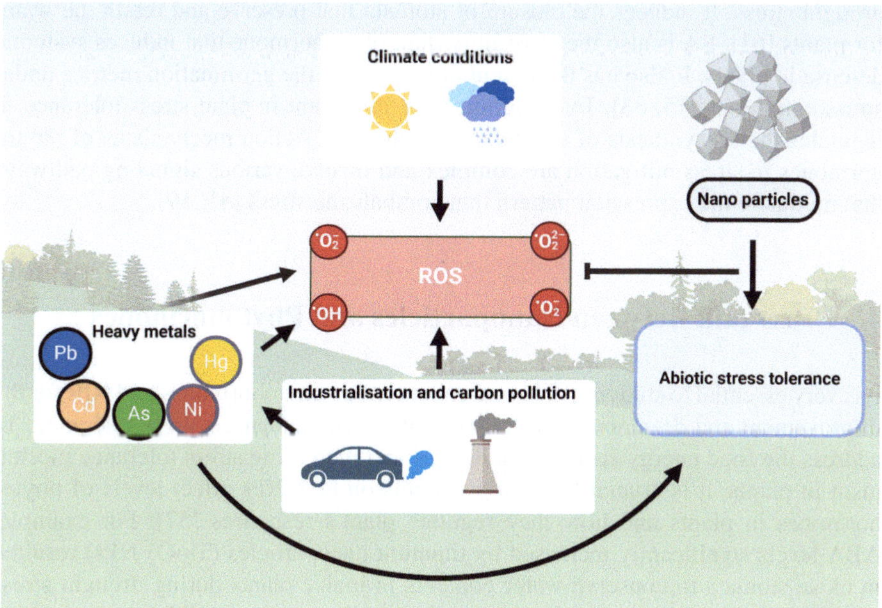

Fig. 1 Production of ROS via several abiotic factors that enhance the oxidation burst. Production of ROS enhances the oxidation burst; however application of NPs reduces the ROS and provides the abiotic stress tolerance

3 Phytohormones and Abiotic Stress Tolerance

Phytohormones are commonly known plant hormones that play vital roles in modulating plant growth, development, reproductive processes, and responses to various types of ecological stress conditions [8, 39]. They work as signaling molecules and are produced in specific tissue under specific environmental conditions and transported to different parts of the plant, where they exhibit their effects even at low concentrations [11]. Key phytohormones including IAA, CKs, ABA, JA, and GAs play essential roles in stress responses. IAA is significantly involved in cell elongation, and division and plays a key role in directing growth responses to enable plant species to optimize their exposure to nutrients and light [64]. CKs are essential to influencing plant growth and development by promoting cell differentiation and division, particularly in young shoots and roots. They also help to maintain ion homeostasis in salt stress and enhance antioxidant systems to reduce oxidative damage caused by ROS. In HM stress, they facilitate metal sequestration into vacuoles to reduce their toxicity [42]. ABA is an important plant hormone that regulates the stress mechanism of plants. It is also known as stress hormones upon exposure to drought stress. It induces the closure of stomata that preserve and retain the water for plants [61]. SA is also the most important plant hormone that induces systemic defense in plants. It also has the potential to regulate the germination metrics under stress conditions [15, 33]. JA is significantly important in plant stress tolerance, it regulates the biosynthesis of secondary metabolites. Action mechanisms of phytohormones in stress mitigation are complex and involve various signaling pathways that modulate the expression pattern that combats the stress [41, 59].

4 Cross-talk Between Nanoparticles and Phytohormones

It is very essential to studying the cross-talk between phytohormones and NPs for the improvement and development of plant varities with enhanced stress tolerance. To address the food energy security challenges and increasing stress tolerance mechanism in plants, it is crucial to getting insights on how NPs affect levels of phytohormones in plants and how they regulate plant's responses [57]. For example, ABA levels significantly increased by titanium nanoparticles ($TioO_2$ NPs) resulted in close stomata to conserve water contents in maize plants during drought stress conditions [31]. In sustainable agriculture approaches, the cross-talk between phytohormones and NPs is effectively used to develop advanced agro-technologies and substrates such as NP-based fertilizers and novel stress mitigation strategies [56]. For example, silicon nanoparticles (Si NPs) have been used in combination with cytokinin which promotes shoot and root growth significantly and improves plant resilience and yield under salt stress conditions [29]. The novel strategies based on cross-talk between phytohormones and NPs could significantly improve essential nutrient uptake, enhance stress signaling pathways, and promote sustainable agriculture practices. Furthermore, a thorough understanding of the mechanisms involved in the cross-talk between the phytohormones and NPs will help develop specialized NPs with targeted efficacies in stress mitigation processes and contribute to the production of crop varieties with enhanced stress tolerance capabilities [53]. For example, selenium nanoparticles (Se NPs) have been used to increase the production of gibberellins in plants like barley which leads to improved seed germination and plant growth, especially under stress conditions [10].

Currently, studies are emphasizing the impact of phytohormones and NPs on plant growth and development. For example, metal NPs such as gold nanoparticles (Au NPs) are found to increase the synthesis of ABA and lead to improving drought stress tolerance in plant species. Moreover, NPs have been found to modulate the activity of phytohormones by influencing their signaling pathways and ultimately affecting various physiological processes in plants [30, 57]. For example, copper nanoparticles (Cu NPs) are found to be involved in the regulation of auxin activity and lead to influence the growth and development of roots and the efficiency of nutrient uptake from soil [62]. Recent studies have investigated that phytohormones affect the accumulation and transportation of nanoparticles within plants. To optimize the use of NPs in agriculture, it is crucial to studying and understanding the associated signaling pathways, as the reciprocal cross-talk revealed. For example, according to a recent study, phytohormones can increase the bioavailability of NPs by enhancing their capacity to promote plant growth, development, and adaptation to stress conditions [3]. According to research studies, SA boost the antimicrobial activity and overall health of plants by enhancing the uptake of silver nanoparticles (Ag NPs) in plants such as wheat [7]. Studies have investigated the functional role of ethylene in the accumulation and internalization of iron oxide (Fe_2O_3 NPs) in plants by increasing the nutritional status of the plant under deficient conditions [60]. In Egypt, a study was conducted on chitosan NPs, where they were applied under drought stress in tomatoes. Two different concentrations of synthesized chitosan NPs were given (60 and 90 ppm. The results showed that applying these NPs significantly enhanced the drought stress tolerance in tomatoes when genetic expressions were performed. The results were interesting, the application of these NPs enhances the upregulation of *NCED* gene that is responsible for the biosynthesis of ABA. This study confirms the involvement and cross-talk of phytohormones and their biosynthesis genes [36]. Previously it was believed that exposure to metallic NPs has negative effects on the plants. In 2020, China, a study was conducted to see the effectiveness of metallic oxide NPs on modulation of auxin via expression analysis of the FT-INTERACTING PROTEIN 7 (*OsFTIP7*). The results were interesting such that upon exposure to the CuO and ZnO NPs, there is endogenous production of the IAA. The results are interesting and give us the direction that applying NPs modulates IAA production [28]. A most recent study demonstrates that in *Arabidopsis thaliana*, upon exposure to UV-B stress Cerium oxide nanoparticles (CeO_2 NPs) enhance stress tolerance via accumulation of auxin. Moreover, the NPs also enhance the antioxidant defense system. The endogenous accumulation of auxin was confirmed via expression analysis of PIN1, 2, and 7 [55]. In another study, Ca NPs were applied to *Brassica napus* under

Table 1 Impact of nanoparticles on different plant species under different conditions with endogenous interaction with phytohormones or their biosynthesis genes

Phytohormone/ biosynthesis gene	Nanoparticles	Experimental plant	Stress	Key references
OsFTIP7 (IAA)	CuO NPs	*Oryza sativa*	Exposure to metal oxide	[28]
NECD (ABA)	Chitosan	*Solanum lycopersicum*	Drought	[36]
ABA, JA, SA	Ca NPs	*Brassica napus*	Drought	[5]
PIN1, PIN2, and PIN7 (IAA)	CeO_2 NPs	*Arabidopsis thaliana*	UV-B	[55]

drought stress. The results revealed it enhanced the drought stress tolerance of the plant via several different mechanisms. It also modulates the level of three important endogenous phytohormones ABA, SA, and JA that enhance the antioxidant defense in plants [5]. Some examples of the interaction of phytohormones or their relative gene in response to stress are given in Table 1.

5 Mechanism of Interaction

In the interaction of the NPs with phytohormones, the surface charge or coating is important in the entire cross-talk. Usually positively charged NPs enhance the uptake of phytohormones while negatively charged NPs inhibit their uptake [34]. Moreover, due to their unique smaller size, they penetrate cells and influence the hormonal signaling pathways. Some studies demonstrate that metallic NPs generate ROS which act as signaling molecules that interact with the phytohormones and modulate the stress response of plants [30, 37, 55]. The mechanism of NP cross-talk with phytohormones is given in Fig. 2.

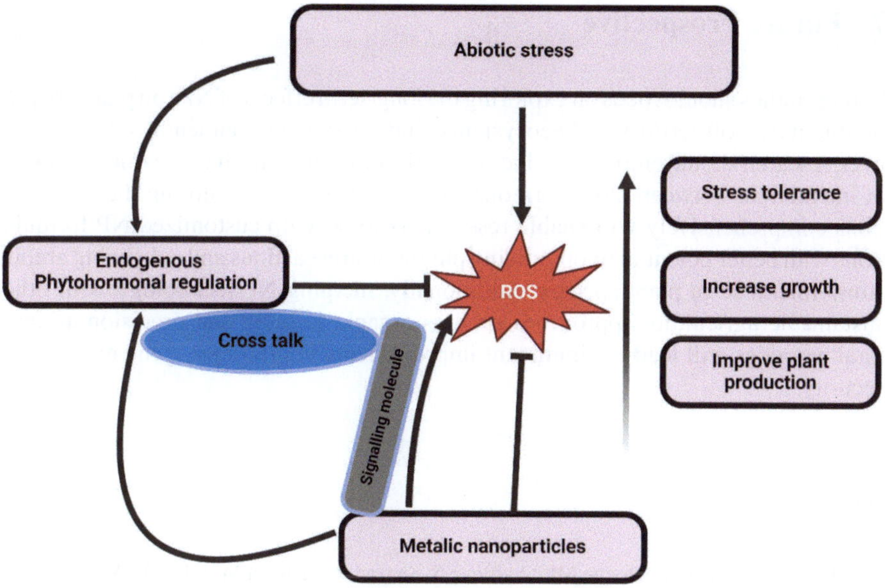

Fig. 2 Production of ROS through exposure to abiotic stress; however, NPs also enhance the ROS production that acts as a signaling molecule that may interact with the phytohormones to enhance plant stress tolerance and increase growth

6 Conclusion

This chapter explores the essential functional roles of NPs in improving plant resilience to various abiotic stresses through their interactions with phytohormones. The reviewed literature revealed that NPs significantly influence the production of phytohormones and play essential roles in improving stress tolerance in plants. Phytohormones and NP cross-talk or interactions open new windows for researchers to develop novel agriculture strategies for effective mitigation of adverse effects of various stress factors. It is required to comprehensively study the molecular mechanisms involved in these interactions to develop and optimize applications of NPs in agriculture. Keeping the significant advantages and potential of NPs in mind, it is also important to identify and address the risks and challenges associated with their applications. In summary, exploring the cross-talk between the key phytohormones and NPs is an elementary need that holds significant promise in sustainable agriculture for the improvement of stress tolerance in crops and to ensure food security for the ever-increasing population worldwide.

7 Future Prospective

Future studies should focus on exploring the long-term effects of NPs on plant growth, productivity, soil fertility and ecosystems, and overall environmental safety. Moreover, research should emphasize elucidating the molecular mechanisms and pathways of interactions between phytohormones and nanoparticles. Exploring these interactions comprehensively will enable researchers to develop customized NP formulations with better potential of optimizing phytohormone actions and enhancing abiotic stress tolerance in plant species. Additionally, merging NP technology with other sustainable agriculture approaches such as organic farming and precision agricultural practices will lead to significant improvements in crop management and food security.

References

1. Ahsan SM, Imran M, Hoque MI, Shaffique S, Shazad R, Rahman MM, Lee IJ, Woo CH, Ray RL (2024) Nanotechnology in crop protection. In: Shahzad R, Fiaz S, Qayyum A, Ul Islam M, Lee I-J (eds) Revolutionizing agriculture: a comprehensive exploration of agri-nanotechnology. Springer, Switzerland pp 101–120. https://doi.org/10.1007/978-3-031-76000-6_5
2. Aizaz M, Khan I, Lubna, Asaf S, Bilal S, Jan R, Khan AL, Kim KM, AL-Harrasi A (2023). Enhanced physiological and biochemical performance of mung bean and maize under saline and heavy metal stress through application of endophytic fungal strain SL3 and exogenous IAA. Cells 12(15):1960
3. Al-Khayri JM, Rashmi R, Surya Ulhas R, Sudheer WN, Banadka A, Nagella P, Aldaej MI, Rezk AA-S, Shehata WF, Almaghasla MI (2023) The role of nanoparticles in response of plants to abiotic stress at physiological, biochemical, and molecular levels. Plants 12(2):292
4. Altammar KA (2023) A review on nanoparticles: characteristics, synthesis, applications, and challenges. Front Microbiol 14:1155622
5. Ayyaz A, Zhou Y, Batool I, Hannan F, Huang Q, Zhang K, Shahzad K, Sun Y, Farooq MA, Zhou W (2024) Calcium nanoparticles and abscisic acid improve drought tolerance, mineral nutrients uptake and inhibitor-mediated photosystem II performance in *Brassica napus*. J Plant Growth Regul 43(2):516–537
6. Banerjee A, Roychoudhury A (2022) Explicating the cross-talks between nanoparticles, signaling pathways and nutrient homeostasis during environmental stresses and xenobiotic toxicity for sustainable cultivation of cereals. Chemosphere 286:131827
7. Bano A (2020) Interactive effects of Ag-nanoparticles, salicylic acid, and plant growth promoting rhizobacteria on the physiology of wheat infected with yellow rust. J Plant Pathol 102(4):1215–1225
8. Ciura J, Kruk J (2018) Phytohormones as targets for improving plant productivity and stress tolerance. J Plant Physiol 229:32–40
9. Cramer GR, Urano K, Delrot S, Pezzotti M, Shinozaki K (2011) Effects of abiotic stress on plants: a systems biology perspective. BMC Plant Biol 11:1–14
10. Dilnawaz F, Misra AN, Apostolova E (2023) Involvement of nanoparticles in mitigating plant's abiotic stress. Plant Stress 100280
11. EL Sabagh A, Islam MS, Hossain A, Iqbal MA, Mubeen M, Waleed M, Reginato M, Battaglia M, Ahmed S, Rehman A, Arif M (2022) Phytohormones as growth regulators during abiotic stress tolerance in plants. Front Agron 4:765068
12. Fahad S, Nie L, Chen Y, Wu C, Xiong D, Saud S, Hongyan L, Cui K, Huang J (2015) Crop plant hormones and environmental stress. Sustain Agric Rev 15:371–400

13. Farooq SS, Sultana S, Dilshad R, Nisar R, Muneer S, Sarwar G, Shaffique S (2025) Integrating computational approaches and phytohormones for enhancing drought and salt stress tolerance in marginal conditions. Pak J Bot 57:4
14. Fraire-Velázquez S, Balderas-Hernández VE (2013) Abiotic stress in plants and metabolic responses. In: Abiotic stress-plant responses and applications in agriculture, pp 25–48
15. Horváth E, Szalai G, Janda T (2007) Induction of abiotic stress tolerance by salicylic acid signaling. J Plant Growth Regul 26:290–300
16. Husen A (2021a) The harsh environment and resilient plants: an overview. In: Husen A (ed) Harsh environment and plant resilience. Springer, Cham, pp 1–23. Gewerbestrasse 11:6330. https://doi.org/10.1007/978-3-030-65912-7_1
17. Husen A (2021b) Morpho-anatomical, physiological, biochemical and molecular responses of plants to air pollution. In: Husen A (ed) Harsh environment and plant resilience. Springer, Cham, pp. 203–234. Gewerbestrasse 11:6330 https://doi.org/10.1007/978-3-030-65912-7_9
18. Husen A (2021c) Cross talk between autophagy and hormones for abiotic stress tolerance in plants. In: Husen A (eds) Plant performance under environmental stress. Springer, Cham, pp 1–15. Gewerbestrasse 11:6330. https://doi.org/10.1007/978-3-030-78521-5_1
19. Husen A (2025a) Agricultural crop improvement: plant and soil relationships. Taylor & Francis Group, NW. https://doi.org/10.1201/9781032630366
20. Husen A (2023) Plants and Their Interaction to Environmental Pollution (Damage detection, adaptation, tolerance, physiological and molecular responses). Elsevier Inc., Cambridge. https://doi.org/10.1016/C2020-0-03319-0
21. Husen A (2025b) Roles of osmolytes in plant systems under changing environment. Elsevier Inc., Cambridge
22. Husen A (2025c) Role of antioxidants in plant stress mitigation. Elsevier Inc., Cambridge
23. Husen A, Iqbal M, Aref IM (2016) IAA-induced alteration in growth and photosynthesis of pea (*Pisum sativum* L.) plants grown under salt stress. J Environ Biol 37:421–429
24. Husen A, Iqbal M, Aref IM (2017) Plant growth and foliar characteristics of faba bean (*Vicia faba* L.) as affected by indole-acetic acid under water-sufficient and water-deficient conditions. J Environ Biol 38:179–186
25. Husen A, Iqbal M, Khanum N, Aref IM, Sohrab SS, Meshresa G (2019) Modulation of salt-stress tolerance of Niger (*Guizotia abyssinica*), an oilseed plant, by application of salicylic acid. J Environ Biol 40:94–104
26. Husen A, Iqbal M, Sohrab SS, Ansari MKA (2018) Salicylic acid alleviates salinity-caused damage to foliar functions, plant growth and antioxidant system in Ethiopian mustard (*Brassica carinata* A. Br.) Agric Food Secur 7(44):1–14
27. Husen A, Zhang W (2023) Hormonal cross-talk, plant defense and development. Elsevier Inc., Cambridge
28. Jiang M, Wang J, Rui M, Yang L, Shen J, Chu H, Song S, Chen Y (2021) OsFTIP7 determines metallic oxide nanoparticles response and tolerance by regulating auxin biosynthesis in rice. J Hazard Mater 403:123946
29. Jin W, Li L, He W, Wei Z (2024) Application of silica nanoparticles improved the growth, yield, and grain quality of two salt-tolerant rice varieties under saline irrigation. Plants 13(17):2452
30. Karabulut F (2024) The impact of nanoparticles on plant growth, development, and stress tolerance through regulating phytohormones. In: Nanoparticles in plant biotic stress management. Springer, pp 87–105
31. Karvar M, Azari A, Rahimi A, Maddah-Hosseini S, Ahmadi-Lahijani MJ (2022) Titanium dioxide nanoparticles (TiO_2-NPs) enhance drought tolerance and grain yield of sweet corn (*Zea mays* L.) under deficit irrigation regimes. Acta Physiologiae Plantarum 44(2):14
32. Khan I, Zhang Y, Akbar F, Khan J (2022) Abiotic stress tolerance in cereals through genome editing. In: Omics approach to manage abiotic stress in cereals. Springer, pp 295–319
33. Khan MIR, Fatma M, Per TS, Anjum NA, Khan NA (2015) Salicylic acid-induced abiotic stress tolerance and underlying mechanisms in plants. Front Plant Sci 6:462
34. Mascarenhas R, Mathur T, Maheshwari J, Nagella P (2023) Interaction of nanomaterials with plant macromolecules: nucleic acid, proteins and hormones. In: Nanomaterial interactions

with plant cellular mechanisms and macromolecules and agricultural implications. Springer, pp 231–271
35. Minhas PS, Rane J, Pasala RK (2017) Abiotic stress management for resilient agriculture. Springer
36. Mohamed NG, Abdel-Hakeem MA (2023) Chitosan nanoparticles enhance drought tolerance in tomatoes (*Solanum lycopersicum* L) via gene expression modulation. Plant Gene 34:100406
37. Mukarram M, Petrik P, Mushtaq Z, Khan MMA, Gulfishan M, Lux A (2022) Silicon nanoparticles in higher plants: uptake, action, stress tolerance, and crosstalk with phytohormones, antioxidants, and other signalling molecules. Environ Pollut 310:119855
38. Mukherjee A, Gaurav AK, Singh S, Yadav S, Bhowmick S, Abeysinghe S, Verma JP (2022) The bioactive potential of phytohormones: a review. Biotechnology Reports 35:e00748
39. Nadeem SM, Ahmad M, Zahir ZA, Kharal MA (2016) Role of phytohormones in stress tolerance of plants. In: Plant, soil and microbes: volume 2: mechanisms and molecular interactions, pp 385–421
40. Pestovsky YS, Martínez-Antonio A (2017) The use of nanoparticles and nanoformulations in agriculture. J Nanosci Nanotechnol 17(12):8699–8730
41. Raza A, Charagh S, Zahid Z, Mubarik MS, Javed R, Siddiqui MH, Hasanuzzaman M (2021) Jasmonic acid: a key frontier in conferring abiotic stress tolerance in plants. Plant Cell Rep 40(8):1513–1541
42. Rehman M, Saeed MS, Fan X, Salam A, Munir R, Yasin MU, Khan AR, Muhammad S, Ali B, Ali I (2023) The multifaceted role of jasmonic acid in plant stress mitigation: an overview. Plants 12(23):3982
43. Roy TK, Islam MS, Mahiddin NA, Hossain SA, Biswas T, Antu UB, Serity SA, Miti JF, Akter S, Roy S (2025) Application of nanoparticles (NPs) to ameliorate abiotic stress in economically important crop species: a potential review. J Crop Health 77(1):1–20
44. Shaffique S, Farooq M, Kang S-M, Lee I-J (2024) Recent advances in biochemical reprogramming network under drought stress in soybean. J Soil Sci Plant Nutr 1–12
45. Shaffique S, Injamum-Ul-Hoque M, Husen A, Lee I-J (2024) Revolutionizing heat stress tolerance in Glycine max: exploring the latest advances in microbial application. Plant Stress 100725
46. Shaffique S, Odongkara P, Imran M, Injamum-Ul-Hoque M, Ahsan SM, Shah AA, Khan MA, Kang SM, Khan AL, Lee IJ (2024) Role of green synthesized nanoparticles in agriculture technology. In: Shahzad R, Fiaz S, Qayyum A, Ul Islam M, Lee I-J (eds) Revolutionizing agriculture: a comprehensive exploration of agri-nanotechnology. Springer, Switzerland, pp 369–378. https://doi.org/10.1007/978-3-031-76000-6_16
47. Sharma R, Kumar V (2024) Nano enabled agriculture for sustainable soil. Waste Manag Bull
48. Siddiqi KS, Husen A (2017) Plant response to strigolactones: current developments and emerging trends. Appl Soil Ecol 120:247–253
49. Siddiqi KS, Husen A (2019) Plant response to jasmonates: current developments and their role in changing environment. Bull Natl Res Cent 43(153):1–11
50. Siddiqi KS, Husen A (2021) Significance of brassinosteroids and their derivatives in the development and protection of plants under abiotic stress. Biologia 76:2837–2857. https://doi.org/10.1007/s11756-021-00853-3
51. Singh S, Husen A (2020) Behavior of agricultural crops in relation to nanomaterials under adverse environmental conditions. In: Husen A Jawaid M (eds) Nanomaterials for agriculture and forestry applications. Elsevier Inc., Cambridge, pp 219–256 https://doi.org/10.1016/B978-0-12-817852-2.00009-3
52. Singh S, Husen A (2019) Role of nanomaterials in the mitigation of abiotic stress in plants. In: Husen A, Iqbal M (eds) N*anomaterials* and plant potential. Springer, Cham, pp 441–471. Gewerbestrasse 11:6330. https://doi.org/10.1007/978-3-030-05569-1_18
53. Singh A, Sharma A, Singh O, Rajput VD, Movsesyan HS, Minkina T, Alexiou A, Papadakis M, Singh RK, Singh S (2024) In-depth exploration of nanoparticles for enhanced nutrient use efficiency and abiotic stresses management: present insights and future horizons. Plant Stress 100576

54. Sonkar S, Sharma L, Singh RK, Pandey B, Rathore SS, Singh AK, Porwal P, Singh SP (2021) Plant stress hormones nanobiotechnology. In: Nanobiotechnology: mitigation of abiotic stress in plants, pp 349–373
55. Sun C, Zhao C, Wang G, Han R (2025) Cerium oxide nanoparticles ameliorate *Arabidopsis thaliana* root damage under UV-B stress by modulating the cell cycle and auxin pathways. Protoplasma 1–19
56. Thabet SG, Alqudah AM (2024) Unraveling the role of nanoparticles in improving plant resilience under environmental stress condition. Plant Soil 1–18
57. Tripathi D, Singh M, Pandey-Rai S (2022) Crosstalk of nanoparticles and phytohormones regulate plant growth and metabolism under abiotic and biotic stress. Plant Stress 6:100107
58. Wang J, Song L, Gong X, Xu J, Li M (2020) Functions of jasmonic acid in plant regulation and response to abiotic stress. Int J Mol Sci 21(4):1446
59. Wang Y, Wang S, Xu M, Xiao L, Dai Z, Li J (2019) The impacts of γ-Fe_2O_3 and Fe_3O_4 nanoparticles on the physiology and fruit quality of muskmelon (*Cucumis melo*) plants. Environ Pollut 249:1011–1018
60. Wani SH, Kumar V (2015) Plant stress tolerance: engineering ABA: a potent phytohormone. Transcriptomics 3(2):1000113
61. Wu Q, Shi J, Jiang X, Wu H (2021) Regulatory mechanism of copper oxide nanoparticles on uptake of different species of arsenic in rice. Nanomaterials 11(9):2228
62. Yadav S, Modi P, Dave A, Vijapura A, Patel D, Patel M (2020) Effect of abiotic stress on crops. Sustain Crop Prod 3(17):5–16
63. Zhang H, Zhu J, Gong Z, Zhu J-K (2022) Abiotic stress responses in plants. Nat Rev Genet 23(2):104–119
64. Zhang Q, Gong M, Xu X, Li H, Deng W (2022) Roles of auxin in the growth, development, and stress tolerance of horticultural plants. Cells 11(17):2761

Synergistic Impacts of Nanoparticles and PGPR Under Various Abiotic Stresses for Higher Performance of Agricultural Crops

Ana Aslam, Sana Aslam, Akasha Amjad, Emaan Amjad, and Allah Ditta

Abstract Globally, abiotic stresses such as salt, drought, and severe temperatures have a major impact on agricultural output and food security. Researchers have investigated the possibility of plant growth-promoting rhizobacteria (PGPR) and nanoparticles (NPs) as novel strategies to lessen these impacts. This chapter summarizes the synergistic impacts of NPs and PGPR on agricultural crops under different abiotic stresses. We go over how NPs and PGPR work together to improve plant production, growth, and stress tolerance. The chapter also discusses the possible advantages and difficulties of incorporating PGPR and NPs into farming methods. We also discuss possible uses of this synergistic technique in sustainable agriculture as well as future research possibilities.

Keywords Nanoparticles · PGPR · Abiotic stresses · Agricultural crops · Synergistic impacts · Sustainable agriculture

A. Aslam
Soil Chemistry Section, Institute of Soil Chemistry and Environmental Sciences, Ayub Agricultural Research Institute, Faisalabad, Pakistan

S. Aslam
Department of Botany, University of Agriculture, Faisalabad, Pakistan

A. Amjad · E. Amjad
Institute of Soil and Environmental Sciences, University of Agriculture, Faisalabad, Pakistan

A. Ditta (✉)
Department of Environmental Sciences, Shaheed BB, University Sheringel, Khyber Pakhtunkhwa, Pakistan
e-mail: allah.ditta@sbbu.edu.pk

School of Biological Sciences, The University of Western Australia, Perth, WA, Australia

A. Husen (ed.), *Nanobiotechnology for Abiotic Stress Adaptation and Mitigation in Agricultural Crops*, Smart Nanomaterials Technology,
https://doi.org/10.1007/978-981-96-9298-9_10

1 Introduction

Climate change is increasingly threatening global food security. The projected impacts of climate variability on crops have become a serious concern in meeting the food demand of the growing human population. The profuse use of agrochemicals in agriculture to address the growing food demand leads to environmental pollution, soil degradation, and deleterious health effects on humans and the ecosystem. This led to the inefficiency of traditional agricultural systems. Hence, there is a need for a sustainable approach to meet the demand for food and reduce health and pollution hazards. Nanotechnology can enhance the quality of agriculture, improve food productivity, and reduce the harmful effects of agriculture toxins on human health and the environment [1–5]. Nanostructured materials in association with agricultural inputs (nanopesticides and nanofertilizers) can be used to increase the targeted delivery and release to plants at low doses per application and decrease the byproducts which could persist in soil and disturb the ecosystem [6–8]. Furthermore, plant growth-promoting rhizobacteria are eco-friendly and sustainable alternatives to synthetic agrochemicals for promoting healthier crops. They have attained vital importance for the capability to increase productivity, profitability, and sustainability when rural livelihood and food security are the main priority [9, 10]). They improve plant growth and health in various ways [11]. Integration of nanoparticles and PGPR into mainstream agriculture practices is necessary for reducing reliance on agrochemicals.

2 Nanoparticles

Nanoparticles are ultrafine particles that have a size range of 1–100 nm. These can be made by single elements such as silver or a mixture of elements composing oxides (ZnO, SiO_2, TiO_2) [12, 13]. Metal base nanoparticles possess at least one metallic element. Many of them have various characteristics in comparison to bulk metals because of their large surface-to-volume ratio, high catalytic activity, and unique electromagnetic behavior [14]. Nanoparticles can be organic such as hydrogels and carbon nanotubes, and inorganic like metallic nanoparticles [7, 15, 16]. Organic, i.e., mostly natural product NPs, inorganic, i.e., metal and metal oxide NPs, and combined, i.e., clay have been tested in the food industry [17]. Most of the nanoparticles are of synthetic origin, while a few may be of natural origin such as magnetite [7]. Physical, chemical, and biological are the synthetic methods [18] used for the preparation of nanoparticles.

Physical techniques include size reduction by laser sputtering, grinding, and evaporation–condensation. Chemical techniques include spray pyrolysis, precipitation, microemulsion, precipitation, reduction, and microwave combustion. Green synthesis using material drawn out from various plants and parts of plants and using microbes are the biological methods [19–22]. Sputtering [23], mechanical milling

[24], sonication [25], chemical etching [26], electron beam evaporation [27], lithography [28], ultraviolet radiation [29] and electro explosion [30], and laser ablation [31] are the methods used for the preparation of nanoparticles. The physical method used for the synthesis of NPs is the "Top-down approach". Chemical and biological processes used for the synthesis of NPs is the "Bottom-up approach" [32]. Top-down synthesis is a destructive approach, in which larger molecules decompose into smaller units, and then these smaller units transform into appropriate nanoparticles [7]. The bottom-up approach is in reverse, in which NPs are synthesized from simpler substances, hence this is called the building-up approach [7]. Nanoparticles can be used as nanopesticides and nanofertilizers [33, 34].

3 Plant Growth-Promoting Rhizobacteria

Improving food safety and agricultural productivity is crucial due to the growing population and rising living standards [35, 36]. To improve crop production and achieve food security, in recent years, plant growth-promoting rhizobacteria (PGPR) have been recognized as a component of the fresh green revolution [14]. Plant growth-promoting rhizobacteria are becoming eco-friendly and promising techniques to minimize the usage of synthetic agrochemicals to improve sustainable agricultural systems [37]. These are soil bacteria that colonize the roots upon inoculation to seed and enhance plant growth first defined by Kloepper and Schroth [38]. Plant growth-promoting rhizobacteria are associated with host plants autistically, where both get benefits [39, 40]. These are groups of bacteria present in the soil surrounding the plant roots called the rhizosphere [41]. Plants supply habitat for microbes due to root exudation of organic compounds which are essential for their metabolism [42]. Plant growth-promoting rhizobacteria belong to various genera such as *Bacillus, Azotobacter, Azospirillum, Pseudomonas, Arthrobacter, Serratia, Burkholderia, Rhodococcus, Pantoae, Azoarcus, Paenibacillus, Alcaligenes Lactobacillius, Beijerinck, Stenotrophomonas* [43]; *Acinetobacter, Agrobacterium, Rhizobium, Bradyrhizobium, Frankia, Thiobacillus, Bacillus* [44]; *Staphylococcus, Gluconacetobacter, Methylobacterium, Achromobacter, Exiguobacterium, Flavobacterium, Herbaspirillum* [45]. Plant growth-promoting rhizobacteria (rhizosphere-colonizing bacteria) can improve the growth of plants in many direct and indirect mechanisms [46]. These bacteria improve plant growth directly by producing growth-enhancing hormones, biological nitrogen fixation, solubilizing minerals, and producing siderophores [47]. Plant growth is also improved indirectly by induced systemic resistance, and by decreasing the harmful effect of phytopathogenic organisms by generating antagonist materials such as antibiotics, bacteriocins, and siderophores [45]. Figure 1 shows direct and indirect mechanisms of growth promotion by plant growth-promoting rhizobacteria.

The bacteria that reside in the rhizospheric zone are extremely useful for plants as they stimulate plant growth by producing phytohormones such as auxin, gibberellins, cytokinin, 1-amonocyclopropaane-1-carboxylase deaminase, many lytic enzymes,

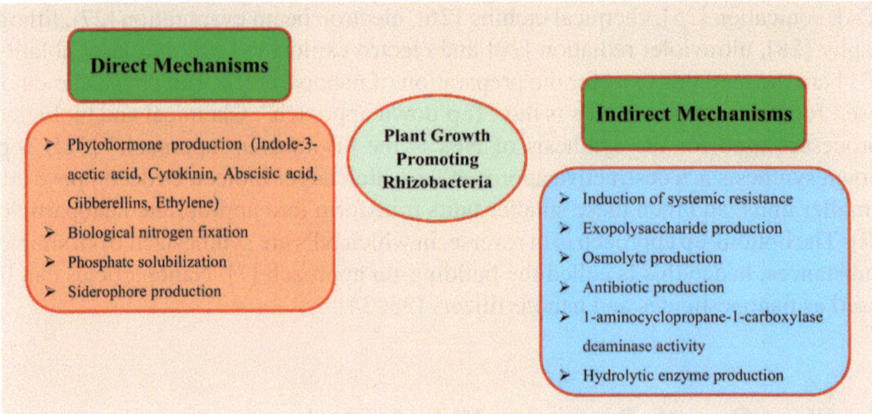

Fig. 1 Direct and indirect mechanisms of growth promotion by plant growth-promoting rhizobacteria

ammonia, and hydrogen cyanide [45] producing volatile compounds [48]. These bacteria exhibit synergistic or antagonistic interaction with rhizosphere microbes and soil which improve or increase the growth rate of plants [48, 49]. These bacteria can affect the growth of plants with one or more mechanisms, and they also describe exceptional characteristics to enhance the growth of plants at different growth stages during their life cycle [50]. Many PGPR are used either by biofertilizers or biocontrol agents commercially, nowadays [16, 51, 52]. Many PGPR enhance plant growth and assist plants in mitigating different biotic and abiotic stresses through the production of phytohormones such as cytokinin, auxin, and many volatile organic compounds [53].

4 Abiotic Stress

Abiotic stresses (salinity, high temperature, drought, and heavy metals) are the most serious harmful factors for modern agriculture, severely affecting crop productivity and sustainable agriculture throughout the world [54–57]. These stresses have affected crop growth and productivity either in single or combined form [58]. Approximately 50% of plant productivity is affected by abiotic stress [59]. Plant growth is compromised under abiotic stress mainly due to impaired biochemical and physiological processes, which cause nutrient imbalance, photosynthesis, respiration disturbance, and cellular organelle dysfunction [33, 60–69]. Drought stress drastically affects the yield of crops, significantly decreasing the productivity of food crops such as rice, wheat, maize, pea and faba bean [70–73]. Drought stress decreases plant height, leaf area, and plant development. Drought severity, duration, plant species, age, and growth stages are the key factors that determine the plant's response to drought [74]. About 21 and 40% of wheat and maize yield reduction have been reported due to drought [75]. Yield losses under drought vary between crop species and typically range from 30 to 90% in the field [76]. Salt stress is a major problem in all arid and semiarid regions. It leads to a 50% reduction in crop standard yield [77]. Salinity stress considerably reduced rice yield with an annual estimation of 30–50% [78]. Salinity stress adversely affects cell development, systolic metabolism, and membrane function, and enhances the production of reactive oxygen species [58, 79–83]. It causes ion toxicity and osmotic stress by escalating Na^+ assimilation and reducing the Na^+/K^+ ratio due to decreased osmotic potential in the roots of the plant [84]. It has been expected that a 20% yield loss of crops has occurred due to salinity stress [85]. Salinity stress adversely affects seed germination, development, and vigor of seedling, flowering, and fruit setting [86]. Heavy metal stress caused reduced seed germination, yellowing of leaves, chlorosis, and necrosis, shortened roots, and slowed growth [87]. Lead, cadmium, arsenic, zinc, copper, chromium, and nickel are the major heavy metals [88]. Over 20 million hectares of land are influenced by heavy metals including lead, mercury, cadmium, arsenic, nickel, zinc, copper, and chromium [87, 89]. Heavy metal-induced denaturation and inactivation of intracellular enzymes, modification in protein structures, and disruption of membrane integrity ultimately affect metabolic processes (respiration, photosynthesis, and transpiration) of plants [90]. Temperature stress is a serious issue affecting the distribution of plant and crop yield [91]. These stresses hampered the physiological, biochemical, and molecular processes in plants that ultimately influence the growth and development of plants [40]. Extraordinarily low temperature also has a considerable impact on the growth of plants through morphological, anatomical, and physiological modifications which modify the cell division, root architecture, water transport, nutrient uptake, phytohormonal signaling, and photosynthesis [92]. Increased heat stress is generally caused because of high average temperatures [93]. Heat shocks, due to increased atmospheric temperature, are becoming a vital limiting factor to the productivity of crops worldwide (Fahad et al. 2017). Plant growth and development are affected by heat stress, either heat stress chronic or abrupt [94]. It has been estimated that wheat production will decrease by 6% due to every degree (Celsius) increase in temperature [95]. Heat stress decreases photosynthetic efficiency, shortens the life cycle, and reduces crop productivity [58, 96]. Abnormal temperature and suboptimal water supply negatively affect the growth and yield of crops due to biochemical changes, physiological disruptions, and physical damage [97]. Heat stress exposure to crops at grain filling and reproductive stages disturbs the source-sink activities and creates unwanted biochemical changes in the seed [98]. Impacts of different abiotic stresses on plants are summarized in Table 1.

5 Nanoparticles and PGPR Under Drought Stress

Drought is one of the factors that limits agriculture growth and productivity throughout the world [111, 112]. It stopped the early development and production of crops due to poor water absorption and impeded the performance of enzymes

Table 1 Impacts of abiotic stress on plants

Abiotic stresses	Impacts on plants	Key references
Drought	Impaired physiological, biochemical, molecular, ecological, and morphological traits and processes	[99]
	Decrease cell division and enlargement, differentiation of root, shoot length, alteration of stomatal movement, mineral nutrition, decrease yield, and water use efficiency	[100]
	Reduction in leaf area, shoot length, plant height, leaf rolling, stomata closure, decrease in photosynthesis and transpiration rate, decrease in chlorophyll content, leaf water potential, production of reactive oxygen species, oxidative damage, accumulation of abscisic acid and proline	[101]
Salinity	Disrupt water balance, water uptake, gaseous exchange, photosynthesis, transpiration, ionic homeostasis, growth pattern, morphology, and anatomy of the plant	[102]
	Inhibit plant growth by osmotic stress, ionic imbalance, oxidative stress, and decreased nutrient uptake	[103]
	Delay seedling germination, lower seedling growth, and disperse seedling metabolism	[104]
Heavy metals	Chlorosis, decreased photosynthetic efficiency, disturbance in enzymatic function and nutrient uptake, stunted growth	[105]
	Overproduction of reactive oxygen species leading to oxidative stress	[106]
	Accumulation of metal ions, disturbance of cellular ionic homeostasis	[107]
Heat	Membrane damage, protein denaturation, decrease in biomass, and inactivation of enzyme	[108, 109]
	Damage to the photosynthetic machinery such as denaturation of the electron transport chain and chloroplast protein decreases photosynthesis	[96]
	Decrease seed germination, seedling growth, water use efficiency, cell turgidity, disturb cellular functions, leaf senescence, deactivation of photosynthetic enzyme, reduce grain number, size, oxidative damage to chloroplast	[110]

[113]. It negatively influences all aspects of the plant-water relationship such as water potential, leaf relative water content, osmotic potential, rate of photosynthesis, respiration and transpiration, and nutrient uptake [114]. For enhancing drought resilience in crops, an eco-friendly and suitable approach is needed. Nanotechnology is a promising approach due to the unique properties of nanomaterials. Nanoparticles can be used in transporting water and nutrients more efficiently to the roots of plants, and can reduce water loss by forming protective coatings on surfaces of leaves and by improving photosynthesis under stressed conditions [115, 116]. Moreover, leveraging microbes in agriculture can promote sustainable farming practices and increase crop resilience to water-deficit conditions. The integration of microbes and

nanotechnology holds the potential to improve drought resilience in crops synergistically [117]. Nanoparticles can improve the water-holding capacity of soil due to their higher surface area-to-volume ratio. In comparison to the sole application of NPs and PGPR, the co-application of NPs and PGPR increases the water-holding capacity of the soil [118]. In nanoparticle-amended soil, a large surface area-to-volume ratio of NPs increases water-holding capacity under drought stress [119]. Different studies have reported synergistic impact via the co-application of nanoparticles and bacteria under drought stress (Table 2).

The combined application of bacterial strain and nano-Si induced considerable tolerance to water-deficit stress in wheat and canola. Application of 100 mg nano-Si kg^{-1} and *Pseudomonas sp.* 19 performed better in canola and 200 mg nano-Si kg^{-1} and *Pseudomonas sp.*19 in wheat. The combined application of bacteria and Si considerably reduces the damaging effect of drought on the nutrient content (N, P, K, Si) of canola and wheat. A significant increase in the activity of superoxide dismutase glutathione peroxidase and ascorbate peroxidase and a decrease in MDA content also reinforced the importance of bacterial strain and PGPB in water-deficit stress. Nano-Si improved the morphophysiological and nutritional responses of wheat and canola under water stress [121]. Nano-silicon may increase water uptake and its transport in plants. Small size and large surface area-to-weight ratio are the distinctive properties that can increase its solubility and transport in plants [123]. Simultaneous application of graphene oxide (GO) nanosheets with *Rhizobium sp.* proved effective in coping with the drought-induced stress on maize by providing antioxidant and osmotic protection through GO and mitigating the effect of GO on plant biochemistry through *Rhizobium* sp. E20-8. Graphene oxide nanosheet and *Rhizobium Sp.* E20-*8*-based biostimulant formulation not only improved the performance of the plant and provided a high degree of protection to shoot but also increased crop production under drought stress [120]. Combining silicon dioxide nanoparticles (SiO_2 NPs) along with PGPB (*Pseudomonas fluorescens* p-187, *Pseudomonas putida* p-168) effectively increases drought tolerance in wheat by increasing N, P, K, and Si content in the

Table 2 Co application of nanoparticles and bacteria under drought stress

Stress	Nanoparticles	Bacteria	Plant	Key references
Drought	Graphene oxide (GO) nanosheets 4 mg mL^{-1}, soil (Sand) application	*Rhizobium sp.* E20-8	*Zea mays*	[120]
Water deficit	Nano-Silicon (100 mg nano-Si kg^{-1}, 200 mg nano-Si kg^{-1})	PGPB (Pseudomonas sp. 19, *Bacillus sp.* 76, and consortium of 19 and 76 strains)	Canola and wheat	[121]
Water deficit	Silicon dioxide nanoparticle (SiO_2 NP) soil application of 100 and 200 $mgkg^{-1}$, and foliar application of 200 $mgkg^{-1}$	*Pseudomonas fluorescens* p-187 *Pseudomonas putida* p-168	Wheat (*Triticum aestivum* L.)	[122]

shoot, decreasing malondialdehyde content, and boosting activities of glutathione peroxidase and superoxide dismutase in wheat [122].

6 Nanoparticles and PGPR Under Salt Stress

Salinity stress dramatically impairs global crop production. Salinity stress enhances the intracellular osmotic pressure and accumulates sodium to a toxic level [124]. It considerably decreases the surface area of the root by lowering root hair length and root hair density which is directly related to the uptake of nutrients [125]. It affects plant development including germination, vegetative growth, and reproductive development. It induces osmotic stress, ion toxicity, oxidative stress, and nutrient deficiency and limits the uptake of water from the soil [126]. Plant growth-promoting rhizobacteria and nanomaterials are used at present to alleviate the deleterious impacts of stresses on the growth and development of plants [127]. Combined use of zinc oxide nanoparticles and plant growth-promoting bacteria (*Bacillus subtilis, Bacillus pumilus,* and *Lactobacillus casei*) decreased the deleterious impact of salinity, reduced molecular disorders, and improved the morphological traits of tomato. Taking together, ZnO-NPs and bacteria had a positive impact on tomato seedlings under salt stress by decreasing genetic impairment (Table 3). Co-application of PGPB and ZnO-NPs plays a protective role against salt stress on DNA hypermethylation and DNA damage [128]. Use of copper (ii) oxide nanoparticles (CuO-NPs) with PGPB (*Bacillus subtilis, Bacillus pumilus, Lactobacillus casei*) improved morphological traits, increased genomic template stability, and reduced genetic impairment relieve the adverse effect of salinity and in wheat. Co-application of PGPB and CuO-NPs provides a strong anti-genotoxic effect against salt stress in wheat [128]. Higher colonization efficiency was observed when NPs and PGPR were applied together compared to the sole application of NP and PGPR [98]. Combined application of plant growth-promoting rhizobacteria (*Pseudomonas koreensis* MG209738 and *Bacillus coagulans* NCAIM B.01123) and silica nanoparticles (Si-NP) not only lessen the harmful effect of co-stressors (irrigation with saline water or soil salinity) but also improve morpho-physiological characteristics and productivity of sugarbeet. Coupled use of Si-NP and PGPR lessens the oxidative stress indicators such as lipid peroxidation, hydrogen peroxide, and sodium ion and enhances the activities of catalase, superoxide dismutase peroxidase, and potassium ion [129].

7 Nanoparticles and PGPR Under Heavy Metal Stress

Nanoparticles have a high affinity for heavy metals and arsenic due to great surface area and active binding sites and can be used for securing crops from contamination [111, 130–132]. Combined application of graphene nanoplatelets (GNPs) and

Table 3 Co-application of nanoparticles and bacteria under salinity stress

Stress	Nanoparticles	Bacteria	Plant	Key references
Salinity	Silica nanoparticles (Si-NP), 12.5 mg L^{-1}, Foliar spray	*Pseudomonas koreensis* MG209738 and *Bacillus coagulans* NCAIM B.01123	Sugar beet (*Beta vulgaris L.*)	[129]
Salinity	Zinc oxide nanoparticles (ZnO-NPs) (0, 20, and 40 mg L^{-1}), Soil application	*Bacillus subtilis, Bacillus pumilus, Lactobacillus casei*	Tomato *Solanum lycopersicum* L. Linda'	[128]
Salinity-Sodicity	Nano-gypsum (NG) soil application, Nano-calcium oxide (NCaO) 50 mg NCaO L^{-1} Nano-zinc oxide (NZnO) 50 mg NZnO L^{-1}, and nano-silica ($NSiO_2$) 300 mg $NSiO_2$ L^{-1} Foliar application	Plant growth-promoting bacteria	Wheat (*Triticum aestivum* L.)	[22]
Salinity	Copper (II) oxide nanoparticles (CuO-NPs) (0, 50, and 100 mg L^{-1}); Soil Application	*Bacillus subtilis, Bacillus pumilis, Lactobacillus casei*	Wheat (*Triticum aestivum* L.)	[128]

Cd tolerant *Serratia sp.* promotes the biomass of ryegrass and accumulation of Cd in plants. The PGPB-GNP combination decreased the Cd-induced stress injury by enhancing antioxidant production. Graphene nanoplatelets increased plant growth-promoting bacterial survival in the rhizosphere. These bacteria enhanced the enzyme activity such as urease, catalase, dehydrogenase, and phosphatase in soil. Increased activities of urease and phosphatase facilitate the uptake of nitrogen and phosphorus by reviving their transformation in soil. Inoculation of bacteria in combination with NPs decreased the H_2O_2 and MDA content and enhanced the antioxidant enzymes. Adding GNPs to soil together with bacteria increases Cd uptake due to increased levels of exchangeable Cd [133]. The GNPs are expected to be eco-friendly, cost-effective, and very effective absorbent for metal ions [134]. Combined application of cerium dioxide (CeO_2) nanoparticles with *Staphylococcus aureus* increases the performance of sunflowers under Cr stress by prompting the activation of the enzymatic defense system, and improving the metabolic process [135]. Co-application of plant growth-promoting rhizobacteria (*Pseudomonas fluorescens*) and titanium dioxide nanoparticles (TiO_2 NPs) enhanced plant growth of *Trifolium repens* that may be due to increased symbiosis between plant root rhizospheric microbes reduced toxicity of Cd in the presence of bacteria Additionally, co-application of *Pseudomonas fluorescens* and TiO_2 NPs enhanced the chlorophyll content, uptake, and accumulation capacity of Cd in plant [136]. Integrated application of chromium-resistant bacterium (*Staphylococcus aureus*) and iron nanoparticles (Fe NPs) proved a good approach for increasing plant ability to combat metal-induced toxicity and reduce chromium accumulation and toxicity in plants from contaminated soil (Table 4). The use of Fe NPs considerably increased the growth, biomass, photosynthetic activity, and yield by increasing chlorophyll content and lessening the oxidative damage. Application of Fe NPs also decreases the acquisition and accumulation of Cr in plants by enhancing micronutrient bioavailability to plants. Moreover, Fe NPs increased enzymatic and non-enzymatic activity and decreased oxidative damage. The positive effect of Fe NPs is further enhanced by the inoculation of rice plants with Cr-resistant bacteria S. aureus by converting the toxic form of Cr (Cr^{6+}) into a less toxic form (Cr^{3+}). Inoculation of bacteria also reduces Cr uptake by adsorption of Cr ion which decreases the bioavailability of Cr [137]. Co-application of iron oxide nanoparticles (IONPs) and *Bacillus subtilis* S4 increased the activity of superoxide dismutase and peroxidase and decreased the level of malondialdehyde, hydrogen peroxide, and electrolyte leakage under arsenic stress. The synergistic association between bacteria and IONPs increases the stomatal conductance photosynthetic rate, transportation rate, and synthesis of stress-combating polyamines such as putrescine and spermidine in arsenic-stressed seedlings [130]. Application of *Providencia vermicola*, zinc oxide nanoparticles (ZnO-NPs) and oxalic acid (OA) reduce the arsenic stress on *Luffa acutangula* by improving biomass, phytohormones, antioxidants, nutrients, and proline content. Combined application of *P. vermicola*, ZnO-NPs, and OA increased growth and improved defense mechanism against As stress. Furthermore, improved soil phosphatase and invertase exhibit effective amelioration of As [138]. Combined application of zinc oxide nanoparticles (ZnO-NPs) with plant growth-promoting rhizobacteria (PGPR) *Klebsiella sp.* SBP-8 alleviates the Cr stress in *Brassica juncea* (L.) and escalates the antioxidant enzyme (catalase, superoxide dismutase, and ascorbate peroxidase activities [139]).

8 Nanoparticles and PGPR Under Temperature Stress

Crop productivity is highly affected by high and low temperatures. Temperature extremes such as high temperature, freezing, and chilling can lead to metabolic and physiological disorders in plants by increasing chlorophyll decomposition and influencing photosynthetic pigment synthesis [141]. To combat the impacts of extreme weather on plants, several measures have been adopted to enhance the adaptation of plants to temperature [142]. Therefore, a new approach is needed to cope with these stresses and attain sustainability. A combination of NPs and PGPR has gained the expectation to attain sustainable productivity under temperature stress. Certain

Table 4 Co-application of nanoparticles and bacteria under heavy metals stress

Heavy metal stress	Nanoparticles	Bacteria	Plant	Key reference
Cadmium	Graphene nanoplatelets (GNPs), and graphitic carbon nitride (g-C_3N_4) 100 mg kg^{-1}, Soil application	*Serratia bozhouensis* and *Serratia marcescens*	Ryegrass (*Lolium perenne*)	[133]
Cadmium	Titanium dioxide TiO_2-NPs (0, 100, 250, 500 and 1000 mg kg^{-1}), soil application	*Pseudomonas fluorescens*	White clover (*Trifolium repens* L.)	[136]
Chromium	Cerium dioxide nanoparticles CeO2-NPs (0, 25, and 50 mg L^{-1}), Foliar spray	*Staphylococcus aureus*	Sunflower (*Helianthus annuus* L.)	[135]
Chromium	Fe nanoparticles, Fe NPs (0, 10, and 20 mg L^{-1}), Foliar spray	*Staphylococcus aureus*	Rice *Oryza sativa* L	[137]
Cadmium	Zinc oxide nanoparticles 20 mg kg^{-1}, Soil application	*Bacillus fortis* IAGS 223	*Cucumis melo*	[140]
Chromium	Zinc oxide nanoparticles (ZnO-NPs) 60 ppm, Foliar spray	*Klebsiella sp.* SBP-8	*Brassica juncea*	[139]
Arsenic	Iron oxide nanoparticles (IONPs) (0, 5, 10, 15, and 20 mg L^{-1}), seed priming	*Bacillus subtilis* S4	*Cucurbita moschata*	[130]
Arsenic enriched soil	Zinc oxide nanoparticles ZnO-NPs (10 ppm), soil application	*Providencia vermicola*	Ridge gourd *Luffa acutangula*	[138]

nanoparticles have been reported that enhance the ability of plants to combat temperature extremes as stress mitigators. These NPs may expedite an increase in photosynthetic efficiency, enhance water and nutrient uptake, and strengthen the antioxidant defenses, thus aiding the plant in combating temperature-induced challenges [40]. The application of NP assists the plant in boosting its defense system against high temperatures [143]. Plant growth-promoting rhizobacteria enhance plant growth under cold stress in the temperature range 5–14 °C and under heat stress 40–50 °C. Plant growth-promoting rhizobacteria increases the capability of plants to combat temperature stress by promoting plant growth and development and reducing the temperature stress-induced damage to plants [91].

9 Mechanisms Responsible for Synergistic Effects of Nanoparticles and PGPR Under Various Abiotic Stress

In many studies, the co-application of NPs and PGPR has been used to improve agronomic productivity and soil quality in various scenarios [144]. The clear conjecture from these studies is that NPs would. It has been assumed that NPs would increase nutrient availability and create a favorable habitat for the establishment of PGPR. In response, PGPR would perform their specific tasks such as nutrient solubilization and plant hormone production at the highest levels [43]. Combined applications of various PGPR and NPs have proposed fruitful ways to enhance soil quality. Nanoparticles provide the substrate with a high surface area to PGPR, enhanced nutrition for survival of PGPR, hence their presence may increase the efficiency of PGPR [129, 145]. In comparison to a single application of PGPR and NPs, co-application of NPs and PGPR usually increases the quantity of mineral nutrients in rhizospheric soil [146]. ZnO and TiO_2 NP are effective ways of delivering Ti and Zn to plants. Zn is necessary for the synthesis of Auxin and IAA regulation [147]. Zinc also facilitates the oxidative stress reduction induced by reactive oxygen species, resulting in high activity of antioxidants including ascorbates, peroxidases, superoxide dismutase, and catalase [148]. Silicon dioxide nanoparticles have distinct characteristics due to their high surface area and small size [149]. These nanoparticles increase the uptake of nutrients and growth of plants by improving root development stress tolerance mechanism and photosynthesis [150, 151]. One key benefit of silicon dioxide nanoparticles is their capability to increase nutrient uptake, interact with soil nutrients, and make complexes that are easily taken by the roots of plants. Moreover, silicon dioxide nanoparticles increase the activity of plant growth-promoting rhizobacteria which further increases the availability of nutrients to plants [122]. These bacteria (PGPR) play a key role in facilitating the plant against adverse conditions and increasing plant development. The bacteria increase plant growth in various ways [27]. These bacteria colonize the root zone and increase the availability of nutrients to plants, solubilize phosphorus, fix nitrogen, and make it available for absorption of the root. These bacteria also release phytohormone and siderophores which increase the availability of nutrients and plant growth for the plants [139]. These bacteria also produce volatile signaling molecules which increase immunity against the pathogen, enhance actions of plant enzymes, indulge in the uptake of nutrients, and enhance other metabolic processes and stress tolerance [152]. Plant growth-promoting rhizobacteria promotes the mobility and bioavailability of heavy metals in the soil. Consequently, increases in their access to plants promote and affect the phytoremediation process [153].

10 Conclusions and Future Prospects

Abiotic stresses such as drought, salinity, low and high temperatures, and heavy metals impair physiological and biochemical processes which in turn affect crop productivity. The combined application of nanoparticles and plant growth-promoting rhizobacteria significantly influence plant growth, development, and responses under different abiotic stresses. These findings also reveal that the co-application of nanoparticles and PGPR is an effective strategy for improving plant growth and stress mitigation for combating the negative effect of stresses on crops. Future studies are required to investigate the long-term effect of nanoparticles and plant growth-promoting rhizobacteria on plant growth, development, and soil health under different abiotic stresses (drought, salinity, temperature, and heavy metals). Studies are also needed to reveal the specific mechanisms involved in stress tolerance by co-applied nanoparticles and PGPR. For their practical application in agriculture, work should focus on the standardization of nanoparticles in crops to avoid the contamination of nanoparticles and to sustain the ecosystem.

References

1. Bahrulolum H, Nooraei S, Javanshir N, Tarrahimofrad H, Mirbagheri VS, Easton AJ, Ahmadian G (2021) Green synthesis of metal nanoparticles using microorganisms and their application in the agri-food sector. J Nanobiotechnol 19:1–26
2. Ditta A (2012) How helpful is nanotechnology in agriculture? Adv Nat Sci: Nano Sci Nanotechnol 3:033002
3. Husen A (2024a) Plant response to gold nanoparticles. Springer Nature Singapore Pte Ltd., 152 Beach Road, #21-01/04 Gateway East, Singapore 189721, Singapore. https://doi.org/10.1007/978-981-97-7174-5
4. Husen A (2025a) Plant response to silver nanoparticles. Springer Nature Singapore Pte Ltd., 152 Beach Road, #21-01/04 Gateway East, Singapore 189721, Singapore. https://doi.org/10.1007/978-981-97-7352-7
5. Husen A (2025b) Emerging carbon nanomaterials for sustainable agricultural practices (synthesis, plant growth, performance, production and protection). Springer Nature Singapore Pte Ltd., 152 Beach Road, #21-01/04 Gateway East, Singapore 189721, Singapore
6. Faheem M, Ditta A, Du J (2024) Nanomaterials in industrial chemistry, 1st edn. CRC Press, Taylor & Francis Group, Boca Raton, Florida 33487, USA
7. Khan I, Saeed K, Khan I (2019) Nanoparticles: properties, applications and toxicities. Arab J Chem 12(7):908–931
8. Varma A, Uma, Khanuja M (2017) Role of nanoparticles on plant growth with special emphasis on Piriformospora indica: a review. Nanosci Plant–Soil Syst 387–403
9. Jeyanthi V, Kanimozhi S (2018) Plant growth promoting rhizobacteria (PGPR)-perspective and mechanisms: a review. J Pure Appl Microbiol 12(2):733–749
10. Kanwal S, Naeem HK, Batool F, Mirza A, Abdelrahman EA, Sharif G, Maqsood F, Mustaqeem M, Ditta A (2023) Adsorption potential of orange rind-based nanosorbents for the removal of cadmium (II) and chromium (VI) from contaminated water. Environ Sci Pollut Res 30:110658 110673
11. Ehinmitan E, Losenge T, Mamati E, Ngumi V, Juma P, Siamalube B (2024) BioSolutions for green agriculture: unveiling the diverse roles of plant growth-promoting Rhizobacteria. Int J Microbiol 2024(1):6181491
12. Dinesh R, Anandaraj M, Srinivasan V, Hamza S (2012) Engineered nanoparticles in the soil and their potential implications to microbial activity. Geoderma 1(173):19–27
13. Ditta A, Arshad M (2016) Applications and perspectives of using nanomaterials for sustainable plant nutrition. Nanotechnol Rev 2(5):209–229
14. Gao C, Lyu F, Yin Y (2020) Encapsulated metal nanoparticles for catalysis. Chem Rev 121(2):834–881
15. Ditta A (2019) Role of nanoclay polymers in agriculture: applications and perspectives. In: Sharma SK (ed) Nanohybrids in environmental and biomedical applications. Taylor and Francis (CRC Press), USA, pp 323–334
16. Ditta A, Arshad M, Ibrahim M (2015) Nanoparticles in sustainable agricultural crop production: applications and perspectives. In: Siddiqui MH, Al-Whaibi MH, Mohammad F (eds) Nanotechnology and plant sciences—nanoparticles and their impact on plants. Springer, Switzerland, pp 55–75

17. He X, Jia HX, Deng Hua DH, Hwang HueyMin HH (2019) The current application of nanotechnology in food and agriculture. J Food Drug Anal 27(1):1–21
18. Ansari M, Shahzadi K, Ahmed S (2020) Nanotechnology: a breakthrough in agronomy. Nanoagronomy 1–21
19. Akhtar MS, Fiaz S, Aslam S, Chung S, Ditta A, Irshad MA, Al-Mohaimeed AM, Iqbal R, Al-onazi WA, Rizwan M, Nakashima Y (2024) Green synthesis of magnetite iron oxide nanoparticles using Azadirachta indica leaf extract loaded on reduced graphene oxide and degradation of methylene blue. Sci Rep 14:18172
20. Husen A (2023b) Secondary metabolites based green synthesis of nanomaterials and their applications. Springer Nature Singapore Pte Ltd., 152 Beach Road, #21-01/04 Gateway East, Singapore 189721, Singapore. https://doi.org/10.1007/978-981-99-0927-8
21. Husen A (2023c) Nanomaterials from agricultural and horticultural products. Springer Nature Singapore Pte Ltd., 152 Beach Road, #21-01/04 Gateway East, Singapore 189721, Singapore. https://link.springer.com/book/9789819934348
22. Saleh MA, Metwally SM, Ayman M, Abdelghany AM (2024) Synergistic approaches for wheat productivity in saline-sodic soils: nano-gypsum, bacillus inoculation, and nanomaterial foliar treatments. Biochar Compost Technol 1(1):36–55
23. Orozco-Montes V, Caillard A, Brault P, Chamorro-Coral W, Bigarre J, Sauldubois A, Andreazza P, Cuynet S, Baranton S, Coutanceau C (2021) Synthesis of platinum nanoparticles by plasma sputtering onto glycerol: effect of argon pressure on their physicochemical properties. J Phys Chem C 125(5):3169–3179
24. Shojaei M, Shokuhfar A, Zolriasatein A (2021) Synthesis and characterization of CuAlS2 nanoparticles by mechanical milling. Mater Today Commun 1(27):102243
25. Tang SY, Qiao R, Lin Y, Li Y, Zhao Q, Yuan D, Yun G, Guo J, Dickey MD, Huang TJ, Davis TP (2019) Functional liquid metal nanoparticles produced by liquid-based nebulization. Adv Mater Technol 4(2):1800420
26. Butterfield AG, Steimle BC, Schaak RE (2020) Retrosynthetic design of morphologically complex metal sulfide nanoparticles using sequential partial cation exchange and chemical etching. ACS Mater Lett 2(9):1106–1114
27. Nomoev AV, Bardakhanov SP (2012) Synthesis and structure of Ag–Si nanoparticles obtained by the electron-beam evaporation/condensation method. Tech Phys Lett 38:375–378
28. Fu X, Cai J, Zhang X, Li WD, Ge H, Hu Y (2018) Top-down fabrication of shape-controlled, monodisperse nanoparticles for biomedical applications. Adv Drug Deliv Rev 1(132):169–187
29. Henglein A (1999) Radiolytic preparation of ultrafine colloidal gold particles in aqueous solution: optical spectrum, controlled growth, and some chemical reactions. Langmuir 15(20):6738–6744
30. Lozhkomoev AS, Pervikov AV, Kazantsev SO, Sharipova AF, Rodkevich NG, Toropkov NE, Suliz KV, Svarovskaya NV, Kondranova AM, Lerner MI (2021) Synthesis of Fe/Fe_3O_4 core-shell nanoparticles by electrical explosion of the iron wire in an oxygen-containing atmosphere. J Nanopart Res 23:1–2
31. Menazea AA (2020) Femtosecond laser ablation-assisted synthesis of silver nanoparticles in organic and inorganic liquids medium and their antibacterial efficiency. Radiat Phys Chem 1(168):108616
32. Álvarez-Chimal R, Arenas-Alatorre JÁ (2023) Green synthesis of nanoparticles. Biol Approach
33. Krasensky J, Jonak C (2012) Drought, salt, and temperature stress-induced metabolic rearrangements and regulatory networks. J Exp Bot 63(4):1593–1608
34. Okey-Onyesolu CF, Hassanisaadi M, Bilal M, Barani M, Rahdar A, Iqbal J, Kyzas GZ (2021) Nanomaterials as nanofertilizers and nanopesticides: an overview. ChemistrySelect 6:8645–8663 [Internet]
35. Ditta A, Ullah N, Imtiaz M, Li X, Jan AU, Mehmood S, Rizwan MS, Rizwan M (2022) Zn biofortification in crops through Zn-solubilizing plant growth promoting rhizobacteria. In: Mahmood Q (ed) Sustainable plant nutrition under contaminated environments. Springer Nature Switzerland AG, pp 115–133

36. Roth JA, Galyon J (2024) Food security: the ultimate one-health challenge. One Health 28:100864
37. Agrawal R, Satlewal A, Varma A (2015) Characterization of plant growth-promoting rhizobacteria (PGPR): a perspective of conventional versus recent techniques. In: Heavy metal contamination of soils: monitoring and remediation, pp 471-485
38. Kloepper JW, Schroth MN (1978) Plant growth promoting rhizobacteria on radishes. In: Proceedings of the 4th international conference on plant pathogenic bacteria. Angers, France, pp 879–882
39. Ahmad M, Hussain A, Dar A, Luqman M, Ditta A, Iqbal Z, Ahmad HT, Nazli F, Soufan W, Almutairi K, Sabagh AEl (2023) Combating iron and zinc malnutrition through mineral biofortification in maize through plant growth promoting Bacillus and Paenibacillus species. Front Plant Sci 13:1094551
40. Sidhu AK, Sharma M, Bhickchand Agrawal S, Pradip Bhavsar P, Samota MK (2024) Nanomaterial strategies for enhancing plant resilience in the face of temperature stress. CABI Agric Biosci 5(1):60
41. Santoyo G, Urtis-Flores CA, Loeza-Lara PD, Orozco-Mosqueda MD, Glick BR (2021) Rhizosphere colonization determinants by plant growth-promoting rhizobacteria (PGPR). Biology 10(6):475
42. Canarini A, Kaiser C, Merchant A, Richter A, Wanek W (2019) Root exudation of primary metabolites: mechanisms and their roles in plant responses to environmental stimuli. Front Plant Sci 21(10):157
43. Vega-Celedón P, Bravo G, Velásquez A, Cid FP, Valenzuela M, Ramírez I, Vasconez IN, Alvarez I, Jorquera MA, Seeger M (2021) Microbial diversity of psychrotolerant bacteria isolated from wild flora of andes mountains and Patagonia of chile towards the selection of plant growth-promoting bacterial consortia to alleviate cold stress in plants. Microorganisms 9(3):538
44. Cordero AP, Vergara DE, Mendoza YA (2023) Production of 1-aminocyclopropane-1-carboxylic acid deaminase (ACC) by Burkholderia cepacia as an indicator of cadmium contamination. J Positive Psychol Wellbeing 7(1):942–949
45. Kour D, Rana KL, Yadav N, Yadav AN, Kumar A, Meena VS, Singh B, Chauhan VS, Dhaliwal HS, Saxena AK (2019) Rhizospheric microbiomes: biodiversity, mechanisms of plant growth promotion, and biotechnological applications for sustainable agriculture. Plant growth promoting rhizobacteria for agricultural sustainability: from theory to practices, pp 19–65
46. Hartman K, Tringe SG (2019) Interactions between plants and soil shape the root microbiome under abiotic stress. Biochem J 476(19):2705–2724
47. Nasir G, Batool F, Iqbal S, Akbar J, Noreen S, Munawar KS, Iqbal T, Ditta A (2024) Biosynthesis of lead oxide nanoparticles using mulberry leaf extract for adsorptive removal of Diazine black dye. Biomass Convers Biorefinery. https://doi.org/10.1007/s13399-024-06208-7
48. Kapoor D, Sharma P, Sharma MMM, Yadav S, Husen A (2024) Exploring soil microbiota and their role in plant growth, stress tolerance, disease control and nutrient immobilizer. Biocatal Agric Biotechnol 61:103358. https://doi.org/10.1016/j.bcab.2024.103358
49. Ashraf M, Athar HR, Harris PJ, Kwon TR (2008) Some prospective strategies for improving crop salt tolerance. Adv Agron 1(97):45–110
50. Glick BR, Patten CL, Holguin G, Penrose DM (1999) Biochemical and genetic mechanisms used by plant growth-promoting bacteria
51. Ditta A, Muhammad J, Imtiaz M, Mehmood S, Qian Z, Tu S (2018) Application of rock phosphate enriched composts increases nodulation, growth and yield of chickpea. Int J Recycl Org Waste Agric 7(1):33–40
52. Olanrewaju OS, Glick BR, Babalola OO (2017) Mechanisms of action of plant growth promoting bacteria. World J Microbiol Biotechnol 33:1–6
53. Orozco-Mosqueda MD, Santoyo G, Glick BR (2023) Recent advances in the bacterial phytohormone modulation of plant growth. Plants 12(3):606

54. Abbas G, Batool F, Ahmed A, Wagi S, Gondal HY, Maqsood F, Abdelrahman EA, Naeem HK, Kanwal S, Mustaqeem M, Ditta A (2024) Green synthesized silver nanoparticles: characterization, phytostimulatory impacts, and degradation potential for organic pollutants. Biocatal Agric Biotechnol 55:102993
55. Abdelrahman M, Burritt DJ, Tran LS (2018) The use of metabolomic quantitative trait locus mapping and osmotic adjustment traits for the improvement of crop yields under environmental stresses. In: Seminars in cell and developmental biology, vol 83. Academic Press, pp 86–94
56. Jha Y, Mohamed HI (2023) Enhancement of disease resistance, growth potential, and biochemical markers in maize plants by inoculation with plant growth-promoting bacteria under biotic stress. J Plant Pathol 105(3):729–748
57. Pandey P, Irulappan V, Bagavathiannan MV, Senthil-Kumar M (2017) Impact of combined abiotic and biotic stresses on plant growth and avenues for crop improvement by exploiting physio-morphological traits. Front Plant Sci 18(8):537
58. Wani SH, Kumar V (eds) (2020) Heat stress tolerance in plants: physiological, molecular and genetic perspectives. Wiley
59. Yadav S, Modi P, Dave A, Vijapura A, Patel D, Patel M (2020) Effect of abiotic stress on crops. Sustain Crop Prod 3(17):5–16
60. Husen A (2021a) Harsh environment and plant resilience, vol 11. Springer International Publishing AG, Gewerbestrasse, 6330 Cham, pp 1–23. https://doi.org/10.1007/978-3-030-65912-7_1
61. Husen A (2021b) Plant performance under environmental stress, vol 11. Springer International Publishing AG, Gewerbestrasse, 6330 Cham, pp 1–15. https://doi.org/10.1007/978-3-030-78521-5_1
62. Husen A (2022) Environmental pollution and medicinal plants. CRC Press 6000 Broken Sound Parkway NW, Suite 300, Boca Raton, FL, pp 33487–2742. https://doi.org/10.1201/9781003178866
63. Husen A (2023a) Plants and their interaction to environmental pollution (Damage detection, adaptation, tolerance, physiological and molecular responses). Elsevier Inc., 50 Hampshire St., 5th Floor, Cambridge, MA 02139, USA. https://doi.org/10.1016/C2020-0-03319-0
64. Husen A (2024b) Biostimulants in plant protection and performance. Elsevier Inc., 50 Hampshire St., 5th Floor, Cambridge, MA 02139, USA. https://doi.org/10.1016/C2022-0-02207-8
65. Husen A (2024c) Essential minerals in plant-soil system (Coordination, signalling, and interaction under adverse situations). Elsevier Inc., 50 Hampshire St., 5th Floor, Cambridge, MA 02139, USA. https://doi.org/10.1016/C2022-0-02437-5
66. Husen A (2025c) Agricultural crop improvement: plant and soil relationships. Taylor and Francis Group, LLC, 6000 Broken Sound Pkwy. NW. https://doi.org/10.1201/9781032630366
67. Husen A (2025d) Role of Osmolytes in changing environment. Elsevier Inc., 50 Hampshire St., 5th Floor, Cambridge, MA 02139, USA
68. Husen A (2025e) Role of antioxidants in mitigating plant stress. Elsevier Inc., 50 Hampshire St., 5th Floor, Cambridge, MA 02139, USA
69. Srivastava AK, Penna S, Nguyen DV, Tran LS (2016) Multifaceted roles of aquaporins as molecular conduits in plant responses to abiotic stresses. Crit Rev Biotechnol 36(3):389–398
70. Husen A, Iqbal M, Aref IM (2016) IAA-induced alteration in growth and photosynthesis of pea (*Pisum sativum* L.) plants grown under salt stress. J Environ Biol
71. Husen A, Iqbal M, Aref IM (2017) Plant growth and foliar characteristics of faba bean (*Vicia faba* L.) as affected by indole-acetic acid under water-sufficient and water-deficient conditions. J Environ Biol 38:179–186
72. Qiao M, Hong C, Jiao Y, Hou S, Gao H (2024) Impacts of drought on photosynthesis in major food crops and the related mechanisms of plant responses to drought. Plants 13(13):1808
73. Tuteja N, Gill SS (eds) (2013) Climate changes and plant abiotic stress tolerance. Wiley

74. Gray SB, Brady SM (2016) Plant developmental responses to climate change. Dev Biol 419(1):64–77
75. Daryanto S, Wang L, Jacinthe PA (2016) Global synthesis of drought effects on maize and wheat production. PLoS ONE 11(5):e0156362
76. Hussain S, Hussain S, Qadir T, Khaliq A, Ashraf U, Parveen A, Saqib M, Rafiq M (2019) Drought stress in plants: an overview on implications, tolerance mechanisms and agronomic mitigation strategies. Plant Sci Today 6(4):389–402
77. Ilyas M, Maqsood MF, Shahbaz M, Zulfiqar U, Ahmad K, Naz N, Ali MF, Ahmad M, Ali Q, Yong JW, Ali HM (2024) Alleviating salinity stress in canola (*Brassica napus L.*) through exogenous application of salicylic acid. BMC Plant Biology 24(1):611
78. Eynard A, Lal R, Wiebe K (2005) Crop response in salt-affected soils. J Sustain Agric 27(1):5
79. Embiale A, Hussein M, Husen A, Sahile S, Mohammed K (2016) Differential Sensitivity of *Pisum sativum* L. cultivars to water-deficit stress: changes in growth, water status, chlorophyll fluorescence and gas exchange attributes. J Agron 15:45–57
80. Getnet Z, Husen A, Fetene M, Yemata G (2015) Growth, water status, physiological, biochemical and yield response of stay green sorghum {*Sorghum bicolor* (L.) Moench} varieties-a field trial under drought-prone area in Amhara regional state, Ethiopia. J Agron 14:188–202. https://doi.org/10.3923/ja.2015.188.202
81. Hafeez MB, Ghaffar A, Zahra N, Ahmad N, Al-Qahtani WH, Sharma P, Husen A, Li J (2024) Exploring the efficacy of plant growth promoters in nutrient acquisition and antioxidant defense responses of late sown wheat under saline conditions. S Afr J Bot 171:812–824. https://doi.org/10.1016/j.sajb.2024.06.042
82. Zahra N, Kausar A, Abdelghani HTM, Singh S, Vashishth DS, Bachheti A, Bachheti RK, Husen A (2024b) Serotonin improves plant growth, foliar functions and antioxidant defence system in Ethiopian mustard (*Brassica carinata* A. Br.). South Afr J Bot 170:1–9 https://doi.org/10.1016/j.sajb.2024.05.002
83. Zahra N, Kausar A, Sharma P, Wabaidur SM, Husen A (2024a) Serotonin application to Guizotia abyssinica improves growth, physiological functions, antioxidant system and alleviates the deleterious effects induced by salt stress. Plant Biosyst—Int J Dealing Aspects Plant Biol 1–9. https://doi.org/10.1080/11263504.2024.2384972
84. El Sabagh A, Islam MS, Skalicky M, Ali Raza M, Singh K, Anwar Hossain M, Hossain A, Mahboob W, Iqbal MA, Ratnasekera D, Singhal RK (2021) Salinity stress in wheat (*Triticum aestivum L.*) in the changing climate: adaptation and management strategies. Front Agron 3:661932
85. Ashraf M, Harris PJC (2005) Abiotic stresses. Plant resistance through breeding and molecular approaches
86. Irik HA, Bikmaz G (2024) Effect of different salinity on seed germination, growth parameters and biochemical contents of pumpkin (*Cucurbita pepo L.*) seeds cultivars. Sci Rep 14(1):6929
87. Liu L, Li W, Song W, Guo M (2018) Remediation techniques for heavy metal-contaminated soils: principles and applicability. Sci Total Environ 15(633):206–219
88. Angon PB, Islam MS, Kc S, Das A, Anjum N, Poudel A, Suchi SA (2024) Sources, effects and present perspectives of heavy metals contamination: soil, plants, and human food chain. Heliyon
89. Gang M, Munir A, Murtaza G, Iqbal R, Majeed A, Shoaib M, Vyas S, Ahmed I, Akram MI, Ejaz Z, Ditta A, Alsakkaf WAA, Alkahtani J (2024) Agro-waste-based bio-nanocomposite for the mitigation of Pb (II) ions from water. Pol J Environ Stud 33(5):5635–5646
90. Liu S, Yang R, Pan Y, Ma M, Pan J, Zhao Y, Cheng Q, Wu M, Wang M, Zhang L (2015) Nitric oxide contributes to minerals absorption, proton pumps, and hormone equilibrium under cadmium excess in *Trifolium repens L.* plants. Ecotoxicol Environ Saf 119:35–46
91. Zhang X, Yang Z, Wang L, Yue Y, Wang L, Yang X (2023) The effects of plant growth-promoting rhizobacteria on plants under temperature stress: a meta-analysis. Rhizosphere 26:100788
92. Ahmad W, Coffman L, Weerasooriya AD, Crawford K, Khan AL (2024) Silicon regulates microbiome diversity and plant defenses during cold stress in Glycine max L. Front Plant Sci 10(14):1280251

93. Sun Q, Miao C, Hanel M, Borthwick AG, Duan Q, Ji D, Li H (2019) Global heat stress on health, wildfires, and crops under different levels of climate warming. Environ Int 1(128):125–136
94. Mishra S, Spaccarotella K, Gido J, Samanta I, Chowdhary G (2023) Effects of heat stress on plant-nutrient relations: an update on nutrient uptake, transport, and assimilation. Int J Mol Sci 24(21):15670
95. Zhao C, Liu B, Piao S, Wang X, Lobell DB, Huang Y, Huang M, Yao Y, Bassu S, Ciais P, Durand JL (2017) Temperature increase reduces global yields of major crops in four independent estimates. Proc Natl Acad Sci 114(35):9326–9331
96. Hu S, Ding Y, Zhu C (2020) Sensitivity and responses of chloroplasts to heat stress in plants. Front Plant Sci 2(11):375
97. Fahad S, Bajwa AA, Nazir U, Anjum SA, Farooq A, Zohaib A, Sadia S, Nasim W, Adkins S, Saud S, Ihsan MZ (2017) Crop production under drought and heat stress: plant responses and management options. Front Plant Sci 8:1147. https://doi.org/10.1186/s11671-021-03612-0
98. Kumar S, Bhushan B, Wakchaure GC, Dutta R, Jat BS, Meena KK, Rakshit S, Pathak H (2023) Unveiling the impact of heat stress on seed biochemical composition of major cereal crops: implications for crop resilience and nutritional value. Plant Stress 1(9):100183
99. Ortiz N, Armada E, Duque E, Roldán A, Azcón R (2015) Contribution of arbuscular mycorrhizal fungi and/or bacteria to enhancing plant drought tolerance under natural soil conditions: effectiveness of autochthonous or allochthonous strains. J Plant Physiol 1(174):87–96
100. Kumawat KC, Sharma B, Nagpal S, Kumar A, Tiwari S (2022) RM nair plant growth-promoting rhizobacteria: salt stress alleviators to improve crop productivity for sustainable agriculture development, p 13. https://doi.org/10.3389/fpls.2022:1101862
101. Seleiman MF, Al-Suhaibani N, Ali N, Akmal M, Alotaibi M, Refay Y, Dindaroglu T, Abdul-Wajid HH, Battaglia ML (2021) Drought stress impacts plants and different approaches to alleviate its adverse effects. Plants 10(2):259
102. Ismail AM, Horie T (2017) Genomics, physiology, and molecular breeding approaches for improving salt tolerance. Annu Rev Plant Biol 68(1):405–434
103. Atta K, Mondal S, Gorai S, Singh AP, Kumari A, Ghosh T, Roy A, Hembram S, Gaikwad DJ, Mondal S, Bhattacharya S (2023) Impacts of salinity stress on crop plants: improving salt tolerance through genetic and molecular dissection. Front Plant Sci 15(14):1241736
104. El Sabagh A, Hossain A, Barutçular C, Iqbal MA, Islam MS, Fahad S, Sytar O, Çiğ F, Meena RS, Erman M (2020) Consequences of salinity stress on the quality of crops and its mitigation strategies for sustainable crop production: an outlook of arid and semi-arid regions. Environ, Clim, Plant Vegetation Growth 503–33
105. Zulfiqar U, Ayub A, Hussain S, Waraich EA, El-Esawi MA, Ishfaq M, Ahmad M, Ali N, Maqsood MF (2022) Cadmium toxicity in plants: recent progress on morpho-physiological effects and remediation strategies. J Soil Sci Plant Nutr 22(1):212–269
106. Romero-Puertas MC, Terron-Camero LC, Pelaez-Vico MA, Olmedilla A, Sandalio LM (2019) Reactive oxygen and nitrogen species as key indicators of plant responses to Cd stress. Environ Exp Bot 1(161):107–119
107. Yadav SK (2010) Heavy metals toxicity in plants: an overview on the role of glutathione and phytochelatins in heavy metal stress tolerance of plants. S Afr J Bot 76(2):167–179
108. Heckathorn SA, Giri A, Mishra S, Bista D (2013) Heat stress and roots. In: Tuteja NN, Gill S (eds) Climate change and plant abiotic stress tolerance. Wiley Blackwell; Weinheim, Germany, pp 109–136
109. Thind S, Hussain I, Rasheed R, Ashraf MA, Perveen A, Ditta A, Hussain S, Khalil N, Ullah Z, Mahmood Q (2021) Alleviation of Cd stress by silicon nanoparticles during different phenological stages of Ujala wheat variety. Arab J Geosci 14:1028
110. Akter N, Rafiqul IM (2017) Heat stress effects and management in wheat. A review. Agron Sustain Dev 37:1–7
111. Teklić T, Parađiković N, Špoljarević M, Zeljković S, Lončarić Z, Lisjak M (2021) Linking abiotic stress, plant metabolites, biostimulants and functional food. Ann Appl Biol 178(2):169–191

112. Hosseinpour A, Ilhan E, Özkan G, Öztürk Hİ, Haliloglu K, Cinisli KT (2022) Plant growth-promoting bacteria (PGPBs) and copper (II) oxide (CuO) nanoparticle ameliorates DNA damage and DNA Methylation in wheat (*Triticum aestivum L.*) exposed to NaCl stress. J Plant Biochem Biotechnol 31(4):751–64
113. Lamaoui M, Jemo M, Datla R, Bekkaoui F (2018) Heat and drought stresses in crops and approaches for their mitigation. Front Chem 19(6):26
114. Franco-Navarro JD, Díaz-Rueda P, Rivero-Núñez CM, Brumós J, Rubio-Casal AE, de Cires A, Colmenero-Flores JM, Rosales MA (2021) Chloride nutrition improves drought resistance by enhancing water deficit avoidance and tolerance mechanisms. J Exp Bot 72(14):5246–5261
115. Ditta A, Ullah N (2022) An update on nanotechnology and sustainable agriculture. In: Ahmed W, Nourafkan E (eds) Science and applications of nanoparticles. Jenny Stanford Publishing Pte. Ltd., Singapore, pp 159–197
116. Ditta A, Mehmood S, Imtiaz M, Rizwan MS, Islam I (2020) Soil fertility and nutrient management with the help of nanotechnology. In: Husen A, Jawaid M (eds) Nanomaterials for agriculture and forestry applications, woodhead publishing Ltd. Elsevier UK (Invited book chapter) pp 273–288
117. Meel S, Saharan BS (2024) Enhancing crop resilience towards drought: by integrating nanotechnology, microbiomes, and growth-promoting rhizobacteria. Discover Agric 2(1):112
118. Nayana AR, Joseph BJ, Jose A, Radhakrishnan EK (2020) Nanotechnological advances with PGPR applications. Sustainable agriculture reviews 41: nanotechnology for plant growth and development, pp 163–80
119. El-Saadony MT, Saad AM, Soliman SM, Salem HM, Desoky ES, Babalghith AO, El-Tahan AM, Ibrahim OM, Ebrahim AA, Abd El-Mageed TA, Elrys AS (2022) Role of nanoparticles in enhancing crop tolerance to abiotic stress: a comprehensive review. Front Plant Sci 2(13):946717
120. Lopes T, Cruz C, Cardoso P, Pinto R, Marques PA, Figueira E (2021) A multifactorial approach to untangle graphene oxide (GO) nanosheets effects on plants: plant growth-promoting bacteria inoculation, bacterial survival, and drought. Nanomaterials 11(3):771
121. Valizadeh-Rad K, Motesharezadeh B, Alikhani HA, Jalali M, Etesami H, Javadzarin I (2023) Morphophysiological and nutritional responses of canola and wheat to water deficit stress by the application of plant growth-promoting bacteria, nano-silicon, and silicon. J Plant Growth Regul 42(6):3615–3631
122. Davoudi F, Jalali M, Valizadeh-Rad K, Etesami H (2024) Enhancing water deficit stress tolerance in wheat: synergistic effects of silicon nanoparticles and plant growth-promoting bacteria. SILICON 14:1–6
123. Salajegheh M, Yavarzadeh M, Payandeh A, Akbarian MM (2020) Effects of titanium and silicon nanoparticles on antioxidant enzymes activity and some biochemical properties of *Cuminum cyminum L.* under drought stress. Banat's J Biotechnol 11(21):19–25
124. Zhao S, Zhang Q, Liu M, Zhou H, Ma C, Wang P (2021) Regulation of plant responses to salt stress. Int J Mol Sci 22(9):4609
125. Arif MR, Islam MT, Robin AH (2019) Salinity stress alters root morphology and root hair traits in Brassica napus. Plants 8(7):192
126. Shrivastava P, Kumar R (2015) Soil salinity: a serious environmental issue and plant growth promoting bacteria as one of the tools for its alleviation. Saudi J Biol Sci 22(2):123–131
127. Ditta A, Arshad M, Zahir ZA, Jamil A (2015) Comparative efficacy of rock phosphate enriched organic fertilizer versus mineral phosphatic fertilizer for nodulation, growth and yield of lentil. Int J Agric Biol 17:589–595
128. Hosseinpour A, Haliloglu K, Tolga Cinisli K, Ozkan G, Ozturk HI, Pour-Aboughadareh A, Poczai P (2020) Application of zinc oxide nanoparticles and plant growth promoting bacteria reduces genetic impairment under salt stress in tomato (*Solanum lycopersicum L.* 'Linda'). Agriculture 10(11):521
129. Alharbi K, Hafez E, Omara AE, Awadalla A, Nehela Y (2022) Plant growth promoting rhizobacteria and silica nanoparticles stimulate sugar beet resilience to irrigation with saline water in salt-affected soils. Plants 11(22):3117

130. Mushtaq T, Shah AA, Akram W, Yasin NA (2020) Synergistic ameliorative effect of iron oxide nanoparticles and Bacillus subtilis S4 against arsenic toxicity in Cucurbita moschata: polyamines, antioxidants, and physiochemical studies. Int J Phytorem 22(13):1408–1419
131. Rai PK, Kumar V, Lee S, Raza N, Kim KH, Ok YS, Tsang DC (2018a) Nanoparticle-plant interaction: Implications in energy, environment, and agriculture. Environ Int 1(119):1–9
132. Rai PK, Lee J, Kailasa SK, Kwon EE, Tsang YF, Ok YS, Kim KH (2018b) A critical review of ferrate (VI)-based remediation of soil and groundwater. Environ Res 1(160):420–448
133. Farooqi A, ul Haq E, Johansen A, Ellegaard-Jensen L, Iqbal M, Yousaf S, Lackner M (2025) Plant growth-promoting bacteria and nanomaterials synergism to enhance Lolium perenne growth and phytoremediation in cadmium-contaminated soil. Chem Eng J Adv 21:100695
134. Lyu D, Backer R, Subramanian S, Smith DL (2020) Phytomicrobiome coordination signals hold potential for climate change-resilient agriculture. Front Plant Sci 25(11):634
135. Ma J, Alshaya H, Okla MK, Alwasel YA, Chen F, Adrees M, Hussain A, Hameed S, Shahid MJ (2022) Application of cerium dioxide nanoparticles and chromium-resistant bacteria reduced chromium toxicity in sunflower plants. Front Plant Sci 4(13):876119
136. Zand AD, Mikaeili Tabrizi A, Vaezi HA (2020) Application of titanium dioxide nanoparticles to promote phytoremediation of Cd-polluted soil: contribution of PGPR inoculation. Bioremediat J 24(2–3):171–189
137. Alharby HF, Ali S (2022) Combined role of Fe nanoparticles (Fe NPs) and Staphylococcus aureus L. in the alleviation of chromium stress in rice plants. Life 12(3):338
138. Tanveer Y, Yasmin H, Nosheen A, Ali S, Ahmad A (2022) Ameliorative effects of plant growth promoting bacteria, zinc oxide nanoparticles and oxalic acid on Luffa acutangula grown on arsenic enriched soil. Environ Pollut 1(300):118889
139. Shah T, Khan Z, Alahmadi TA, Imran A, Asad M, Khan SR, Ansari MJ (2024) Mycorrhizosphere bacteria inhibit chromium uptake and phytotoxicity by regulating proline metabolism, antioxidant defense system, and aquaporin gene expression in tomato. Environ Sci Pollut Res 31(17):24836-24850
140. Shah A, Nazari M, Antar M, Msimbira LA, Naamala J, Lyu D, Rabileh M, Zajonc J, Smith DL (2021) PGPR in agriculture: a sustainable approach to increasing climate change resilience. Front Sustain Food Syst 5:667546
141. Dhankher OP, Foyer CH (2018) Climate resilient crops for improving global food security and safety. Plant, Cell Environ 41(5):877–884
142. Baier M, Bittner A, Prescher A, van Buer J (2019) Preparing plants for improved cold tolerance by priming. Plant, Cell Environ 42(3):782–800
143. Khalid MF, Iqbal Khan R, Jawaid MZ, Shafqat W, Hussain S, Ahmed T, Rizwan M, Ercisli S, Pop OL, Alina MR (2022) Nanoparticles: the plant savior under abiotic stresses. Nanomaterials 12(21):3915
144. Caradonia F, Francia E, Morcia C, Ghizzoni R, Moulin L, Terzi V, Ronga D (2019) Arbuscular mycorrhizal fungi and plant growth promoting rhizobacteria avoid processing tomato leaf damage during chilling stress. Agronomy 9(6):299
145. Rajput VD, Singh A, Minkina TM, Verma KK, Singh AK (2023b) Nanotechnology for sustainable agriculture: an innovative and eco-friendly approach, 1st edn. Apple Academic Press, Burlington, Canada. https://doi.org/10.1201/9781003333128
146. Tang L, Shi Y, Zhang Y, Yang D, Guo C (2023) Effects of plant-growth-promoting rhizobacteria on soil bacterial community, soil physicochemical properties, and soil enzyme activities in the rhizosphere of alfalfa under field conditions. Diversity 15(4):537
147. Wang J, Yang S (2021) Dose-dependent responses of Arabidopsis thaliana to zinc are mediated by auxin homeostasis and transport. Environ Exp Bot 1(189):104554
148. Ahmed S, Ahmad M, Swami BL, Ikram S (2016) A review on plant extract mediated synthesis of silver nanoparticles for antimicrobial applications: a green expertise. J Adv Res 7(1):17–28
149. Al-Selwey WA, Alsadon AA, Ibrahim AA, Labis JP, Seleiman MF (2023) Effects of zinc oxide and silicon dioxide nanoparticles on physiological, yield, and water use efficiency traits of potato grown under water deficit. Plants 12(1):218

150. Goswami P, Mathur J, Srivastava N (2022) Silica nanoparticles as novel sustainable approach for plant growth and crop protection. Heliyon 8(7)
151. Mahawar L, Ramasamy KP, Suhel M, Prasad SM, Živčák M, Brestic M, Rastogi A, Skalický M (2023) Silicon nanoparticles: comprehensive review on biogenic synthesis and applications in agriculture. Environ Res 1(232):116292
152. de Andrade LA, Santos CH, Frezarin ET, Sales LR, Rigobelo EC (2023) Plant growth-promoting rhizobacteria for sustainable agricultural production. Microorganisms 11(4):1088
153. Liu A, Wang W, Zheng X, Chen X, Fu W, Wang G, Ji J, Jin C, Guan C (2022) Improvement of the Cd and Zn phytoremediation efficiency of rice (Oryza sativa) through the inoculation of a metal-resistant PGPR strain. Chemosphere 1(302):134900

Nanoparticle Interactions with Beneficial Microorganisms: Enhancing Abiotic Stress Adaptation and Mitigation in Agricultural Crops

Karuvelan Murugan, Hemamalini Palvannan, G. Hariharan, Udayadasan Sathishkumar, Ramachandran Chelliah, and Deog-Hwan Oh

Abstract Nanotechnology plays a crucial role in addressing environmental challenges in agriculture, particularly in mitigating abiotic stress in crops. The interaction between nanoparticles (NPs) and beneficial microorganisms has shown promise in enhancing plant resilience. Soil microorganisms play a vital role in nutrient cycling, pollutant decomposition, and soil-quality enhancement, with plant growth-promoting *rhizobacteria* such as *Bacillus, Azospirillum*, *Pseudomonas*, and *Rhizobium* stabilizing nutrients in the rhizosphere and enhancing plant resilience. In this connection, it has been noticed that the increasing application of NPs in agriculture, including nano-fertilizers, nano-pesticides, and nanoremediation agents, raises concerns about their interactions with soil microbiota. NPs such as Fe, ZnO, $TiO_{0.}$, CuO, and Ag have demonstrated both beneficial and harmful effects on microbial communities. While some NPs promote microbial activity and enhance nutrient bioavailability, others induce toxicity by disrupting cell membranes, generating reactive oxygen species, and interfering with metabolic functions. The extent of nanoparticle impact depends on factors like particle size, surface charge, capping agents, and the composition of bacterial cell walls. Taken together, this chapter discusses the mechanisms

---

K. Murugan · G. Hariharan · U. Sathishkumar
National Centre for Sustainable Coastal Management (NCSCM), Ministry of Environment, Forest and Climate Change, Chennai, India

H. Palvannan
Department of Microbiology, Vels Institute of Science, Technology and Advanced Studies (VISTAS), Pallavaram, Chennai, TamilNadu, India

R. Chelliah (✉) · D.-H. Oh
Department of Food Science and Biotechnology, College of Agriculture and Life Sciences, Kangwon National University, Chuncheon, South Korea
e-mail: ramachandran865@gmail.com

K. Murugan · H. Palvannan · G. Hariharan · U. Sathishkumar · R. Chelliah · D.-H. Oh
Saveetha School of Engineering, (SIMATS) University, Sriperumbudur, Tamil Nadu, India

A. Husen (ed.), *Nanobiotechnology for Abiotic Stress Adaptation and Mitigation in Agricultural Crops*, Smart Nanomaterials Technology,
https://doi.org/10.1007/978-981-96-9298-9_11

by which NPs interact with beneficial soil microbes to support plant growth, alleviate abiotic stress, and promote nutrient uptake. Additionally, the potential risks and bioaccumulation concerns of NPs in agricultural systems are explored.

Keywords Nanotechnology · Plants · Crop · Agriculture · Nanoparticle · Stress

1 Introduction

Over the past decade, nanotechnology has experienced significant growth, revolutionizing various fields, including biology, commerce, ecology, and computational modeling. It involves studying minute materials at the nanoscale and their applications across diverse disciplines [163]. Nanoparticles, ranging from 1 to 100 nm in size, possess a high surface-to-volume ratio, giving them unique electrical, optical, magnetic, and mechanical properties that distinguish them from bulk materials [174]. These particles exist in multiple forms, including fullerenes, carbon nanotubes, metals (Ag, Cu, Ni, Fe, Au), metal oxides (CeO_2, TiO_2, ZnO, NiO, CuO), and polymeric structures [136]. Their high surface area makes them valuable for environmental applications, such as wastewater treatment [189], functional food additives [23], agriculture, and medicine. While nanoparticles naturally exist in minimal quantities, the rise of synthetic nanoparticles in industrial and commercial applications has led to concerns about their environmental impact. Despite ongoing efforts to regulate their release, the widespread use of nanoparticles continues to raise questions about their potential hazards. Understanding their interactions in ecosystems is essential to preventing unintended consequences and ensure responsible usage.

In agriculture, nanoparticles serve as an alternative to traditional chemicals, pesticides, and fungicides. Their compact size, long shelf life, high efficiency, and ease of handling make them a preferred choice over conventional approaches, helping to reduce biodiversity loss and maintain soil fertility [162]. Metals like silver, zinc, copper, iron, silicon, titanium, magnesium, and manganese have been applied to crops in various forms, improving plant growth, yield, and resistance to disease. Nanoparticles also act as carriers for nutrients, plant growth regulators, nano-fertilizers, and insecticides, as well as tools for phytopathogen defense and biosensing, thereby contributing to enhanced agricultural productivity and sustainability [7–156].

2 Streamlines that Facilitate Nanoparticles Penetration into Natural Ecosystems

The increasing use of nanoparticles in various industrial and commercial products, such as coatings, cosmetics, and catalysts, has led to concerns about their environmental impact. Nanoparticles enter natural ecosystems through different pathways, including emissions during manufacturing, usage, and disposal. These particles can reach the environment directly or indirectly via industrial waste streams, landfills, and wastewater treatment plants (WWTPs), where their fate is influenced by factors such as surface modifications and structural alterations. Understanding these emission pathways is crucial for assessing the long-term impact of nanoparticles on agricultural soils and their interactions with beneficial microorganisms. Once released into the environment, nanoparticles undergo complex transformations, including agglomeration, redox reactions, and surface interactions with biomolecules. These changes influence their toxicity, mobility, and potential effects on soil microbial communities. Studies indicate that most nanoparticle pollution arises not during production but rather through their usage and disposal, highlighting the need for effective waste management strategies. In agricultural settings, the presence of nanoparticles in soil can affect microbial functions, influencing nutrient cycling, plant growth, and abiotic stress responses [52].

Nanoparticles interact with soil microorganisms through processes such as biofilm formation, enzymatic activity modulation, and nutrient bioavailability changes. Natural organic matter (NOM) and extracellular polymeric substances secreted by microbes play a significant role in nanoparticle stabilization or aggregation. The formation of a "protein corona" around nanoparticles alters their behavior, potentially affecting microbial colonization and plant–microbe interactions. These interactions can either enhance plant stress tolerance by promoting beneficial microbial activity or disrupt microbial ecosystems, leading to unintended ecological consequences. To ensure sustainable agricultural applications, it is essential to understanding how nanoparticles influence beneficial microorganisms that aid in crop resilience against abiotic stress. Future research should focus on nanoparticle dynamics in real agricultural conditions, assessing their interactions with microbial communities and potential risks. Developing eco-friendly nanomaterials and controlled release mechanisms can help minimize negative environmental impacts while leveraging nanotechnology for improved plant adaptation and stress mitigation [150].

2.1 Role of Nanoparticles in Ecosystem Dynamics

Nanoparticles play a crucial role in environmental applications, particularly in mitigating soil contamination. Their use in fertilizers, herbicides, and growth promoters enhances efficiency by reducing chemical inputs and minimizing environmental pollution [182]. Nanomaterials such as Zn and Al coatings regulate nutrient availability, while nanocomposite membranes protect pesticides from degradation, ensuring controlled release and prolonged effectiveness [111]. These advancements contribute to sustainable agricultural practices by improving nutrient uptake and reducing excess chemical accumulation in the environment.

In aquatic ecosystems, nanoparticles exhibit complex behaviors influenced by particle size, surface characteristics, and interactions with organic and inorganic substances. Natural waters, including rivers, groundwater, and atmospheric humidity, inherently contain nanoparticles [98]. These colloidal particles play a significant role in pollutant transport, with humic and fulvic acids acting as binding agents for contaminants [93]. Engineered nanoparticles, used in wastewater treatment and industrial applications, pose challenges as they may accumulate in aquatic environments, altering chemical compositions and impacting aquatic life.

The widespread use of nano-fertilizers and synthetic nanoparticles in agriculture has raised concerns regarding nutrient runoff into water bodies, leading to marine habitat disruptions and potential risks to human health [102]. While nanoparticles hold promise for environmental remediation, excessive accumulation can lead to ecological imbalances. Research highlights the need for controlled applications to prevent unintended consequences, emphasizing the importance of understanding nanoparticle behavior in water systems and their interactions with natural and artificial contaminants.

Nanoparticles in aquatic environments originate from both natural and anthropogenic sources, with mining, industrial discharge, and accidental release contributing to their presence [99]. While engineered nanoparticles like Fe and TiO_2 are beneficial for groundwater and wastewater treatment, their long-term ecological impact remains uncertain [92]. The study of nanoparticle transport, transformation, and interactions within aquatic systems is critical for predicting future contamination scenarios and developing sustainable remediation strategies.

2.2 *Impact of Nanoparticles on Aquatic Microorganisms*

Nanoparticles, particularly metal oxides like titanium dioxide (TiO_2) and zinc oxide (ZnO), have become pervasive contaminants in marine and freshwater ecosystems due to their widespread use in industrial and consumer applications [94]. These nanoparticles enter aquatic environments through wastewater discharge, industrial effluents, and surface runoff, where they interact with microbial communities, often leading to significant ecological consequences [49]. The primary mechanisms of nanoparticle toxicity in aquatic microorganisms involve the generation of reactive oxygen species (ROS) and the release of metal ions. ROS production by TiO_2 and ZnO nanoparticles induces oxidative stress, resulting in DNA damage, lipid peroxidation, and enzyme inhibition in microorganisms [122, 184]. ZnO nanoparticles, due to their higher solubility in water, pose an even greater risk, as the dissolution of Zn^{2+} ions disrupts microbial metabolism and cellular functions. Studies have shown that free Zn^{2+} ions are particularly toxic to various aquatic microorganisms, including bacteria (Escherichia coli), freshwater algae (*Pseudokirchneriella subcapitata*), and marine phytoplankton (*Thalassiosira pseudonana*) [9, 151].

Nanoparticle interactions in aqueous systems are influenced by several environmental factors, including ionic strength, natural organic matter (NOM), and pH, which affect their aggregation, dispersion, and bioavailability. While TiO_2 is relatively insoluble and tends to aggregate, ZnO nanoparticles display variable solubility based on pH and particle size, impacting their toxic effects [112]. NOM and extracellular polymeric substances secreted by microbes can further modify nanoparticle behavior, either stabilizing them or promoting their sedimentation. These transformations determine the extent of nanoparticle exposure to microbial communities and their long-term ecological consequences. Although direct interactions between nanoparticles and microorganisms may be limited due to aggregation and sedimentation, their long-term persistence in aquatic environments raises concerns about potential disruptions to microbial ecosystems. Nanoparticles can alter microbial community structures, affect nutrient cycling, and interfere with symbiotic relationships crucial for maintaining ecosystem balance. Given their increasing presence in aquatic environments, further research is necessary to assess their ecological risks and develop sustainable strategies for mitigating their impact.

2.3 Impact of Nanoparticles on Soil Microorganisms

Soil microorganisms play a vital role in nutrient cycling, pollutant decomposition, and soil-quality enhancement, with plant growth-promoting *rhizobacteria* (PGPR) such as *Bacillus, Azospirillum, Pseudomonas*, and *Rhizobium* stabilizing nutrients in the rhizosphere and enhancing plant resilience [31–86]. However, the increasing application of nanoparticles in agriculture, including nano-fertilizers, nano-pesticides, and nanoremediation agents, raises concerns about their interactions with soil microbiota. Nanoparticles such as Fe, ZnO, TiO_2, CuO, and Ag have demonstrated both beneficial and harmful effects on microbial communities. While some nanoparticles promote microbial activity and enhance nutrient bioavailability, others induce toxicity by disrupting cell membranes, generating reactive oxygen species (ROS), and interfering with metabolic functions [34–132]. The extent of nanoparticle impact depends on factors like particle size, surface charge, capping agents, and the composition of bacterial cell walls. For instance, Fe and TiO_2 nanoparticles, commonly used in water treatment, have shown both inhibitory and stimulatory effects on microbial populations, whereas nano-zerovalent iron (nZVI) has exhibited predominantly toxic effects [30, 117]. Given the growing integration of nanotechnology in agriculture, further research is essential to assessing the long-term ecological consequences of nanoparticle exposure on soil microbial diversity, functionality, and plant–microbe interactions.

2.4 Toxicity Effects of Nanoparticles on Beneficial Microorganisms

Nanoparticles induce toxicity by generating reactive oxygen species (ROS), leading to oxidative stress that disrupts biological processes. Their oxidative potential, catalytic interactions, and redox-active surfaces contribute to cellular damage. In algae and plants, nanoparticle exposure alters photosynthesis, pigment content, and metabolism, affecting growth and ecosystem balance. In soil, they impact microbial communities by disrupting carbon fixation and water dynamics. The release of toxic metal ions further intensifies their harmful effects, varying across species and environments. Understanding these interactions is crucial for minimizing ecological risks.

2.4.1 Nanoparticle-Induced Cellular Damage and Bioactive Molecule Disruption in Beneficial Microorganisms

Nanoparticles impact beneficial soil microorganisms by altering cell morphology, inhibiting bioactive molecule production, and disrupting metabolic processes. Their toxicity depends on factors such as particle size, molecular structure, concentration, and exposure duration. While some *rhizobacteria* like *Sinorhizobium meliloti*, *Bacillus thuringiensis*, *Azotobacter chroococcum*, and *Pseudomonas mosselii* tolerate high levels of CuO, TiO_2, and Al_2O_3 nanoparticles, they exhibit sensitivity to Ag and ZnO nanoparticles due to increased oxidative stress, membrane disruption, and reduced microbial adhesion [6]. Mechanisms of resistance include structural modifications, efflux pumps, and physiological adaptations, whereas Ag and ZnO nanoparticles enhance cellular toxicity by inducing oxidative damage, reducing respiration, and impairing biofilm formation [26, 191]. Furthermore, nanoparticles interfere with the synthesis of essential bioactive compounds like chlorophyll, indole-3-acetic acid (IAA), and carotenoids, which are critical for plant–microbe interactions. Studies show that Cu and ZnO nanoparticles significantly suppress plant growth, reduce chlorophyll synthesis, and increase oxidative stress, negatively affecting root elongation and nutrient uptake [180]. The heightened reactivity of nanoparticles compared to bulk materials amplifies their toxicity, with higher surface area and metal ion release contributing to microbial stress and growth inhibition.

2.4.2 Nanoparticle-Induced Molecular Alterations in Microorganisms

Recent advancements in omics technologies, including proteomics, genomics, metabolomics, and transcriptomics, have significantly enhanced nano-ecotoxicology research by providing insights into nanoparticle-induced molecular stress in microorganisms. Studies have shown that metal and non-metal nanoparticles alter microbial proteomes and gene expression, impacting key cellular processes. For instance, exposure to silver nanoparticles caused significant changes in protein clusters of Articulospora tetracladia, with alterations in stress-related proteins like acetyl-CoA and lipid peroxidases [17]. Similarly, transcriptomic analysis of Fusarium solani revealed that silver nanoparticle exposure led to differential expression of over 2500 genes, affecting metabolism, genetic signaling, and energy pathways [7]. These findings indicate that nanoparticles disrupt fundamental biological processes, leading to impaired microbial growth and metabolic dysfunction.

Nanoparticles also affect bacterial transcriptomes, altering gene expression and metabolic pathways in soil and aquatic bacteria. Silver and Al_2O_3 nanoparticles influenced stress-response genes in Pseudomonas stutzeri and Bacillus cereus, with P. stutzeri exhibiting an 8.2-fold increase in catalase-related gene expression to counter oxidative stress [40]. Similarly, exposure to CuO nanoparticles in Desulfovibrio vulgaris upregulated genes related to protein and ATP synthesis while downregulating genes for flagellar assembly and respiration, leading to reduced mobility and disrupted energy metabolism [25]. These molecular alterations highlight the capacity of nanoparticles to interfere with cellular functions, necessitating further research into their long-term ecological effects on microbial communities.

3 Abiotic Stress in Plants and Its Agricultural Impact

Plants face numerous environmental challenges, including biotic stress (caused by bacteria, fungi, insects, nematodes, and weeds) and abiotic stress (such as salinity, drought, extreme temperatures, heavy metals, and UV radiation). The impact of these stressors can be categorized into three main types: single stress, multiple individual stresses (occurring at separate times without overlap), and combined stresses (occurring simultaneously with overlap). Biotic and abiotic stresses often interact, particularly under global warming and changing climate conditions, significantly affecting plant biomass and secondary metabolite production [127, 135]. Abiotic stress alone can prevent crops from reaching their full genetic potential, leading to substantial global yield losses. These stressors are complex traits that impair plant growth, development, and productivity. Plant hormones and nanoparticles can influence plants' stress responses, either synergistically or antagonistically. As adaptable regulators of growth, plant hormones provide a promising area for exploring plant–nanoparticle interactions. Nanotechnology is emerging as a powerful tool in agriculture, offering methods to enhance soil quality, water relations, nutrient uptake, plant growth, and productivity, while also protecting crops from various biotic and abiotic stresses (Fig. 1, Table 1).

3.1 Salt Stress and Its Effects on Crop Growth

Nanoparticles have emerged as a promising approach to mitigate the effects of salt stress on plant growth and performance [163, 164]. Globally, around 62 million hectares (20% of cultivated land) are affected by salinity, a challenge that continues to grow, particularly in arid and semi-arid regions. Salinity, defined as soil with salt concentrations exceeding 4 dS m^{-1}, significantly impacts plant life by reducing

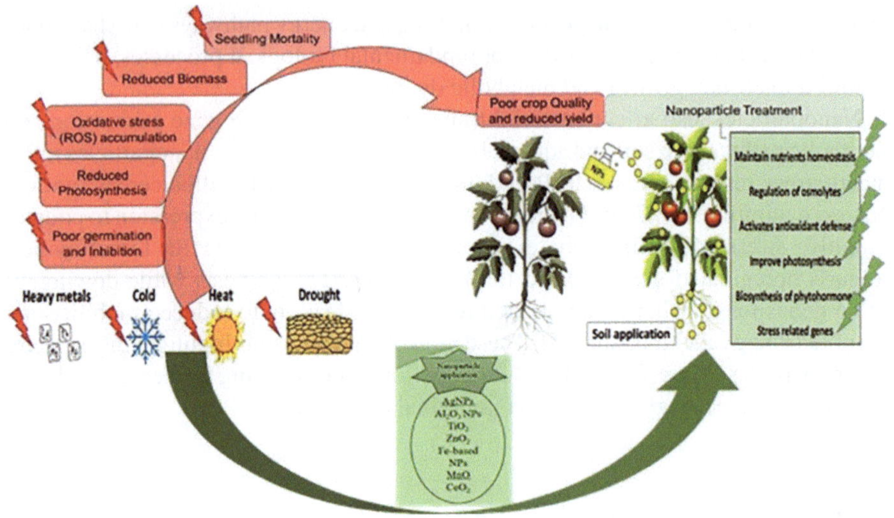

Fig. 1 Nanoparticles used for Abiotic stress: The types of abiotic stresses (e.g., temperature, drought, salinity, heavy metals, and climatic changes) and their relevance in stress management for plants

seed germination, seedling development, vegetative growth, and flowering. It causes osmotic imbalance, nutrient deficiencies, and ionic toxicity, leading to plant mortality when salt concentrations exceed 200 mM [4].

Salinity is one of the greatest threats to modern agriculture, stunting plant growth and reducing yield [53–71]. Nanoparticles, including Zn, Ag, SiO_2, Cu, Fe, Mn, C, Ti, Ce, and K, have been shown to alleviate salt stress. NPs are primarily absorbed through roots and leaves, influencing plant morphology, biochemical and physiological functions, and molecular activities, with effects depending on the NP concentration and plant species [91]. For example, Ag NPs improve germination and seedling development in Lathyrus sativus under salt stress by enhancing osmotic regulation and reducing stress effects [55]. Cu NPs alleviate oxidative stress in wheat, enhancing plant growth and yield [124], while ZnO NPs increase the dry weight of sunflower under salinity. CeO_2 NPs enhance photosynthetic activity and mineral uptake in Brassica napus, boosting biomass under salt stress [142].

Ag NPs also reduce oxidative damage, enhance antioxidant enzyme activity, and improve salt tolerance in wheat [89]. In rice, low concentrations of Ag NPs stimulate growth, whereas higher concentrations damage root structures. Similarly, Ag NPs enhance salt resistance in S. hortensis and cumin plants [118, 121]. Gold NPs (Au NPs) increase seed germination, vegetative growth, and free radical scavenging in Arabidopsis thaliana under salt stress [97], while Fe_3O_4 NPs protect mint plants from oxidative stress caused by high NaCl levels [10]. These findings highlight the potential of nanoparticles to reduce salinity's detrimental effects and enhance plant adaptation in agriculture.

Table 1 NPs used for abiotic stress and plant response to nanoparticles

Type of NPs	Plant and abiotic stress	Plant response to NPs	Key references
Silicon-Zinc oxide NPs	Pea, salinity	Improve the efficiency of photosynthesis	
Calcium oxide NPs	Barley, heavy metals	Promotes plant growth, photosynthesis, and antioxidant enzyme activity	[75]
Iron NPs	Wheat, heavy-metal, and drought stress	Enhances photosynthesis while reducing oxidative stress	
Copper NPs	Wheat, heavy metal	Promotes plant growth while decreasing chromium bioavailability	
Zinc oxide NPs	Wheat, heavy-metal, and drought stress	Enhances tissue and dry weight, while reducing cadmium accumulation	
Silica NPs	Rice, heavy metal, and drought	Enhanced plant growth and development	
Maghemite NPs	Sunflower, droughts stress	Mitigates the harmful effects of drought	
Titanium oxide NPs	Tomato, heat	Boosts plant growth and photosynthetic efficiency	[131]
Selenium NPs	Wheat, drought	Promotes plant growth and development	
Silver NPs	Rockcress, cold	Enhances the expression of genes associated with antioxidant activity	
Chitosan NPs	Corn, salinity	Reduces the harmful impacts of salinity	

3.2 Drought Stress and Water Deficiency Management

Drought is a persistent abiotic stress that severely impacts plant growth and development, contributing to approximately 70% of global crop yield losses [186]. Insufficient rainfall, low soil moisture, and reduced groundwater levels disrupt agriculture and forestry worldwide. The effects of drought are exacerbated by global warming, leading to more frequent and severe water shortages, posing significant challenges to agriculture in changing climates. Drought occurs when a plant's water demand exceeds available moisture due to factors like high transpiration rates, insufficient precipitation, or poor soil water retention [148].

Drought stress alters key physiological and biochemical processes in plants, including photosynthesis, respiration, enzymatic activity, and nutrient metabolism, ultimately reducing crop yield and quality [71–158]. Limited CO_2 uptake from stomatal closure decreases CO_2 reduction activity and impacts pigment composition, depending on the plant's tolerance to water deficiency [80]. Reduced gaseous exchange, transpiration, and carbon assimilation further exacerbate the effects of drought. Nanoparticles (NPs) are emerging as an effective tool to mitigating drought-induced stress by enhancing crop tolerance mechanisms. NPs improve photosynthetic activity, reduce water loss through stomatal closure by increasing ABA accumulation, and alleviate oxidative stress by reducing reactive oxygen species (ROS) and activating antioxidant defenses. NPs also boost antioxidant enzyme activity, regulate phytohormone levels, and improve physiological properties in plants.

For instance, analcite NPs enhance wheat germination in hot, dry soils. Metal-oxide NPs like Fe_3O_4, ZnO, and TiO_2 improve plant physiological and metabolic processes under drought stress. ZnO NPs increase germination rates in soybean under arid conditions [152]. Cu and Zn NPs in wheat improve antioxidant enzyme activity, maintain photosynthetic pigment levels, and reduce stress effects [172]. Green ZnO NPs enhance drought tolerance in tomato plants at lower concentrations [35]. Silica NPs strengthen physical barriers by depositing on cell walls, promote lodging tolerance, and prime defense responses to oxidative stress. NPs induce morphological, physiological, and biochemical changes in plants, increasing hydraulic conductance in roots, improving water uptake, and modulating proteins involved in ROS detoxification, stress signaling, and hormone pathways, thereby enhancing drought resistance [85]. These applications demonstrate the potential of NPs to improve crop resilience and productivity under drought conditions (Table 2).

Additionally, green-synthesized nano-fertilizers, such as Fe_3O_4 NPs, have proven effective in mitigating drought stress in fenugreek plants. Foliar application of SiO_2 and TiO_2 NPs has also been reported to reduce the adverse effects of drought stress on cotton and chickpea plants under arid conditions. Applying NPs to soil under drought stress significantly enhances leaf area, plant height, biomass, and relative water content, promoting overall plant resilience and growth.

3.2.1 Heavy-Metal Contamination and Toxicity in Plants

Heavy-metal (HM) stress is a significant toxic factor reducing agricultural productivity. Rapid industrialization and anthropogenic activities have introduced heavy metals into soil, water, and air, disrupting ecosystems and harming crops [77]. Toxic heavy metals such as arsenic (As), lead (Pb), cadmium (Cd), mercury (Hg), chromium (Cr), aluminum (Al), and beryllium (Be) severely impact plant growth and significantly reduce yields when present beyond safe levels [161]. Additionally, modern agricultural practices, such as the use of chemical pesticides and fertilizers, contribute to heavy-metal stress. Heavy metals like cadmium, zinc, and chromium inhibit metabolic activity, while copper, mercury, and nickel generate oxidative stress and reactive oxygen species (ROS), further reducing photosynthetic activity [81].

Table 2 Impact of NPs on plants under drought stress

NPs	Plant	Effect	Reference
ZnO	*Triticum aestivum*	Enhances the growth of plant and mineral content in grains	[31]
B NPs, SiO_2 NPs	*Triticum aestivum*	Mitigates drought stress and enhances the protein contents	[3]
ZnO NPs	*Zea mays*	Enhances crop yield and ameliorates antioxidative behavior	[169]
Nano-Cu NPs	*Zea mays*	Under drought conditions, NPs upgrade the protective mechanism of maize crops	[123]
Nano-Si NPs	*Tanacetum parthenium*	Improves water and phosphorus absorbing capabilities in plants and provides general growth	[38]
ZnO NPs	*Solanum lycopersicum*	Increases ascorbic acid and free phenol concentrations while enhancing the activity of antioxidant enzymes	[35]
Zn, Cu, and Fe NPs	*Glycine max* Merrill	Glycine max Merrill modulates the expression of drought-responsive genes	

These toxic elements can bioaccumulate in the food chain, causing chronic illnesses such as kidney and liver damage in humans and animals, as well as morphological abnormalities and metabolic disorders in plants [185].

Nanoparticles (NPs) have been shown to mitigate heavy-metal stress by absorbing and transforming toxic metals in the soil, reducing bioaccumulation and mobility. NPs activate the antioxidant system in plants, alleviating ROS-induced stress [183]. For instance, Fe_3O_4 NPs reduce Cd in soil [179], while TiO_2 NPs decrease Cd toxicity in maize and enhance superoxide dismutase (SOD) and glutathione levels [101]. SiO_2 NPs promote organic acid production, mitigating heavy-metal damage [29]. The application of nano-SiO_2 (30 nm and 50 nm) to soybean reduces mercury accumulation, minimizes growth inhibition, and increases chlorophyll content and antioxidant enzyme activity. ZnO NPs lower arsenic (As) and Cd accumulation in rice crops, reducing phytotoxicity. Fe_2O_3 NPs limit As uptake in *Vigna radiata* while enhancing antioxidant capacity by increasing SOD and catalase (CAT) activity [153]. CuO NPs in rice significantly reduce As uptake, improve plant biomass, and enhance antioxidant activity [181]. Nano-hydroxyapatite alleviates lead (Pb) toxicity, boosting ryegrass biomass production and antioxidant enzyme activity. Chromium (Cr) stress reduces photosynthetic pigments and chloro phyll fluorescence, but citrate-coated magnetite NPs in wheat alleviate Cd and Cr toxicity, promoting root growth [105]. Similarly, graphite carbon nitride reduces Cd toxicity in rice while increasing plant biomass. These findings demonstrate that NPs are effective tools for mitigating heavy-metal stress and improving plant resilience and productivity (Table 3).

Table 3 Impact of NPs on plant under heavy-metal stress

Plant species	NPs	Heavy Metal	Treatment	Reference
Wheat crop (*Triticum aestivum*)	Iron oxide	Cadmium	Reduces cadmium toxicity while enhancing growth, yield, and chlorophyll content in crops	[76]
Rice crop (*Oryza sativa*)	Zinc oxide	Arsenic	Reduces cadmium and arsenic concentrations in roots, shoots, and leaves	
Maize crop (*Zea mays*)	Titanium dioxide	Cadmium	Reduces cadmium concentration in leaves and shoots while enhancing antioxidant enzyme activity	[101]
Coriander crop (*Coriandrum sativum*)	Silicon nanomaterials	Lead	Enhances plant growth while reducing lead toxicity	[41]
Wheat crop (*Triticum aestivum*)	Zinc oxide	Cadmium	Reduces electrolyte leakage while boosting antioxidant activity	[75]
Pea crop (*Pisum sativum*)	Silicon nanomaterials	Chromium	Reduces chromium phytotoxicity, accumulation, and oxidative stress markers	[175]
Soybean crop (*Glycine max*)	Titanium dioxide	Cadmium	Enhances photosynthesis rate and promotes growth	[161]
Corn plants (*Zea mays*) and broad bean seed (*Vicia faba*)	Magnesium oxide	Chromium, cobalt, lead, cadmium, and nickel	Reduces heavy-metal toxicity while promoting plant growth	[42]
Cabbage crop (*brassica*)	Copper	Cadmium	Enhances the rate of photosynthesis and boosts SOD, POD, and CAT enzyme activity	[178]
Rice crop (*Oryza sativa*)	Iron(II) oxide	Arsenic	Reduces arsenic accumulation, enhances iron uptake, and restores photosynthetic pigments	

3.3 High-temperature Stress and Heat Shock Responses

Global warming and rising temperatures negatively impact the environment, reducing crop yields and causing severe damage to agriculture and tree populations [67, 69, 166]. Extreme temperature fluctuations and drought exacerbate these challenges. Nanoparticles (NPs) sourced from heavy metals can affect soil, water, and plant systems; however, they also play a vital role in mitigating abiotic stresses, including heat stress [130, 140]. NPs translocate through lateral roots to plant tissues, traveling via the xylem to the entire plant, with their size influencing their efficiency. At low concentrations, NPs exhibit antioxidative properties, helping plants combat abiotic stresses. Under heat stress, plants synthesize molecular chaperones and heat shock proteins (HSPs). For instance, CeO_2 NPs in maize induce H_2O_2 generation and upregulate HSP70, reducing heat stress by regulating stomatal opening [178]. Green-synthesized selenium NPs (100 μg/mL) enhance growth, photosynthesis, and gas exchange at elevated temperatures [34]. Nano-ZnO improves chlorophyll content, gas exchange, enzymatic activity, and pod yield in mung bean under heat stress [87].

In wheat seedlings, ZnO and TiO_2 NPs enhance membrane stability and antioxidant defense in roots and shoots [173]. Nano-ZnO regulates osmotic potential, minimizes thylakoid damage, boosts antioxidant and osmolyte production, and curtails lipid peroxidation under heat stress. Its foliar application increases chlorophyll content, gas exchange, and agronomic traits, improving yield even in late-sown crops [87]. In sorghum, Se NPs mitigate heat stress by activating antioxidant defenses, reducing membrane damage, and enhancing pollen germination and yield [32]. Ag NPs improve wheat growth under heat stress [78], while Zn NPs boost wheat heat tolerance by enhancing antioxidant enzymes and reducing lipid peroxidation [51]. Foliar application of Si NPs helps plants cope with heat stress by activating defense mechanisms, particularly in crops like tomatoes. Nanotechnology offers a promising solution to protect crops from the detrimental effects of rising temperatures, enhancing plant resilience, growth, and productivity (Table 4).

3.4 Cold Stress and Its Impact on Metabolic Processes

Climatic changes significantly impact plant growth and adaptation, with global climate change contributing to cold stress, which hinders plant development. Cold stress includes chilling (0–15 °C) and freezing, depending on the plant species and their temperature tolerance. Factors like air temperature and wind speed further influence the impact of chilling. Cold temperatures reduce protein and enzyme activity, slow crop growth, and disrupt secondary metabolism, respiration, defense mechanisms, and nucleic acid synthesis [11, 193]. Cold stress affects plants' physiological, biochemical, molecular, and metabolic processes, reducing agricultural yields. Nanoparticles such as TiO_2 and chitosan NPs have shown promise in mitigating cold stress. TiO_2 NPs improve electrolyte leakage, photosynthetic activity, and membrane stability in chickpea plants under cold stress [8, 115]. In rice, ZnO NPs enhance seedling growth by activating the antioxidative system and chilling-response transcription factors [165]. Similarly, Si NPs improve photosynthesis in sugarcane under chilling stress [36] (Fig. 2).

Table 4 Impact of NPs on plants under temperature stress

NPs	Stress	Plant	Effect	Reference
Ag NPs	Heat stress	*Triticum aestivum*	Enhances the plant's morphological characteristics	[173]
ZnO and TiO_2	Heat stress	*Triticum aestivum*	Enhances plant morphology and antioxidant defense system (SOD, GPX) while reducing H_2O_2 content	[46]
TiO_2	Cold stress	*Cicer arietinum*	Reduces membrane damage indices and improves redox status	[50]
Zn NPs, Fe NPs	Heat stress	*Triticum aestivum*	Enhances crop yield and antioxidant enzyme activity	[51]
ZnO NPs	Heat stress	*Triticum aestivum*	Enhances plant biomass, photosynthetic pigments, soluble sugars, protein, and indole acetic acid (IAA) content	[13]
ZnO NPs	Chilling stress	*Oryza Sativa*	Promotes crop growth, alleviates oxidative stress, and upregulates gene expression of the antioxidative system	[165]
Chitosan CH NPs	Chilling stress	*Musa acuminata var*	Enhances crop growth, nutrient content, photosynthesis, and antioxidant enzyme activities	[179]

4 Plant-Based Nanoparticles and Their Role in Agricultural Sustainability

Carbon-based nanoparticles, metallic nanoparticles (e.g., gold, silver, zinc, copper), and oxidized nanoparticles (e.g., MgO, CaO, TiO_2 NMs) are extensively used in plant science due to their diverse applications. Nanogels are also employed to enhance plant growth and physiology [113]. Research on plant–nanomaterial interactions has

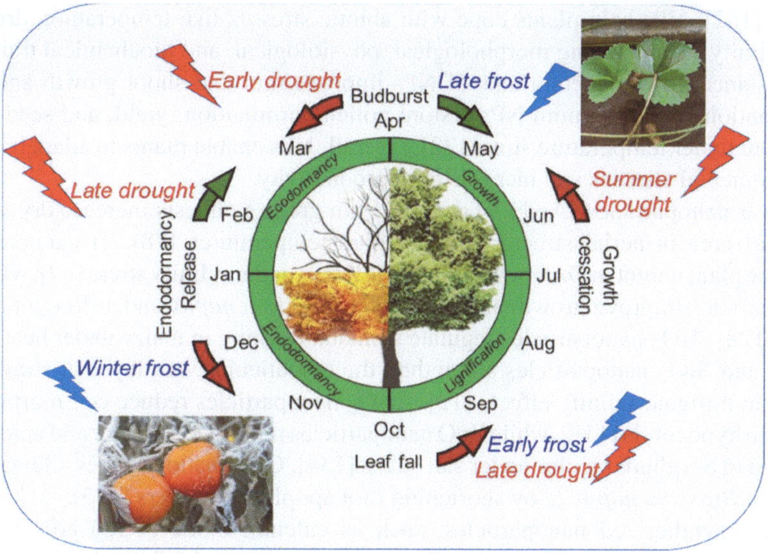

Fig. 2 Cold stress in plants affect growth and reduces yield: the adverse effects of cold stress on plants. On the left, healthy plants exhibit normal growth, while the right image shows frost-damaged plants, highlighting reduced growth and productivity caused by low-temperature stress

revealed both positive and negative effects on plant growth in agriculture. Nanotechnology holds great potential in plant science, particularly in using nano-products as carriers for biomolecules and agrochemicals to boost plant production.

4.1 Interaction of Nanoparticles with Plant

Nanoparticles (NPs) are materials within the nano-size range, occurring naturally or released into the environment through widespread applications, including metal-based NPs [110]. Their interaction with plants involves uptake, translocation, and accumulation in plant tissues [7–63, 138]. Effective delivery of NPs into plants is crucial for optimal tissue functioning and is achieved through methods like seed coating, soil wetting, and foliar spraying. Transport within plants depends on membrane pore size and involves three main pathways: deposition on surfaces (leaves, roots, stems), penetration through the cuticle and epidermis, and translocation via the vascular system. NPs move actively through osmotic pressure, capillary forces, cell wall pores, and *plasmodesmata* [177]. Smaller NPs pass through cell walls, while larger NPs travel via stomata, stigma, or hydathodes [56]. Factors influencing NP transport include cell wall pore size, sap composition, and sap flow dynamics. Common NPs like carbon nanotubes, metal-based NPs (Ag, Au), and metal-oxide NPs (TiO_2, ZnO, SiO_2, FeO) stimulate molecular mechanisms in plants [162]. NPs help plants cope with abiotic stresses like temperature, drought, and salinity by enhancing morphological, physiological, and biochemical traits [5]. For instance, low concentrations of NPs improve root and shoot growth and seed germination [52]. Selenium NPs restore pollen germination, yield, and seed set in sorghum under temperature stress [32]. Overall, NPs enable plants to adapt better to environmental stresses and increase crop productivity.

Silver nanoparticles (Ag NPs) derived from green synthesis increase dry weight and leaf area in aerial structures under high temperatures [78]. Ti nanoparticles enhance plant height in *Dracocephalum moldavica* under salinity stress [47], while Fe nanoparticles improve growth and biomass in *Brassica napus* under drought conditions [128]. TiO_2 nanoparticles regulate stomatal opening in maize under heat stress [131], and SiO_2 nanoparticles strengthen the epicuticular wax layer in strawberry plants to mitigate salinity effects [12]. Al_2O_3 nanoparticles reduce cell mortality in soybean hypocotyls [118], while ZnO nanoparticles minimize vascular and epidermal damage in Sorghum bicolor under salt stress [134]. CeO_2 nanoparticles alleviate salt stress in *Brassica napus L.* by shortening root apoplastic barriers [143].

Green-synthesized nanoparticles, such as calcium oxide (CaO) NPs from *P. granatum* fruit extract, effectively alleviate salinity stress in Triticale callus [187]. Zinc nanoparticles from Sorghum bicolor leaf extract improve growth and photosynthesis in *Abelmoschus esculentus* during salinity stress by upregulating antioxidant defenses [188]. Similarly, TiO_2 nanoparticles from *Buddleja asiatica* extracts enhance morphological, physiological, and biochemical traits in *Triticum aestivum* under salinity stress [119]. Positively charged gold nanoparticles (Au NPs) induce hairless root development in barley, while neutral Au NPs vacuolize cortical and rhizodermal cells at root tips [97] (Kumar et al. 2022). Under abiotic stress, NPs help maintain carbohydrate content, protect proteins and enzymes from denaturation, and stabilize membrane integrity, reducing stress impacts.

Plants respond to abiotic stress by producing enzymes and secondary metabolites like lignin, anthocyanins, flavonoids, phenolic acids, and other molecules to mitigate stressors and manage oxidative damage. Foliar application of Zn nanoparticles enhances proline, glycine betaine, free sugars, and amino acids under drought conditions. CeO_2 nanoparticles reduce ROS and H_2O_2 concentrations in germinating rice seedlings in a dose-dependent manner [139]. At low concentrations, CeO_2 NPs suppress ROS production in *M. arborea*, while higher concentrations mimic SOD activity [48]. In *A. thaliana*, CeO_2 NPs also increased lipid peroxidation by 2.5-fold [107]. Spinach seeds treated with rutile and anatase-form TiO_2 NPs reduced oxidative stress under UV-B irradiation. However, graphene NPs and MWCNTs applied to cabbage, tomato, red spinach, and rice suspension cells increased H_2O_2 and ROS levels, resulting in cell viability loss [18, 171]. These findings highlight the dual effects of nanoparticles, with their efficacy depending on type, concentration, and plant species.

4.1.1 Mechanisms of Nanoparticle Uptake and Transport in Plants

Nanoparticles enter plant cells and tissues through various mechanisms, including ion channel transport, passive diffusion, aquaporin-mediated water molecule transport, carrier proteins, endocytosis, pore formation, and interactions with organic matter. Among crops, onion shows the highest uptake, accumulation, and toxicity of nanoparticles, followed by tomato, wheat [43], cucumber, zucchini/pumpkin, soybean, lettuce, and rice. Cucumber and zucchini exhibit higher nanoparticle uptake and translocation due to greater water absorption and larger vascular bundles.

The biological activity of NPs in plants depends on their concentration, properties, and application methods, such as seed treatment, foliar sprays, and soil application. NPs penetrate plants through leaf cuticles, stomata, trichomes, hydrathodes, and cell cytoplasm, binding with organelles and interfering with metabolic processes [154, 190]. In seeds, NPs are absorbed through parenchymatic intercellular spaces in the seed coat. The uptake, accumulation, and translocation of NPs vary based on plant species, NP size, type, stability, and interactions with roots, soil, and microbes. Gold nanoparticles (0.5–100 nm) have been detected in plant tissues, with smaller sizes (e.g., 3.5 nm Au NPs) entering tobacco root cytoplasm, while larger sizes (18 nm) do not [143–157]. However, the release of silver nanoparticles can negatively impact naturally occurring microorganisms (Fig. 5).

4.1.2 Effects of Nanoparticles on Crop Physiology and Productivity

Biogenic copper nanoparticles synthesized via green methods using *Shigella flexneri* effectively reduce cadmium (Cd) translocation from soil to wheat plants. Cu nanoparticle treatment in Cd-contaminated soil enhances biomass production, nutrient uptake, and decreases acropetal Cd translocation and sodium content in wheat. Mung beans (*Phaseolus radiatus*) remain unaffected by Ag nanoparticles in soil, while sorghum exhibits vulnerability at concentrations of 200 mg/kg or higher. Similarly, flax, ryegrass, and barley show minimal or no growth impact at Ag NP concentrations of 20–100 mg/kg. Ag NPs, at sub-lethal levels (around 0.1 mg/L), promote growth in poplar and Arabidopsis in liquid systems. In contrast, ZnO NPs display mixed effects: they reduce cucumber biomass in hydroponic cultures and suppress growth in wheat, corn, beans, and ryegrass in sand or liquid systems, while also altering wheat root architecture by stimulating lateral root growth. CuO nanoparticles negatively affect growth and metabolism in crops like rice, causing oxidative damage and reduced germination. Similarly, cucumber growth declines under CuO NP exposure, accompanied by increased stress-response enzyme activity. Radish, ryegrass, and buckwheat studies further reveal growth, yield, and DNA integrity reductions. TiO_2 nanoparticles inhibit leaf growth and transpiration in maize, as well as root, stem elongation, and biomass in tomato crops. However, TiO_2 NPs mitigate root-knot nematode infestations in tomatoes, demonstrating dual effects depending on context.

4.1.3 Entry Routes of Nanoparticles into Plants

NPs in the atmosphere, water, and soil interact with plants, potentially causing harm and entering the food chain through plant uptake, influencing their environmental fate [133]. Plant roots initially contact NPs present in soil or wastewater effluents used for crop growth [49, 28]. Root exudates, containing biomolecules like amino acids, organic acids, fatty acids, and polysaccharides, form the rhizosphere and attract NPs, which adhere to or deposit on root surfaces [43, 14]. Absorbed NPs undergo physicochemical modifications through interactions with root exudates and humic acids [138]. For instance, CuO NPs in maize translocate from roots to shoots via xylem water uptake, as observed through TEM and EDX analysis. The size of root pores influences NP absorption, with crops like corn, tomato, cucumber, and alfalfa absorbing CeO_2 NPs larger than 7 nm in diameter [106]. Cucumber roots can absorb CeO_2 NPs up to 25 nm, translocating them to shoots. When applied as foliar sprays or aerosols, NPs deposit on leaf surfaces and penetrate plant systems through their nanoscale size and interaction with gaseous uptake. This versatility highlights multiple NP entry routes into plants (Fig. 3).

Silicon (Si) Nanoparticles: Enhancing Stress Resistance

The plant response to silica nanoparticles (Si NPs) depends on factors like size, shape, application method, and their chemical and physical properties [137]. Si NPs are biologically synthesized from plant sources, including bamboo stem, culm, sugar beet bagasse, and rice husk, which contain 60% silica. These three-dimensional nanoparticles have a large surface area and superior absorption properties, enhancing molecular transport [174]. Mesoporous Si NPs (~20 nm) are applied to plants via soil supplementation or foliar spray. Soil application success depends on factors like soil texture, pH, salt content, and agrochemical release duration, while foliar application involves spraying liquid Si NPs on leaves, where they are absorbed through stomata or epidermal cells [167]. Once absorbed, Si NPs form a protective binary film at the epidermal cell wall, potentially adding structural color to plants. Soil-applied Si NPs have been found more effective than foliar-applied ones. In agriculture, Si nanoparticles (Si NPs) are utilized to mitigate heavy-metal toxicity [29], UVB stress [176], salinity stress [1], and dehydration [84]. Si NPs (~12 nm) improve germination in tomato plants, enhancing fresh and dry seedling weight. Application of SiO_2 nanoparticles increases potato tuber mass, root and sprout length, and chlorophyll and carotenoid content. Similarly, maize seed germination and root elongation benefit from Si NPs [88]. Si NPs also serve as fertilizers, pesticides, herbicides, genetic and drug transfer agents, soil improvers, and soil analysis sensors, showcasing their potential to enhance crops and support sustainable agriculture.

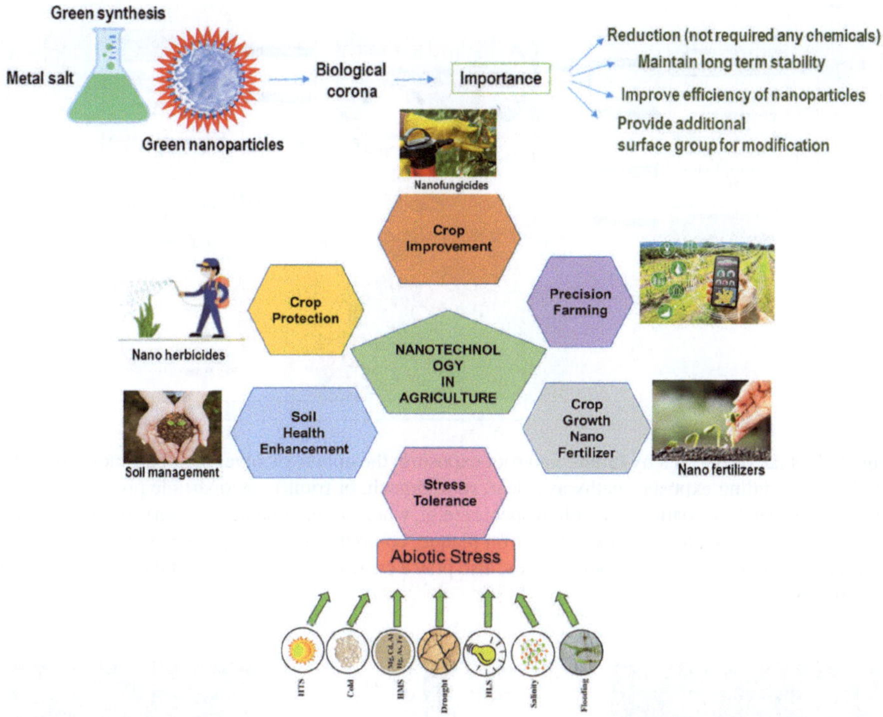

Fig. 3 Methods used for the application of nanoparticles in the agricultural field: various methods for applying nanoparticles in agriculture, including biolistic guns, spraying, injection, hydroponics, irrigation, and incubation. These techniques enhance the efficient delivery of nano-fertilizers to plants, optimizing growth and productivity while addressing specific agricultural challenges

Silver (Ag) Nanoparticles: Antimicrobial and Growth-Promoting Effects

The green synthesis of silver nanoparticles (Ag NPs) using plant extracts and microorganisms offers several advantages over chemical and physical methods. This synthesis can be classified into three types: (a) use of plant extracts; (b) use of microbes like yeasts, bacteria, viruses, and actinomycetes; and (c) use of viral membranes, DNA, and RNA. Plant extracts derived from leaves, flowers, seeds, bark, and roots act as both reducing and capping agents during AgNP synthesis. Ag NPs exhibit unique physiological and chemical properties, including antimicrobial and antioxidant activities [114, 129]. Medicago sativa shoots were the first live plant system used for AgNP synthesis [44]. Similarly, *Acalypha lanceolata* leaf extract has been employed for AgNP production, highlighting the potential of plant-based green synthesis [95]. *Aloe vera* (Xanthorrhoeaceae) has been used to synthesize silver nanoparticles (Ag NPs) of varying sizes and shapes [104]. Similarly, green spherical Ag NPs have been produced using *Sambucus nigra* (Adoxaceae) and *Syzygium sp.* fruit extracts (Myrtaceae), with demonstrated stress resistance

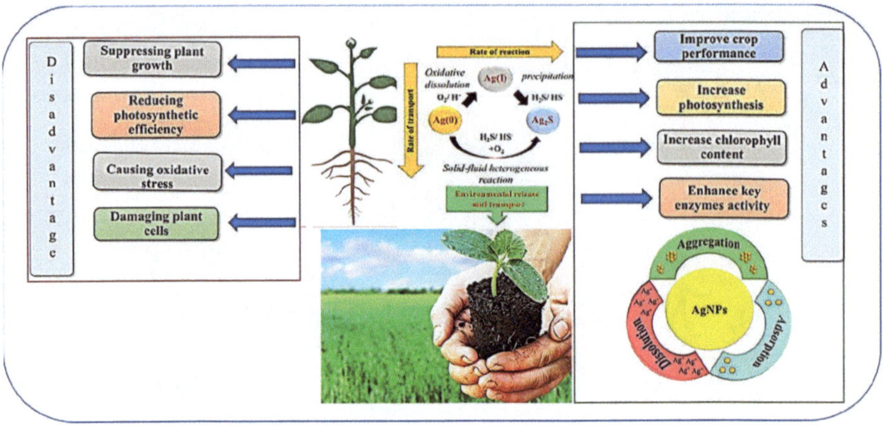

Fig. 4 Uptake of Ag NPs by failure and root exposure: the uptake of silver nanoparticles (Ag NPs) in plants, including exposure pathways (leaf, root, branch, or trunk), nanoparticle properties (size, morphology, surface coating), and plant species (e.g., wheat, rice, tomato, soybean). Soil properties, such as pH, clay content, root exudates, iron plaque, co-contaminants (e.g., heavy metals, organic pollutants), and symbiotic microorganisms, also play a critical role in Ag NP uptake and interaction with plants

Fig. 5 Nanotechnology in agriculture utilized, including nano-fertilizers for improved nutrient delivery, nano-herbicides for weed management, nano-pesticides for crop protection, and nanosensors for precise monitoring of agricultural conditions

and increased tolerance via antioxidant activity [150, 4, 144]. These plant-based Ag NPs exhibit antibacterial, antidiabetic, anti-inflammatory, antioxidant, anti-HIV, antifungal, snake venom neutralization, and larvicidal activities. Biologically synthesized Ag NPs also enhance seed sprouting and sapling growth in plants such as *Boswellia ovalifoliolata* (*Burseraceae*) and improve growth parameters like plant length and leaf area in Zea mays (Poaceae), *Phaseolus vulgaris* (*Fabaceae*), and *Brassica juncea* (*Brassicaceae*) [146, 155] (Fig. 4).

Biogenic silver nanoparticles (Ag NPs) have gained attention in agriculture for their stress-tolerant capabilities. These nanomaterials enhance stress defense mechanisms by improving nutrient absorption, enzyme reactions, and root interactions with growth-promoting bacteria during abiotic stress [90]. Biogenic Ag NPs help maintain water balance in lentils under drought conditions, improving plant growth, length, and weight, as well as enhancing germination processes. In *Triticum aestivum* (wheat), Ag NPs improve physiological growth by enhancing membrane stability, chlorophyll content, relative water content, and antioxidant enzyme activity under drought and salt stress. Low concentrations of Ag NPs also increase fresh and dry weight, soluble sugar content, and chlorophyll levels, and improve germination rates at various NaCl concentrations (5, 10, 15, and 20 dS/m) [114, 54].

Zinc (Zn) Nanoparticles: Role in Nutrient Fortification

Plant-mediated synthesis is an effective biological method for fabricating ZnO nanoparticles [22]. Phytochemicals in plant extracts act as reducing agents, transforming metal precursors into nanoparticles [159]. Plant-secreted compounds, like organic acids, influence nanoparticle size and morphology. Compared to microbial systems, plant-based methods offer multiple advantages, such as reduced costs, simpler processes, and sustainable, renewable resources for nanoparticle synthesis. The size of nanoparticles is crucial for their ability to permeate plant tissues effectively. Smaller nanoparticles can pass through pores in plant tissues, while larger nanoparticles may enter through two mechanisms: the creation of slightly larger pores or cell wall relaxation induced by reactive oxygen species (ROS). Root and foliar applications are the most common methods for delivering ZnO nanoparticles to plants. ZnO nanoparticles can decompose in the environment, releasing ions or forming smaller particles that are absorbed by plant tissues. Upon exposure to plant tissues, nanoparticles penetrate the root epidermis and cortex, travel through the vascular bundle (xylem), and move toward the stele. Foliar application involves spraying ZnO nanoparticles on leaf surfaces, where they are absorbed through stomata and cuticles, traveling further via phloem sieve tubes for whole-plant distribution. Zhu et al. demonstrated that ZnO nanoparticles penetrate stomata, accumulate in plastid exosomes, and translocate to chloroplasts. Proper nanoparticle sizing is critical to avoiding blocking stomata, which can disrupt transpiration. ZnO nanoparticles enhance drought tolerance by mitigating its negative effects on crop yield and biomass accumulation. Dimkpa et al. studied the application of ZnO nanoparticles on sorghum, demonstrating their role in enhancing plant development, nutrient uptake (N, K, Zn), and grain fortification. ZnO nanoparticles improve crop yield, expedite growth, and augment Zn content in sorghum. Additionally, their utilization in wheat enhances antioxidant enzyme activity, chlorophyll content, total proteins, soluble sugars, lysine, and arginine levels. At the molecular level, ZnO nanoparticles upregulate drought-related genes such as Wdhn13, CAT1, P5CS, and DREB2, mitigating drought stress and improving wheat yield. Rai-Kalal et al. reported that ZnO nanoparticle pretreatment in wheat seeds prevents chlorophyll degradation under drought stress, enhances photosynthesis, and promotes plant growth. ZnO nanoparticles significantly improve PSII photochemical efficiency under drought stress, reduce reactive oxygen species production, and protect the photosynthetic apparatus from oxidative damage.

The application of ZnO nanoparticles in agriculture has been shown to enhance heat stress tolerance in crops like alfalfa, mung bean, chickpea, and wheat. Under heat stress, ZnO nanoparticles improve water relations, increase zinc ion concentrations in leaves, elevate free proline levels, boost antioxidant enzyme activities (SOD, MDA, H_2O_2, and APX), and regulate PSII efficiency, enhancing photosynthesis and growth. ZnO nanoparticles also strengthen membrane integrity, scavenge ROS, regulate cell division, and improve nutrient transport, carbohydrate levels, protein metabolism, and photosynthetic pigments, increasing salt tolerance in crops such as sunflower, wheat, tomato, pea, rice, and rapeseed. Foliar application of ZnO nanoparticles in salt-stressed rice improves phosphorus (P), potassium (K), and zinc (Zn) uptake while reducing sodium ion absorption, enhancing agronomic traits like plant height, panicle length, tiller number, and grain weight. ZnO nanoparticles also regulate stress-responsive genes and hormone signaling pathways, making them a cost-effective and eco-friendly strategy to boost agricultural productivity and mitigate stress impacts.

5 Interaction of Nanoparticles with Soil Microbiota and Their Implications

The application of nanoparticles in the environment can threaten beneficial soil microorganisms, including bacteria and fungi. Nanoparticles influence microbial activity by affecting soil respiration and enzymatic functions [160]. Studies indicate that TiO_2 and CuO nanoparticles reduce microbial biomass, enzymatic activities, and alter community structures in flooded paddy soil, while higher concentrations of Fe_3O_4 nanoparticles significantly lower bacterial populations. ZnO and CeO_2 nanoparticles impact Azotobacter, phosphorus- and potassium-solubilizing bacteria, and inhibit plant enzyme activities [21]. Copper nanoparticles release ions that cause toxicity to both pathogenic and beneficial bacteria. CuO nanoparticles have been observed to cause membrane degradation, cellular collapse, and cell lysis in native soil bacteria. Additionally, Pradhan et al. reported that exposure to CuO and Ag nanoparticles reduces leaf microbial decomposition rates, further highlighting their potential ecological impact.

6 Engineered Nanoparticles: Tailoring Properties for Agricultural Applications

Engineered nanomaterials (ENPs), synthesized using inorganic, polymeric, and lipid nanoparticles through techniques like emulsification, ion gelation, polymerization, and oxido-reduction, aim to enhance agricultural productivity. These ENPs are tailored for specific physical (size, shape) and electrical properties (surface characteristics), providing unique catalytic activity, improved electrical and thermal conductivity, and controlled host molecule distribution. This enhances nutrient uptake, reduces eutrophication, and minimizes nutrient leaching [102]. Recent advancements, including improved pesticide solubility, resistance to activity loss, and targeted delivery, have increased agricultural safety and minimized off-site effects [126]. ENPs like hydrogels, nanoclays, and nano-zeolites have also been explored to enhance soil water retention and slow-release capacity, aiding crop sustainability in water-scarce and degraded lands. ENPs are primarily applied to roots or vegetative parts, particularly leaves, although uptake in soils is more complex than aerial applications [149].

Numerous studies have demonstrated that carbon nanotubes, foliar sprays, irrigation, and specific low-dose ENPs significantly enhance plant growth, photosynthesis, chlorophyll a and b, carotenoid content, antioxidants, and tolerance to biotic and abiotic stresses. The impact of ENPs on plants largely depends on factors like size, surface charge, composition, concentration, and physicochemical properties of the nanoparticles, as well as the plant species' susceptibility. For instance, Zn nanomaterials promote chlorophyll content, fertilization, pollen function, seed germination, and drought stress tolerance. ENP effects vary based on plant growth stages and exposure duration. Soil symbiotic bacteria and fungi interact with ENPs, influencing plant root behavior. While these microbes increase heavy-metal nanoparticle accumulation in turfgrasses, they reduce the uptake of nano-Ag and nano-FeO in legumes. To better understand ENP interactions with plants and microflora, advanced techniques like magnetic resonance imaging (MRI), microscopy, and fluorescence spectroscopy could be employed for more accurate conclusions.

7 The Role of Nanotechnology in Agriculture: Innovations and Applications

Nanotechnology offers innovative tools to enhance agricultural productivity, with "nano-fertilizers" emerging as a promising solution to meet the demands of a growing population. Nano-fertilizers are nanomaterials designed to supply one or more essential nutrients to crops, improving growth and yield while enhancing the efficiency of conventional fertilizers [102]. This approach has the potential to significantly boost agricultural output and sustainability (Fig. 5).

Nanotechnology is a rapidly advancing field with applications across industries, including electronics, medicine, energy, cosmetics, agriculture, and more [174–170]. In agriculture, it plays a crucial role in the development of nano-pesticides, nano-herbicides, nano-fertilizers, and nanosensors. Nano-fertilizers, categorized based on plant nutrient uptake and requirements, are essential for providing macronutrients like nitrogen (N), phosphorus (P), potassium (K), magnesium (Mg), sulfur (S), and calcium (Ca), which are vital for plant growth. Conventional farming practices often result in significant nutrient runoff into groundwater and surface water, posing risks to human health, aquatic life, and ecosystem balance due to chemical fertilizers and heavy metals. Nano-fertilizers, developed using nanomaterials, offer an eco-friendly and sustainable alternative for nutrient delivery. Additionally, micronutrients like iron (Fe), manganese (Mn), zinc (Zn), copper (Cu), molybdenum (Mo), nickel (Ni), boron (B), and chlorine (Cl), required in trace amounts for plant metabolic activities, are integrated into composite fertilizers at concentrations like 5 ppm, enhancing crop yield and growth efficiency.

7.1 Nano-fertilizers: Enhancing Nutrient Use Efficiency

Nano-fertilizers are synthesized or modified versions of traditional fertilizers using nanotechnology. These fertilizers provide essential nutrients to crops, boosting growth and productivity [102]. Fabricated using "bottom-up" or "top-down" approaches, nano-fertilizers are characterized by their nano-sized particles (1–100 nm), which enhance nutrient uptake, photosynthesis, and biomass production due to their high surface area-to-volume ratio. They require smaller quantities, have a slow-release mechanism, and reduce soil salt accumulation compared to conventional fertilizers. Nano-fertilizers include elements like Fe, Mn, Zn, Cu, Mo, Ti, and carbon nanotubes, offering efficient and cost-effective nutrient delivery. Nano-fertilizers also improve crop nutrient use efficiency (NUE) and nutritional quality through biofortification. For example, balanced NPK application with S, Zn, B, and Mo has been shown to increase yields in red and lateritic soils. Nano-fertilizers like nanoporous zeolites prevent nutrient loss to the environment, enhancing bioavailability and meeting crop nutrient requirements. While their application significantly improves physiological and biochemical indices, overuse can lead to phytotoxicity and environmental harm. Challenges like cost-effectiveness, recyclability, and biodegradability remain critical for manufacturers. The integration of nano-fertilizers with smart delivery systems, including nano-insecticides, nano-herbicides, and nanosensors, marks a revolutionary step in modern agriculture. These systems release active ingredients gradually, enhancing the efficiency and effectiveness of fertilizers while addressing public concerns about the adverse effects of chemical fertilizers on ecosystems.

7.2 *Nano-pesticides: Targeted Pest and Disease Management*

Nano-pesticides are formulated using nanomaterials to enhance agricultural applications, whether fixed on hybrid substrates, encapsulated in matrices (Fig. 6), or carried by functionalized nanocarriers responsive to external stimuli or enzymes. These nanosized particles, with specific shapes and properties, enable innovative pesticide activities using materials such as silica, lipids, polymers, ceramics, metals, and carbon. Common nano-pesticides include Ag, Cu, SiO_2, and Zn nanoparticles, which offer greater efficiency and cost savings in agricultural use. By reducing the reliance on chemical fertilizers and pesticides, nano-pesticides help maintain ecosystem balance disrupted by synthetic chemicals. They improve agrochemicals' water solubility, bioavailability, and resistance to environmental degradation, potentially transforming crop disease, weed, and insect management. Specifically, silicon nanoparticles (Si NPs) are utilized in two main ways: as mesoporous nanocarriers that release commercial pesticides to enhance their effectiveness, or directly applied in the field to kill insects and larvae. In order to optimize pesticide efficacy, targeted delivery is crucial, and mesoporous silica nanoparticles (Si NPs) enable controlled release of bioformulations [24, 100]. Studies show that mesoporous nanosilica increases both effectiveness and longevity of commercial insecticides, indicating that Si NPs can be applied alone or in combination with other pesticides for enhanced transport. Silicon nanocarriers embedded with herbicides (e.g., chloroacetanilide, anilide, benzimidazole) in diatom frustules release the active herbicidal component into the field [103]. However, copper-based nano-pesticides may inadvertently disrupt non-target soil microbial communities due to their broad-spectrum antimicrobial properties. Moreover, runoff from agricultural lands, industrial wastewater discharge, and soil leaching can transport nano-pesticides into water supplies, raising concerns about soil quality, human exposure, and ecosystem health. Due to their biomimetic properties, nanoparticles have shown potential toxicological effects, high distribution capability, and bioaccumulation in soil, water systems, food, and ultimately in mammals [20, 53].

7.3 *Nano-herbicides: Precision Weed Control for Sustainable Farming*

Herbicides are commonly used to control weeds in agriculture, but they can compete with crops for light, water, and nutrients. Encapsulating herbicides in nanomaterials improves selectivity, prolongs their release, and reduces losses. Various nanocarriers, including polymeric nanoparticles, nanocapsules, and nanospheres (e.g., alginate, chitosan, pectin, poly (ε-caprolactone), poly(methyl methacrylate), and poly(lactic-co-glycolic acid)), have been used to formulate herbicides like paraquat, 2,4-D, diuron, ametryn, atrazine, and simazine. Other carriers (e.g., rice husk nano sorbents, mesoporous silver nanoparticles, nanoclay) also serve as efficient systems

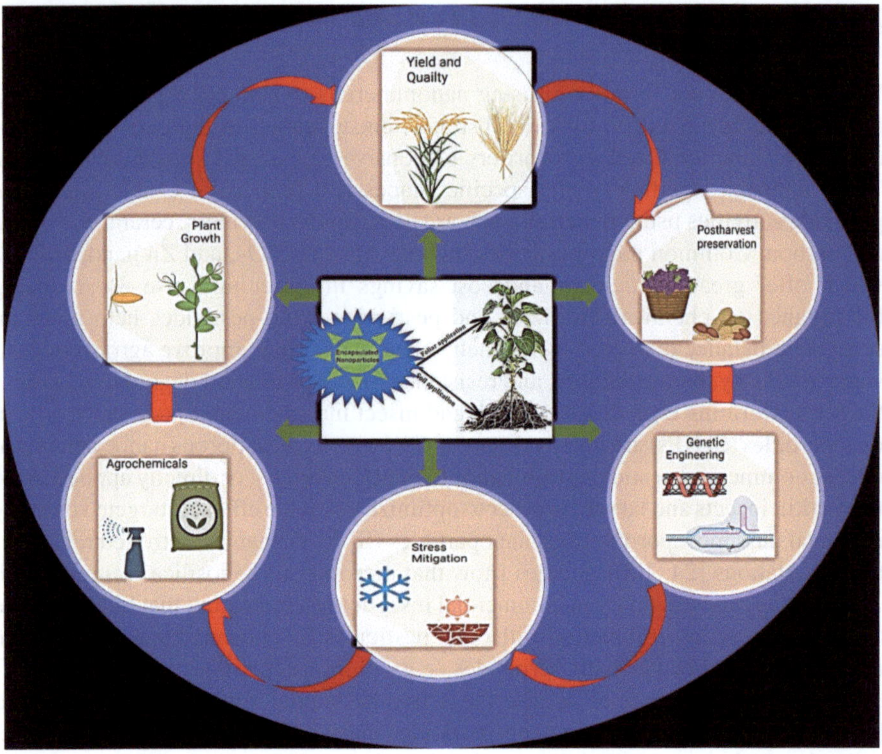

Fig. 6 Application of encapsulated nanoparticles (NPs) in plants demonstrates benefits such as sustainable microbial activity, weed control with significantly lower dosages, enhanced sustainable plant growth, and reduced toxicity to the environment and humans, and cost-effectiveness

for herbicide transport. Nano-herbicides effectively control various weeds, including *Echinochloa crus-galli, Chenopodium album, Bidens pilosa, Amaranthus viridis*, and *Raphanus raphanistrum*, offering an alternative to conventional herbicide formulations. Chitosan, acting as both a stabilizer and capping agent, is widely applied for controlled herbicide release, and has additional agricultural applications such as plant defense, growth stimulation, seed coating, frost protection, and sustained release of insecticides, fertilizers, and nutrients [192].

Nano-herbicides, with regulated release and gradual disintegration under natural conditions, can reduce herbicide toxicity and boost bioactive compound efficacy [19]. They remain effective at doses up to ten times lower than traditional herbicides, posing fewer environmental risks [125]. According to [97], nano-herbicides offer benefits such as improved chemical stability, solubility, bioavailability, photodecomposition, and soil adsorption. By incorporating biodegradable carriers like chitosan, these formulations can sustain weed control, reduce herbicide application frequency, and minimize residue in the environment. However, prolonged pesticide use has led to resistance in weeds and ecological contamination [27, 120]. Nano-herbicides present an effective alternative, enabling lower doses for weed management in wheat and other crops while mitigating environmental risks and enhancing herbicide efficiency.

7.4 Nanosensors in Agriculture: Smart Farming and Precision Monitoring

Nano-biosensors are transforming farming by providing precise, efficient, and cost-effective diagnostic tools for addressing a range of food, agricultural, and environmental issues. They enable rapid monitoring of temperature, traceability, humidity, heavy-metal ions, contaminants, microbial loads, and pathogens. In particular, high-surface-area carbon nanotubes and thin nanosensors are used to measure soil moisture, pesticide residues, and plant nutrient requirements under various environmental conditions. Their compact design, small size, high sensitivity, and affordability make nanosensors indispensable for modern agriculture. One notable application involves positioning sensors on plant leaves to detect hydrogen peroxide (H_2O_2) signaling waves. H_2O_2 acts as an internal messenger in plants, prompting leaf cells to produce defensive compounds against pests. Examples of nanosensors include films, polymer-based nanomaterials, nanoscale wires, and semiconducting carbon nanotubes, which can detect conductance changes upon exposure to specific chemicals. Silica nanoparticles (Si NPs) are also highly effective as nanosensors, offering superior sensor stability, accuracy, and sensitivity due to their core–shell structures [168]. When combined with other nanoparticles such as silver, silica enhances optical stability, and sensor sensitivity, thin-film configurations further stabilize colloidal silver [79]. These cutting-edge nano-biosensors deliver rapid and reliable data, helping farmers predict and mitigate potential crop losses while fostering sustainable agricultural practices (Fig. 7).

7.4.1 Potential Applications of Nanosensors in the Agri-food Sector

Nanosensors and related nanotechnologies have numerous agricultural applications, including monitoring soil conditions (moisture, pH, pesticides, herbicides, fertilizers, insecticides, pathogens) and crop growth, detecting food-borne contaminants, and tracking environmental conditions via nanochips. Nanocapsules facilitate the controlled release of pesticides, herbicides, fertilizers, and vaccines, while nano-based smart delivery systems optimize the use of water, nutrients, and chemicals in precision farming. Nanoparticles can carry growth hormones or DNA to plants, acting as smart nanosensors that provide early warnings of changing conditions. Aptasensors, meanwhile, are employed to detect pesticides, insecticides, antibiotics, drugs, and heavy metals such as silver (Ag), silicon (Si), and copper (Cu^{2+}) in plants, thereby safeguarding both crop yield and food safety.

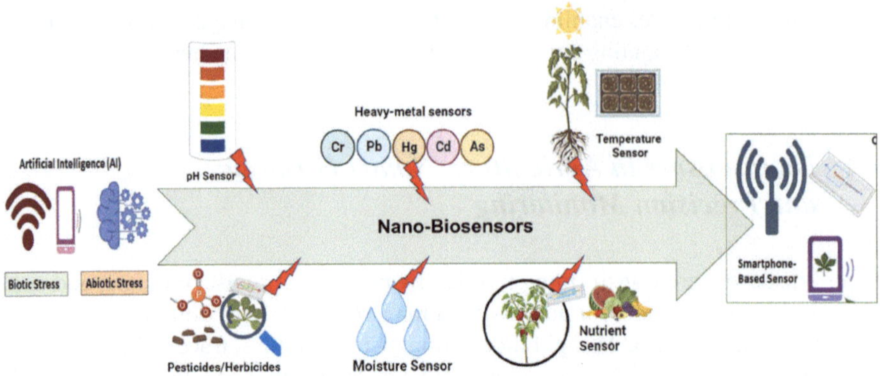

Fig. 7 Application of biosensors in agricultural systems to monitor plant health and environmental conditions. By collecting real-time data, these sensors inform decisions about applying nano-fertilizers and nano-pesticides, adjusting to changes in environmental factors, and ultimately improving crop management and productivity

8 Future Perspectives on Nanoparticles for Crop Enhancement

Nanotechnology holds significant promise for improving agricultural outcomes, particularly by enhancing nutrient management, reducing fertilizer loss, and minimizing pesticide usage. Engineered nanomaterials (ENMs) can influence plant growth, biomass production, and physiological parameters, ultimately affecting yield and crop quality. Researchers are especially interested in applying nanoparticles to plant science due to their beneficial effects—such as targeted fertilizer delivery, soil-quality monitoring via sensors, and efficient pest and disease management. However, while certain nanoparticles may exhibit toxicity or reactivity, others have demonstrated positive impacts on plant biomass and yield. To fully realize the potential of nano-biotechnology in agriculture, a deeper understanding of the molecular mechanisms activated by ENMs particularly in seed germination and plant growth will be essential for guiding future innovations.

9 Practical Applications of Nanoparticles in Agriculture

The agri-food sector has emerged as a major recipient of global investment, with over 300 nano-food products currently on the international market. According to a study by the German firm Helmuth Kaiser Consultancy, the market for nano-foods was projected to grow from USD 2.6 billion in 2004 to USD 20 billion by 2010, with Asia led by China expected to dominate this growth. Nanotechnology contributes to agriculture through crop enhancement, improved fertilizer and pesticide efficiency, soil management, plant disease detection, water management, gene expression analysis and regulation, and post-harvest technologies. Ranking second among potential nanotechnology applications in developing nations for achieving the Millennium Development Goals (after energy conversion/storage and ahead of water treatment), enhancing agricultural productivity underscores its global importance [147]. Furthermore, nanotechnology aids in environmental preservation by reducing pollution, remediating existing contaminants, and leveraging catalysts or renewable resources. Future agricultural systems may become "smart" with nanoscale devices that detect plant diseases before they become visible, initiating remedial actions autonomously and accelerating agricultural growth.

10 Conclusion

Nanotechnology stands out as a promising interdisciplinary field with applications across electronics, medicine, and particularly agriculture. In the face of global population growth and mounting environmental challenges like salinity, extreme temperatures, heavy-metal contamination, and water scarcity, nanotechnology-based approaches have emerged to enhance plant resilience. Numerous studies show that nanoparticles (NPs) can substantially bolster abiotic stress tolerance by modulating key physiological, biochemical, and molecular pathways in plants. Moreover, NPs can help mitigate heavy-metal pollution limiting bioaccumulation and reducing entry into the food chain. Although nano-pesticides, nano-herbicides, nanosensors, and nano-fertilizers demonstrate high reactivity and efficient plant uptake, their agronomic use remains primarily at the laboratory level. Future efforts must focus on designing self-degradable, non-toxic, cost-effective, and eco-friendly NPs to transition from controlled experiments to practical field applications. Such innovations will foster sustainable solutions to weed management, pathogenic threats, drought, and crop yield limitations. While the physicochemical properties of diverse NPs, along with environmental and crop-specific factors, can restrict large-scale deployment, continued research will refine nanotechnologies for safe and efficient agricultural use. Ultimately, judicious application of NPs can improve crop productivity, nutritional quality, and long-term environmental stewardship.

References

1. Abdel-Haliem ME, Hegazy HS, Hassan NS, Naguib DM (2017) Effect of silica ions and nano silica on rice plants under salinity stress. Ecol Eng 99:282–289
2. Acharya UR, Oh SL, Hagiwara Y, Tan JH, Adeli H (2018) Deep convolutional neural network for the automated detection and diagnosis of seizure using EEG signals. Comput Biol Med 100:270–278
3. Ahmadian K, Jalilian J, Pirzad A (2021) Nano-fertilizers improved drought tolerance in wheat under deficit irrigation. Agric Water Manag 244:106544

4. Ahmed Q, Gupta N, Kumar A, Nimesh S (2017) Antibacterial efficacy of silver nanoparticles synthesized employing *Terminalia arjuna* bark extract. Artif Cells, Nanomed, Biotechnol 45(6):1192–1200
5. Ahmed T, Noman M, Ijaz M, Ali S, Rizwan M, Ijaz U, Hameed A, Ahmad U, Wang Y, Sun G, Li B (2021) Current trends and future prospective in nanoremediation of heavy metals contaminated soils: a way forward towards sustainable agriculture. Ecotoxicol Environ Saf 227:112888
6. Ahmed B, Rizvi A, Ali K, Lee J, Zaidi A, Khan MS, Musarrat J (2021) Nanoparticles in the soil–plant system: a review. Environ Chem Lett 19:1545–1609
7. Ameen F, Alsamhary K, Alabdullatif JA, ALNadhari S (2021) A review on metal-based nanoparticles and their toxicity to beneficial soil bacteria and fungi. Ecotoxicol Environ Saf 213:112027
8. Amini S, Maali-Amiri R, Mohammadi R, Kazemi-Shahandashti SS (2017) CDNA-AFLP analysis of transcripts induced in chickpea plants by TiO_2 nanoparticles during cold stress. Plant Physiol Biochem 111:39–49
9. Aruoja V, Dubourguier HC, Kasemets K, Kahru A (2009) Toxicity of nanoparticles of CuO, ZnO and TiO_2 to microalgae Pseudokirchneriella subcapitata. Sci Total Environ 407(4):1461–1468
10. Askary M, Talebi SM, Amini F, Bangan ADB (2017) Effects of iron nanoparticles on *Mentha piperita L.* under salinity stress. Biologija 63(1)
11. Aslam M, Fakher B, Ashraf MA, Cheng Y, Wang B, Qin Y (2022) Plant low-temperature stress: signaling and response. Agronomy 12(3):702
12. Avestan S, Ghasemnezhad M, Esfahani M, Byrt CS (2019) Application of nano-silicon dioxide improves salt stress tolerance in strawberry plants. Agronomy 9(5):246
13. Azmat A, Tanveer Y, Yasmin H, Hassan MN, Shahzad A, Reddy M, Ahmad A (2022) Coactive role of zinc oxide nanoparticles and plant growth promoting *rhizobacteria* for mitigation of synchronized effects of heat and drought stress in wheat plants. Chemosphere 297:133982
14. Bais HP, Weir TL, Perry LG, Gilroy S, Vivanco JM (2006) The role of root exudates in rhizosphere interactions with plants and other organisms. Annu Rev Plant Biol 57(1):233–266
15. Bali R, Harris AT (2010) Biogenic synthesis of Au nanoparticles using vascular plants. Ind Eng Chem Res 49(24):12762–12772
16. Barot M, Modi A, Husen A (2025) Engineering soil microbiota for crop plant improvement. In: Husen A (ed) Agricultural crop improvement: plant and soil relationships. Taylor and Francis Group, LLC, 6000 Broken Sound Pkwy. NW, pp 56–63 https://doi.org/10.1201/978 1032630366-4
17. Barros D, Pradhan A, Pascoal C, Cássio F (2021) Transcriptomics reveals the action mechanisms and cellular targets of citrate-coated silver nanoparticles in a ubiquitous aquatic fungus. Environ Pollut 268:115913
18. Begum P, Ikhtiari R, Fugetsu B (2011) Graphene phytotoxicity in the seedling stage of cabbage, tomato, red spinach, and lettuce. Carbon 49(12):3907–3919
19. Bhaskar M, Kumar A, Rani R (2023) Application of nano formulations in agriculture. Biocatal Agric Biotechnol 102934
20. Bombo AB, Pereira AES, Lusa MG, de Medeiros Oliveira E, de Oliveira JL, Campos EVR, de Jesus MB, Oliveira HC, Fraceto LF, Mayer JLS (2019) A mechanistic view of interactions of a nanoherbicide with target organism. J Agric Food Chem 67(16):4453–4462
21. Chai H, Yao J, Sun J, Zhang C, Liu W, Zhu M, Ceccanti B (2015) The effect of metal oxide nanoparticles on functional bacteria and metabolic profiles in agricultural soil. Bull Environ Contam Toxicol 94:490–495
22. Chandra H, Patel D, Kumari P, Jangwan JS, Yadav S (2019) Phyto-mediated synthesis of zinc oxide nanoparticles of *Berberis aristata*: characterization, antioxidant activity and antibacterial activity with special reference to urinary tract pathogens. Mater Sci Eng, C 102:212–220
23. Chen J, Guo Y, Zhang X, Liu J, Gong P, Su Z, Fan L, Li G (2023) Emerging nanoparticles in food: sources, application, and safety. J Agric Food Chem 71(8):3564–3582

24. Chen J, Wang W, Xu Y, Zhang X (2011) Slow-release formulation of a new biological pesticide, pyoluteorin, with mesoporous silica. J Agric Food Chem 59(1):307–311
25. Chen F, Xiao Z, Yue L, Wang J, Feng Y, Zhu X, Wang Z, Xing B (2019) Algae response to engineered nanoparticles: current understanding, mechanisms and implications. Environ Sci Nano 6(4):1026–1042
26. Choi HS, Ashitate Y, Lee JH, Kim SH, Matsui A, Insin N, Bawendi MG, Semmler-Behnke M, Frangioni JV, Tsuda A (2010) Rapid translocation of nanoparticles from the lung airspaces to the body. Nat Biotechnol 28(12):1300–1303
27. Colbach N, Petit S, Chauvel B, Deytieux V, Lechenet M, Munier-Jolain N, Cordeau S (2020) The pitfalls of relating weeds, herbicide use, and crop yield: don't fall into the trap! a critical review. Front Agron 2:615470
28. Cox A, Venkatachalam P, Sahi S, Sharma N (2017) Reprint of: Silver and titanium dioxide nanoparticle toxicity in plants: a review of current research. Plant Physiol Biochem 110:33–49
29. Cui J, Liu T, Li F, Yi J, Liu C, Yu H (2017) Silica nanoparticles alleviate cadmium toxicity in rice cells: mechanisms and size effects. Environ Pollut 228:363–369
30. Cullen LG, Tilston EL, Mitchell GR, Collins CD, Shaw LJ (2011) Assessing the impact of nano-and micro-scale zerovalent iron particles on soil microbial activities: particle reactivity interferes with assay conditions and interpretation of genuine microbial effects. Chemosphere 82(11):1675–1682
31. Dimkpa CO, Andrews J, Sanabria J, Bindraban PS, Singh U, Elmer WH, Gardea-Torresdey JL, White JC (2020) Interactive effects of drought, organic fertilizer, and zinc oxide nanoscale and bulk particles on wheat performance and grain nutrient accumulation. Sci Total Environ 722:137808
32. Djanaguiraman M, Belliraj N, Bossmann SH, Prasad PV (2018) High-temperature stress alleviation by selenium nanoparticle treatment in grain sorghum. ACS Omega 3(3):2479–2491
33. Duhan JS, Kumar R, Kumar N, Kaur P, Nehra K, Duhan S (2017) Nanotechnology: the new perspective in precision agriculture. Biotechnol Rep 15:11–23
34. El-Saadony MT, Saad AM, Najjar AA, Alzahrani SO, Alkhatib FM, Shafi ME, Selem E, Desoky ESM, Fouda SE, El-Tahan AM, Hassan MA (2021) The use of biological selenium nanoparticles to suppress *Triticum aestivum L.* crown and root rot diseases induced by Fusarium species and improve yield under drought and heat stress. Saudi J Biol Sci 28(8):4461–4471
35. El-Zohri M, Al-Wadaani NA, Bafeel SO (2021) Foliar sprayed green zinc oxide nanoparticles mitigate drought-induced oxidative stress in tomato. Plants 10(11):2400
36. Elsheery NI, Sunoj VSJ, Wen Y, Zhu JJ, Muralidharan G, Cao KF (2020) Foliar application of nanoparticles mitigates the chilling effect on photosynthesis and photoprotection in sugarcane. Plant Physiol Biochem 149:50–60
37. Embiale A, Hussein M, Husen A, Sahile S, Mohammed K (2016) Differential sensitivity of *Pisum sativum* L. cultivars to water-deficit stress: changes in growth, water status, chlorophyll fluorescence and gas exchange attributes. J Agron 15:45–57
38. Esmaili S, Tavallali V, Amiri B (2021) Nano-silicon complexes enhance growth, yield, water relations and mineral composition in *Tanacetum parthenium* under water deficit stress. SILICON 13(8):2493–2508
39. Etesami H, Maheshwari DK (2018) Use of plant growth promoting rhizobacteria (PGPRs) with multiple plant growth promoting traits in stress agriculture: action mechanisms and future prospects. Ecotoxicol Environ Saf 156:225–246
40. Fajardo C, Martinez-Rodriguez G, Blasco J, Mancera JM, Thomas B, De Donato M (2022) Nanotechnology in aquaculture: applications, perspectives and regulatory challenges. Aquac Fish 7(2):185–200
41. Fatemi H, Esmaiel Pour B, Rizwan M (2021) Foliar application of silicon nanoparticles affected the growth, vitamin C, flavonoid, and antioxidant enzyme activities of coriander (*Coriandrum sativum L.*) plants grown in lead (Pb)-spiked soil. Environ Sci Pollut Res 28:1417–1425

42. Fouda A, Hassan SED, Saied E, Hamza MF (2021) Photocatalytic degradation of real textile and tannery effluent using biosynthesized magnesium oxide nanoparticles (MgO-NPs), heavy metal adsorption, phytotoxicity, and antimicrobial activity. J Environ Chem Eng 9(4):105346
43. Gao X, Avellan A, Laughton S, Vaidya R, Rodrigues SM, Casman EA, Lowry GV (2018) CuO nanoparticle dissolution and toxicity to wheat (*Triticum aestivum*) in rhizosphere soil. Environ Sci Technol 52(5):2888–2897
44. Gardea-Torresdey JL, Parsons JG, Gomez E, Peralta-Videa J, Troiani HE, Santiago P, Yacaman MJ (2002) Formation and growth of Au nanoparticles inside live *alfalfa* plants. Nano Lett 2(4):397–401
45. Getnet Z, Husen A, Fetene M, Yemata G (2015) Growth, water status, physiological, biochemical and yield response of stay green sorghum {*Sorghum bicolor* (L.) Moench varieties-a field trial under drought-prone area in Amhara regional state, Ethiopia. J Agron 14:188–202. https://doi.org/10.3923/ja.2015.188.202
46. Ghabel VK, Karamian R (2020) Effects of TiO_2 nanoparticles and spermine on antioxidant responses of *Glycyrrhiza glabra L.* to cold stress. Acta Botanica Croatica 79(2):137–147
47. Gohari G, Mohammadi A, Akbari A, Panahirad S, Dadpour MR, Fotopoulos V, Kimura S (2020) Titanium dioxide nanoparticles (TiO_2 NPs) promote growth and ameliorate salinity stress effects on essential oil profile and biochemical attributes of *Dracocephalum moldavica.* Sci Rep 10(1):912
48. Gomez-Garay A, Pintos B, Manzanera JA, Lobo C, Villalobos N, Martín L (2014) Uptake of CeO_2 nanoparticles and its effect on growth of *Medicago arborea* in vitro plantlets. Biol Trace Elem Res 161:143–150
49. Gottschalk F, Sonderer T, Scholz RW, Nowack B (2009) Modeled environmental concentrations of engineered nanomaterials (TiO_2, ZnO, Ag, CNT, fullerenes) for different regions. Environ Sci Technol 43(24):9216–9222
50. Hasanpour H, Maali-Amir R, Zeinali H (2015) Effect of TiO_2 nanoparticles on metabolic limitations to photosynthesis under cold in chickpea. Russ J Plant Physiol 62:779–787
51. Hassan NS, Salah El Din TA, Hendawey MH, Borai IH, Mahdi AA (2018) Magnetite and zinc oxide nanoparticles alleviated heat stress in wheat plants. Curr Nanomater 3(1):32–43
52. Hassanisaadi M, Barani M, Rahdar A, Heidary M, Thysiadou A, Kyzas GZ (2022) Role of agrochemical-based nanomaterials in plants: biotic and abiotic stress with germination improvement of seeds. Plant Growth Regul 97(2):375–418
53. Hayles J, Johnson L, Worthley C, Losic D (2017) Nanopesticides: a review of current research and perspectives. New pesticides and soil sensors, pp 193–225
54. Hojjat SS, Kamyab M (2017) The effect of silver nanoparticle on Fenugreek seed germination under salinity levels. Russ Agric Sci 43:61–65
55. Hojjat SS, Mozumder C, Bora T, Hornyak GL (2019) Polyvinylpyrrolidone-coated silver nanoparticle mitigation of salinity on germination and seedling parameters of bitter vetch (*Vicia ervilia L.*) plants. Nanotechnol Russ 14:582–587
56. Hossain Z, Mustafa G, Sakata K, Komatsu S (2016) Insights into the proteomic response of soybean towards Al_2O_3, ZnO, and Ag nanoparticles stress. J Hazard Mater 304:291–305
57. Husen A (2010) Growth characteristics, physiological and metabolic responses of teak (*Tectona grandis* Linn. f.) clones differing in rejuvenation capacity subjected to drought stress. Silvae Genetica 59:124–136. https://doi.org/10.1515/sg-2010-0015
58. Husen A, Iqbal M, Aref MI (2014) Growth, water status and leaf characteristics of *Brassica carinata* under drought stress and rehydration conditions. Braz J Botany 37:217–227. https://doi.org/10.1007/s40415-014-0066-1
59. Husen A, Iqbal M, Aref IM (2016) IAA-induced alteration in growth and photosynthesis of pea (*Pisum sativum* L.) plants grown under salt stress. J Environ Biol 37:421–429
60. Husen A, Iqbal M, Aref IM (2017) Plant growth and foliar characteristics of faba bean (*Vicia faba* L.) as affected by indole-acetic acid under water-sufficient and water-deficient conditions. J Environ Biol 38:179–186
61. Husen A, Iqbal M, Khanum N, Aref IM, Sohrab SS, Meshresa G (2019) Modulation of salt-stress tolerance of niger (*Guizotia abyssinica*), an oilseed plant, by application of salicylic acid. J Environ Biol 40:94–104

62. Husen A (2025b) Emerging carbon nanomaterials for sustainable agricultural practices (Synthesis, plant growth, performance, production and protection). Springer Nature Singapore Pte Ltd., 152 Beach Road, #21-01/04 Gateway East, Singapore 189721, Singapore. https://doi.org/10.1007/978-981-97-5104-4
63. Husen A (2025c) Nanosensors for agricultural and other related sectors (Towards sustainable use and future prospects) Springer Nature Singapore Pte Ltd., 152 Beach Road, #21-01/04 Gateway East, Singapore 189721, Singapore
64. Husen A, Iqbal M (2019) Nanomaterials and plant potential. Springer International Publishing AG, Gewerbestrasse 11, 6330 Cham, Switzerland. https://doi.org/10.1007/978-3-030-05569-1
65. Husen A, Siddiqi KS (2023) Advances in smart nanomaterials and their applications. Elsevier Inc., 50 Hampshire St., 5th Floor, Cambridge, MA 02139, USA. https://doi.org/10.1016/C2021-0-02202-1
66. Husen A (2023) Nanomaterials from agricultural and horticultural products. Springer Nature Singapore Pte Ltd., 152 Beach Road, #21-01/04 Gateway East, Singapore 189721, Singapore. https://doi.org/10.1007/978-981-99-3435-5
67. Husen A (2021a) Harsh environment and plant resilience (*Molecular and functional aspects*). Springer Nature Switzerland AG, Gewerbestrasse 11, 6330 Cham, Switzerland. https://doi.org/10.1007/978-3-030-65912-7
68. Husen A (2021b) The harsh environment and resilient plants: an overview. In: Husen A (ed) Harsh environment and plant resilience. Springer International Publishing AG, Gewerbestrasse 11, 6330 Cham, pp 1–23. https://doi.org/10.1007/978-3-030-65912-7_1
69. Husen A (2021c) Plant performance under environmental stress (*Hormones, biostimulants and sustainable plant growth management*). Springer Nature Switzerland AG, Gewerbestrasse 11, 6330 Cham, Switzerland. https://doi.org/10.1007/978-3-030-78521-5
70. Husen A (2021d) Cross talk between autophagy and hormones for abiotic stress tolerance in plants. In: Husen A (eds) Plant performance under environmental stress. Springer International Publishing AG, Gewerbestrasse 11, 6330 Cham, pp 1–15. https://doi.org/10.1007/978-3-030-78521-5_1
71. Husen A, Iqbal M, Sohrab SS, Ansari MKA (2018) Salicylic acid alleviates salinity-caused damage to foliar functions, plant growth and antioxidant system in Ethiopian mustard (*Brassica carinata* A. Br.). Agric Food Secur 7(44):1–14
72. Husen A (2022) Engineered nanomaterials for sustainable agricultural production, soil improvement and stress management. Elsevier Inc., 50 Hampshire St., 5th Floor, Cambridge, MA 02139, USA. https://doi.org/10.1016/C2021-0-00054-7
73. Husen A (2024) Plant response to gold nanoparticles. Springer Nature Singapore Pte Ltd., 152 Beach Road, #21-01/04 Gateway East, Singapore 189721, Singapore. https://doi.org/10.1007/978-981-97-7174-5
74. Husen A (2025a) Plant response to silver nanoparticles. Springer Nature Singapore Pte Ltd., 152 Beach Road, #21-01/04 Gateway East, Singapore 189721, Singapore. https://doi.org/10.1007/978-981-97-7352-7
75. Hussain A, Ali S, Rizwan M, ur Rehman MZ, Javed MR, Imran M, Chatha SAS, Nazir R (2018) Zinc oxide nanoparticles alter the wheat physiological response and reduce the cadmium uptake by plants. Environ Pollut 242:1518–1526
76. Hussain A, Ali S, Rizwan M, ur Rehman MZ, Qayyum MF, Wang H, Rinklebe J (2019) Responses of wheat (Triticum aestivum) plants grown in a Cd contaminated soil to the application of iron oxide nanoparticles. Ecotoxicol Environ Saf 173:156–164
77. Iqbal M, Ahmad A, Ansari MKA, Qureshi MI, Aref IM, Khan PR, Hegazy SS, El-Atta H, Husen A, Hakeem KR (2015) Improving the phytoextraction capacity of plants to scavenge metal(loid)-contaminated sites. Environ Rev 23:44–65. https://doi.org/10.1139/er-2014-0043
78. Iqbal M, Raja NI, Mashwani ZUR, Hussain M, Ejaz M, Yasmeen F (2019) Effect of silver nanoparticles on growth of wheat under heat stress. Iran J Sci Technol, Trans A: Sci 43:387–395

79. Jean RD, Chiu KC, Chen TH, Chen CH, Liu DM (2010) Functionalized silica nanoparticles by nanometallic Ag decoration for optical sensing of organic molecule. J Phys Chem C 114(37):15633–15639
80. Jeyasri R, Muthuramalingam P, Satish L, Pandian SK, Chen JT, Ahmar S, Wang X, Mora-Poblete F, Ramesh M (2021) An overview of abiotic stress in cereal crops: negative impacts, regulation, biotechnology and integrated omics. Plants 10(7):1472
81. Jin C, Fan J, Liu R, Sun R (2015) Single and joint toxicity of sulfamonomethoxine and cadmium on three agricultural crops. Soil Sediment Contam: Int J 24(4):454–470
82. Jin-Chul K, Madhusudhan A, Husen A (2021) Smart nanomaterials in biomedical applications. Springer Nature Switzerland AG, Gewerbestrasse 11, 6330 Cham, Switzerland. https://doi.org/10.1007/978-3-030-84262-8
83. Judy JD, Unrine JM, Rao W, Wirick S, Bertsch PM (2012) Bioavailability of gold nanomaterials to plants: importance of particle size and surface coating. Environ Sci Technol 46(15):8467–8474
84. Jullok N, Van Hooghten R, Luis P, Volodin A, Van Haesendonck C, Vermant J, Van der Bruggen B (2016) Effect of silica nanoparticles in mixed matrix membranes for pervaporation dehydration of acetic acid aqueous solution: plant-inspired dewatering systems. J Clean Prod 112:4879–4889
85. Kandhol N, Jain M, Tripathi DK (2022) Nanoparticles as potential hallmarks of drought stress tolerance in plants. Physiol Plant 174(2):e13665
86. Kapoor D, Sharma P, Sharma MMM, Yadav S, Husen A (2024) Exploring soil microbiota and their role in plant growth, stress tolerance, disease control and nutrient immobilizer. Biocatal Agric Biotechnol 61:103358. https://doi.org/10.1016/j.bcab.2024.103358
87. Kareem HA, Saleem MF, Saleem S, Rather SA, Wani SH, Siddiqui MH et al (2022) Zinc oxide nanoparticles interplay with physiological and biochemical attributes in terminal heat stress alleviation in mungbean (*Vigna radiata L.*). Front Plant Sci 13:842349
88. Karunakaran G, Suriyaprabha R, Rajendran V, Kannan N (2016) Influence of ZrO_2, SiO_2, Al_2O_3 and TiO_2 nanoparticles on maize seed germination under different growth conditions. IET Nanobiotechnol 10(4):171–177
89. Kashyap PL, Xiang X, Heiden P (2015) Chitosan nanoparticle based delivery systems for sustainable agriculture. Int J Biol Macromol 77:36–51
90. Khan A, Anwar Y, Hasan MM, Iqbal A, Ali M, Alharby HF, Hakeem KR, Hasanuzzaman M (2017) Attenuation of drought stress in Brassica seedlings with exogenous application of Ca2+ and H_2O_2. Plants 6(2):20
91. Khan I, Raza MA, Awan SA, Shah GA, Rizwan M, Ali B, Tariq R, Hassan MJ, Alyemeni MN, Brestic M, Zhang X (2020) Amelioration of salt induced toxicity in pearl millet by seed priming with silver nanoparticles (AgNPs): the oxidative damage, antioxidant enzymes and ions uptake are major determinants of salt tolerant capacity. Plant Physiol Biochem 156:221–232
92. Kirschling TL, Gregory KB, Minkley EG Jr, Lowry GV, Tilton RD (2010) Impact of nanoscale zero valent iron on geochemistry and microbial populations in trichloroethylene contaminated aquifer materials. Environ Sci Technol 44(9):3474–3480
93. Klaine SJ, Alvarez PJJ, Batley GE, Fernandes TF, Handy RD, Lyon DY et al (2008) Nanomaterials in the environment: behavior, fate, bioavailability, and effects. Environ Toxicol Chem 27(9):1825–1851
94. Klaine SJ, Alvarez PJJ, Batley GE, Fernandes TF, Handy RD, Lyon DY, Mahendra S, McLaughlin MJ, Lead JR (2008) Nanomaterials in the environment: behavior, fate, bioavailability, and effects. Environ Toxicol Chem 27(9):1825–1851
95. Krishnaraj C, Jagan EG, Rajasekar S, Selvakumar P, Kalaichelvan PT, Mohan NJCSBB (2010) Synthesis of silver nanoparticles using *Acalypha indica* leaf extracts and its antibacterial activity against water borne pathogens. Colloids Surf, B 76(1):50–56
96. Kumar A, Choudhary A, Kaur H, Mehta S, Husen A (2021) Metal-based nanoparticles, sensors and their multifaceted application in food packaging. J Nanobiotechnol 19(256):1–25. https://doi.org/10.1186/s12951-021-00996-0

97. Kumar S, Nehra M, Dilbaghi N, Marrazza G, Hassan AA, Kim KH (2019) Nano-based smart pesticide formulations: emerging opportunities for agriculture. J Control Release 294:131–153
98. Lead JR, Wilkinson KJ (2006) Aquatic colloids and nanoparticles: current knowledge and future trends. Environ Chem 3(3):159–171
99. Lenz HD, Haller E, Melzer E, Kober K, Wurster K, Stahl M, Bassham DC, Vierstra RD, Parker JE, Bautor J, Molina A (2011) Autophagy differentially controls plant basal immunity to biotrophic and necrotrophic pathogens. Plant J 66(5):818–830
100. Li ZZ, Chen JF, Liu F, Liu AQ, Wang Q, Sun HY, Wen LX (2007) Study of UV-shielding properties of novel porous hollow silica nanoparticle carriers for avermectin. Pest Manage Sci: Formerly Pesticide Sci 63(3):241–246
101. Lian J, Zhao L, Wu J, Xiong H, Bao Y, Zeb A, Tang J, Liu W (2020) Foliar spray of TiO_2 nanoparticles prevails over root application in reducing Cd accumulation and mitigating Cd-induced phytotoxicity in maize (*Zea mays L.*). Chemosphere 239:124794
102. Liu R, Lal R (2015) Potentials of engineered nanoparticles as fertilizers for increasing agronomic productions. Sci Total Environ 514:131–139
103. Lodriche SS, Soltani S, Mirzazadeh R (2012) U.S. patent application no. 13/406,538
104. Logaranjan K, Raiza AJ, Gopinath SC, Chen Y, Pandian K (2016) Shape-and size-controlled synthesis of silver nanoparticles using *Aloe vera* plant extract and their antimicrobial activity. Nanoscale Res Lett 11:1–9
105. López-Luna J, Silva-Silva MJ, Martinez-Vargas S, Mijangos-Ricardez OF, González-Chávez MC, Solís-Domínguez FA, Cuevas-Díaz MC (2016) Magnetite nanoparticle (NP) uptake by wheat plants and its effect on cadmium and chromium toxicological behavior. Sci Total Environ 565:941–950
106. López-Moreno ML, de la Rosa G, Hernández-Viezcas JA, Peralta-Videa JR, Gardea-Torresdey JL (2010) X-ray absorption spectroscopy (XAS) corroboration of the uptake and storage of CeO_2 nanoparticles and assessment of their differential toxicity in four edible plant species. J Agric Food Chem 58(6):3689–3693
107. Ma X, Geiser-Lee J, Deng Y, Kolmakov A (2010) Interactions between engineered nanoparticles (ENPs) and plants: phytotoxicity, uptake and accumulation. Sci Total Environ 408(16):3053–3061
108. Madhusudhan A, Purohit SD, Prasad R, Husen A (2024) Functional smart nanomaterials and their theranostics approaches. Springer Nature Singapore Pte Ltd., 152 Beach Road, #21-01/04 Gateway East, Singapore 189721, Singapore
109. Mandal AK, Ghorai S, Husen A (2023) Functionalized smart nanomaterials for point-of-care testing. Springer Nature Singapore Pte Ltd., 152 Beach Road, #21-01/04 Gateway East, Singapore 189721, Singapore
110. Maurer-Jones MA, Gunsolus IL, Murphy CJ, Haynes CL (2013) Toxicity of engineered nanoparticles in the environment. Anal Chem 85(6):3036–3049
111. Medina-Pérez G, Fernández-Luqueño F, Vazquez-Nuñez E, López-Valdez F, Prieto-Mendez J, Madariaga-Navarrete A, Miranda-Arámbula M (2019) Remediating polluted soils using nanotechnologies: environmental benefits and risks. Pol J Environ Stud 28(3):1013–1030
112. Miller GAD, Suzuki N, Ciftci-Yilmaz SULTAN, Mittler RON (2010) Reactive oxygen species homeostasis and signalling during drought and salinity stresses. Plant, Cell Environ 33(4):453–467
113. Mishra V, Mishra RK, Dikshit A, Pandey AC (2014) Interactions of nanoparticles with plants: an emerging prospective in the agriculture industry. In: Emerging technologies and management of crop stress tolerance. Academic press, pp 159–180
114. Mohamed AKS, Qayyum MF, Abdel-Hadi AM, Rehman RA, Ali S, Rizwan M (2017) Interactive effect of salinity and silver nanoparticles on photosynthetic and biochemical parameters of wheat. Arch Agron Soil Sci 63(12):1736–1747
115. Mohammadi R, Maali-Amiri R, Abbasi A (2013) Effect of TiO_2 nanoparticles on chickpea response to cold stress. Biol Trace Elem Res 152:403–410
116. Mondal R, Dam P, Chakraborty J, Paret ML, Katı A, Altuntas S, Sarkar R, Ghorai S, Mandal AK, Husen A (2022) Potential of nanobiosensor in sustainable agriculture: the state-of-art. Heliyon 8(12):e12207. https://doi.org/10.2139/ssrn.4134873

117. Mueller NC, Nowack B (2010) Nanoparticles for remediation: solving big problems with little particles. Elements 6(6):395–400
118. Mustafa G, Sakata K, Komatsu S (2015) Proteomic analysis of flooded soybean root exposed to aluminum oxide nanoparticles. J Proteomics 128:280–297
119. Mustafa N, Raja NI, Ilyas N, Ikram M, Mashwani ZUR, Ehsan M (2021) Foliar applications of plant-based titanium dioxide nanoparticles to improve agronomic and physiological attributes of wheat (*Triticum aestivum L.)* plants under salinity stress. Green Proc Synth 10(1):246–257
120. Nath CP, Singh RG, Choudhary VK, Datta D, Nandan R, Singh SS (2024) Challenges and alternatives of herbicide-based weed management. Agronomy 14(1):126
121. Nejatzadeh F (2021) Effect of silver nanoparticles on salt tolerance of *Satureja hortensis l.* during in vitro and in vivo germination tests. Heliyon 7(2)
122. Nel A, Xia T, Madler L, Li N (2006) Toxic potential of materials at the nanolevel. Science 311(5761):622–627
123. Van Nguyen D, Nguyen HM, Le NT, Nguyen KH, Nguyen HT, Le HM, Nguyen AT, Dinh NTT, Hoang SA, Van Ha C (2022) Copper nanoparticle application enhances plant growth and grain yield in maize under drought stress conditions. J Plant Growth Regul 41(1):364–375
124. Noman M, Ahmed T, Hussain S, Niazi MBK, Shahid M, Song F (2020) Biogenic copper nanoparticles synthesized by using a copper-resistant strain *Shigella flexneri* SNT22 reduced the translocation of cadmium from soil to wheat plants. J Hazard Mater 398:123175
125. Nsairat H, Lafi Z, Al-Sulaibi M, Gharaibeh L, Alshaer W (2023) Impact of nanotechnology on the oral delivery of phyto-bioactive compounds. Food Chem 424:136438
126. Nuruzzaman MD, Rahman MM, Liu Y, Naidu R (2016) Nanoencapsulation, nano-guard for pesticides: a new window for safe application. J Agric Food Chem 64(7):1447–1483
127. Onyekachi OG, Boniface OO, Gemlack NF, Nicholas N (2019) The effect of climate change on abiotic plant stress: a review. Abiotic Biotic Stress Plants 17
128. Palmqvist NM, Seisenbaeva GA, Svedlindh P, Kessler VG (2017) Maghemite nanoparticles acts as nanozymes, improving growth and abiotic stress tolerance in *Brassica napus.* Nanoscale Res Lett 12:1–9
129. Panyuta O, Belava V, Fomaidi S, Kalinichenko O, Volkogon M, Taran N (2016) The effect of pre-sowing seed treatment with metal nanoparticles on the formation of the defensive reaction of wheat seedlings infected with the eyespot causal agent. Nanoscale Res Lett 11:1–5
130. Punia P, Naagar M, Chalia S, Dhar R, Ravelo B, Thakur P, Thakur A (2021) Recent advances in synthesis, characterization, and applications of nanoparticles for contaminated water treatment—a review. Ceram Int 47(2):1526–1550
131. Qi M, Liu Y, Li T (2013) Nano-TiO_2 improve the photosynthesis of tomato leaves under mild heat stress. Biol Trace Elem Res 156:323–328
132. Raffi MM, Husen A (2019) Impact of fabricated nanoparticles on the rhizospheric microorganisms and soil environment. In: Husen A, Iqbal M (eds) Nanomaterials and plant potential. Springer International Publishing AG, Gewerbestrasse 11, 6330 Cham, pp 529–552. https://doi.org/10.1007/978-3-030-05569-1_21
133. Rai PK, Kumar V, Lee S, Raza N, Kim KH, Ok YS, Tsang DC (2018) Nanoparticle-plant interaction: implications in energy, environment, and agriculture. Environ Int 119:1–19
134. Rakgotho T, Ndou N, Mulaudzi T, Iwuoha E, Mayedwa N, Ajayi RF (2022) Green-synthesized zinc oxide nanoparticles mitigate salt stress in *Sorghum* bicolor. Agriculture 12(5):597
135. Ramegowda V, Senthil-Kumar M (2015) The interactive effects of simultaneous biotic and abiotic stresses on plants: mechanistic understanding from drought and pathogen combination. J Plant Physiol 176:47–54
136. Rassaei L, Marken F, Sillanpää M, Amiri M, Cirtiu CM, Sillanpää M (2011) Nanoparticles in electrochemical sensors for environmental monitoring. TrAC, Trends Anal Chem 30(11):1704–1715
137. Rastogi A, Zivcak M, Sytar O, Kalaji HM, He X, Mbarki S, Brestic M (2017) Impact of metal and metal oxide nanoparticles on plant: a critical review. Front Chem 5:78
138. Rico CM, Majumdar S, Duarte-Gardea M, Peralta-Videa JR, Gardea-Torresdey JL (2011) Interaction of nanoparticles with edible plants and their possible implications in the food chain. J Agric Food Chem 59(8):3485–3498

139. Rico CM, Morales MI, McCreary R, Castillo-Michel H, Barrios AC, Hong J, Tafoya A, Lee WY, Varela-Ramirez A, Peralta-Videa JR, Gardea-Torresdey JL (2013) Cerium oxide nanoparticles modify the antioxidative stress enzyme activities and macromolecule composition in rice seedlings. Environ Sci Technol 47(24):14110–14118
140. Rizwan M, Ali S, ur Rehman MZ, Riaz M, Adrees M, Hussain A, Zahir ZA, Rinklebe J (2021) Effects of nanoparticles on trace element uptake and toxicity in plants: a review. Ecotoxicol Environ Saf 221:112437
141. Rodriguez E, Parsons JG, Peralta-Videa JR, Cruz-Jimenez G, Romero-Gonzalez J, Sanchez-Salcido BE et al (2007) Potential of *Chilopsis linearis* for gold phytomining: using XAS to determine gold reduction and nanoparticle formation within plant tissues. Int J Phytorem 9(2):133–147
142. Rossi L, Zhang W, Lombardini L, Ma X (2016) The impact of cerium oxide nanoparticles on the salt stress responses of *Brassica napus L.* Environ Pollut 219:28–36
143. Rossi L, Zhang W, Ma X (2017) Cerium oxide nanoparticles alter the salt stress tolerance of *Brassica napus L.* by modifying the formation of root apoplastic barriers. Environ Pollut 229:132–138
144. Roy K, Sarkar CK, Ghosh CK (2015) Photocatalytic activity of biogenic silver nanoparticles synthesized using yeast (*Saccharomyces cerevisiae*) extract. Appl Nanosci 5:953–959
145. Sabo-Attwood T, Unrine JM, Stone JW, Murphy CJ, Ghoshroy S, Blom D, Bertsch PM, Newman LA (2012) Uptake, distribution and toxicity of gold nanoparticles in tobacco (*Nicotiana xanthi*) seedlings. Nanotoxicology 6(4):353–360
146. Salama HM (2012) Effects of silver nanoparticles in some crop plants, common bean (*Phaseolus vulgaris L.*) and corn (*Zea mays L.*). Int Res J Biotechnol 3(10):190–197
147. Salamanca-Buentello F, Persad DL, Court EB, Martin DK, Daar AS, Singer PA (2005) Nanotechnology and the developing world. PLoS Med 2(5):e97
148. Salehi-Lisar SY, Bakhshayeshan-Agdam H (2016) Drought stress in plants: causes, consequences, and tolerance. Drought stress tolerance in plants, vol 1: physiology and biochemistry, pp 1–16
149. Sanzari I, Leone A, Ambrosone A (2019) Nanotechnology in plant science: to make a long story short. Front Bioeng Biotechnol 7:120
150. Saravanakumar A, Peng MM, Ganesh M, Jayaprakash J, Mohankumar M, Jang HT (2017) Low-cost and eco-friendly green synthesis of silver nanoparticles using *Prunus japonica* (*Rosaceae*) leaf extract and their antibacterial, antioxidant properties. Artif Cells, Nanomed, Biotechnol 45(6):1165–1171
151. Schiavo S, Oliviero M, Miglietta M, Rametta G, Manzo S (2016) Genotoxic and cytotoxic effects of ZnO nanoparticles for Dunaliella tertiolecta and comparison with SiO_2 and TiO_2 effects at population growth inhibition levels. Sci Total Environ 550:619–627
152. Sedghi M, Mitra H, Sahar T (2013) Effect of nano zinc oxide on the germination of soybean seeds under drought stress. Ann West Univ Timiş Ser Biol 16:73–78
153. Shabnam N, Kim M, Kim H (2019) Iron (III) oxide nanoparticles alleviate arsenic induced stunting in *Vigna radiata*. Ecotoxicol Environ Saf 183:109496
154. Sharif F, Westerhoff P, Herckes P (2013) Sorption of trace organics and engineered nanomaterials onto wetland plant material. Environ Sci Process Impacts 15(1):267–274
155. Sharma P, Jha AB, Dubey RS, Pessarakli M (2012) Reactive oxygen species, oxidative damage, and antioxidative defense mechanism in plants under stressful conditions. J Bot 2012(1):217037
156. Sharma P, Pandey V, Sharma MMM, Patra A, Singh B, Mehta S, Husen A (2021) A review on biosensors and nanosensors application in agroecosystems. Nanoscale Res Lett 16(136):1–24. https://doi.org/10.1186/s11671-021-03593-0
157. Sharma NC, Sahi SV, Nath S, Parsons JG, Gardea-Torresde JL, Pal T (2007) Synthesis of plant-mediated gold nanoparticles and catalytic role of biomatrix-embedded nanomaterials. Environ Sci Technol 41(14):5137–5142
158. Sharma MMM, Kapoor D, Loyal A, Kumar R, Sharma P, Husen A (2024) Mitigation of drought stress in plants using gold nanoparticles. In: Husen A (eds) Plant response to gold

nanoparticles. Smart nanomaterials technology. Springer, Singapore. https://doi.org/10.1007/978-981-97-7174-5_10
159. Shayegan Mehr E, Sorbiun M, Ramazani A, Taghavi Fardood S (2018) Plant-mediated synthesis of zinc oxide and copper oxide nanoparticles by using *ferulago angulata* (*schlecht*) boiss extract and comparison of their photocatalytic degradation of Rhodamine B (RhB) under visible light irradiation. J Mater Sci: Mater Electron 29:1333–1340
160. Simonin M, Richaume A (2015) Impact of engineered nanoparticles on the activity, abundance, and diversity of soil microbial communities: a review. Environ Sci Pollut Res 22:13710–13723
161. Singh J, Lee BK (2016) Influence of nano-TiO_2 particles on the bioaccumulation of Cd in soybean plants (Glycine max): a possible mechanism for the removal of Cd from the contaminated soil. J Environ Manage 170:88–96
162. Singh A, Tiwari S, Pandey J, Lata C, Singh IK (2021) Role of nanoparticles in crop improvement and abiotic stress management. J Biotechnol 337:57–70
163. Singh S, Husen A (2020) Behavior of agricultural crops in relation to nanomaterials under adverse environmental conditions. In: Husen, A, Jawaid M (eds) Nanomaterials for agriculture and forestry applications. Elsevier Inc. 50 Hampshire St., 5th Floor, Cambridge, MA 02139, USA, pp 219–256. https://doi.org/10.1016/B978-0-12-817852-2.00009-3
164. Singh S, Husen A (2019) Role of nanomaterials in the mitigation of abiotic stress in plants. In: Husen A, Iqbal M (eds) N*anomaterials* and plant potential. Springer International Publishing AG, Gewerbestrasse 11, 6330 Cham, pp 441–471. https://doi.org/10.1007/978-3-030-05569-1_18
165. Song Y, Jiang M, Zhang H, Li R (2021) Zinc oxide nanoparticles alleviate chilling stress in rice (*Oryza Sativa L.*) by regulating antioxidative system and chilling response transcription factors. Molecules 26(8):2196
166. Steiner JL, Briske DD, Brown DP, Rottler CM (2018) Vulnerability of southern plains agriculture to climate change. Clim Change 146:201–218
167. Strout G, Russell SD, Pulsifer DP, Erten S, Lakhtakia A, Lee DW (2013) Silica nanoparticles aid in structural leaf coloration in the Malaysian tropical rainforest understorey herb *Mapania caudata*. Ann Bot 112(6):1141–1148
168. Sun D, Hussain HI, Yi Z, Rookes JE, Kong L, Cahill DM (2016) Mesoporous silica nanoparticles enhance seedling growth and photosynthesis in wheat and lupin. Chemosphere 152:81–91
169. Sun L, Song F, Guo J, Zhu X, Liu S, Liu F, Li X (2020) Nano-ZnO-induced drought tolerance is associated with melatonin synthesis and metabolism in maize. Int J Mol Sci 21(3):782
170. Taghiyari HR, Morrell JF, Husen A (2023) Emerging nanomaterials (Opportunities and challenges in forestry sectors). Springer Nature Switzerland AG, Gewerbestrasse 11, 6330 Cham, Switzerland. https://doi.org/10.1007/978-3-031-17378-3
171. Tan XM, Lin C, Fugetsu B (2009) Studies on toxicity of multi-walled carbon nanotubes on suspension rice cells. Carbon 47(15):3479–3487
172. Taran N, Storozhenko V, Svietlova N, Batsmanova L, Shvartau V, Kovalenko M (2017) Effect of zinc and copper nanoparticles on drought resistance of wheat seedlings. Nanoscale Res Lett 12:1–6
173. Thakur S, Asthir B, Kaur G, Kalia A, Sharma A (2021) Zinc oxide and titanium dioxide nanoparticles influence heat stress tolerance mediated by antioxidant defense system in wheat. Cereal Res Commun 1–12
174. Tiwari JN, Tiwari RN, Kim KS (2012) Zero-dimensional, one-dimensional, two-dimensional and three-dimensional nanostructured materials for advanced electrochemical energy devices. Prog Mater Sci 57(4):724–803
175. Tripathi DK, Singh VP, Prasad SM, Chauhan DK, Dubey NK (2015) Silicon nanoparticles (SiNp) alleviate chromium (VI) phytotoxicity in *Pisum sativum* (L.) seedlings. Plant Physiol Biochem 96:189–198
176. Tripathi DK, Singh S, Singh VP, Prasad SM, Dubey NK, Chauhan DK (2017) Silicon nanoparticles more effectively alleviated UV-B stress than silicon in wheat (*Triticum aestivum*) seedlings. Plant Physiol Biochem 110:70–81

177. Usman M, Farooq M, Wakeel A, Nawaz A, Cheema SA, ur Rehman H, Ashraf I, Sanaullah M (2020) Nanotechnology in agriculture: current status, challenges and future opportunities. Sci Total Environ 721:137778
178. Wang S, Fu Y, Zheng S, Xu Y, Sun Y (2022) Phytotoxicity and accumulation of copper-based nanoparticles in *brassica* under cadmium stress. Nanomaterials 12(9):1497
179. Wang Y, Liu Y, Zhan W, Zheng K, Lian M, Zhang C et al (2020) Long-term stabilization of Cd in agricultural soil using mercapto-functionalized nano-silica (MPTS/nano-silica): a three-year field study. Ecotoxicol Environ Saf 197:110600
180. Wang X, Yang X, Chen S, Li Q, Wang W, Hou C et al (2016) Zinc oxide nanoparticles affect biomass accumulation and photosynthesis in Arabidopsis. Front Plant Sci 6:1243
181. Wang Y, Lin Y, Xu Y, Yin Y, Guo H, Du W (2019) Divergence in response of lettuce *(var. ramosa Hort.)* to copper oxide nanoparticles/microparticles as potential agricultural fertilizer. Environ Pollut Bioavailab 31(1):80–84
182. Weijie X, Shuzhen H, AmanKhan M, Yu C, Linlin X, Zebin R, Liu H, Zhenhua C, Shengwei C, Ye Z, Liu D (2021) Effect of water and fertilization management on Cd immobilization and bioavailability in Cd-polluted paddy soil. Chemosphere 276:130168
183. Wu H, Tito N, Giraldo JP (2017) Anionic cerium oxide nanoparticles protect plant photosynthesis from abiotic stress by scavenging reactive oxygen species. ACS Nano 11(11):11283–11297
184. Xia T, Kovochich M, Liong M, Madler L, Gilbert B, Shi H, Yeh JI, Zink JI, Nel AE (2008) Comparison of the mechanism of toxicity of zinc oxide and cerium oxide nanoparticles based on dissolution and oxidative stress properties. ACS Nano 2(10):2121–2134
185. Xiong T, Leveque T, Shahid M, Foucault Y, Mombo S, Dumat C (2014) Lead and cadmium phytoavailability and human bioaccessibility for vegetables exposed to soil or atmospheric pollution by process ultrafine particles. J Environ Qual 43(5):1593–1600
186. Yadav T, Kumar A, Yadav RK, Yadav G, Kumar R, Kushwaha M (2020) Salicylic acid and thiourea mitigate the salinity and drought stress on physiological traits governing yield in pearl millet-wheat. Saudi J Biol Sci 27(8):2010–2017
187. Yazıcılar B, Böke F, Alaylı A, Nadaroglu H, Gedikli S, Bezirganoglu I (2021) In vitro effects of CaO nanoparticles on Triticale callus exposed to short and long-term salt stress. Plant Cell Rep 40:29–42
188. Zafar S, Hasnain Z, Aslam N, Mumtaz S, Jaafar HZ, Wahab PEM et al (2021) Impact of Zn nanoparticles synthesized via green and chemical approach on okra (*Abelmoschus esculentus L.*) growth under salt stress. Sustainability 13(7):3694
189. Zahra Z, Habib Z, Chung S, Badshah MA (2020) Exposure route of TiO_2 NPs from industrial applications to wastewater treatment and their impacts on the agro-environment. Nanomaterials 10(8):1469
190. Zhang LW, Monteiro-Riviere NA (2009) Mechanisms of quantum dot nanoparticle cellular uptake. Toxicol Sci 110(1):138–155
191. Zhang W, Xiao B, Fang T (2018) Chemical transformation of silver nanoparticles in aquatic environments: mechanism, morphology and toxicity. Chemosphere 191:324–334
192. Zhang M, Zhang F, Li C, An H, Wan T, Zhang P (2022) Application of chitosan and its derivative polymers in clinical medicine and agriculture. Polymers 14(5):958
193. Zhang H, Zhao Y, Zhu JK (2020) Thriving under stress: how plants balance growth and the stress response. Dev Cell 55(5):529–543

Comparative Assessment of Crop Performance Due to Exposure to Chemically and Green-Synthesized Nanoparticles Under Abiotic Stress

Karuvelan Murugan, Kaliyamoorthy Suganya, G. Hariharan, Ramachandran Chelliah, and Deog-Hwan Oh

Abstract Abiotic stress significantly affects agricultural productivity, with factors such as salinity, toxic metals, drought, and temperature variations reducing crop yields. Green nanotechnology offers an environmentally sustainable approach to mitigate these stresses while minimizing ecological harm. Unlike traditional physical and chemical nanoparticle (NP) synthesis methods, green synthesis leverages biological resources, making it a safer and more effective alternative. This chapter explores the role of green-synthesized NPs in enhancing abiotic stress resilience, improving plant defense mechanisms, and sustaining crop growth and yield.

Keywords Chemical synthesis · Green synthesis · Abiotic stresses · Nanoparticles · Stress resilience · Crop performance

K. Murugan · G. Hariharan
National Centre for Sustainable Coastal Management (NCSCM), Ministry of Environment, Forest and Climate Change, Chennai, India

K. Suganya
Department of Microbiology, Vels Institute of Science, Technology & Advanced Studies (VISTAS), Pallavaram, Chennai, Tamil Nadu, India

R. Chelliah (✉) · D.-H. Oh
Department of Food Science and Biotechnology, College of Agriculture and Life Sciences, Kangwon National University, Chuncheon, South Korea
e-mail: ramachandran865@gmail.com

R. Chelliah
Saveetha School of Engineering, SIMATS University, Sriperumbudur, Tamil Nadu, India

A. Husen (ed.), *Nanobiotechnology for Abiotic Stress Adaptation and Mitigation in Agricultural Crops*, Smart Nanomaterials Technology,
https://doi.org/10.1007/978-981-96-9298-9_12

1 Introduction

Abiotic stresses impact biomass production and crop viability, with environmental factors such as salinity, water scarcity, and temperature extremes leading to oxidative stress through reactive oxygen species (ROS) generation. These conditions degrade soil fertility, disrupt photosynthesis, and reduce agricultural output. Sustainable approaches, including nanotechnology-based solutions, are emerging to counteract these challenges. The abiotic stresses are inherent in organic and instinctual farming systems, yet the physical environment can amplify their unexpected and indirect impacts on plant life, affecting biomass, crop yield, and viability [28]. Major environmental challenges include salinity, water scarcity, cold stress, toxic chemicals, atmospheric pollutants, and extreme temperatures [76–68, 72, 73, 90]. Plants exhibit dynamic, multifaceted responses to these overlapping stressors, often generating reactive oxygen species (ROS). Unmitigated ROS can degrade soil quality and fertility, while also disrupting processes like photosynthesis, carbon adaptation, and membrane permeability, ultimately reducing crop yield [79, 91]. The Earth's atmosphere has experienced notable warming, limiting plant productivity and complicating efforts to meet global food demands. By 2025, the global population is projected to reach 9 billion, necessitating a 70% increase in food production [121]. Employing environmentally sustainable technologies that help plants mitigate abiotic stress is essential to boosting crop yields, curtailing dependence on chemical pesticides and fertilizers, and addressing environmental challenges. Agricultural and physiological methods are used to reduce stress impacts and enhance plant resilience, while the integration of nanotechnology into farming holds significant promise for boosting crop output and offering unprecedented advancements [80].

Nanoparticles (1–100 nm) exhibit unique properties such as high surface area, enhanced reactivity, and controlled release mechanisms, making them effective for agricultural applications. Their use in seed coating, foliar spray, and soil amendments enhances crop resilience to abiotic stresses by modulating physiological and biochemical processes. This chapter evaluates the comparative effects of chemically and green-synthesized nanoparticles in improving crop performance under stress conditions [146]. Notably, nanoparticle dosage can elicit both beneficial and detrimental biological outcomes: low nanomolar concentrations enhance plant growth and development, whereas higher doses disrupt metabolic functions [180]. In addition, research highlights the significant impact of nanoparticles [76, 77], heavy metals [190], salinity [193], and drought [117, 171] on crop responses to diverse environmental conditions. Over the years, various methodologies, including chemical, physical, and biological methods, have been used to synthesize metallic nanoparticles [13, 12, 132]. Studies have shown that nanoparticles can alter the biochemical, structural, physiological, and genetic processes in plants stressed by factors such as salt [148], drought [116, 115], and heavy metals [148], improving their ability to withstand these abiotic stresses [165, 166]. It has been suggested that nanoparticles are significantly more effective than other methods in enhancing plant resistance

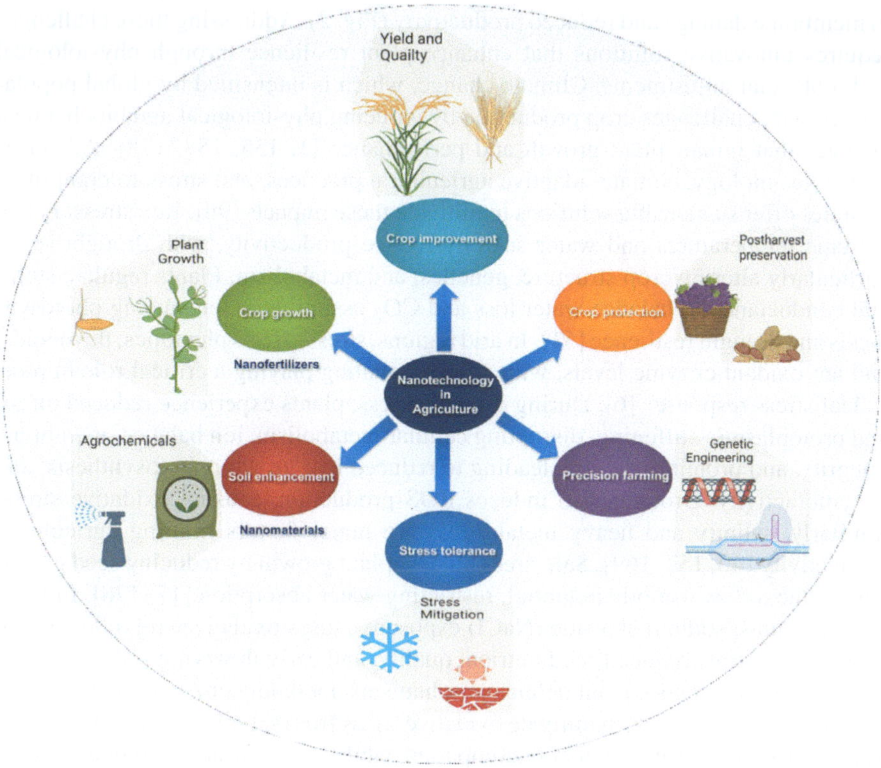

Fig. 1 Exploration on prospective correlation of nanotechnology within the agricultural sector. Heavy metals, commonly abbreviated as HM

to abiotic stress, with longer lasting effects. Nanoparticles possess unique properties, including a high response capacity, large surface area, varied pore sizes, and diverse morphologies, offering new research avenues in agriculture. Recent studies highlight an increase in nanoparticle-mediated therapies aimed at alleviating abiotic stress and optimizing plant growth, thus improving crop yields. The primary methods for nanoparticle application include seed coating, foliar spray, and soil drenching, all optimized to enhance nanoparticle uptake by plants [37]. The following section explores the crucial roles nanoparticles play in plant systems under abiotic stress conditions (Fig. 1).

2 Abiotic Stress and Its Impact on Crops

Abiotic stressors, including drought, salinity, heavy metal toxicity, and extreme temperatures, disrupt plant metabolism and growth [76–68]. These stresses alter photosynthesis, respiration, and nutrient uptake, triggering oxidative stress that leads to membrane damage and reduced productivity (Fig. 2). Addressing these challenges requires innovative solutions that enhance plant resilience through physiological and molecular adjustments. Climate change, which is intensified by global population growth, challenges crop production by inducing physiological and biochemical changes that impair plant growth and performance [1, 155, 158, 178]. Advances in nanotechnology, climate-adaptive agricultural practices, and stress-tolerant plant varieties offer sustainable solutions to mitigate these impacts [96]. Key stressors like extreme temperatures and water scarcity reduce productivity, with drought stress particularly altering crop structure, genetics, and metabolism. Plants regulate stomatal conductance to balance water loss and CO_2 assimilation, enhancing photosynthesis and drought resilience [51]. In arid regions, stress affects phenolics, flavonoids, and antioxidant enzyme levels, with xylem signaling playing a critical role in root-to-leaf stress responses [6]. During drought stress, plants experience reduced turgor and protoplasmic stiffening, disrupting cellular metabolism, ion balance, membrane integrity, and protein structure, leading to reduced leaf growth, photosynthesis, and enzyme activity. Drought also induces ROS production, causing oxidative stress. Similarly, salinity and heavy metal stress are major factors limiting agricultural productivity [40, 157, 169]. Salt stress hinders plant growth by reducing seed germination due to low osmotic potential, restricting water absorption [17, 178]. In leguminous plants, sodium chloride (NaCl) exposure causes oxidative stress, leading to growth inhibition, reduced seed nutrient quality, and early flowering [10].

Plants employ antioxidant defense mechanisms, including enzyme-mediated and nonenzymatic pathways, to mitigate oxidative stress from salinity. Proline (Pro) plays a critical role as an osmoregulatory compound, while ascorbate and carotenoids act as

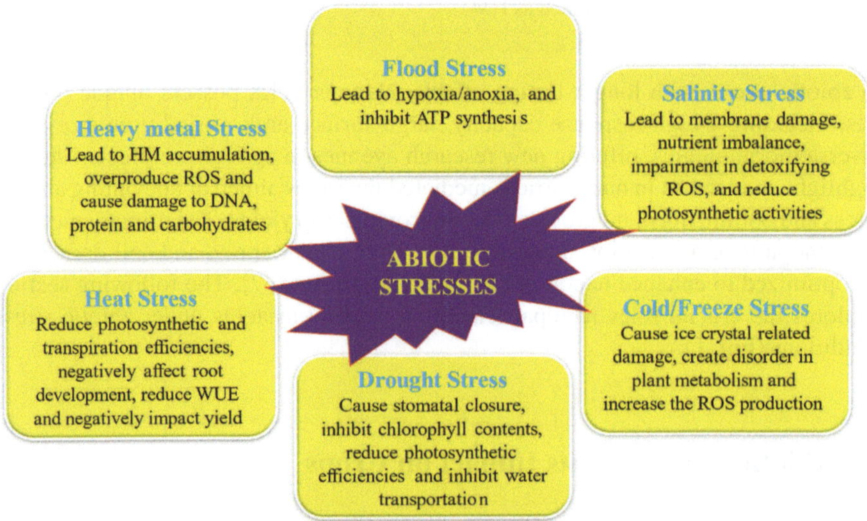

Fig. 2 The harmful effects of abiotic stress on plants. Reactive oxygen species (ROS); water usage efficiency (WUE); and heavy metals (HM)

nonenzymatic defenses against salinity stress [19]. Essential trace elements such as cobalt (Co), copper (Cu), iron (Fe), manganese (Mn), molybdenum (Mo), nickel (Ni), and zinc (Zn) support plant growth, while heavy metals like cadmium (Cd), chromium (Cr), lead (Pb), and mercury (Hg) are toxic at high concentrations. These toxic metals disrupt metabolic pathways, destabilize protein structures through binding to sulfhydryl groups, impair membrane integrity [52], and inhibit photosynthesis, respiration, and enzymatic activity [64].

3 Role of Nanoparticles in Alleviating Abiotic Stress

Nanoparticles (NPs) are minuscule particles that can penetrate plant tissues through above-ground routes (e.g., cuticle, epidermis, stomata, hydathodes) and below-ground pathways (e.g., root tips, cortex, lateral roots, injuries). Their effects on plant growth and physiology depend on factors such as plant type, developmental stage, application method, exposure level, and duration [147]. According to the pressure flow hypothesis, NPs entering stomata are transported through the phloem, driven by pressure differences between leaves and roots [186]. NPs influence physiological processes like sprouting, antioxidant activity, nutrient levels, chlorophyll content, chloroplast count, and photosynthetic efficiency [32]. In Arabidopsis, NPs affect root signaling via ethylene production [177]. NPs pass through the cell wall and membrane, enabling distribution to plant parts like leaves, xylem, and epidermis, progressing from passive transport through the epidermis to the central cylinder [82, 184].

3.1 Mechanisms of Nanoparticle Uptake and Transport in Plants

Nanoparticles penetrate plant tissues through root and foliar pathways, influencing stress responses at cellular levels. Their interactions with antioxidant enzymes, ROS scavenging mechanisms, and phytohormone signaling pathways improve stress tolerance and crop productivity. The active transport of nanoparticles in plants is significantly influenced by osmotic pressure, capillary tension, cell wall pore size, and plasma membrane structure in the root system, especially along the symplastic pathway [187]. Nanoparticles can bind to protein carriers like aquaporins or ion channels, disrupting plasma membrane integrity and forming pores for cellular entry. Their ability to cross membranes depends on size—smaller nanoparticles penetrate more easily [54], while larger ones enter through openings like hydathodes and stigmas [63]. Nanoparticles between 15 and 40 nm can pass through stomatal openings and move via the phloem to other plant parts [43, 184]. Table 1 highlights nanoparticles commonly used in agriculture, showcasing their role in enhancing crop resilience against abiotic stressors, as illustrated in Fig. 3.

During seed coating, nanoparticles penetrate cavities within the parenchymatous tissue cells of the seed coat, with water channels regulating their influx [99]. The high surface area-to-volume ratio of nanomaterials has advanced their applications in agriculture, with size and structure playing critical roles in their functionality [185]. As biofertilizers, nanoparticles enable controlled nutrient release, requiring smaller quantities than traditional fertilizers and reducing environmental risks [187]. The excessive use of conventional fertilizers and pesticides has spurred a shift toward nano-based alternatives to enhance crop yields and reduce environmental impacts [26]. Nano-fertilizers, sized 30–40 nm, pass through stomatal openings and interact with ions to facilitate nutrient release [23, 32]. Studies show that nano-fertilizers improve plant growth, hormone production, and resilience to stressors [32, 44, 87, 150]. Their production employs chemical, biological, and physical methods, supporting sustainable agricultural innovation.

3.2 Green-Synthesized Nanoparticles Versus Chemically Synthesized Nanoparticles

Chemical synthesis methods, including precipitation and microemulsion techniques, often involve toxic reagents that pose environmental risks. In contrast, green synthesis utilizes plant extracts, microorganisms, or biopolymers as reducing and stabilizing agents, reducing toxicity and ensuring sustainable nanoparticle production. Biogenic nanoparticles exhibit enhanced biocompatibility and greater efficiency in modulating plant stress responses. Nano-based pesticides containing elements such as silver (Ag), copper (Cu), silica (SiO_2), and zinc oxide (ZnO) have demonstrated superior efficacy compared to conventional pesticides, offering enhanced control of plant diseases and pests [31]. Additionally, chitosan–metal oxide nanoparticles have been employed to improve fertilizer uptake, with chitosan boosting enzymatic activities like nitrate reductase, protease, and glutamine synthetase in nitrogen metabolism, thereby promoting plant growth and development [23].

Studies have shown that nanoparticles can significantly influence seed germination and seedling development, with effects varying based on concentration, plant species, and nanoparticle type. For instance, ZnO nanoparticles enhanced seed germination and seedling growth in peanut and maize [137, 168]. However, high concentrations, such as 2000 mg L^{-1} of 60-nm aluminum (Al) nanoparticles, reduced root length in maize seedlings but did not affect plants like *Lolium perenne*, *Raphanus sativus*, and *Cucumis sativus* [197]. Similarly, Zn nanoparticles at the same concentration inhibited maize root growth and entirely arrested it in *Brassica oleracea*, *C. sativus*, *Daucus carota*, and *Glycine max* [103]. Despite these adverse effects, most studies suggest that nanoparticles, particularly SiO_2 and AgNPs, generally enhance germination rates, seedling vigor, and biomass in various crops, including

Table 1 The various types of nanoparticles and their effects on enhancing crop tolerance to different abiotic stresses

	Plant species	Nano particles	Administration	Impact	Key references
Salinity	*Lupins termis*	Zn	A suspension of nano-ZnO was introduced into half of these pots	The treatment stimulated plant growth, restored the concentrations of photosynthetic pigments, organic solutes, total phenols, ascorbic acid, and antioxidant enzymes, while simultaneously reducing the levels of malondialdehyde (MDA)	[4]
Salinity	Tomato	Zn	The application of foliar spray to the leaf area involves directly spraying a liquid solution onto the plant's leaves, allowing for the absorption of nutrients, chemicals	Antioxidant enzyme levels were significantly elevated, promoting improved root and shoot development. This, in turn, led to an increase in biomass and enhanced levels of photosynthetic pigments	[49]
Salinity	*Mangifera indica* L.	Zn	Application of foliar spray to the leaf area	The treatment resulted in increased concentrations of proline (Pro), total sugars, and antioxidant enzymes, contributing to improved stress tolerance and enhanced physiological responses in the plants	[48]
Drought	*Zea mays L. cv. Jidan 27*	Zn	A nano-ZnO suspension was introduced into half of the pots seven days after emergence, allowing for the assessment of its effects on plant growth and development	The production of melatonin was enhanced, leading to an improvement in the operation of the antioxidant system	[174]

(continued)

Table 1 (continued)

	Plant species	Nano particles	Administration	Impact	Key references
Mineral nutrient	Eggplant	Zn	Foliar application	The treatment resulted in enhanced relative water content and photosynthetic pigment levels, which contributed to improved fruit production and growth metrics	[159]
Salinity	*Prunus mahaleb*	Si	Added in water used for irrigation	The treatment enhanced seed germination and growth, improved photosynthetic metrics, and boosted antioxidant enzyme activity. It also led to decreased levels of MDA and H_2O_2, reduced chlorophyll degradation, and minimized oxidative damage	[21]
Cd toxicity	Wheat	Si	Seed priming	The treatment reduced oxidative stress, enhanced antioxidant enzyme activity, and diminished cadmium (Cd) content, particularly in the grains. Additionally, it led to an increase in silicon (Si) concentration within the plants	[74, 75]
Hg toxicity	Soybean	Si	Added in water used for irrigation	Inactivated and immobilized Hg	[101]
Cold	Wheat grass	Si	Incorporated into petri dishes holding seeds	Prevented seed dormancy, enhanced seedling weight and germination	[22]
Heat	Wheat	Ag	Added in water used for irrigation	Better plant development and resistance to high temperature	[76, 77]
Flood	Soybean	Ag	Nutrient solution has been incorporated into the seedling environment	Reduced cytotoxic glycolysis byproducts, increased stress-related protein abundance, and seedling development	[119]

(continued)

Table 1 (continued)

	Plant species	Nano particles	Administration	Impact	Key references
Drought	Linseed or flax	Ag	Added in water used for irrigation	Increased levels of pigments used for photosynthesis, improved development of crops, enhanced yield output, and reduced concentrations of hydrogen peroxide and MDA were observed	[8]
HMs	*Vigna radiata* L.	TiO_2	Exogenous application	The treatment induced antioxidant activity, resulting in decreased levels of MDA. It also enhanced plant growth and elevated biomass production	[86]
Cold	Licorice	TiO_2	Added in nutrient solution	The treatment led to a reduction in lipid peroxidation and hydrogen peroxide levels, while simultaneously elevating the levels of phenolics, total protein, and osmolytes	
Heat	Tomato	TiO_2	Tomato spraying	Regulating energy dissipation and increasing stomatal openness led to increased photosynthesis and leaf cooling	[138]
Salinity	Cucumber	Mn_3O_4	Foliar application	Enhanced pigments used for photosynthesis and net photosynthesis contributed to changes in biomass and metabolomes	[107]
Flooding	Soybean	Al_2O_3	Seed soaking	The ascorbate–glutathione process, the ability of the membranes to let things pass through, and the tricarboxylic acid cycle's functions were all controlled	[118]

(continued)

Table 1 (continued)

	Plant species	Nano particles	Administration	Impact	Key references
Salinity	Grape	Fe	In nutrient media	The overall amount of protein exhibited an increase, whereas the levels of Pro content, antioxidant enzyme action, and hydrogen per oxide demonstrated a decline. Decreased sodium ions and increased potassium ions	[116]
Salinity	Bell peppers	Mn	Foliar application	The treatment enhanced seed germination and root development, while also modifying the expression of specific genes related to growth and stress response	

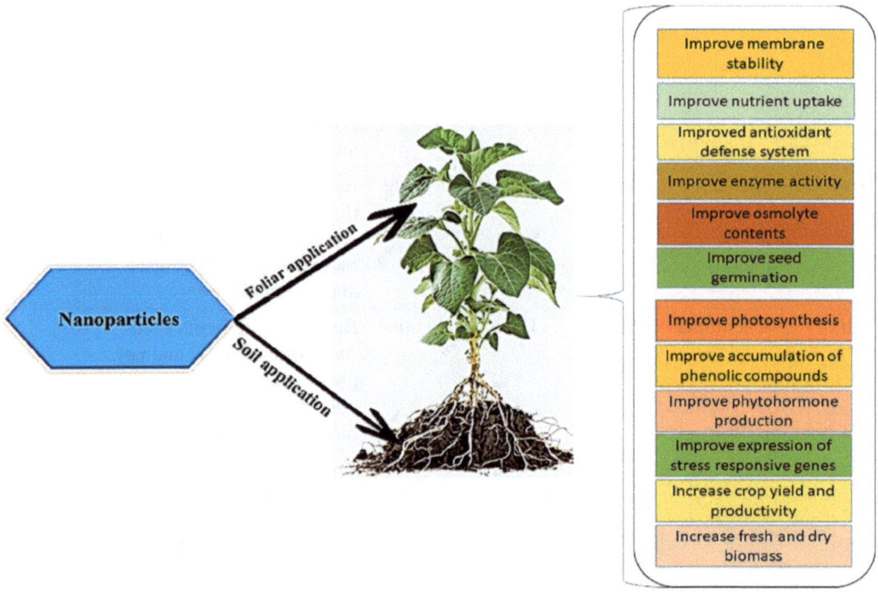

Fig. 3 The impact of nanoparticles on various physiological indicators, growth performance, and overall plant health, highlighting their potential to enhance plant well-being. One key parameter, Water Use Efficiency (WUE), is also represented, demonstrating how nanoparticles may influence water management in plants, contributing to more efficient water usage during growth

Agropyron elongatum [22], *Thymus kotschyanus* [3], and *Pennisetum glaucum* [135]. However, AgNPs have negatively affected germination and growth in some species, such as *Brassica nigra*, *Medicago sativa*, and *Ocimum basilicum* [18, 46]. These findings highlight the need for optimized concentrations and targeted applications to maximize benefits while minimizing potential toxic effects.

The application of SiO_2 nanoparticles in *Onobrychis sativa* promoted shoot length, while TiO_2 nanoparticles improved germination rate and percentage [113]. Fe_2O_3 nanoparticles accelerated *L. perenne* germination [195] and influenced mycorrhizal biomass, as well as shoot and root length in *Trifolium repens*, *Satureja hortensis*, and *H. vulgare* and *L. perenne* [46]. TiO_2 nanoparticles positively impacted germination in *Foeniculum vulgare* and *Petroselinum crispum* [34], while AgNPs enhanced shoot length and chlorophyll content in *Brassica juncea* and *Sorghum bicolor* [161]. SiO_2 nanoparticles also stimulated root growth in *T. kotschyanus* and *Alopecurus textilis* [2] and increased plant height, tiller count, yield, FW, DW, chlorophyll, and carotenoid content in *M. sativa* [108]. However, inhibitory effects on shoot and root growth were observed in *S. bicolor*, *Stipa hohenackeriana*, and *Secale montanum* [112]. ZnO nanoparticles have been widely reported to enhance biomass, root and shoot length, and chlorophyll content [198], while TiO_2 nanoparticles increased essential oil content and yield in medicinal plants [11]. Conversely, CuO nanoparticles adversely affected the morphology, physiology, and biochemistry of *H. vulgare*, *L. perenne*, *M. sativa*, and *Triticum aestivum* [62].

4 Classification of Nanoparticles

Nanoparticles are classified into three main categories based on their composition: organic, carbon-based, and inorganic nanoparticles [41].

4.1 Nanoparticles of Organic Matter

Organic nanoparticles, composed of polymers, lipids, proteins, carbohydrates, and other organic compounds, include examples like dendrimers, liposomes, micelles, and complex proteins such as ferritin. These nanoparticles are often environmentally friendly, biodegradable, and feature a hollow core structure akin to liposomes [81]. They exhibit high sensitivity to electromagnetic and thermal radiation and increased brittleness [126]. Their chemical composition, surface morphology, rigidity, and transport capacity play critical roles in determining their applications.

4.2 Nanoparticles Composed of Carbon

Carbon-based nanoparticles include fullerenes, carbon black nanoparticles, and carbon quantum dots. Fullerenes, particularly C60, feature a symmetrical closed-cage structure of 60 carbon atoms in a football-like shape, with other variants like C70 and C540 also identified [106]. Carbon black nanoparticles consist of strongly fused spherical aggregates resembling grape clusters [198]. Carbon quantum dots, measuring less than 10 nm, represent another notable type of carbon nanoparticle. Carbon-based nanoparticles have a wide range of applications, including drug delivery, energy storage [104], and bioimaging [29]. Their unique properties, such as high strength and strong electron affinity [89], make them highly versatile. Advanced forms like nanodiamonds and carbon nano onions are distinguished by their low toxicity and excellent biocompatibility, making them ideal for tissue engineering and drug delivery systems [9].

4.3 Inorganic Nanoparticles

Inorganic nanoparticles lack carbon or biological components and include semiconductor, metallic, and ceramic nanoparticles. Metallic nanoparticles (MNPs), made from metal precursors, can be monometallic, bimetallic, or polymetallic [124]. Bimetallic nanoparticles are synthesized as alloys or layered structures, such as core–shell configurations, exhibiting unique optical and electrical properties due to localized surface plasmon resonances [89]. Additionally, some MNPs possess distinct biological, magnetic, and thermal characteristics, making them essential in nanodevice development [41, 114]. Precise control over size, shape, and surface facets is critical for their synthesis in advanced material applications [38]. Semiconductor nanoparticles, distinct from bulk materials due to their tunable wide bandgaps, exhibit unique metallic and non-metallic properties, enabling significant property modifications [89]. They are pivotal in photocatalysis and widely applied in optical and electrical devices [60, 175]. Ceramic nanoparticles, composed of inorganic solids like metal and metalloid oxides (e.g., calcium, titanium), carbides, phosphates, and carbonates, feature various structural forms—amorphous, polycrystalline, dense, porous, or hollow—achieved through heating and controlled cooling [182, 89]. Their high stability and load capacity make them ideal for biomedical applications and versatile in catalysis, dye degradation, photonics, and optoelectronics [182].

5 Approaches to Nanoparticle Synthesis

Nanoparticles are synthesized using top-down and bottom-up approaches. The top-down method, including mechanical and mechanochemical synthesis, reduces bulk materials into nanoparticles using ball milling or pulverization, though it often results in irregular shapes, wide size distributions, and impurities [98, 196]. In contrast, the bottom-up approach synthesizes nanoparticles in gas or liquid phases. Gas-phase methods can produce toxic byproducts, while liquid-phase techniques, such as precipitation, microemulsion, and solution–gel synthesis, are more controlled. The Stöber method [172], widely used for mesoporous nanoparticles, employs tetraethyl orthosilicate, ammonia, and alcohol-based solvents. Additionally, biosynthesis using microorganisms, algae, and plant extracts offers an eco-friendly alternative [61, 199]. The structural features of nanoparticles, such as porosity and size, are strongly influenced by the synthesis method [33]. In the solution–gel method, adjusting the reaction solution's pH controls nanoparticle size [172]. These structural properties impact nanoparticle uptake and distribution in biological systems [102, 200]. Pore size is determined by the surfactants used during liquid-phase synthesis, with longer chain surfactants producing larger pores and shorter ones yielding smaller pores [55, 123]. Higher porosity accelerates nanoparticle dissolution, which affects their degradation rate and subsequent efficacy. Rapidly dissolving nanoparticles have been shown to improve plant biomass and fruit yield [59, 83].

5.1 Environmentally Sustainable Synthesis of Nanoparticles Using Plant Extracts

Green nanotechnology, or phyto-nanotechnology, is rapidly advancing across scientific disciplines, focusing on environmentally friendly nanoparticle synthesis to minimize ecological impact [35]. Plants can produce diverse nanoparticles, including gold (Au), palladium (Pd), copper (Cu), iron (Fe), silver (Ag), platinum (Pt), selenium (Se), and zinc (Zn) (Table 2). These nanoparticles are distributed across different plant structures, such as stems, leaves, roots, flowers, bark, fruits, epidermis, and seeds [80–71]. Green nanoparticles (NPs), primarily synthesized from botanical and microbial sources, are gaining prominence for their accessibility and eco-friendliness. These methods enable cost-efficient production while reducing reliance on mineral ores and minimizing adverse environmental and human health effects.

5.1.1 Role of Roots in Nanoparticle Synthesis and Their Unique Bio-reducing Agents

Roots play a significant role in green nanoparticle synthesis due to their biochemical properties, which include reducing and stabilizing agents. Various root extracts

Table 2 Application of botanical methods in the production of various nanoparticles

Nanoparticles	Utilized components of the plant	Plant species	The dimensions and morphology of the nanoparticle	Key references
FeO	Flower	*Avicennia marina*	30–100 nm	[105]
Al_2O_3	Leaves	*Aerva lanata*	50–70 nm spherical	[39]
Fe_3O_4	Fruit	*Couroupita guianensis*	17 ± 10 nm	[154]
ZnO	Flower	*Moringa oleifera*	13.2 nm triangle shape	[127]
FeO	Leaves	*Ramalina sinensis*	Spherical 20–70 nm	[20]
ZnO	Root	*Rubus fairholmianus*	Spherical 11.34 nm	[149]
Al_2O_3	Seeds	*Terminalia chebula*	50–100 nm spherical	[39]
Pt	Leaves	*Ocimum sanctum*	23 nm irregular	[78]
CeO_2	Leaves	*Azadirachta indica*	10–50 nm	[143]
NiO	Leaves	*Rauwolfia serpentina*	40 nm	[139]
Samarium oxide (Sm_2O_3)	Leaves	*Andrographis paniculata*	30–50 nm cubic structure	[120]
Ag	Fruit	*Terminalia chebula*	25 nm spherical	[42]
Au	Leaves	*Combretum erythrophyllum*	Not specified	[50]
NiO	Leaves	*Rauwolfia serpentina*	40 nm	[139]
Ag	Leaves	*Eclipta prostrate*	35–60 nm hexagons	[140]

contain phytochemicals like oxalic acid, ascorbic acid, phenylpropanoids, and zingerone, which facilitate nanoparticle synthesis. For instance, root extracts of *Zingiber officinale* are rich in bioactive compounds that enable the formation of silver (AgNPs) and gold (AuNPs) nanoparticles [189]. Likewise, *Withania somnifera* root extracts have been used for the biosynthesis of titanium dioxide nanoparticles (TiO_2NPs), as confirmed by energy-dispersive X-ray spectral (EDS) analysis [16]. Other root-based biosynthesized nanoparticles include AgNPs from *Rheum turkestanicum* and Fe_0O_4NPs from *Mimosa pudica*, which have been characterized using X-ray diffraction (XRD) and Fourier transform infrared (FTIR) spectroscopy [129, 151].

5.1.2 Leaf-Mediated Nanoparticle Synthesis: A Rich Source of Bioactive Compounds

Leaves serve as an excellent biological template for nanoparticle synthesis due to their high content of bioactive compounds. Leaf extracts of various plant species have been utilized to synthesize a wide range of nanoparticles with distinct properties. For example, *Diospyros kaki* leaf extract has been used to produce platinum nanoparticles (PtNPs) [170]. Additionally, the adsorption properties of *Ocimum sanctum* leaf extract have facilitated the synthesis of nickel oxide nanoparticles (NiONPs), which have proven effective in the removal of anionic pollutants from wastewater [134]. Leaves have also been instrumental in producing rare-earth nanoparticles such as lanthanum oxide (La_2O_3NPs) with *Eucalyptus globulus* and neodymium oxide (Nd_2O_3NPs) with *Andrographis paniculata*, demonstrating their versatility in nanoparticle synthesis [109, 176].

5.1.3 Contribution of Stems to Nanoparticle Biosynthesis and Their Industrial Applications

Stems, being a structural component of plants, are a valuable source for nanoparticle biosynthesis. The rich bioactive composition of stems enables the synthesis of metallic and bimetallic nanoparticles. Silver nanoparticles (AgNPs) have been synthesized from *Salvadora persica*, *Momordica charantia*, and *Coleus aromaticus* stem extracts [15, 188]. Furthermore, bimetallic nanoparticles such as Fe_3O_4-AgNPs have been derived from *Vitis vinifera* stem extracts, exhibiting strong magnetic properties and potential for industrial applications [191]. Selenium nanoparticles (SeNPs) synthesized from *Leucas lavandulifolia* stem extract have shown unique morphological features and high stability [93] (Fig. 4).

5.1.4 Bud-Derived Nanoparticles and Their Unique Biochemical Properties

Bud extracts, known for their high bioactive content, play a significant role in nanoparticle synthesis. These extracts act as both oxidizing and reducing agents, facilitating controlled nanoparticle formation. *Syzygium aromaticum* buds have been extensively used for synthesizing nanoparticles such as gold (AuNPs), silver (AgNPs) [97], palladium (PdNPs) [160], and copper (CuNPs) [141]. *Couroupita guianensis* bud extracts, rich in phenolics, flavonoids, and proteins, have also demonstrated effective AgNP synthesis [95, 128]. Additionally, bud-derived *Tussilago farfara* extracts have been successfully used to synthesize spherical AgNPs and AuNPs [100].

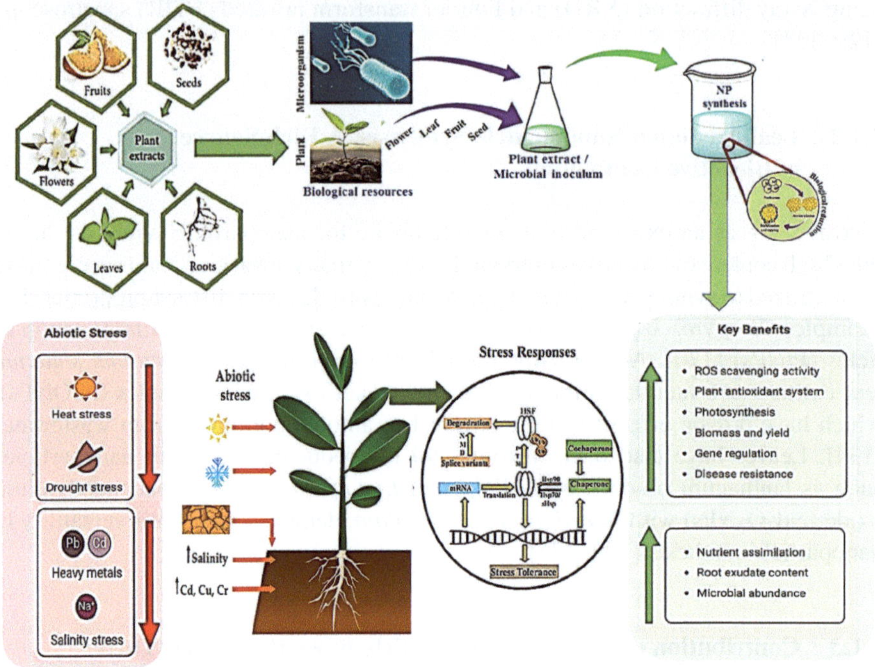

Fig. 4 Schematic representation of eco-friendly production of nanoparticles from a variety of biological sources and their positive impacts on plants experiencing abiotic stress

5.1.5 Flower-Based Nanoparticle Synthesis for Agricultural and Environmental Applications

Flowers are widely recognized for their abundance of bioactive compounds, including flavonoids, anthocyanins, carotenoids, and phenolics, making them ideal candidates for green nanoparticle synthesis. Flower-derived nanoparticles exhibit remarkable antimicrobial and antioxidant properties, enhancing plant tolerance to abiotic stress. Flower-mediated synthesis has been extensively used to produce gold (AuNPs) and silver nanoparticles (AgNPs) [7]. Kumar et al. [94] proposed a model for flower-mediated nanoparticle synthesis, underscoring its potential for sustainable agricultural application.

5.1.6 Fruit-Extracted Nanoparticles and Their Structural Versatility

Fruits, rich in bioactive compounds, have been effectively used in the synthesis of nanoparticles with diverse structural properties. Citrus extracts from *Citrus limon*, *Citrus reticulata*, and *Citrus sinensis* have enabled the production of various gold nanostructures, including nanospheres, nanotriangles, and nanoprisms [173]. Similarly, *Emblica officinalis* extracts have shown potential in synthesizing silver nanoparticles (AgNPs) [144]. These studies highlight the suitability of fruit extracts for the eco-friendly production of nanoparticles with diverse shapes and applications.

5.1.7 Seeds

Seed extracts offer a non-toxic, cost-effective approach to nanoparticle synthesis. The phytochemicals present in seeds, such as reducing sugars, phenolic compounds, and carbohydrates, facilitate nanoparticle formation and stabilization [36]. Seed extract concentrations can be adjusted to control nanoparticle size and morphology. Silver nanoparticles (AgNPs) have been synthesized from *Coffea italica* seed extracts and *Sinapis arvensis* seeds [92]. Additionally, zinc oxide nanoparticles (ZnONPs) from *Peganum harmala* seeds and iron oxide nanoparticles (Fe_2O_3NPs) from *Punica granatum* seeds have demonstrated significant photocatalytic activity and heavy metal adsorption potential [53]. These findings emphasize the adaptability of seed extracts for sustainable nanoparticle production.

6 Nanoparticle Applications in Mitigating Specific Abiotic Stresses

6.1 Salinity-Induced Stress

Salinity disrupts water uptake and nutrient balance, leading to osmotic stress and ion toxicity. Soil salinization, driven by ecological degradation, climate change, and poor irrigation practices, affects approximately 20% of irrigated agricultural land. Salt stress disrupts molecular, physiological, and biochemical processes, reducing both food quality and quantity. It causes osmotic and ionic imbalances, leading to increased reactive oxygen species (ROS) and toxic ion accumulation, which damage biomolecules, induce lipid peroxidation, and result in cell death [24].

Green-synthesized ZnO and Se nanoparticles improve salt tolerance by enhancing antioxidant enzyme activity, maintaining ionic homeostasis, and reducing oxidative stress. Recent studies emphasize the potential of green-synthesized nanoparticles (NPs) for enhancing salt stress tolerance in plants. Vegetative priming with ZnO nanoparticles (5–10 mg/L) from *Agathosma betulina* improved shoot length, fresh weight, and physiological health in *S. bicolor* exposed to 400 mM NaCl, reducing the Na/K ratio and oxidative stress [142]. Similarly, green-synthesized gold nanoparticles (AuNPs) enhanced salt stress recovery in *T. aestivum* through improved K/Na ratios, nitrogen assimilation, chlorophyll content, stomatal regulation, and ROS/RNS management [194]. Selenium nanoparticles (SeNPs) from *Allamanda cathartica* increased germination rate (31%), shoot length (92%), root length (78%), and chlorophyll content (49%) in *Brassica campestris* under 200 mM NaCl stress [153]. Additionally, SeNPs from fresh grape extracts (30 mg/L) boosted antioxidant enzyme activities (SOD, CAT, APX, POX) and phenolic and flavonoid contents in *B. juncea*, enhancing salt tolerance [153].

Biosynthesized nanoparticles (NPs) have shown remarkable potential in alleviating salt stress in plants. SeNPs and ZnONPs improved growth, antioxidant activity, and phytochemical accumulation in garlic under high salinity (14 dS m). Similarly, AgNPs derived from *Rosmarinus officinalis* enhanced photosynthetic pigments, soluble sugars, proline, and proteins in *Linum usitatissimum*, while reducing hydrogen peroxide and malondialdehyde levels [88]. Foliar application of ZnONPs from *Moringa* leaf extract improved salt tolerance in *Vicia faba* by increasing phenolic content, proline levels, and antioxidant enzyme activity (PPO, APX, CAT), alongside better chlorophyll content and reduced Na levels [57]. Green SeNPs synthesized by *Bacillus cereus* TAH boosted wheat germination, vigor index, shoot length, and dry weight in high electrical conductivity soil [57]. Collectively, green-synthesized NPs, including SeNPs, ZnONPs, and AuNPs, enhance growth, antioxidant defense, and oxidative damage regulation, offering sustainable solutions to improve crop resilience in saline-affected regions.

6.2 Drought Stress

Drought affects photosynthetic efficiency and water regulation. Drought stress, driven by global warming, monsoon shifts, and irregular precipitation, significantly hampers plant growth and agricultural productivity by affecting structural, physiological, chemical, and genomic traits, ultimately reducing photosynthetic capacity [155, 156]. Drought-induced effects such as stomatal closure, membrane damage, disrupted enzyme activities, reduced ATP synthesis, and altered water balance exacerbate plant wilting and resilience challenges [24]. Green nanoparticles offer an eco-friendly solution to mitigate drought stress by maintaining cellular homeostasis and enhancing plant tolerance. For example, foliar application of green ZnO nanoparticles (25–50 mg/L) on tomato plants increased shoot and root biomass, improved ascorbic acid and free phenol levels, boosted antioxidant enzyme activities (SOD, APX, CAT), and reduced oxidative stress markers such as H_2O_2 and MDA by 25% under severe drought conditions [47]. Green FeO and ZnO nanoparticles mitigate drought stress by enhancing chlorophyll content, improving root water uptake, and modulating stress-responsive proteins.

In hydroponic systems, green-synthesized ZnO nanoparticles (75 nm) from *Lawsonia inermis* demonstrated superior stress alleviation compared to chemically synthesized ZnO nanoparticles, as evidenced by enhanced antioxidant enzyme activities, biomass, and shoot and root lengths [163]. Similarly, applying green ZnO_x nanoparticles (100 ppm) derived from *Camellia sinensis* leaf extract to *Coriandrum sativum* increased chlorophyll and proline levels, reduced MDA levels, and strengthened antioxidant defenses [90]. Biosynthesized selenium nanoparticles (SeNPs) from

Allium sativum buds improved wheat drought resistance by enhancing plant height, leaf surface area, and biomass under water-limited conditions. Additionally, FeONPs derived from *Z. officinale* and *Cuminum cyminum* seed extracts increased wheat drought tolerance by boosting proline, total iron, soluble sugar, and chlorophyll content while reducing lipid peroxidation and electrolyte leakage [133].

Magnetite nanoparticles (Fe_0O_4) synthesized from *Polyalthia longifolia* leaves improved growth, photosynthetic pigments, and overall photosynthetic performance in *Trigonella foenum-graecum* under drought stress [25]. Similarly, nano-fertilizers made from FeONPs using marine algae (*Chaetomorpha antennina*) enhanced drought-stressed *Setaria italica* by increasing iron, chlorophyll, and soluble carbohydrate content [171]. Phyto-synthesized FeONPs (100 ppm) from *Prosopis cineraria* leaf extracts, applied via nano-priming and foliar methods, significantly improved wheat morphological and yield traits, providing drought resilience in arid regions [167]. These studies highlight the potential of green nanoparticles to enhance drought tolerance by boosting growth, photosynthesis, and antioxidant defenses while reducing oxidative damage, offering a sustainable solution for agriculture in water-limited environments.

6.3 Heavy Metal Toxicity

Heavy metals such as cadmium (Cd), nickel (Ni), mercury (Hg), and arsenic (As) accumulate in plant tissues, causing oxidative damage, which pose serious risks to plant health despite some being essential micronutrients [58]. Prolonged heavy metal exposure induces oxidative stress by overproducing reactive oxygen species (ROS), damaging membranes, proteins, pigments, and nucleic acids while disrupting antioxidant systems. Cost-effective, eco-friendly nanoparticles have shown promise in mitigating heavy metal toxicity in crops [56, 111]. Green nanoparticles reduce metal toxicity by facilitating immobilization, enhancing antioxidant responses, and promoting nutrient uptake; for instance, silver (AgNPs) and copper oxide nanoparticles (CuONPs) synthesized from *Catharanthus roseus* leaf extract reduced chromium and cadmium contamination in soil and water [192]. Additionally, biogenic hydroxyapatite nanoparticles (HAP-NPs) from *Aspergillus niger* fermentation significantly lowered cadmium levels in mung bean seedlings, reducing hydrogen peroxide (H_2O_2) and malondialdehyde (MDA) concentrations while enhancing antioxidant enzyme activities, mitigating oxidative stress, and promoting seedling regrowth [162].

Biogenic nanoparticles have shown significant potential in mitigating heavy metal stress in plants. MgO nanoparticles derived from *Enterobacter sp.* strain RTN2 enhanced rice resistance to arsenic, improving biomass, antioxidant enzyme activity, and reducing ROS and arsenic translocation at 200 mg/kg [13, 12]. Similarly, SiNPs synthesized from potassium silica fluoride and *Aspergillus tubingensis* AM11 improved *Phaseolus vulgaris* growth, photosynthesis, and nutrient uptake while reducing electrolyte leakage, MDA, H_2O_2, and heavy metal levels, including Na, Pb, Cd, and Ni [45]. Green CuNPs from *Klebsiella pneumoniae* alleviated chromium stress in wheat, enhancing growth, biomass, and antioxidant defenses while reducing ROS and Cr translocation at 25–50 mg/kg [130]. Additionally, CuNPs (100 mg/kg) synthesized from *Shigella flexneri* SNT22 reduced cadmium stress in wheat, increasing plant length, dry weight, nutrient content, and ion balance while decreasing Na accumulation and immobilizing Cd in soil [131]. FeONPs from *Adiantum lunulatum* leaf extract aided rice in coping with arsenic stress, improving germination, seedling growth, and reducing arsenic uptake and oxidative stress [30]. Molybdenum nanoparticles (MoNPs) combined with *Bacillus* sp. strain ZH16 reduced arsenic uptake by 30.3%, enhancing nutrient composition, ionic balance, and growth under arsenic stress [14]. These findings underscore the role of biogenic nanoparticles in mitigating heavy metal stress, promoting plant resilience, and enhancing growth in contaminated environments.

6.4 Temperature Extremes

Extreme temperatures, whether excessive heat or cold, pose significant challenges to plant growth and productivity. High temperatures disrupt enzymatic activities, denature proteins, and impair metabolic processes, leading to reduced photosynthetic efficiency, increased transpiration, and oxidative stress. On the other hand, low temperatures cause membrane rigidity, slow enzymatic reactions, and lead to the accumulation of harmful reactive oxygen species (ROS), which can damage cellular structures. Biosynthesized silver (Ag) and titanium dioxide (TiO_2) nanoparticles have shown promise in enhancing plant thermotolerance. These nanoparticles contribute to stress mitigation by stabilizing cell membranes, protecting photosynthetic pigments, and modulating the expression of heat shock proteins (HSPs) and cold-responsive genes. Ag nanoparticles, due to their antimicrobial properties, also help prevent secondary infections that can occur under extreme temperature stress. TiO_2 nanoparticles, known for their photocatalytic properties, enhance light absorption and energy conversion efficiency, thereby improving photosynthetic performance under high temperatures. Additionally, these nanoparticles have been reported to enhance antioxidant enzyme activities, including superoxide dismutase (SOD) and catalase (CAT), which scavenge excess ROS and reduce oxidative damage in plant cells. By improving cellular homeostasis, maintaining osmotic balance, and reducing protein degradation, biosynthesized Ag and TiO_2 nanoparticles provide a robust protective mechanism against temperature extremes, ultimately leading to better crop resilience and sustained agricultural productivity.

6.5 Additional Abiotic Factors

Green-synthesized nanoparticles have shown promise in mitigating a range of abiotic stressors beyond drought, salinity, and heavy metal stress. For example, wheat plants treated with 75 mg/L of biosynthesized silver nanoparticles (AgNPs) derived from *Moringa oleifera* extract under heat stress conditions (35–40 °C for 3 h daily over three days) exhibited notable improvements in growth parameters. Root length increased by 5.4%, shoot length by 26.1%, root number by 7.5%, fresh weight by 2%, dry weight by 0.6%, leaf area by 33.8%, leaf number by 4.8%, fresh leaf weight by 0.15%, and dry leaf weight by 0.18% compared to untreated controls [76, 77]. These findings highlight the potential of green nanoparticles in enhancing plant resilience to heat and other abiotic stresses.

Green-synthesized nanoparticles have demonstrated versatility in mitigating various abiotic stressors, offering eco-friendly solutions to enhance plant resilience and productivity under adverse conditions. For example, biogenic nanoparticles synthesized from green tea (20 mg per plant) applied via foliar spraying and soil watering reduced oxidative stress in *Lactuca sativa* by lowering ascorbate peroxidase (APX) and catalase (CAT) levels while increasing R-S-nitroso thiols (RSNOs), phenolic compounds, and total flavonoid content, thereby improving stress tolerance. Similarly, foliar application of 25 mg/L of biogenic AgNPs was as effective as ethylene diurea (EDU), a conventional anti-ozonate agent, in alleviating the effects of tropospheric ozone (O_3) stress in wheat cultivars HD-2967 and DBW-17. This treatment enhanced growth, physiology, antioxidant defenses, and yield during vegetative and reproductive stages, showcasing its potential as a promising ozone-protectant agent [84].

7 Environmental and Ecotoxicological Considerations

Despite their promising applications, the widespread use of nanoparticles (NPs) in agriculture raises significant environmental and ecotoxicological concerns. One of the primary issues is the accumulation of nanoparticles in soil and water systems, which can lead to unintended consequences for microbial communities, soil health, and aquatic ecosystems. The persistence and mobility of nanoparticles depend on their composition, size, surface charge, and interaction with organic and inorganic matter in the environment. In soil, nanoparticles may undergo various transformations, including aggregation, dissolution, and interactions with soil organic matter, which can influence their bioavailability and toxicity. Studies suggest that metal-based nanoparticles such as silver (Ag), zinc oxide (ZnO), and titanium dioxide (TiO_2) can alter soil microbial diversity, affecting the composition of essential bacterial and fungal populations. This disruption in microbial communities can impair nutrient cycling processes, such as nitrogen fixation and organic matter decomposition, ultimately reducing soil fertility. Additionally, certain nanoparticles have been found to induce oxidative stress in earthworms and other soil organisms, leading to potential disruptions in soil ecosystems. In aquatic environments, nanoparticles can enter water systems through agricultural runoff, leaching, and atmospheric deposition. Their interaction with aquatic organisms depends on factors such as particle size, surface coating, and exposure duration. Nanoparticles can accumulate in sediments, where they may undergo transformations that affect their stability and toxicity. Some studies have reported bioaccumulation of nanoparticles in fish, invertebrates, and algae, raising concerns about their potential bio-magnification along the food chain. Furthermore, nanoparticles with antimicrobial properties, such as silver and copper oxide (CuO), could negatively impact beneficial aquatic microbes, which play a crucial role in maintaining water quality and ecosystem stability.

To address these challenges, future research should prioritize the development of environmentally safe nanoparticle formulations that minimize toxicity while maintaining efficacy. Studies must focus on understanding the fate, transformation, and degradation of nanoparticles in different environmental matrices, including soil, water, and plant tissues. Long-term field studies are essential to assessing the cumulative effects of nanoparticles on soil health, water quality, and non-target organisms. Moreover, regulatory frameworks and safety guidelines must be established to ensure the responsible use of nanoparticles in agriculture. Standardized protocols for toxicity testing, risk assessment, and environmental monitoring should be implemented to mitigate potential hazards. It is also essential to exploring green synthesis methods that utilize biodegradable stabilizers and plant-based reducing agents to create more environmentally friendly nanoparticles. Encouraging interdisciplinary collaborations among chemists, biologists, agronomists, environmental scientists, and policymakers will be crucial in designing sustainable nanotechnology applications that balance agricultural benefits with ecological safety. These measures will be essential to harnessing the advantages of nanotechnology while preventing unintended environmental and ecological consequences.

8 The Reprogramming of Plant Traits Through the Mediation of Green-Synthesized Nanoparticles

Research on phytogenic nanoparticles has highlighted their ability to enhance plant traits, particularly during early growth stages such as seed germination. Titanium dioxide nanoparticles (TiO_2NPs) synthesized from *C. cyminum* seed extract improved germination indices in *Vigna italica* at 50 μg/mL [110]. Similarly, gold nanoparticles (AuNPs) derived from *Tiliacora triandra* leaves at 2000 mg/L enhanced *Oryza sativa* germination [125], while *Allium cepa*-based AuNPs at 5.4 $\mu g/mL^3$ increased seedling growth and yield [5]. Cerium oxide nanoparticles (Ce_2O_3NPs) from *Elaeagnus angustifolia* leaves promoted growth and metabolism in *Solanum lycopersicum* at 20–100 mg/L [164]. Similarly, copper oxide nanoparticles (CuONPs) from *Sesbania italica* leaves (25–30 mg/100 mL) improved *B. nigra* growth [179].

Zinc oxide (ZnONPs) and magnesium oxide (MgONPs) nanoparticles from *Aloe barbadensis* leaves enhanced growth, root, and shoot lengths in *Vigna radiata* and *Cajanus cajan* at 5 mg [145]. Phytogenic nanoparticles also impact nutrient uptake, photosynthetic pigment biosynthesis, and antioxidant defenses. Sulfur nanoparticles (SNPs) from *P. granatum* increased nutrient content in *S. lycopersicum* fruits at 200 ppm [152], while *Cinnamomum zeylanicum*-derived SNPs improved osmolyte levels and reduced stress indicators like malondialdehyde (MDA) and hydrogen peroxide (H_2O_2) in *L. sativa* [122]. Additionally, ZnONPs (50 mg/L) from *Lemna minor* enhanced *Zea mays* growth and oxidative defense mechanisms [27]. These findings underscore the potential of green nanoparticles in reprogramming plant traits to improve growth, stress tolerance, and crop productivity.

Nanoparticles regulate gene expression associated with plant defense and secondary metabolite production. For example, silver nanoparticles (AgNPs) synthesized from *Vitex negundo* upregulated anthocyanin regulatory genes (PAP1, ANS, PAL) in *Brassica rapa*, resulting in increased anthocyanin accumulation. Similarly, *Withania coagulans*-derived AgNPs (1 mg/L) elevated mRNA expression levels of antioxidant genes, including SOD, CAT, glutathione reductase (GR), and ascorbate peroxidase (APX) [183]. Iron oxide nanoparticles (Fe_2O_3NPs) biosynthesized from *Hydrangea paniculata* flowers enhanced antioxidant enzyme activities (POX, CAT) and seed vigor in *L. usitatissimum* at 1000 mg/L. Additionally, copper oxide nanoparticles (CuONPs) synthesized from *A. lunulatum* extracts increased phenolic and flavonoid content, boosting antioxidant enzyme activity in *Lens culinaris* [153]. These findings highlight the transformative potential of phytogenic nanoparticles in reprogramming plant traits by enhancing growth, stress tolerance, secondary metabolite production, and antioxidant defenses. This offers sustainable solutions for improving crop productivity and resilience under adverse environmental conditions.

9 Nanoparticles Influence Phytohormonal Activity in Response to Abiotic Stress Conditions

Phytohormones, essential chemical messengers, regulate metabolic pathways and coordinate cellular functions, impacting plant growth and development [136]. Green-synthesized nanoparticles (NPs) have been shown to modulate phytohormonal activity under abiotic stress, offering promising applications in sustainable agriculture [181]. Recent studies reveal that NPs interact with phytohormones, influencing signaling pathways and plant physiological traits. For instance, copper (CuNPs) and silver nanoparticles (AgNPs) derived from *Justicia spicigera* leaf extracts (250–1000 ppm) enhanced root growth in *Annona muricata* by modulating indole-3-acetic acid (IAA) and gibberellic acid (GA_3) levels. Similarly, iron nanoparticles synthesized from *A. cepa* extract (20–160 mg/L) increased photosynthetic pigments and nonenzymatic antioxidants in *Citrullus lanatus*, boosting jasmonic acid (JA) biosynthesis via enhanced uptake of its precursor, 12-oxophytodienic acid (OPDA) [85].

Under salinity stress, AgNPs (300 ppm) from *T. aestivum* leaf extract improved proline metabolism, nitrogen assimilation, chlorophyll content, and stomatal regulation by reducing abscisic acid (ABA) levels, thereby enhancing antioxidant defenses [193]. AgNPs (1 mg/L) synthesized from *Capparis spinosa* further mitigated salinity stress in *T. aestivum* by decreasing ABA levels while increasing indole-3-butyric acid (IBA), 1-naphthalene acetic acid (NAA), and 6-benzylaminopurine (BAP), improving photosynthetic efficiency and reducing oxidative damage. These findings highlight the potential of nanoparticles in modulating phytohormonal activity to enhance plant resilience under stress. The application of TiO_2 nanoparticles (40 ppm) to *T. aestivum* improved phytohormonal interactions with IAA and GA, enhancing photosynthesis, amino acid and carbohydrate metabolism, and antioxidant enzyme activities (SOD, POX, CAT), thereby promoting drought resilience [117]. Green-synthesized nanoparticles consistently influence endogenous phytohormones under both stress and optimal conditions. However, further research is essential to unraveling the complex interactions between nanoparticles and phytohormones. Such studies could advance the strategic use of nanoparticles to enhance crop resilience and productivity, fostering more sustainable agricultural practices.

10 Conclusion and Future Perspectives

Nanotechnology presents a promising avenue for enhancing crop resilience against abiotic stresses. Green-synthesized nanoparticles, due to their biocompatibility and sustainability, are preferable over chemically synthesized alternatives. However, further research is needed to bridge the gap between laboratory findings and large-scale agricultural applications. Emphasis should be placed on optimizing formulation strategies, minimizing environmental risks, and ensuring regulatory compliance for sustainable implementation. The study of abiotic stresses and their impact on agricultural yields has become a critical area of research. This chapter emphasizes the potential of nanoparticles to mitigate abiotic stressors by enhancing antioxidant capacity, nutrient absorption, photosynthetic efficiency, and regulating biochemical and molecular processes in plants. Nano-fertilizers, in particular, provide essential nutrients to strengthen plant resilience against such stresses. Despite these advancements, concerns regarding the ecological implications of nanoparticles persist, necessitating further research to assess their interactions with ecosystems, residual effects, and impacts on plant and soil systems. Studies on nanoparticle longevity, ecotoxicological risks, and their effects on the metabolome, proteome, metagenome, and transcriptome are crucial. Bridging the gap between controlled laboratory results and field applications remains a major challenge, requiring optimized formulations and long-term effect evaluations for real-world agricultural settings. Scaling up nanoparticle production through the development of manufacturing industries, educating farmers about nano-formulations, and enhancing application methodologies are vital for widespread adoption. Addressing environmental considerations will ensure the safe and effective integration of nanoparticles in agriculture. Advancing research and fostering collaboration between academia and industry will enable the development of robust nano-enabled cultivation practices, addressing global food security challenges and improving crop resilience in the face of climate change. Nanotechnology holds the potential to revolutionize sustainable agriculture and contribute significantly to overcoming abiotic stressors.

References

1. Abasi F, Raja NI, Mashwani ZUR, Amjad MS, Ehsan M, Mustafa N, Haroon M, Proćków J (2022) Biogenic silver nanoparticles as a stress alleviator in plants: a mechanistic overview. Molecules 27(11):3378
2. Abbasi Khalaki M, Ghorbani A, Dadjou F (2019) Influence of nanopriming on *Festuca ovina* seed germination and early seedling traits under drought stress in laboratory condition. Ecopersia 7:133–139
3. Abbasi Khalaki M, Ghorbani A, Moameri M (2016) Effects of silica and silver nanoparticles on seed germination traits of *Thymus kotschyanus* in laboratory conditions. J Rangel Sci 6:221–231
4. Abdel Latef AAH, Alhmad MFA, Abdelfattah KE (2017) The possible roles of priming with ZnO nanoparticles in mitigation of salinity stress in lupine (*Lupinus termis*) plants. J Plant Growth Regul 36:60–70
5. Acharya P, Jayaprakasha GK, Crosby KM, Jifon JL, Patil BS (2019) Green-synthesized nanoparticles enhanced seedling growth, yield, and quality of onion (*Allium cepa L.*). ACS Sustain Chem Eng 7(17):14580–14590
6. Afshari M, Pazoki A, Sadeghipour O (2021) Foliar-applied silicon and its nanoparticles stimulates physio-chemical changes to improve growth, yield and active constituents of coriander (*Coriandrum sativum L.*) essential oil under different irrigation regimes. SILICON 13:4177–4188
7. Agarwal H, Kumar SV, Rajeshkumar S (2017) A review on green synthesis of zinc oxide nanoparticles—an eco-friendly approach. Resource-Efficient Technol 3(4):406–413
8. Aghdam MTB, Mohammadi H, Ghorbanpour M (2016) Effects of nanoparticulate anatase titanium dioxide on physiological and biochemical performance of *Linum usitatissimum (Linaceae)* under well-watered and drought stress conditions. Braz J Bot 39:139–146
9. Ahlawat J, Asil SM, Barroso GG, Nurunnabi M, Narayan M (2021) Application of carbon nano onions in the biomedical field: recent advances and challenges. Biomater Sci 9(3):626–644
10. Ahmad P, John R, Sarwat M, Umar S (2008) Responses of proline, lipid peroxidation and antioxidative enzymes in two varieties of *Pisum sativum L.* under salt stress. Int J Plant Prod 2:353–366
11. Ahmad B, Shabbir A, Jaleel H, Khan MM, Sadiq Y (2018) Efficacy of titanium dioxide nanoparticles in modulating photosynthesis, peltate glandular trichomes and essential oil production and quality in *Mentha piperita L.* Curr Plant Biol 13:6–15
12. Ahmed T, Noman M, Manzoor N, Ali S, Rizwan M, Ijaz M, Allemailem KS, BinShaya AS, Alhumaydhi FA, Li B (2021) Recent advances in nanoparticles associated ecological harms and their biodegradation: global environmental safety from nano-invaders. J Environ Chem Eng 9(5):106093
13. Ahmed T, Noman M, Manzoor N, Shahid M, Hussaini KM, Rizwan M, Ali S, Maqsood A, Li B (2021) Green magnesium oxide nanoparticles-based modulation of cellular oxidative repair mechanisms to reduce arsenic uptake and translocation in rice (*Oryza sativa L.*) plants. Environ Pollut 288:117785
14. Ahmed T, Noman M, Rizwan M, Ali S, Ijaz U, Nazir MM, ALHaithloul HAS, Alghanem SM, Abdulmajeed AM, Bin Li (2022) Green molybdenum nanoparticles-mediated bio-stimulation

of *Bacillus sp.* strain ZH16 improved the wheat growth by managing in planta nutrients supply, ionic homeostasis and arsenic accumulation. J Hazard Mater 423:127024
15. Akinsiku AA, Dare EO, Ajanaku KO, Adekoya JA, Ayo-Ajayi J (2018) Green synthesized optically active organically capped silver nanoparticles using stem extract of African cucumber (*Momordica charantia*). J Mater Environ Sci 3:902–908
16. Al-Shabib NA, Husain FM, Qais FA, Ahmad N, Khan A, Alyousef AA, Arshad M, Noor S, Khan JM, Alam P, Albalawi TH (2020) Phyto-mediated synthesis of porous titanium dioxide nanoparticles from *Withania somnifera* root extract: broad-spectrum attenuation of biofilm and cytotoxic properties against HepG2 cell lines. Front Microbiol 11:1680
17. Alkharabsheh HM, Seleiman MF, Hewedy OA, Battaglia ML, Jalal RS, Alhammad BA et al (2021) Field crop responses and management strategies to mitigate soil salinity in modern agriculture: a review. Agronomy 11:2299
18. Amooaghaie R, Tabatabaei F, Ahadi AM (2015) Role of hematin and sodium nitroprusside in regulating *Brassica nigra* seed germination under nanosilver and silver nitrate stresses. Ecotoxicol Environ Saf 113:259–270
19. Anoop N, Gupta AK (2003) Transgenic indica rice cv IR-50 overexpressing *Vigna aconitifolia* 11-pyrroline-5-carboxylate synthetase cDNA shows tolerance to high salt. J Plant Biochem Biotechnol 12:109–116
20. Arjaghi SK, Alasl MK, Sajjadi N, Fataei E, Rajaei GE (2021) Retracted article: green synthesis of iron oxide nanoparticles by RS Lichen extract and its application in removing heavy metals of lead and cadmium. Biol Trace Elem Res 199:763–768
21. Ashkavand P, Zarafshar M, Tabari M, Mirzaie J, Nikpour A, Bordbar SK et al (2018) Application of SiO_2 nanoparticles as pretreatment alleviates the impact of drought on the physiological performance of *Prunus mahaleb (Rosaceae)*. Bol Soc Argent Bot 53:207–219
22. Azimi R, Borzelabad MJ, Feizi H, Azimi A (2014) Interaction of SiO_2 nanoparticles with seed prechilling on germination and early seedling growth of tall wheatgrass (*Agropyron elongatum L.*). Pol J Chem Technol 16:25–29
23. Bal A (2019) Effect of foliar application of nanoparticles of chitosan, iron oxide and chitosan-iron oxide complex on secondary metabolites of *Hypericum triquetrifolium* Turra. Dicle University, Graduate School of Natural and Applied Sciences, Department of Biology, Master's Thesis, Diyarbakir
24. Bhattacharya S, Saha J (2023) Green synthesis and application of biogenic nanomaterials as a blueprint in mitigation of abiotic stress in crop plants: a conceptual review. Advancement of GI-science and sustainable agriculture: a multi-dimensional approach, pp 155–182
25. Bisht S, Sharma V, Kumari N (2022) Biosynthesized magnetite nanoparticles from *Polyalthia longifolia* leaves improve photosynthetic performance and yield of *Trigonella foenum-graecum* under drought stress. Plant Stress 5:100090
26. Bratovcic A, Hikal WM, Said-Al Ahl HA, Tkachenko KG, Baeshen RS, Sabra AS et al (2021) Nanopesticides and nanofertilizers and agricultural development: scopes, advances and applications. Open Ecol J 11:301–316
27. Del Buono D, Di Michele A, Costantino F, Trevisan M, Lucini L (2021) Biogenic ZnO nanoparticles synthesized using a novel plant extract: application to enhance physiological and biochemical traits in maize. Nanomaterials 11(5):1270
28. Calanca PP (2017) Effects of abiotic stress in crop production. Quantification of climate variability, adaptation and mitigation for agricultural sustainability, pp 165–180
29. Chandra S, Das P, Bag S, Laha D, Pramanik P (2011) Synthesis, functionalization and bioimaging applications of highly fluorescent carbon nanoparticles. Nanoscale 3(4):1533–1540
30. Chatterjee A, Mridha D, Banerjee J, Chanda S, Ray K, Acharya K, Das M, Roychowdhury T, Sarkar J (2021) Green synthesis of iron oxide nanoparticles and their ameliorative effect on arsenic stress relief in *Oryza sativa* seedlings. Biocatal Agric Biotechnol 38:102207
31. Chhipa H (2017) Nanofertilizers and nanopesticides for agriculture. Environ Chem Lett 15:15–22. https://doi.org/10.1007/s10311-016-0600-4

32. Cinisli KT, Uçar S, Dikbaş N (2019) Use of nanomaterials in agriculture. Yuzuncu Yil Univ J Agric Sci 29:817–831. https://doi.org/10.29133/yyutbd.595658
33. Croissant JG, Butler KS, Zink JI, Brinker CJ (2020) Synthetic amorphous silica nanoparticles: toxicity, biomedical and environmental implications. Nat Rev Mater 5(12):886–909
34. Dehkourdi EH, Mosavi M (2013) Effect of anatase nanoparticles (TiO_2) on parsley seed germination (*Petroselinum crispum*) in vitro. Biol Trace Elem Res 155:283–286. https://doi.org/10.1007/s12011-013-9788-3
35. Devatha CP, Thalla AK (2018) Green synthesis of nanomaterials. In: Synthesis of inorganic nanomaterials. Woodhead Publishing, pp 169–184
36. Dhand V, Soumya L, Bharadwaj S, Chakra S, Bhatt D, Sreedhar BJMS (2016) Green synthesis of silver nanoparticles using *Coffea arabica* seed extract and its antibacterial activity. Mater Sci Eng C 58:36–43
37. Dilnawaz F, Misra AN, Apostolova E (2023) Involvement of nanoparticles in mitigating plant's abiotic stress. *Plant Stress* 100280
38. Dreaden EC, Alkilany AM, Huang X, Murphy CJ, El-Sayed MA (2012) The golden age: gold nanoparticles for biomedicine. Chem Soc Rev 41(7):2740–2779
39. Duraisamy P (2018) Green synthesis of aluminium oxide nanoparticles by using Aerva lanta and *Terminalia chebula* extracts
40. Dustgeer Z, Seleiman MF, Imran KHAN, Chattha MU, Alhammad BA, Jalal RS, Refay Y, Hassan MU (2021) Glycine-betaine induced salinity tolerance in maize by regulating the physiological attributes, antioxidant defense system and ionic homeostasis. Not Bot Horti Agrobot Cluj-Napoca 49(1):12248–12248
41. Ealia SAM, Saravanakumar MP (2017) A review on the classification, characterisation, synthesis of nanoparticles and their application. In: IOP conference series: materials science and engineering, vol 263, no 3, p 032019. IOP Publishing
42. Edison TJI, Sethuraman MG (2012) Instant green synthesis of silver nanoparticles using *Terminalia chebula* fruit extract and evaluation of their catalytic activity on reduction of methylene blue. Process Biochem 47(9):1351–1357
43. Eichert T, Kurtz A, Steiner U, Goldbach HE (2008) Size exclusion limits and lateral heterogeneity of the *stomatal foliar* uptake pathway for aqueous solutes and water suspended nanoparticles. Physiol Plant 134:151–160
44. El-Ashry RM, El-Saadony MT, El-Sobki AE, El-Tahan AM, Al-Otaibi S, El-Shehawi AM et al (2022) Biological silicon nanoparticles maximize the efficiency of nematicides against biotic stress induced by *Meloidogyne incognita* in eggplant. Saudi J Biol Sci 29:920–932
45. El-Saadony MT, Desoky E-SM, Saad AM, Eid RSM, Selem E, Elrys AS (2021) Biological silicon nanoparticles improve *Phaseolus vulgaris L.* yield and minimize its contaminant contents on a heavy metal contaminated saline soil. J Environ Sci 106:1–14
46. El-Temsah YS, Joner EJ (2010) Impact of Fe and Ag nanoparticles on seed germination and differences in bioavailability during exposure in aqueous suspension and soil. Environ Toxicol 27:42–49
47. El-Zohri M, Al-Wadaani NA, Bafeel SO (2021) Foliar sprayed green zinc oxide nanoparticles mitigate drought-induced oxidative stress in tomato. Plants 10:2400
48. Elsheery NI, Helaly MN, El-Hoseiny HM, Alam-Eldein SM (2020) Zinc oxide and silicone nanoparticles to improve the resistance mechanism and annual productivity of salt-stressed mango trees. Agronomy 10:558
49. Faizan M, Bhat JA, Chen C, Alyemeni MN, Wijaya L, Ahmad P et al (2021) Zinc oxide nanoparticles (Zno-NPs) induce salt tolerance by improving the antioxidant system and photosynthetic machinery in tomato. Plant Physiol Biochem 161:122–130
50. Fanoro OT, Parani S, Maluleke R, Lebepe TC, Varghese JR, Mavumengwana V, Oluwafemi OS (2021) Facile green, room-temperature synthesis of gold nanoparticles using *Combretum erythrophyllum* leaf extract: antibacterial and cell viability studies against normal and cancerous cells. Antibiotics 1:893
51. Faraji J, Sepehri A (2020) Exogenous nitric oxide improves the protective effects of TiO_2 nanoparticles on growth, antioxidant system, and photosynthetic performance of wheat seedlings under drought stress. J Soil Sci Plant Nutr 20:703–714

52. Farid M, Shakoor MB, Ehsan S, Ali S, Zubair M, Hanif MA (2013) Morphological, physiological and biochemical responses of different plant species to Cd stress. Int J Chem Biochem Sci 3:53–60
53. Fazlzadeh M, Khosravi R, Zarei A (2017) Green synthesis of zinc oxide nanoparticles using *Peganum harmala* seed extract, and loaded on *Peganum harmala* seed powdered activated carbon as new adsorbent for removal of Cr (VI) from aqueous solution. Ecol Eng 103:180–190
54. Fleischer A, O'Neill MA, Ehwald R (1999) The pore size of nongraminaceous plant cell walls is rapidly decreased by borate ester crosslinking of the pectic polysaccharide rhamnogalacturonan II. Plant Physiol 12:829–838
55. Ganguly A, Ahmad T, Ganguli AK (2010) Silica mesostructures: control of pore size and surface area using a surfactant-templated hydrothermal process. Langmuir 26(18):14901–14908
56. García-Ovando AE, Ramírez Piña JE, Esquivel Naranjo EU, Cervantes Chávez JA, Esquivel K (2022) Biosynthesized nanoparticles and implications by their use in crops: effects over physiology, action mechanisms, plant stress responses and toxicity. Plant Stress 6:100109
57. Ghazi AA, El-Nahrawy S, El-Ramady H, Ling W (2022) Biosynthesis of nano-selenium and its impact on germination of wheat under salt stress for sustainable production. Sustainability 14:1784
58. Ghori N-H, Ghori T, Hayat MQ, Imadi SR, Gul A, Altay V, Ozturk M (2019) Heavy metal stress and responses in plants. Int J Environ Sci Technol 16:1807–1828
59. Godin B, Gu J, Serda RE, Bhavane R, Tasciotti E, Chiappini C, Liu X, Tanaka T, Decuzzi P, Ferrari M (2010) Tailoring the degradation kinetics of mesoporous silicon structures through PEGylation. J Biomed Mater Res Part A 94(4):1236–1243
60. Gupta SM, Tripathi M (2012) An overview of commonly used semiconductor nanoparticles in photocatalysis. High Energy Chem 46:1–9
61. Habibullah G, Viktorova J, Ruml T (2021) C urrent strategies for noble metal nanoparticle synthesis. Nanoscale Res Lett 16(1):47
62. Hong J, Rico CM, Zhao L, Adeleye AS, Keller AA, Peralta-Videa JR, Gardea-Torresdey JL (2015) Toxic effects of copper-based nanoparticles or compounds to lettuce (*Lactuca sativa*) and alfalfa (*Medicago sativa*). Environ Sci Process Impacts 17(1):177–185
63. Hossain Z, Mustafa G, Sakata K, Komatsu S (2016) Insights into the proteomic response of soybean towards Al_2O_3, ZnO, and Ag nanoparticles stress. J Hazard Mater 304:291–305
64. Hossain MA, Piyatida P, da Silva JAT, Fujita M (2012) Molecular mechanism of heavy metal toxicity and tolerance in plants: central role of glutathione in detoxification of reactive oxygen species and methylglyoxal and in heavy metal chelation. J Bot 2012:872875
65. Husen A (2021) Harsh environment and plant resilience. Molecular and functional aspects. Springer, Cham. https://doi.org/10.1007/978-3-030-65912-7
66. Husen A (2021) The harsh environment and resilient plants: an overview. In: Husen A (ed) Harsh environment and plant resilience. Springer, Cham, pp 1–23 https://doi.org/10.1007/978-3-030-65912-7_1
67. Husen A (2021) Plant performance under environmental stress. Hormones, biostimulants and sustainable plant growth management. Springer, Cham. https://doi.org/10.1007/978-3-030-78521-5
68. Husen A (2021) Cross talk between autophagy and hormones for abiotic stress tolerance in plants. In: Husen A (ed) Plant performance under environmental stress. Springerm Cham, pp 1–15. https://doi.org/10.1007/978-3-030-78521-5_1
69. Husen A (2023) Secondary metabolites based green synthesis of nanomaterials and their applications. Springer, Singapore. https://doi.org/10.1007/978-981-99-0927-8
70. Husen A (2023) Nanomaterials from agricultural and horticultural products. Springer, Singapore. https://link.springer.com/book/9789819934348
71. Husen A, Jawaid M (2020) Nanomaterials for agriculture and forestry applications. Elsevier Inc., Cambridge. https://doi.org/10.1016/C2018-0-02349-X
72. Husen A (2022) Environmental pollution and medicinal plants. CRC Press, Boca Raton. https://doi.org/10.1201/9781003178866

73. Husen A (2023) Plants and their interaction to environmental pollution. Damage detection, adaptation, tolerance, physiological and molecular responses. Elsevier Inc., Cambridge. https://doi.org/10.1016/C2020-0-03319-0
74. Hussain A, Oves M, Alajmi MF, Hussain I, Amir S, Ahmed J et al (2019) Biogenesis of ZnO nanoparticles using *Pandanus odorifer* leaf extract: anticancer and antimicrobial activities. RSC Adv 9:15357–15369
75. Hussain A, Rizwan M, Ali Q, Ali S (2019) Seed priming with silicon nanoparticles improved the biomass and yield while reduced the oxidative stress and cadmium concentration in wheat grains. Environ Sci Pollut Res 26:7579–7588
76. Iqbal M, Raja NI, Mashwani Z-U-R, Hussain M, Ejaz M, Yasmeen F (2019) Effect of silver nanoparticles on growth of wheat under heat stress. Iran J Sci Technol Trans Sci 43:387–395
77. Iqbal M, Raja NI, Mashwani ZUR, Hussain M, Ejaz M, Yasmeen F (2019) Effect of silver nanoparticles on growth of wheat under heat stress. Iranian J Sci Technol Trans A: Sci 43:387–395
78. Jameel MS, Aziz AA, Dheyab MA (2020) Green synthesis: proposed mechanism and factors influencing the synthesis of platinum nanoparticles. Green Process Synth 9(1):386–398
79. Jaspers P, Kangasjärvi J (2010) Reactive oxygen species in abiotic stress signaling. Physiol Plant 138(4):405–413
80. Jeelani PG, Mulay P, Venkat R, Ramalingam C (2020) Multifaceted application of silica nanoparticles. A review. SILICON 12(6):1337–1354
81. Joudeh N, Linke D (2022) Nanoparticle classification, physicochemical properties, characterization, and applications: a comprehensive review for biologists. J Nanobiotechnol 20(1):262
82. Judy JD, Unrine JM, Rao W, Wirick S, Bertsch PM (2012) Bioavailability of gold nanomaterials to plants: importance of particle size and surface coating. Environ Sci Technol 46:8467–8474
83. Kang H, Elmer W, Shen Y, Zuverza-Mena N, Ma C, Botella P, White JC, Haynes CL (2021) Silica nanoparticle dissolution rate controls the suppression of fusarium wilt of watermelon (*Citrullus lanatus*). Environ Sci Technol 55(20):13513–13522
84. Kannaujia R, Singh P, Prasad V, Pandey V (2022) Evaluating impacts of biogenic silver nanoparticles and ethylenediurea on wheat (*Triticum aestivum L.*) against ozone-induced damages. Environ Res 203:111857
85. Kasote DM, Lee JH, Jayaprakasha GK, Patil BS (2019) Seed priming with iron oxide nanoparticles modulate antioxidant potential and defense-linked hormones in watermelon seedlings. ACS Sustain Chem Eng 7(5):5142–5151
86. Katiyar P, Yadu B, Korram J, Satnami ML, Kumar M, Keshavkant S (2020) Titanium nanoparticles attenuates arsenic toxicity by up-regulating expressions of defensive genes in *Vigna radiata L.* J Environ Sci 92:18–27
87. Khairy AM, Tohamy MR, Zayed MA, Mahmoud SF, El-Tahan AM, El-Saadony MT et al (2022) Eco-friendly application of nano-chitosan for controlling potato and tomato bacterial wilt. Saudi J Biol Sci 29:2199–2209
88. Khalofah A, Kilany M, Migdadi H (2021) Phytostimulatory influence of *Comamonas testosteroni* and silver nanoparticles on *Linum usitatissimum L.* under salinity stress. Plants 10:790
89. Khan I, Saeed K, Khan I (2019) Nanoparticles: properties, applications and toxicities. Arab J Chem 12(7):908–931
90. Khan I, Seleiman MF, Chattha MU, Jalal RS, Mahmood F, Hassan FA et al (2021) Enhancing antioxidant defense system of mung bean with a salicylic acid exogenous application to mitigate cadmium toxicity. Not Bot Horti Agrobot Cluj Napoca 49:12303
91. Khan MIR, Khan NA (eds) (2017) Reactive oxygen species and antioxidant systems in plants: role and regulation under abiotic stress. Springer, Singapore
92. Khatami M, Pourseyedi S, Khatami M, Hamidi H, Zaeifi M, Soltani L (2015) Synthesis of silver nanoparticles using seed exudates of *Sinapis arvensis* as a novel bioresource, and evaluation of their antifungal activity. Bioresources Bioprocessing 2:1–7

93. Kirupagaran R, Saritha A, Bhuvaneswari S (2016) Green synthesis of selenium nanoparticles from leaf and stem extract of *Leucas lavandulifolia Sm.* and their application. J Nanosci Technol 224–226
94. Kumar H, Bhardwaj K, Kuča K, Kalia A, Nepovimova E, Verma R, Kumar D (2020) Flower-based green synthesis of metallic nanoparticles: applications beyond fragrance. Nanomaterials 10(4):766
95. Kumar B, Smita K, Cumbal L, Debut A (2016) Ficus carica (Fig) fruit mediated green synthesis of silver nanoparticles and its antioxidant activity: a comparison of thermal and ultrasonication approach. BioNanoScience 6:15–21
96. Kumari S, Khanna RR, Nazir F, Albaqami M, Chhillar H, Wahid I et al (2022) Bio-synthesized nanoparticles in developing plant abiotic stress resilience: a new boon for sustainable approach. Int J Mol Sci 23:4452
97. Lakhan MN, Chen R, Shar AH, Chand K, Shah AH, Ahmed M, Ali I, Ahmed R, Liu J, Takahashi K, Wang J (2020) Eco-friendly green synthesis of clove buds extract functionalized silver nanoparticles and evaluation of antibacterial and antidiatom activity. J Microbiol Methods 173:105934
98. Lam C, Zhang YF, Tang YH, Lee CS, Bello I, Lee ST (2000) Large-scale synthesis of ultrafine Si nanoparticles by ball milling. J Cryst Growth 220(4):466–470
99. Lee CW, Mahendra S, Zodrow K, Li D, Tsai YC, Braam J et al (2010) Developmental phytotoxicity of metal oxide nanoparticles to Arabidopsis thaliana. Environ Toxicol Chem 29:669–675
100. Lee YJ, Song K, Cha SH, Cho S, Kim YS, Park Y (2019) Sesquiterpenoids from *Tussilago farfara* flower bud extract for the eco-friendly synthesis of silver and gold nanoparticles possessing antibacterial and anticancer activities. Nanomaterials 9(6):819
101. Li Y, Zhu N, Liang X, Bai X, Zheng L, Zhao J et al (2020) Silica nanoparticles alleviate mercury toxicity via immobilization and inactivation of Hg(ii) in soybean (*Glycine max*). Environ Sci Nano 7:1807–1817
102. Li L, Liu T, Fu C, Tan L, Meng X, Liu H (2015) Biodistribution, excretion, and toxicity of mesoporous silica nanoparticles after oral administration depend on their shape. Nanomed: Nanotechnol Biol Med 11(8):1915–1924
103. Lin D, Zhing B (2007) Phytotoxicity of nanoparticles: inhibition of seed germination and root growth. Environ Pollut 150:243–250
104. Liu M, Zhao F, Zhu D, Duan H, Lv Y, Li L, Gan L (2018) Ultramicroporous carbon nanoparticles derived from metal–organic framework nanoparticles for high-performance supercapacitors. Mater Chem Phys 211:234–241
105. Lohrasbi S, Kouhbanani MAJ, Beheshtkhoo N, Ghasemi Y, Amani AM, Taghizadeh S (2019) Green synthesis of iron nanoparticles using Plantago major leaf extract and their application as a catalyst for the decolorization of azo dye. BioNanoScience 9:317–322
106. Long CM, Nascarella MA, Valberg PA (2013) Carbon black vs. black carbon and other airborne materials containing elemental carbon: physical and chemical distinctions. Environ Pollut 181:271–286
107. Lu L, Huang M, Huang Y, Corvini PF-X, Ji R, Zhao L (2020) Mn_3O_4 nanozymes boost endogenous antioxidant metabolites in cucumber (*Cucumis sativus*) plant and enhance resistance to salinity stress. Environ Sci Nano 7:1692–1703
108. Ma JF, Yamaji N (2006) Silicon uptake and accumulation in higher plants. Trends Plant Sci 11:392–397
109. Maheshwaran G, Muneeswari RS, Bharathi AN, Kumar MK, Sudhahar S (2021) Eco-friendly synthesis of lanthanum oxide nanoparticles by *Eucalyptus globulus* leaf extracts for effective biomedical applications. Mater Lett 283:128799
110. Mathew SS, Sunny NE, Shanmugam V (2021) Green synthesis of anatase titanium dioxide nanoparticles using *Cuminum cyminum* seed extract; effect on Mung bean (*Vigna radiata*) seed germination. Inorg Chem Commun 126:108485
111. Mathur S, Singh D, Ranjan R (2022) Remediation of heavy metal (loid) contaminated soil through green nanotechnology. Front Sustain Food Syst 6:932424

112. Moameri M, Abbasi Khalaki M (2019) Capability of *Secale montanum* trusted for phytoremediation of lead and cadmium in soils amended with nano-silica and municipal solid waste compost. Environ Sci Pollut Res 26:24315–24322
113. Moameri M, Abbasi Khalaki M, Ghorbani A (2018) Effects of nanopriming and biopriming on growth characteristics of *Onobrychis sativa Lam.* under laboratory conditions. Rangelands 12:101–111
114. Mody VV, Siwale R, Singh A, Mody HR (2010) Introduction to metallic nanoparticles. J Pharmacy Bioallied sci 2(4):282–289
115. Mozafari A, Ghadakchi Asl A, Ghaderi N (2018) Grape response to salinity stress and role of iron nanoparticle and potassium silicate to mitigate salt induced damage under in vitro conditions. Physiol Mol Biol Plants 24:25–35
116. Mozafari AA, Havas F, Ghaderi N (2018) Application of iron nanoparticles and salicylic acid in in vitro culture of strawberries (*Fragaria*× *ananassa* Duch.) to cope with drought stress. Plant Cell Tissue Organ Cult (PCTOC) 132:511–523
117. Mustafa H, Ilyas N, Akhtar N, Raja NI, Zainab T, Shah T, Ahmad A, Ahmad P (2021) Biosynthesis and characterization of titanium dioxide nanoparticles and its effects along with calcium phosphate on physicochemical attributes of wheat under drought stress. Ecotoxicol Environ Saf 223:112519
118. Mustafa G, Komatsu S (2016) Insights into the response of soybean mitochondrial proteins to various sizes of aluminum oxide nanoparticles under flooding stress. J Proteome Res 15:4464–4475
119. Mustafa G, Sakata K, Hossain Z, Komatsu S (2015) Proteomic study on the effects of silver nanoparticles on soybean under flooding stress. J Proteomics 122:100–118
120. Muthulakshmi V, Balaji M, Sundrarajan M (2020) Biomedical applications of ionic liquid mediated samarium oxide nanoparticles by *Andrographis paniculata* leaves extract. Mater Chem Phys 242:122483
121. Mwobobia EG, Sichangi AW, Thiong'o KB (2020) Characterization of wheat production using earth-based observations: a case study of Meru County, Kenya. Model Earth Syst Environ 6(1):13–25
122. Najafi S, Razavi SM, Khoshkam M, Asadi A (2020) Effects of green synthesis of sulfur nanoparticles from *Cinnamomum zeylanicum* barks on physiological and biochemical factors of Lettuce (*Lactuca sativa*). Physiol Mol Biol Plants 26:1055–1066
123. Narayan R, Nayak UY, Raichur AM, Garg S (2018) Mesoporous silica nanoparticles: a comprehensive review on synthesis and recent advances. Pharmaceutics 10(3):118
124. Nascimento MA, Cruz JC, Rodrigues GD, de Oliveira AF, Lopes RP (2018) Synthesis of polymetallic nanoparticles from spent lithium-ion batteries and application in the removal of reactive blue 4 dye. J Clean Prod 202:264–272
125. Ndeh NT, Maensiri S, Maensiri D (2017) The effect of green synthesized gold nanoparticles on rice germination and roots. Adv Nat Sci: Nanosci Nanotechnol 8(3):035008
126. Ng KK, Zheng G (2015) Molecular interactions in organic nanoparticles for phototheranostic applications. Chem Rev 115(19):11012–11042
127. Ngom I, Ngom BD, Sackey J, Khamlich S (2021) Biosynthesis of zinc oxide nanoparticles using extracts of *Moringa oleifera*: structural & optical properties. Mater Today: Proc 36:526–533
128. Nima P, Ganesan V (2015) Green synthesis of silver and gold nanoparticles using flower bud broth of *Couropita guinensis* Aublet
129. Niraimathee VA, Subha V, Ravindran RE, Renganathan S (2016) Green synthesis of iron oxide nanoparticles from *Mimosa pudica* root extract. Int J Environ Sustain Dev 15(3):227–240
130. Noman M, Ahmed T, Hussain S, Niazi MBK, Shahid M, Song F (2020) Biogenic copper nanoparticles synthesized by using a copper-resistant strain *Shigella flexneri* SNT22 reduced the translocation of cadmium from soil to wheat plants. J Hazard Mater 398:123175
131. Noman M, Shahid M, Ahmed T, Tahir M, Naqqash T, Muhammad S, Song F, Abid HMA, Aslam Z (2020) Green copper nanoparticles from a native *Klebsiella pneumoniae* strain alleviated oxidative stress impairment of wheat plants by reducing the chromium bioavailability and increasing the growth. Ecotoxicol Environ Saf 192:110303

132. Noman M, Ahmed T, Ijaz U, Hameed A, Shahid M, Azizullah, Song F et al (2023) Microbe-oriented nanoparticles as phytomedicines for plant health management: an emerging paradigm to achieve global food security. Crit Rev Food Sci Nutr 63(25):7489–7509
133. Noor R, Yasmin H, Ilyas N, Nosheen A, Hassan MN, Mumtaz S, Khan N, Ahmad A, Ahmad P (2022) Comparative analysis of iron oxide nanoparticles synthesized from ginger (*Zingiber officinale*) and cumin seeds (*Cuminum cyminum*) to induce resistance in wheat against drought stress. Chemosphere 292:133201
134. Pandian CJ, Palanivel R, Dhananasekaran S (2015) Green synthesis of nickel nanoparticles using *Ocimum sanctum* and their application in dye and pollutant adsorption. Chin J Chem Eng 23(8):1307–1315
135. Parveen A, Rao S (2015) Effect of nanosilver on seed germination and seedling growth in *Pennisetum glaucum*. J Clust Sci 26:693–701
136. Peleg Z, Blumwald E (2011) Hormone balance and abiotic stress tolerance in crop plants. Curr Opin Plant Biol 14(3):290–295
137. Prasad TNVKV, Sudhakar P, Sreenivasulu Y, Latha P, Munaswamy V, Reddy KR et al (2012) Effect of nanoscale zinc oxide particles on the germination, growth and yield of peanut. J Plant Nutr 35:905–927
138. Qi M, Liu Y, Li T (2013) Nano-TiO_2 improve the photosynthesis of tomato leaves under mild heat stress. Biol Trace Elem Res 156:323–328
139. Rahman MA, Parvin A, Khan MSH, War AR, Lingaraju K, Prasad R, Bhattacharyya A et al (2021) Efficacy of the green synthesized nickel-oxide nanoparticles against pulse beetle, *Callosobruchus maculatus* (F.) in black gram (*Vigna mungo* L.). Int J Pest Manag 67(4):306–314
140. Rajakumar G, Rahuman AA (2011) Larvicidal activity of synthesized silver nanoparticles using *Eclipta prostrata* leaf extract against *filariasis* and malaria vectors. Acta Trop 118(3):196–203
141. Rajesh KM, Ajitha B, Reddy YAK, Suneetha Y, Reddy PS (2018) Assisted green synthesis of copper nanoparticles using *Syzygium aromaticum* bud extract: physical, optical and antimicrobial properties. Optik 154:593–600
142. Rakgotho T, Ndou N, Mulaudzi T, Iwuoha E, Mayedwa N, Ajayi RF (2022) Green-synthesized zinc oxide nanoparticles mitigate salt stress in *Sorghum bicolor*. Agriculture 12:597
143. Ramalingam S, Sankaranarayanan G, Senthil S, Rohith RA, Santosh Kumar R (2023) Effect of cerium oxide nanoparticles derived from biosynthesis of *Azadirachta indica* on stability and performance of a research CI engine powered by Diesel-Lemongrass oil blends. Energy Environ 34(4):886–908
144. Ramesh PS, Kokila T, Geetha D (2015) Plant mediated green synthesis and antibacterial activity of silver nanoparticles using *Emblica officinalis* fruit extract. Spectrochim Acta Part A Mol Biomol Spectrosc 142:339–343
145. Rani P, Kaur G, Rao KV, Singh J, Rawat M (2020) Impact of green synthesized metal oxide nanoparticles on seed germination and seedling growth of *Vigna radiata* (Mung Bean) and *Cajanus cajan* (Red Gram). J Inorg Organomet Polym Mater 30:4053–4062
146. Rastogi A, Zivcak M, Sytar O, Kalaji HM, He X, Mbarki S, Brestic M (2017) Impact of metal and metal oxide nanoparticles on plant: a critical review. Front Chem 5:78
147. Rizwan M, Ali S, Qayyum MF, Ok YS, Adrees M, Ibrahim M et al (2017) Effect of metal and metal oxide nanoparticles on growth and physiology of globally important food crops: a critical review. J Hazard Mater 322:2–16
148. Rossi L, Zhang W, Ma X (2017) Cerium oxide nanoparticles alter the salt stress tolerance of *Brassica napus* L. by modifying the formation of root apoplastic barriers. Environ Pollut 229:132–138
149. Roy SD, Das KC, Dhar SS (2021) Conventional to green synthesis of magnetic iron oxide nanoparticles; its application as catalyst, photocatalyst and toxicity: a short review. Inorg Chem Commun 134:109050
150. Saad AM, El-Saadony MT, El-Tahan AM, Sayed S, Moustafa MA, Taha AE et al (2021) Polyphenolic extracts from pomegranate and watermelon wastes as substrate to fabricate

sustainable silver nanoparticles with larvicidal effect against *Spodoptera littoralis*. Saudi J Biol Sci 28:5674–5683
151. Saheb M, Hosseini HA, Hashemzadeh A, Elahi B, Hasanzadeh L, Oskuee RK, Darroudi M (2019) Photocatalytic and biological attributes of green synthesized nickel oxide nanoparticles by *Rheum turkestanicum* (RT) root extract. ChemistrySelect 4(8):2416–2420
152. Salem NM, Albanna LS, Awwad AM (2016) Green synthesis of sulfur nanoparticles using *Punica granatum* peels and the effects on the growth of tomato by foliar spray applications. Environ Nanotechnol Monit Manag 6:83–87
153. Sarkar RD, Kalita MC (2022) Green synthesized Se nanoparticle-mediated alleviation of salt stress in field mustard, TS-36 variety. J Biotechnol 359:95–107
154. Sathishkumar G, Logeshwaran V, Sarathbabu S, Jha PK, Jeyaraj M, Rajkuberan C, Senthilkumar N, Sivaramakrishnan S (2018) Green synthesis of magnetic Fe_3O_4 nanoparticles using *Couroupita guianensis Aubl.* fruit extract for their antibacterial and cytotoxicity activities. Artif Cells Nanomed Biotechnol 46(3):589–598
155. Seleiman MF, Al-Suhaibani N, Ali N, Akmal M, Alotaibi M, Refay Y, Dindaroglu T, Abdul-Wajid HH, Battaglia ML (2021) Drought stress impacts on plants and different approaches to alleviate its adverse effects. Plants 10:259
156. Seleiman MF, Almutairi KF, Alotaibi M, Shami A, Alhammad BA, Battaglia ML (2021) Nano-fertilization as an emerging fertilization technique: why can modern agriculture benefit from its use? Plants 10:2
157. Seleiman MF, Aslam MT, Alhammad BA, Hassan MU, Maqbool R, Chattha MU et al (2022) Salinity stress in wheat: effects, mechanisms and management strategies. Phyton 91:667
158. Seleiman MF, Santanen A, Mäkelä PS (2020) Recycling sludge on cropland as fertilizer—advantages and risks. Resour Conserv Recycl 155:104647
159. Semida WM, Abdelkhalik A, Mohamed GF, Abd El-Mageed TA, Abd El-Mageed SA, Rady MM et al (2021) Foliar application of zinc oxide nanoparticles promotes drought stress tolerance in eggplant (*Solanum melongena L.*). Plants 10:421
160. Shanthi K, Sreevani V, Vimala K, Kannan S (2017) Cytotoxic effect of palladium nanoparticles synthesized from *Syzygium aromaticum* aqueous extracts and induction of apoptosis in cervical carcinoma. Proc Natl Acad Sci India Sect B: Biol Sci 87:1101–1112
161. Sharma P, Jha AB, Dubey RS, Pessarakli M (2012) Reactive oxygen species, oxidative damage, and antioxidative defense mechanism in plants under stressful conditions. J Bot 2012(1):217037
162. Shen Y, Li J, Zhang S, Jiang X, Liang J, Li T, Guo R, Guan W, Yang L (2022) Cd stress alleviation in mung-bean seedlings with biogenic hydroxyapatite nanoparticles as ecofriendly remediation agents. Environ Sci: Nano 9:3844–3858
163. Shome S, Tewari S, Bhattacharya MK, Panda SK, Upadhyaya H (2022) Phytofunctionalized ZnO nanoparticles ameliorate water stress and its recovery in *Oryza sativa* L. Acta Physiol Plant 44:137
164. Singh A, Hussain I, Singh NB, Singh H (2019) Uptake, translocation and impact of green synthesized nanoceria on growth and antioxidant enzymes activity of *Solanum lycopersicum L.* Ecotoxicol Environ Saf 182:109410
165. Singh S, Husen A (2019) Role of nanomaterials in the mitigation of abiotic stress in plants. In: Husen A, Iqbal M (eds) N*anomaterials* and plant potential. Springer, Cham, pp 441–471. https://doi.org/10.1007/978-3-030-05569-1_18
166. Singh S, Husen A (2020) Behavior of agricultural crops in relation to nanomaterials under adverse environmental conditions. In: Husen A, Jawaid M (eds) Nanomaterials for agriculture and forestry applications. Elsevier Inc., Cambridge, MA, pp 219–256. https://doi.org/10.1016/B978-0-12-817852-2.00009-3
167. Singh A, Basnal N, Shukla G, Chaudhary N, Singh S, Gaurav SS (2022) Evaluation of efficacy of phyto-synthesized iron oxide nanoparticles in contributing drought resilience in wheat (*Triticum aestivum L.*). Nanotechnology 33(48):485101
168. Singh MD, Jayadeva HM, Chirag Gautam Mohan MH (2017) Effects of nano zinc oxide particles on seedling growth of maize (*Zea mays L.*) in germinating paper test. Int J Microbiol Res 9:897–898

169. Sofy MR, Seleiman MF, Alhammad BA, Alharbi BM, Mohamed HI (2020) Minimizing adverse effects of Pb on maize plants by combined treatment with *jasmonic,* salicylic acids and proline. Agronomy 10:699
170. Song JY, Kwon EY, Kim BS (2010) Biological synthesis of platinum nanoparticles using *Diopyros kaki* leaf extract. Bioprocess Biosyst Eng 33:159–164
171. Sreelakshmi B, Induja S, Adarsh PP, Rahul HL, Arya SM, Aswana S, Haripriya R, Aswathy BR, Manoj PK, Vishnudasan D (2021) Drought stress amelioration in plants using green synthesised iron oxide nanoparticles. Mater Today: Proc 41:723–727
172. Stöber W, Fink A, Bohn E (1968) Controlled growth of monodisperse silica spheres in the micron size range. J Colloid Interface Sci 26(1):62–69
173. Sujitha MV, Kannan S (2013) Green synthesis of gold nanoparticles using Citrus fruits (*Citrus limon, Citrus reticulata* and *Citrus sinensis*) aqueous extract and its characterization. Spectrochim Acta Part A Mol Biomol Spectrosc 102:15–23
174. Sun L, Song F, Guo J, Zhu X, Liu S, Liu F et al (2020) Nano-ZnO-induced drought tolerance is associated with melatonin synthesis and metabolism in maize. Int J Mol Sci 21:782
175. Sun S, Murray CB, Weller D, Folks L, Moser A (2000) Monodisperse FePt nanoparticles and ferromagnetic FePt nanocrystal superlattices. Science 287(5460):1989–1992
176. Sundrarajan M, Muthulakshmi V (2021) Green synthesis of ionic liquid mediated neodymium oxide nanoparticles by *Andrographis paniculata* leaves extract for effective bio-medical applications. J Environ Chem Eng 9(1):104716
177. Syu YY, Hung JH, Chen JC, Chuang HW (2014) Impacts of size and shape of silver nanoparticles on Arabidopsis plant growth and gene expression. Plant Physiol Biochem 83:57–64
178. Taha RS, Seleiman MF, Shami A, Alhammad BA, Mahdi AHA (2021) Integrated application of selenium and silicon enhances growth and anatomical structure, antioxidant defense system and yield of wheat grown in salt-stressed soil. Plants 10:1040
179. Tamil Elakkiya V, Meenakshi RV, Senthil Kumar P, Karthik V, Ravi Shankar K, Sureshkumar P, Hanan A (2022) Green synthesis of copper nanoparticles using *Sesbania aculeata* to enhance the plant growth and antimicrobial activities. Int J Environ Sci Technol 1–10
180. Tariq M, Choudhary S, Singh H, Siddiqui MA, Kumar H, Amir A, Kapoor N (2021) Role of nanoparticles in abiotic stress. Technol Agric 323
181. Tayyab QM, Almas MH, Jilani G, Razzaq A (2016) Nanoparticles and plant growth dynamics: a review. J Appl Agric Biotechnol 1(2):14–22
182. Thomas SC, Kumar Mishra P, Talegaonkar S (2015) Ceramic nanoparticles: fabrication methods and applications in drug delivery. Curr Pharm Des 21(42):6165–6188
183. Tripathi D, Rai KK, Pandey-Rai S (2021) Impact of green synthesized WcAgNPs on in-vitro plant regeneration and withanolides production by inducing key biosynthetic genes in *Withania coagulans.* Plant Cell Rep 40:283–299
184. Tripathi DK, Singh S, Singh S, Pandey R, Singh VP, Sharma NC, Prasad SM, Dubey NK, Chauhan DK (2017) An overview on manufactured nanoparticles in plants: uptake, translocation, accumulation and phytotoxicity. Plant Physiol Biochem 110:2–12
185. Tunca E (2015) Nanoparticles as the base of nanotechnology and phytoremediation of nanoparticles. Ordu Univ J Sci Technol 5:23–34
186. Turgeon R (2010) The puzzle of phloem pressure. Plant Physiol 154(2):578–581
187. Usman M, Farooq M, Wakeel A, Nawaz A, Cheema SA, Rehman H et al (2020) Nanotechnology in agriculture: current status, challenges and future opportunities. Sci Total Environ 721:137778
188. Vanaja M, Rajeshkumar S, Paulkumar K, Gnanajobitha G, Malarkodi C, Annadurai G (2013) Phytosynthesis and characterization of silver nanoparticles using stem extract of *Coleus aromaticus.* Int J Mater Biomater Appl 3(1):1–4
189. Velmurugan P, Anbalagan K, Manosathyadevan M, Lee KJ, Cho M, Lee SM, Lee SM, Park JH, Oh SG, Bang KS, Oh BT (2014) Green synthesis of silver and gold nanoparticles using *Zingiber officinale* root extract and antibacterial activity of silver nanoparticles against food pathogens. Bioprocess Biosyst Eng 37:1935–1943

190. Venkatachalam P, Jayaraj M, Manikandan R, Geetha N, Rene ER, Sharma NC, Sahi SV (2017) Zinc oxide nanoparticles (ZnONPs) alleviate heavy metal-induced toxicity in *Leucaena leucocephala* seedlings: a physiochemical analysis. Plant Physiol Biochem 110:59–69
191. Venkateswarlu S, Kumar BN, Prathima B, Anitha K, Jyothi NVV (2015) A novel green synthesis of Fe_3O_4-Ag core shell recyclable nanoparticles using *Vitis vinifera* stem extract and its enhanced antibacterial performance. Physica B 457:30–35
192. Verma A, Bharadvaja N (2022) Plant-mediated synthesis and characterization of silver and copper oxide nanoparticles: antibacterial and heavy metal removal activity. J Clust Sci 33:1697–1712
193. Wahid I, Kumari S, Ahmad R, Hussain SJ, Alamri S, Siddiqui MH, Khan MIR (2020) Silver nanoparticle regulates salt tolerance in wheat through changes in ABA concentration, ion homeostasis, and defense systems. Biomolecules 10(11):1506
194. Wahid I, Rani P, Kumari S, Ahmad R, Hussain SJ, Alamri S, Tripathy N, Khan MIR (2022) Biosynthesized gold nanoparticle-maintained nitrogen metabolism, nitric oxide synthesis, ions balance, and stabilizes the defence systems to improve salt stress tolerance in wheat. Chemosphere 287:132142
195. Wang H, Kou X, Pei Z, Xiao JQ, Shani X, Xing B (2011) Physiological effects of magnetite (Fe_3O_4) nanoparticles on perennial ryegrass (*Lolium perenne L.*) and pumpkin (*Cucurbita mixta*) plants. Nanotoxicology 5:30–42
196. Yadav TP, Yadav RM, Singh DP (2012) Mechanical milling: a top down approach for the synthesis of nanomaterials and nanocomposites. Nanosci Nanotechnol 2(3):22–48
197. Yang L, Watts DJ (2005) Particle surface characteristics may play an important role in phytotoxicity of alumina nanoparticles. Toxicol Lett 158:122–132
198. Yuan J, Chen Y, Li H, Lu J, Zhao H, Liu MM et al (2018) New insights into the cellular responses to iron nanoparticles in *Capsicum annuum*. Sci Rep 8:1–9
199. Zamani H, Jafari A, Mousavi SM, Darezereshki E (2020) Biosynthesis of silica nanoparticle using *Saccharomyces cervisiae* and its application on enhanced oil recovery. J Petrol Sci Eng 190:107002
200. Zhang Q, Wang X, Li PZ, Nguyen KT, Wang XJ, Luo Z, Zhang H, Tan NS, Zhao Y (2014) Biocompatible, uniform, and redispersible mesoporous silica nanoparticles for cancer-targeted drug delivery in vivo. Adv Funct Mater 24(17):2450–2461

Interaction of Arbuscular Mycorrhizal Fungi and Nanoparticles in a Soil Matrix for Enhanced Plant Growth and Production Under Abiotic Stress

Varucha Misra, A. K. Mall, Megha Barot, and Azamal Husen

Abstract Plants are continually exposed to environmental fluctuations and a variety of stresses, which may occur individually or in combination throughout their life cycle. Although they possess natural mechanisms to adapt to adverse conditions, these responses can vary widely even within the same species. Enhancing stress tolerance in plants has therefore become a priority in efforts to achieve sustainable agriculture and secure stable crop yields. In recent years, nanobiotechnology has emerged as a promising tool in this field, attracting significant attention from researchers. Among the innovative approaches, the combined application of arbuscular mycorrhizal fungi (AMF) and nanoparticles (NPs) has shown great potential in strengthening plant resilience to abiotic stresses such as drought, salinity, and heavy metal toxicity. AMF establish symbiotic relationships with plant roots, enhancing nutrient and water uptake while improving tolerance to environmental challenges. Nanoparticles, with their unique physicochemical properties, complement the role of AMF by improving nutrient delivery and influencing soil–plant interactions in a beneficial manner. The synergy between AMF and NPs fosters plant growth and productivity under adverse conditions, reducing dependence on chemical inputs and supporting sustainable agriculture. This chapter explores the mechanisms behind AMF-NP interactions, with a focus on how nanoparticles enhance AMF colonization and functionality, while AMF facilitate the uptake and distribution of nanoparticles within plant systems. Additionally, various types of nanoparticles, including

V. Misra (✉) · A. K. Mall
ICAR-Indian Institute of Sugarcane Research, Lucknow, U.P., India
e-mail: misra.varucha@gmail.com

M. Barot
Research and Development Centre, Parul University, Waghodia, Vadodara, Gujarat, India

Department of Environmental Science, Parul University of Applied Sciences, Parul University, Waghodia, Vadodara, Gujarat, India

A. Husen
Wolaita Sodo University, Wolaita, Ethiopia

Department of Biotechnology, Graphic Era (Deemed to be University), Dehradun, Uttarakhand, India

A. Husen (ed.), *Nanobiotechnology for Abiotic Stress Adaptation and Mitigation in Agricultural Crops*, Smart Nanomaterials Technology,
https://doi.org/10.1007/978-981-96-9298-9_13

metal oxides and carbon-based materials, are examined for their contributions to improving plant stress tolerance. Together, these advancements highlight the potential of AMF-NP applications as a cutting-edge strategy for enhancing agricultural resilience.

Keywords Drought · Salt · Heavy metal · Nanotechnology · Nanoparticles · AFM · Resilience

Abbreviations

AMF	Arbuscular mycorrhizal fungi
AgNPs	Silver nanoparticles
CO_2	Carbon dioxide
CAT	Catalase
nano-CuO	Copper oxide nanoparticles
DOC	Dissolved organic carbon
ENMs	Engineered nanomaterials
FeNPs	Ferrous nanoparticles
HMs	Heavy metals
NPs	Nanoparticles
POD	Peroxidase
P	Phosphorus
PSII	Photosystem II
PGPR	Plant growth-promoting rhizobacteria
K^+	Potassium
ROS	Reactive oxygen species
Na^+	Reduced sodium
Si	Silicon
SiNPs	Silicon nanoparticles
SOD	Superoxide dismutase
TiO_2	Titanium dioxide

1 Introduction

Plants encounter a plethora of abiotic stresses that characterize their natural environment and impact crop production. The world population is expected to hit 10 billion before 2050, so food production is estimated to move up by a factor of two from now [33, 69]. As a result, crop yields have to increase somewhere between 70% and fourfold by mid-century to meet the needs of this growing population [8, 101]. In the past few decades, agricultural production has been on the slide, as farmland has shrunk, owing to both population explosion and environmental aggravations [14, 102]. On a global scale, complete crop failures are anticipated as climate change, increasing global temperatures, and the continued depletion of arable land and water resources [14, 102]. Abiotic stress is one of the leading factors contributing to the impaired growth, development, and productivity of crops [103]. These stresses are caused by numerous environmental conditions, including UV rays, heavy metal contamination, temperature change, drought, salinity, and waterlogging [46–56, 65, 69, 74–73, 97, 98]. As a result, the impact of various sources of abiotic stress on plants is exacerbated by extreme and rapid changes in global climate. Moreover, all agricultural productivity aspects are adversely influenced by these environmental changes [39]. Plant stress induced by a variety of biotic and abiotic agents results in unavoidable crop lowering (grow) in yield. Not only salinity but drought among them is more harmful as they lead to the biochemical and physiological disruptions that can reduce the economic returns of many crops [60]. Stomatal closure, photosynthetic inhibition, reduced leaf area, decreased biomass production, and the production of reactive oxygen species (ROS are examples of these detrimental effects due to drought and salinity stress [47, 58]. Stress conditions induce various physiological and biochemical alterations in plants, which adversely impact their growth, development, and overall productivity [4, 40, 55–44, 57, 95–92, 100, 108, 109]. Increasing crop plants' tolerance to stressful environments has emerged as a priority for plant scientists. Nonetheless, environmental stresses usually affect plants the most during the early growth stages, including germination and seedling growth; this in turn results in decreased agricultural production [27]. This has led to much effort to improve crop yields, mainly through the use of chemical fertilizers. Unfortunately, this practice has had adverse effects on human health and the environment [2]. However, in this modern world, sustainable agriculture is being considered as one of the best solutions to minimize the usage of agrochemicals such as fertilizers, pesticides, and herbicides. As intimately related to plants as they are, soil microbiomes are critical components of soil fertility, plant growth, and environmental stress resilience [41]. The soil microbiome is an incredibly complex system that contains a diverse community of microbes. Importance of arbuscular mycorrhizal fungi (AMF) in the regulation of stress responses in plants is explored [62]. These fungi play a crucial role in improving nutrient absorption, alleviating both biotic and abiotic stresses, shielding plants from pathogens, and boosting crop productivity [32]. AMF are microscopic filamentous organisms that colonize plant root systems and the surrounding rhizosphere, often extending extensively within root cortical cells [1]. Numerous studies have demonstrated the significance of AMF in helping plants adapt to drought conditions while enhancing their morphological and physiological attributes [26]. AMF dynamically assist developmental processes under stress by orchestrating a series of spatiotemporal interactions between themselves and their host plants. This mutualistic relationship supports plant growth by facilitating the efficient exchange of nutrients and photoassimilates derived from the plant's photosynthetic processes. However, the effect of AMF on plants varies depending on the plant species and the particular AMF species colonizing their roots [30].

Besides utilizing the potential of AMF, nanotechnology is also one of the approaches that has been introduced to improve crop productivity [15, 31, 92]. This fast-track science develops new tools to enhance the stress tolerance of plants [87]. Many scientists have shown the potential of nanoparticles (NPs) to enhance the growth and yield of plants under abiotic stress conditions [5, 35, 63]. Outstanding physicochemical properties and high surface area volume ratio of nanoparticles have facilitated their use across agriculture and medicine. Nanoparticles can be prepared by many techniques, including physical, biological, and chemical techniques. Their nanoscale properties impart novel attributes to them which can be applied for numerous agricultural applications. Nanoparticles have been studied for their certain proportions with plant growth, developing properties, and their anticancer activity in dicots and monocots [88]. Modern agriculture is accompanied by many shortcomings, among which is the wasteful use of agrochemicals. This reduced efficacy has prompted researchers to explore nanotechnology as a solution to these problems.

Nanotechnology can be harnessed to facilitate the next agricultural revolution through sustainable, efficient, and resilient farming systems, and food security-readiness [106]. This technology allows for the controlled release of agrochemicals as well as targeted delivery of biomolecules at the site of action in specific plants. The biophysical and biochemical effects of engineered nanomaterials (ENMs) on plants include a range of biochemical and physiological alterations dependent on their size, surface area, dosage, exposure method, chemical profile, plant species, and reactivity [42].

Despite years of research into the use of nanotechnology in agriculture, its industrial implementation lags behind other sectors [36]. Recently, the association involving AMF and ENMs has emerged as a promising research focus for improving crop productivity [107]. Studies suggest that AMF and ENMs can collaborate to improve plant growth under stressful conditions [34]. Current research predominantly focuses on the interaction between plants and ENMs, yet the ecological implications of these interactions are not fully understood. As rhizosphere microbes significantly influence the uptake and transport of ENMs, more attention must be given to understanding how ENMs impact organisms like AMF. This will help elucidate their combined or synergistic effects when applied together in plant systems [95]. Despite the limited studies in this area, this chapter underscores the significance of combining AMF and ENMs to enhance the proven advantages of AMF in promoting plant growth and improving resilience to abiotic stresses.

2 Integrating AMF and ENMs for Improved Plant Resilience to Abiotic Stress

A plant's response to abiotic stress is intricately tied to its developmental stage and the environmental conditions it encounters. The activation of defense mechanisms is largely dependent on the specific stressor at play. These responses are governed by a complex network of signaling pathways uniquely adapted to address each type of stress. Plants adopt various adaptive strategies to counter abiotic stress, including the modulation of gene activity, osmotic regulation, the activation of antioxidant systems, and adjustments in growth patterns [7]. Under stress, changes in aquaporin expression play a critical role in maintaining water balance within the plant [59]. Most plant aquaporins are found either in the plasma membrane or the tonoplast, with a smaller subset potentially residing in other intracellular membranes. The phosphorylation of aquaporin proteins can influence their location, determining whether they remain in the plasma membrane or are redirected to intracellular compartments—a phenomenon supported by sequence-based analyses [61]. This dynamic regulation enhances the plant's ability to transport water and nutrients effectively, bolstering its resilience to challenging environmental conditions.

The interplay between AMF and engineered nanomaterials (ENMs) represents a promising approach to boosting plant resilience against abiotic stresses, achieved in part by influencing aquaporin gene expression. AMF form mutually advantageous relationships with plant roots, enhancing the uptake of water and nutrients. This partnership is driven by the modulation of chloroplast-associated enzymes, which not only reduce chlorophyll degradation but also stimulate its synthesis, thereby enhancing the overall photosynthetic performance of plants [74]. Additionally, AMF contribute significantly to stress tolerance by activating plant antioxidant defense mechanisms, which mitigate oxidative damage (Fig. 1) under adverse environmental conditions [16]. Beyond their fundamental role in nutrient acquisition—encompassing essential elements such as potassium, zinc, nitrogen, and phosphorus—AMF influence a range of vital physiological processes, including photosynthetic efficiency, respiratory metabolism, and disease resistance [18]. Despite these benefits, the performance of AMF can be inconsistent, often influenced by variable environmental conditions and soil properties. Their effective colonization and symbiotic functionality depend on precise management practices, as AMF face competition for space and resources from other soil microorganisms. These rival microbes can outcompete AMF for essential nutrients and water, potentially impeding their ability to colonize host plants and deliver their full spectrum of benefits [6]. In certain scenarios, these competitive interactions can significantly limit AMF efficacy, reducing their impact on plant growth and stress mitigation [85]. The rise of nanotechnology offers innovative solutions to these challenges, with ENMs emerging as a complementary tool to enhance the effectiveness of AMF in agricultural systems. Owing to their large surface area relative to volume, ENMs possess distinctive characteristics that can overcome certain challenges linked to AMF, such as inconsistent performance in varying environments. Recent studies have underscored the

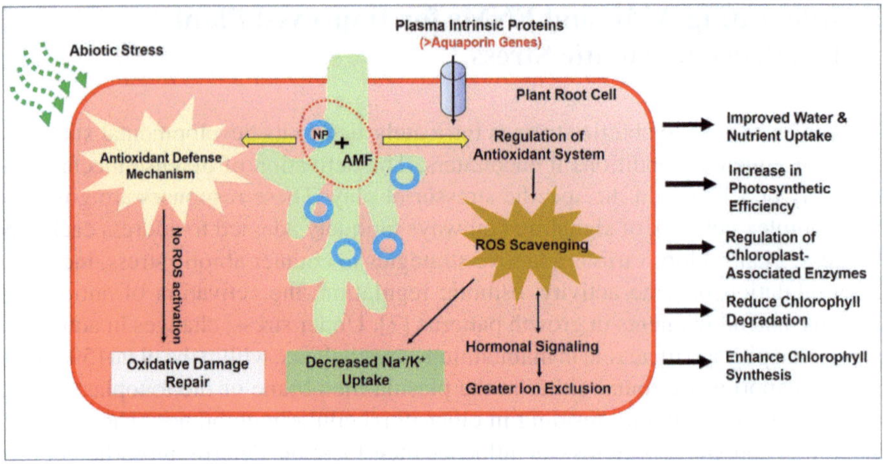

Fig. 1 Probable synergistic mechanism of arbuscular mycorrhizae fungi (AMF) with nanoparticles (NPs) for abiotic stress resilience

synergistic potential of co-applying AMF and ENMs, reporting greater stress tolerance and improved plant performance compared to treatments using either AMF or ENMs alone [34, 76]. This co-application strategy not only amplifies the benefits of both components but also offers a pathway to overcoming some of the bottlenecks that have historically constrained the widespread application of AMF in sustainable agriculture.

2.1 *Synergistic Effects of AMF and Nanoparticles in Mitigating Drought Stress in Crops*

Droughts, becoming increasingly severe and frequent due to climate change, are a major concern for agriculture, particularly in arid and semi-arid regions. These conditions place plants under significant abiotic stress, leading to stunted growth and reduced yields [22]. A lack of soil moisture disrupts critical plant functions, limiting the movement of water and essential nutrients within the plant system. This imbalance in nutrient uptake interferes with ionic homeostasis and gas exchange processes, directly impairing photosynthetic mechanisms. One key consequence is the degradation of photosynthetic pigments and a marked decline in the photochemical efficiency of photosystem II (PSII), a vital component of the photosynthetic apparatus responsible for capturing and converting light energy [64, 66, 80]. Additionally, drought stress triggers the overproduction of reactive oxygen species (ROS), which wreak havoc on cellular structures. ROS damage fundamental biomolecules such as carbohydrates, proteins, lipids, and nucleic acids, disrupting their functionality and creating an environment of oxidative stress [19]. In response, plants attempt to conserve water by regulating their stomata, the microscopic pores on leaves that manage gas exchange. While this adaptation helps minimize water loss, it also restricts the entry of carbon dioxide (CO_2), leading to decreased photosynthesis, lower chlorophyll content, and an accumulation of soluble sugars and proteins as part of the plant's stress response [23]. To counteract these challenges, AM fungi present a sustainable solution for enhancing plant resilience to drought. AM fungi form a symbiotic association with plant roots, enabling plants to better withstand arid conditions. Studies on *Triticum aestivum* grown in controlled environments highlight the effectiveness of AM fungi in alleviating drought stress. Inoculated plants exhibited greater relative water content in both their leaves and surrounding soil, a clear indicator of the fungi's ability to penetrate deeper soil layers in search of moisture. This symbiosis helps plants access water resources that are otherwise unavailable, significantly enhancing drought tolerance. Moreover, AM fungi play a protective role by minimizing the damage caused by drought to the photochemical functions of both PSII and PSI. They help maintain the structural integrity of these photosystems, ensuring that photosynthetic activity continues even under severe water stress [70]. These findings underscore the importance of AM fungi as a natural ally in combating drought stress. Their ability to enhance plant water uptake, maintain photosynthetic efficiency, and protect against oxidative damage makes them an integral component of agricultural strategies aimed at improving productivity in water-limited environments. By integrating AM fungi into crop management practices, farmers can foster more sustainable and resilient agricultural systems in regions most vulnerable to climate variability.

Nanofertilizers are emerging as transformative tools in agriculture, offering significant benefits in enhancing crop growth, yield, and quality [96]. Unlike conventional mineral fertilizers, which may cause nutrient toxicity and inhibit plant growth, nanofertilizers provide nutrients in precise doses, improving growth and productivity without adverse effects [111]. Their high surface area facilitates various metabolic activities, accelerating photosynthesis, increasing dry matter production, and boosting crop yields [96]. However, applying nanofertilizers requires careful optimization of dosage, as excessive concentrations can have inhibitory effects on plants [111]. The synergistic application of AMF with NP for drought tolerance in crops is shown in Table 1. The combined application of carbon nanoparticles, AMF, and compost (Com-AMF-CNPs) has shown promising results in mitigating drought stress in maize. This integrated treatment significantly elevated oxalic and succinic acid levels (by 100%), which are key indicators of enhanced amino acid metabolism. Additionally, it promoted the production of antioxidants and osmoprotectants like proline, alongside increased synthesis of fatty acids such as myristic, palmitic, arachidic, docosanoic, and pentacosanoic acids, with respective increases of 110%, 30%, 100%, and 130%. It also stimulated polyamine biosynthesis pathways, boosting levels of compounds like spermine and spermidine through the upregulation of enzymes such as ornithine decarboxylase and adenosyl methionine decarboxylase [11]. Drought stress significantly hampers plant growth, photosynthetic efficiency, and metabolic stability [12]. However, the synergistic use of AMF and titanium dioxide (TiO_2) nanoparticles has demonstrated potential in alleviating these effects, as seen in *Salvia officinalis* L., where the combination improved essential oil production, plant productivity, and quality under drought conditions [76]. Advanced techniques like transmission electron microscopy and X-ray analysis have revealed the interaction between ferrous nanoparticles (FeNPs) and AMF, showing enhanced photosynthetic efficiency, water-use efficiency, and Fe uptake in seeds and roots. This combination not only promotes plant growth but also significantly enhances drought tolerance [75].

2.2 Synergistic Effects of AMF and Nanoparticles in Mitigating Salinity Stress in Crops

Salt stress represents a major obstacle for global agriculture, negatively impacting plant growth, crop yields, and overall productivity. Soil salinity reduces photosynthetic efficiency, stomatal conductivity, and antioxidant enzyme activity, all of which are critical to plant health and development [84]. To combat these challenges, innovative approaches involving nanomaterials and AMF have emerged as effective solutions to mitigate the detrimental effects of salinity stress on crops [68]. AMF are crucial in boosting crop tolerance to saline conditions by optimizing physiological functions, such as greater plant biomass, improved photosynthetic efficiency, better water retention, and the preferential absorption of vital nutrients [17]. These fungi also bolster the antioxidant defense systems of plants, equipping them with greater tolerance to salinity [110]. AMF support the maintenance of osmotic equilibrium by regulating sodium (Na^+) and potassium (K^+) concentrations and stimulating the accumulation of stress-adaptive compounds such as proline and glycine betaine [105]. However, under extreme salinity, AMF colonization and propagule production tend to decline, necessitating complementary strategies to optimize their effectiveness [21]. The integration of nanotechnology with AMF offers a promising avenue to address these limitations (Fig. 2). For instance, the use of silver nanoparticles (AgNPs) has been shown to enhance the complexity of AMF community co-occurrence networks, thereby improving their functionality under salt stress [28]. Nano-zinc oxide (ZnONPs) in particular has proven effective when used in combination with AMF. This synergistic approach increases K^+ uptake, reduces Na^+ absorption, and significantly boosts crop yield in saline soils [93]. In wheat, the combined application of nano-zinc fertilizers and mycorrhizae not only improved physiological traits and nutrient concentration but also enhanced the K^+/Na^+ ratio and protein content in grains, even under high salinity conditions [9]. Furthermore, the co-application of ZnONPs and AMF amplifies the activity of critical antioxidant enzymes such as catalase (CAT), superoxide dismutase (SOD), and peroxidase (POD), particularly under irrigation with saline water. This protective effect becomes more pronounced as salinity levels increase, underscoring the importance of nanotechnology in stress mitigation [10]. In controlled greenhouse experiments, silicon nanoparticles (SiNPs) have also demonstrated significant potential in enhancing crop resilience to salinity.

Table 1 Co-application of AMF and NPs for abiotic stress resilience in crops

Stress	Application	Plant	Effect	Key references
Drought stress	TiO_2 nanoparticles + AMF	*Salvia officinalis* L. (Common sage)	• Boosted production of essential oil even under drought stress • Enhanced overall plant output • Improved product standards despite challenging condition	[76]
Drought stress	FeNPs + *Glomus intraradices* (AMF)	*Triticum aestivum* L. (Wheat)	• Improved photosynthetic efficiency, leading to better energy capture • Enhanced water-use efficiency for optimal resource management • Increased Fe absorption in seeds and roots • Stimulated robust plant development • Substantially improved tolerance to water scarcity	[75]
Drought stress	Carbon nanoparticles + AMF + compost (Com-AMF-CNPs)	*Zea mays* (Maize)	• Stimulated the production of antioxidants and osmoprotectants, including proline • Enhanced the synthesis of fatty acids like myristic, palmitic, arachidic, docosanoic, and pentacosanoic acids • Myristic acid rose by 110%; Palmitic acid increased by 30%; arachidic acid showed a 100% rise; docosanoic and pentacosanoic acids surged by 130%	[11]

(continued)

Table 1 (continued)

Stress	Application	Plant	Effect	Key references
Salt stress	AgNPs + AMF	*Trifolium repens* (White clover)	• Enhance the complexity of AMF community co-occurrence networks • Improve their functionality	[28]
Salt stress	ZnONPs + AMF	*Triticum aestivum*	• Increases K^+ uptake, reduces Na^+ absorption • Significantly boosts crop yield	[93]
Salt stress	Nano-zinc fertilizers + AMF	*Triticum aestivum* (Wheat)	• Enhanced physiological traits • Significant increase in K+/Na+ ratio • Improved notably in protein content of grains • Improvements persisted even under high salt stress	[9]
Salt stress	SiO_2NPs + AMF	*Triticum aestivum* (Wheat)	• Enhanced physiological parameters like chlorophyll index, anthocyanin levels, leaf area index, chlorophyll fluorescence parameters • Significant higher grain yields • Reduced absorption of Na^+	[3]
Salt stress	TiO_2 ENMs + *Funneliformis mosseae* (AMF)	*Phaseolus vulgaris* (Common bean) roots	• Enhanced plant tolerance to salinity • Improved tolerance at all tested salinity levels • A 10% increase in tolerance was recorded at a salinity level of 100 mM NaCl	[34]

(continued)

Table 1 (continued)

Stress	Application	Plant	Effect	Key references
Salt stress	ZnONPs + *Funneliformis mosseae* (AMF)	*Dracocephalum moldavica*	• Reduced production of ROS • Decrease in ROS production indirectly protected the plant from damage • Zn foliar application boosted the plant's ability to supply and biosynthesize essential metabolites for increasing dry matter	[38]
Copper nano-oxide stress	*Funneliformis mosseae* (AMF)	*Canna indica* (Indian shot)	• Significant stress alleviation when inoculated with AMF	[67]
Heavy metal stress	AgNPs + AMF	*Zea mays* L. (Maize)	• Improved mycorrhizal colonization and biomass • Enhanced P uptake hindered by AgNP toxicity	[25]
Heavy metal stress	Fullerenol nanoparticle + AMF	*Brassica napus* L. (Rapeseed)	• Increase antioxidant enzyme activities • Peroxidase activity rose by 19%; superoxide dismutase activity experienced a 96% enhancement; ascorbate peroxidase activity increased significantly by 200%; glutathione reductase activity improved by 47%; phenylalanine ammonia-lyase activity was elevated by 117%; polyphenol peroxidase activity exhibited a 47% rise	[94]
Arsenic metal stress	CuNPs (100 mM) + AMF	*Elymus sibiricus* (Siberian wild rye)	• Enhanced root growth and shoot growth by declining As content in root tissues and shoot tissues in As-polluted soils	[37]

(continued)

Table 1 (continued)

Stress	Application	Plant	Effect	Key references
Cadmium stress	Molybdenum disulfide (MoS2) nanoparticles (NPs) + AMF	*Coriandrum sativum* (Coriander)	• Significant impact on malondialdehyde content under low-Cd conditions	[29]
Cold stress	ZnONPs + SeNPs + AMF	*Capsicum annuum* L. (Chili)	• Increased peroxidase and nitrogen glutamate dehydrogenase levels, while reducing hydrogen peroxide and malondialdehyde content • Enhanced grana thylakoid thickness and chloroplast count • Improved plant growth and productivity	[89]

Studies using different concentrations of SiNPs (0, 30, and 60 mg/L) in combination with biofertilizers under varying salt concentrations (0, 35, 70, and 105 mM) revealed improved parameters such as plant leaf water potential, soluble protein, and sugar levels, and antioxidant enzyme activities like catalase and peroxidase. Enhanced physiological traits such as stomatal conductance, leaf area index, chlorophyll fluorescence, anthocyanin content, and chlorophyll index were observed, particularly when AMF and plant growth-promoting rhizobacteria (PGPR) were co-inoculated with 60 mg/L SiO_2NPs. This comprehensive approach led to increased uptake of essential elements like silicon (Si), phosphorus (P), and potassium (K^+), reduced sodium (Na^+) absorption, and ultimately resulted in significantly higher grain yields [3]. El-Gazzar et al. [34] observed a significant reduction in hyphal ramification in common bean roots under salinity stress without any engineered nanomaterials (ENMs) applied. However, the inoculation of AMF *Funneliformis mosseae* in the presence of TiO_2 ENMs resulted in a remarkable enhancement of AMF activity. This improvement, evident through increased arbuscular development, was several times greater compared to the control plants and those inoculated with AMF alone under saline stress conditions. The presence of TiO_2 ENMs notably boosted AMF colonization, especially its arbuscule activity (A%), which tripled relative to the control. The positive impact of TiO_2 ENMs on AMF colonization was further corroborated by the enhanced plant dry biomass at both low and high salinity levels. Under low salinity conditions, the efficacy of AMF in promoting plant growth increased more than two-fold compared to the control. At higher salinity levels, the plant growth improvement ranged from 48 to 130%. The plants' reliance on AMF for dry biomass production also saw a four-fold increase in the presence of TiO_2 ENMs. Furthermore,

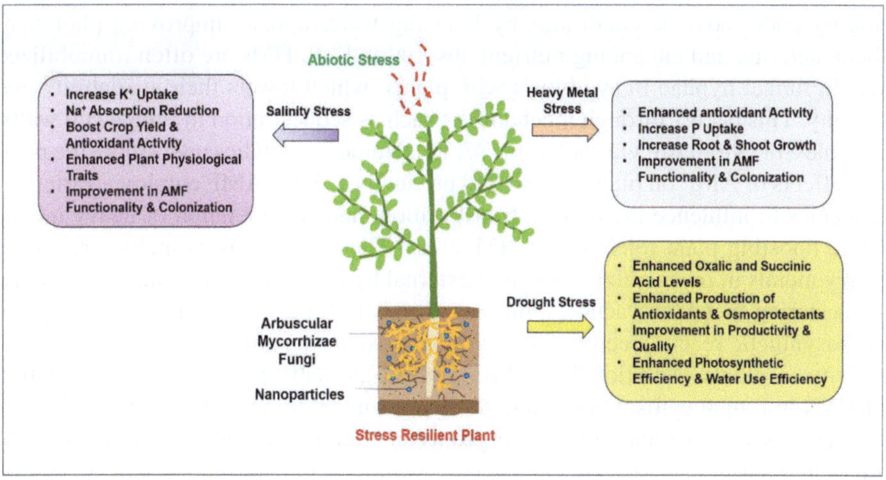

Fig. 2 Impact of AMF and nanoparticles on crop resilience to abiotic stress

co-inoculating AMF with TiO_2 ENMs enhanced the plants' tolerance to salinity across all tested levels, with a 10% increase in tolerance at a salinity level of 100 mM NaCl (Table 1).

2.3 *Synergistic Effects of AMF and Nanoparticles in Mitigating Heavy Metal Stress in Crops*

Improving soils and remediating waste contaminated with heavy metals (HMs) are critical steps toward achieving a clean and sustainable environment. HM contamination not only deteriorates soil quality but also hampers plant growth and threatens ecosystem balance. AMF have emerged as a promising tool to mitigate the adverse effects of pollutants, including HMs, on plant health and development [43]. These symbiotic fungi are integral to enhancing plant tolerance by modulating physiological and biochemical processes under stress conditions. The diversity of AMF spores tends to be significantly lower in soils contaminated with HMs compared to non-polluted environments [78, 81]. AMF colonization in plant roots plays a crucial role in modulating the expression of specific plant genes linked to heavy metal tolerance and detoxification [82, 83]. HMs, once introduced into the soil, pose persistent challenges to plant health, making the role of AMF as biological allies indispensable. These fungi enhance ecosystem functionality by forming structures such as arbuscules, facilitating the exchange of essential inorganic nutrients and minerals critical for plant development. This interaction not only fortifies plants against stress but also acts as a natural filter for heavy metals, mitigating their impact. AMF support plant vitality under adverse conditions by boosting water uptake, improving photosynthetic activity, and enhancing nutrient absorption [20]. HMs are often immobilized within fungal hyphae in symbiosis with plants, which lowers their availability and toxicity. This occurs through mechanisms such as sequestration in fungal cell walls, vacuoles, or cytoplasm via chelation [77, 79]. Numerous studies underscore the positive effects of AMF on plant growth and productivity [99]. AMF employ a variety of strategies to influence the uptake, translocation, and accumulation of heavy metals while boosting plant tolerance to HM stress. These mechanisms include retaining heavy metals in mycorrhizal roots and external hyphae, stimulating nutrient uptake, sequestering metals in vacuoles, binding metals to fungal cell walls, protecting the photosynthetic reaction center, improving gas exchange, and enhancing antioxidant activity in plants. Additionally, AMF-induced glomalin production and chelation of HMs in fungal cytosol contribute to reduced metal toxicity. Furthermore, AMF-mediated phytoremediation has emerged as an effective strategy for mitigating HM stress in plants.

Conversely, the use of nanoparticles (NPs) has also proven effective in removing metals from contaminated sites, with potential applications in phytoremediation strategies [86]. The combined application of AMF and NPs has further enhanced the plants' ability to tolerate heavy metals (Fig. 2). Copper oxide nanoparticles (nano-CuO), widely used across various industries, are a growing environmental concern due to their potential toxicity to plants. For instance, *Canna indica* seedlings exposed to nano-CuO showed significant stress alleviation when inoculated with AMF. This inoculation enhanced antioxidant enzyme activities, reduced reactive oxygen species (ROS) levels in leaves and roots, and led to the upregulation of antioxidant-related genes. Furthermore, AMF helped decrease copper accumulation within the seedlings while increasing the expression of Cu transport and metallothionein genes, which play a role in metal detoxification and homeostasis [67]. The widespread application of ENMs in modern industries poses a significant risk to natural ecosystems, as these materials often end up in soil and water environments. Rhizosphere microorganisms, particularly AMF, are instrumental in mitigating ENM contamination. AMF immobilize ENMs, reduce their phytoaccumulation, and enhance plant growth by triggering systemic protective mechanisms. This dual action of reducing ENM toxicity and promoting plant health underscores the potential of AMF-plant partnerships as a cost-effective and eco-friendly strategy for managing ENM contamination sustainably [104]. Silver nanoparticles (AgNPs), another widely used nanomaterial, have demonstrated varying degrees of phytotoxicity in plants. In *Zea mays* L., the incorporation of AMF significantly alleviated the adverse effects of AgNPs. AMF inoculation not only improved mycorrhizal colonization and biomass but also enhanced phosphorus (P) uptake, which is often hindered by AgNP toxicity. Additionally, AMF played a protective role in preserving soil microbial diversity by reducing the toxic effects of AgNPs on soil bacterial communities. This protection was linked to modulating the expression of potential Ag transporter genes and fostering cooperative interactions between AMF and soil bacteria, leading to a shift in soil bacterial community structure [25]. Copper toxicity, another major environmental challenge, has also been effectively mitigated using AMF. In a study involving spearmint plants treated with $Cu(OH)_2$ nanowires, bulk $Cu(OH)_2$, and ionic $CuSO_4$, AMF inoculation alleviated Cu toxicity. This interaction improved plant growth and productivity while offering environmental advantages by reducing the harmful impact of copper on soil and plant systems. Such findings highlight the critical role of plant–microbe interactions in enhancing crop resilience and productivity while minimizing environmental risks [13]. Iron oxide nanoparticles (Fe_3O_4NPs) present a unique set of challenges to soil ecosystems, particularly in terms of their impact on soil microorganisms and dissolved organic carbon (DOC). In maize plants, AMF inoculation mitigated the adverse effects of Fe_3O_4NPs, showing no significant changes in soil biota or DOC in AMF-treated soils. However, variations in organic matter content were observed under Fe_3O_4NPs treatment alone. This suggests that AMF influence the interaction between Fe_3O_4NPs and soil microorganisms by enhancing plant growth and organic matter release through roots, ultimately stabilizing DOC levels [24]. The integration of carbon-based nanoparticles with AMF inoculation has shown promising potential for mitigating heavy metal stress in crop plants cultivated in contaminated soils [94]. The application of copper nanoparticles (CuNPs) at a concentration of 100 mM, either individually or in combination with AMF, significantly decreased arsenic uptake and mitigated arsenic-induced phytotoxicity in *Elymus sibiricus*. This was achieved through the activation of stress tolerance mechanisms, which ultimately enhanced plant growth parameters. Furthermore, the synergistic interaction between AMF and copper nanoparticles, particularly at elevated concentrations, led to notable improvements in chlorophyll content compared to the effects of either treatment administered separately [37].

3 Conclusion

Agriculture faces the challenge of meeting the increasing global food demand while transitioning to more sustainable farming practices. Abiotic stressors which reduce crop yields worldwide are a major threat to agriculture. Co-application of AMF and ENMs is a promising and cost-effective way to improve plant tolerance to these stressors by amplifying the positive effects of AMF. AMF form mutualistic symbiosis with roots of over 90% of plant species and enhance growth and nutrient uptake. Recent studies have shown the positive effects of combining nanofertilizers with AMF, significant improvement in physiological and chemical properties of plants especially under stress conditions. This combination has been shown to mitigate drought stress, enhance antioxidant activity, and reduce cell death under salinity stress. Nanotechnology especially the interaction between nanoparticles and mycorrhizae has revolutionized plant biology resulting in robust plant growth, biological enrichment, and stronger connection between plants and mycorrhizal networks. But application of nanoparticles at high concentration can sometimes inhibit mycorrhizal growth and negatively affect plant roots. These downsides need to be further investigated to optimize the conditions for these interactions and maximize the benefits. To optimize this co-application, further research is needed to explore more AMF species and ENM types and their interaction under different field conditions. A deeper understanding of the molecular mechanism of the synergistic interaction between AMF and ENMs will provide valuable information on how to apply this strategy across different crops and environments. Moreover, comprehensive cost–benefit analysis should be done to assess the economic viability of using AMF and ENMs together as a sustainable agriculture practice in the future.

References

1. Abd-Alla MH, Nafady NA, Khalaf DM (2016) Assessment of silver nanoparticles contamination on faba bean-*Rhizobium leguminosarum* bv. viciae-Glomusaggrega tum symbiosis: implications for induction of autophagy process in root nodule. Agric Ecosyst Environ 218:163–177. https://doi.org/10.1016/j.agee.2015.11.022
2. Ahanger MA, Hashem A, Abd-Allah EF, Ahmad P (2014) Arbuscular mycorrhiza in crop improvement under environmental stress. In: Ahmad P, Rasool S (eds) Emerging technologies and management of crop stress tolerance. Academic, pp 69–95
3. Ahmadi-Nouraldinvand F, Sharifi RS, Siadat SA, Khalilzadeh R (2024) Fascinating role of silicon dioxide nanoparticles and co-inoculation of mycorrhiza and Rhizobacteria to combat NaCl stress: changes in physiological characteristics, uptake of nutrient elements, and enhancing photosystem II activities in wheat. J Soil Sci Plant Nutr 24:277–294
4. Al-Ashkar I, Al-Suhaibani N, Abdella K, Sallam M, Alotaibi M, Seleiman MF (2021) Combining genetic and multidimensional analyses to identify interpretive traits related to water shortage tolerance as an indirect selection tool for detecting genotypes of drought tolerance in wheat breeding. Plants 10(5):931
5. Al-Khayri JM, Rashmi R, Ulhas RS, Sudheer WN, Banadka A, Nagella P (2023) The role of nanoparticles in response of plants to abiotic stress at physiological, biochemical, and molecular levels. Plants 12(2):292
6. Albornoz FE, Prober SM, Ryan MH, Standish RJ (2022) Ecological interactions among microbial functional guilds in the plant-soil system and implications for ecosystem function. Plant Soil 476(12):301–313
7. Ali S, Gill RA, Ulhassan Z, Zhang N, Hussain S, Zhang K, Huang Q, Sagir M, Tahir MB, Gill MB, Mwamba TM (2023) Exogenously applied melatonin enhanced the tolerance of *Brassica napus* against cobalt toxicity by modulating antioxidant defense, osmotic adjustment, and expression of stress response genes. Ecotoxicol Environ Saf 252:114624
8. Aljabri M (2022) Abiotic stress enhancement tools for improving crop tolerance. Notulae Bot Horti Agrobot Cluj Napoca. https://doi.org/10.15835/nbha50312855
9. Aljenaby HK, Al-Semmak QH, Mohammed BT (2022) The interaction between nano zinc and mycorrhiza on some wheat enzymes activity under saline conditions. Jundishapur J Microbiol 15(2):539–557
10. Aljenaby HK, Mohammed BT, Al-Semmak QH (2022) The interaction between the mineral and nano zinc with mycorrhiza on the concentration of some nutrients, the ratio of K/Na and the proportion of protein in grains of wheat (*Triticum aestivum* L.) irrigated with saline water. NeuroQuantology 20(12):1183
11. Alsherif EA, Almaghrabi O, Elazzazy AM, Abdel-Mawgoud M, Beemster GTS, AbdElgawad H (2023) Carbon nanoparticles improve the effect of compost and arbuscular mycorrhizal fungi in drought-stressed corn cultivation. Plant Physiol Biochem 194:29–40
12. Alsherif EA, Almaghrabi O, Elazzazy AM, Abdel-Mawgoud M, Beemster GTS, Sobrinho RL, AbdElgawad H (2022) How carbon nanoparticles, arbuscular mycorrhiza, and compost mitigate drought stress in maize plant: a growth and biochemical study. Plants 11(23):3324

13. Apodaca SA, Cota-Ruiz K, Hernandez-Viezcas JA, Gardea-Torresdey JL (2022) Arbuscular mycorrhizal fungi alleviate phytotoxic effects of copper-based nanoparticles/compounds in spearmint (*Mentha spicata*). ACS Agric Sci Technol 2(3):661–670
14. Arora NK (2019) Impact of climate change on agriculture production and its sustainable solutions. Environ Sustain 2:95–96. https://doi.org/10.1007/s42398-019-00078-w
15. Ashraf SA, Siddiqui AJ, Elkhalifa AEO, Khan MI, Patel M, Alreshidi M, Moin A, Singh R, Snoussi M, Adnan M (2021) Innovations in nanoscience for the sustainable development of food and agriculture with implications on health and environment. Sci Total Environ 768:144990
16. Begum N, Qin C, Ahanger MA, Raza S, Khan MI, Ashraf M, Ahmed N, Zhang L (2019) Role of arbuscular mycorrhizal fungi in plant growth regulation: implications in abiotic stress tolerance. Front Plant Sci 10:1068. https://doi.org/10.3389/fpls.2019.01068
17. Beltrano J, Ruscitti M, Arango MC, Ronco M (2013) Effects of arbuscular mycorrhiza inoculation on plant growth, biological and physiological parameters and mineral nutrition in pepper grown under different salinity and p levels. J Soil Sci Plant Nutr 13(1):123–141
18. Bindraban PS, Dimkpa C, Nagarajan L, Roy A, Rabbinge R (2015) Revisiting fertilisers and fertilisation strategies for improved nutrient uptake by plants. Biol Fertil Soils 51(8):897–911
19. Birben E, Sahiner UM, Sackesen C, Erzurum S, Kalayci O (2012) Oxidative stress and antioxidant defense. World Allergy Organ J 5:9–19
20. Birhane E, Sterck FJ, Fetene M, Bongers F, Kuyper TW (2012) Arbuscular mycorrhizal fungi enhance photosynthesis, water use efficiency, and growth of frankincense seedlings under pulsed water availability conditions. Oecologia 169:895–904
21. Borde M, Dudhane M, Kulkarni M (2017) Role of arbuscular mycorrhizal fungi (AMF) in salinity tolerance and growth response in plants under salt stress conditions. In: Varma A, Prasad R, Tuteja N (eds) Handbook of mycorrhiza-eco-physiology, secondary metabolites nanomaterials. Springer, Cham, pp 71–86. https://doi.org/10.1007/978-3-319-57849-1_5
22. Boutasknit A, Ait-el-mokhtar M, Fassih B, Ben-laouane R (2024) Effect of arbuscular mycorrhizal fungi and rock phosphate on growth, physiology, and biochemistry of carob under water stress and after rehydration in vermicompost amended soil. Metabolites 14(4):202. https://doi.org/10.3390/metabo14040202
23. Buckley TN (2005) The control of stomata by water balance. New Phytol 168(2):275–292
24. Cao J, Feng Y, Lin X, Wang J (2016) Arbuscular mycorrhizal fungi alleviate the negative effects of iron oxide nanoparticles on bacterial community in rhizospheric soils. Front Environ Sci 4:10
25. Cao J, Feng Y, Lin X, Wang J (2020) A beneficial role of arbuscular mycorrhizal fungi in influencing the effects of silver nanoparticles on plant-microbe systems in a soil matrix. Environ Sci Pollut Res 27(11):11782–11796
26. Cao MJ, Zhang YL, Liu X, Huang H, Zhou XE, Wang WL, Zhu JK (2017) Combining chemical and genetic approaches to increase drought resistance in plants. Nat Commun 8(1):1183. https://doi.org/10.1038/s41467-017-01239-3
27. Chaudhary S, Devi P, Hanumantha Rao B, Jha UC, Sharma KD, Prasad PV, Nayyar H (2022) Physiological and molecular approaches for developing thermotolerance in vegetable crops: a growth, yield and sustenance perspective. Front Plant Sci 13:1892. https://doi.org/10.3389/fpls.2022.878498
28. Chen L, Xiao Y (2023) Silver nanoparticles and arbuscular mycorrhizal fungi influence *Trifolium repen* root associated AMF community structure and its co-occurrence pattern. Scientia Horticulturae (Amsterdam) 320:112232
29. Chen X, Yang J, Zhou Z, Zuo J, Zheng X, Gai J (2024) Effects of arbuscular mycorrhizal inoculation and MoS2 nanoparticles amendment on coriander growth and cadmium uptake in Cd contaminated soil. SSRN https://ssrn.com/abstract=5071285 or https://doi.org/10.2139/ssrn.5071285
30. Deja-Sikora E, Werner K, Hrynkiewicz K (2023) AMF species do matter: *Rhizophagus irregularis* and *Funneliformis mosseae* affect healthy and PVY-infected *Solanum tuberosum* L. in a different way. Front Microbiol 14. https://doi.org/10.3389/fmicb.2023.1127278

31. Desai H, Barot M (2024) Interaction of arbuscular mycorrhizal fungi and silver nanoparticles in a soil matrix for enhanced plant growth and production. In: Plant response to silver nanoparticles: plant growth, development, production, and protection. Springer, Singapore, pp 115–122
32. Diagne N, Ngom M, Djighaly PI, Fall D, Hocher V, Svistoonoff S (2020) Roles of arbuscular mycorrhizal fungi on plant growth and performance: importance in biotic and abiotic stressed regulation. Diversity 12(10):370. https://doi.org/10.3390/d12100370
33. Du J, Liu B, Zhao T, Xu X, Lin H, Ji Y, Li Y, Li Z, Lu C, Li P, Zhao H, Ding X (2022) Silica nanoparticles protect rice against biotic and abiotic stresses. J Nanobiotechnol 20(1):197
34. El-Gazzar N, Almaary K, Ismail A, Polizzi G (2020) Influence of *Funneliformis mosseae* enhanced with titanium dioxide nanoparticles (TiO_2 NPs) on *Phaseolus vulgaris* L. under salinity stress. PLoS One 15(8):e0235355. https://doi.org/10.1371/journal.pone.0235355
35. El-Saadony MT, Saad AM, Soliman SM, Salem HM, Desoky EM, Babalghith AO (2022) Role of nanoparticles in enhancing crop tolerance to abiotic stress: a comprehensive review. Front Plant Sci 13:946717
36. Ezema IC, Ogbobe PO, Omah AD (2014) Initiatives and strategies for development of nanotechnology in nations: a lesson for Africa and other least developed countries. Nanoscale Res Lett 9:1–8
37. Gatasheh MK, Shah AA, Kaleem M, Usman S, Shaffique S (2024) Application of CuNPs and AMF alleviates arsenic stress by encompassing reduced arsenic uptake through metabolomics and inomics alterations in *Elymus sibricus*. BMC Plant Biol 24:667
38. Ghaffari Yaichi Z, Hassanpouraghdam MB, Rasouli F, Aazami MA, Mehrabani LV, Jabbari SF, Asadi M, Esfandiari E, Jimenez-Becker S (2025) Zinc oxide nanoparticles foliar use and arbuscular mycorrhiza inoculation retrieved salinity tolerance in *Dracocephalum moldavica* L. by modulating growth responses and essential oil constituents. Sci Rep 15:492. https://doi.org/10.1038/s41598-024-84198-2
39. Gray E (2021) NASA at your table: climate change and its environmental impacts on crop growth NASA's earth science news team. Washington, DC, USA
40. Hafeez MB, Ghaffar A, Zahra N, Ahmad N, Al-Qahtani WH, Sharma P, Husen A, Li J (2024) Exploring the efficacy of plant growth promoters in nutrient acquisition and antioxidant defense responses of late sown wheat under saline conditions. S Afr J Bot 171:812–824. https://doi.org/10.1016/j.sajb.2024.06.042
41. Hartmann M, Six J (2023) Soil structure and microbiome functions in agroecosystems. Nat Rev Earth Environ 4(1):4–18. https://doi.org/10.1038/s43017-022-00366-w
42. Hatami M, Kariman K, Ghorbanpour M (2016) Engineered nanomaterial-mediated changes in the metabolism of terrestrial plants. Sci Total Environ 571:275–291. https://doi.org/10.1016/j.scitotenv.2016.07.184
43. Herath B, Madushan KWA, Lakmali JPD, Yapa PN (2021) Arbuscular mycorrhizal fungi as a potential tool for bioremediation of heavy metals in contaminated soil. World J Adv Res Rev 10(3):217–228
44. Husen A, Iqbal M, Aref MI (2014) Growth, water status and leaf characteristics of *Brassica carinata* under drought stress and rehydration conditions. Braz J Bot 37:217–227. https://doi.org/10.1007/s40415-014-0066-1
45. Husen A, Iqbal M, Aref IM (2016) IAA-induced alteration in growth and photosynthesis of pea (*Pisum sativum* L.) plants grown under salt stress. J Environ Biol 37:421–429
46. Husen A, Iqbal M, Aref IM (2017) Plant growth and foliar characteristics of faba bean (*Vicia faba* L.) as affected by indole-acetic acid under water-sufficient and water-deficient conditions. J Environ Biol 38:179–186
47. Husen A, Iqbal M, Khanum N, Aref IM, Sohrab SS, Meshresa G (2019) Modulation of salt-stress tolerance of niger (*Guizotia abyssinica*), an oilseed plant, by application of salicylic acid. J Environ Biol 40:94–104
48. Husen A (2021) Harsh environment and plant resilience. Molecular and functional aspects. Springer, Cham. https://doi.org/10.1007/978-3-030-65912-7

49. Husen A (2021) The harsh environment and resilient plants: an overview. In: Husen A (ed) Harsh environment and plant resilience. Springer, Cham, pp 1–23. https://doi.org/10.1007/978-3-030-65912-7_1
50. Husen A (2021) Plant performance under environmental stress. Hormones, biostimulants and sustainable plant growth management. Springer, Cham. https://doi.org/10.1007/978-3-030-78521-5
51. Husen A (2021) Cross talk between autophagy and hormones for abiotic stress tolerance in plants. In: Husen A (ed) Plant performance under environmental stress. Springer, Cham, pp 1–15. https://doi.org/10.1007/978-3-030-78521-5_1
52. Husen A (2024) Current understanding of biostimulants in plant protection and performance. Elsevier, Cambridge, MA
53. Husen A (2024) Essential minerals in plant-soil system (coordination, signalling, and interaction under adverse situations). Elsevier Inc., Cambridge, MA
54. Husen A (2025) Agricultural crop improvement: plant and soil relationships. Taylor & Francis Group, LLC, NW
55. Husen A (2025) Roles of osmolytes in plant systems under changing environment. Elsevier Inc., Cambridge, MA
56. Husen A (2025) Role of antioxidants in plant stress mitigation. Elsevier Inc., Cambridge, MA
57. Husen A, Iqbal M, Sohrab SS, Ansari MKA (2018) Salicylic acid alleviates salinity-caused damage to foliar functions, plant growth and antioxidant system in Ethiopian mustard (*Brassica carinata* A. Br.) Agric Food Secur 7(44):1–14
58. Ibrahim W, Zhu YM, Chen Y, Qiu CW, Zhu S, Wu F, Shuijin Z (2018) Genotypic differences in leaf secondary metabolism, plant hormones and yield under alone and combined stress of drought and salinity in cotton genotypes. Plant Physiol 165:343–355
59. Iwuala E, Odjegba V, Sharma V, Alam A (2020) Drought stress modulates expression of aquaporin gene and photosynthetic efficiency in *Pennisetum glaucum* (L.) R. Br. genotypes. Curr Plant Biol 21:100131. https://doi.org/10.1016/j.cpb.2019.100131
60. Kamran M, Parveen A, Ahmar S, Malik Z, Hussain S, Chattha MS, Saleem MH, Adil M, Heidari P, Chen JT (2019) An overview of hazardous impacts of soil salinity in crops, tolerance mechanisms, and amelioration through selenium supplementation. Int J Mol Sci 21:148
61. Kapilan R, Vaziri M, Zwiazek JJ (2018) Regulation of aquaporins in plants under stress. Biol Res 51:4. https://doi.org/10.1186/s40659-018-0152-0
62. Kapoor D, Sharma P, Sharma MMM, Yadav S, Husen A (2024) Exploring soil microbiota and their role in plant growth, stress tolerance, disease control and nutrient immobilizer. Biocatal Agric Biotechnol 61:103358. https://doi.org/10.1016/j.bcab.2024.103358
63. Khalid MF, Iqbal Khan R, Jawaid MZ, Shafqat W, Hussain S, Ahmed T, Alina Marc R (2022) Nanoparticles: the plant saviour under abiotic stresses. Nanomaterials 12(21):3915
64. Khan F, Siddique AB, Shabala S, Zhou M (2023) Phosphorus plays key roles in regulating plants' physiological responses to abiotic stresses. Plants 12(15):2861. https://doi.org/10.3390/plants12152861
65. Kumar MA, Misra V, Pathak AD (2022) Changing environment and crop breeding. In: Ansari SK, Ansari MI, Husen A (eds) Augmenting crop productivity in stress environment. Springer, Singapore, pp 105–114. https://doi.org/10.1007/978-981-16-6361-1_6. ISBN 978-981-16-6360-4; eISBN: 978-981-16-6361-1
66. Lu C, Zhang J (1999) Effects of water stress on photosystem II photochemistry and its thermostability in wheat plants. J Exp Bot 50(336):1199–1206
67. Luo J, Yan Q, Yang G, Wang Y (2022) Impact of the arbuscular mycorrhizal fungus *Funneliformis mosseae* on the physiological and defence responses of *Canna indica* to copper oxide nanoparticles stress. J Fungi 8(5):513
68. Majeed A, Muhammad Z (2019) Salinity: a major agricultural problem—causes, impacts on crop productivity and management strategies. In: Hasanuzzaman M, Hakeem K, Nahar K, Alharby H (eds) Handbook of plant abiotic stress tolerance. Springer, Cham, pp 83–99. https://doi.org/10.1007/978-3-030-06118-0_3

69. Margaryan G, Singh A, Khachatryan H, Rajput VD, Minkina T, Petropoulos D, Kriemadis A, Alexiou A, Elshikh MS, Mustafa AEZMA, Ghazaryan K (2024) Unveiling the salinity tolerance potential of Armenian Dandur (*Portulaca oleracea* L.) genotypes: enhancing sustainable agriculture and food security. J King Saud Univ Sci 36(8):103332. https://doi.org/10.1016/jjksus2024103332
70. Mathur S, Tomar RS, Jajoo A (2019) Arbuscular mycorrhizal fungi (AMF) protects photosynthetic apparatus of wheat under drought stress. Photosynth Res 139:227–238
71. Misra V, Mall AK, Ansari MI (2022) Role of effective management of harvested crop to increase productivity under stress environment. In: Ansari SK, Ansari MI, Husen A (eds) Augmenting crop productivity in stress environment. Springer, Singapore, pp 223–238. https://doi.org/10.1007/978-981-16-6361-1_14. ISBN: 978-981-16-6360-4; eISBN: 978-981-16-6361-1
72. Misra V, Mall AK, Santeshwari (2022) Productivity losses through bio-deterioration in water-deficit harvested sugarcane. In: Pandey V, Pandey U (eds) Microbial based land restoration handbook, vol 2: Soil and plant health development. Taylor and Francis. CRC Press eBook (chap 12). ISBN: 9781003147077
73. Misra Varucha, Solomon S, Mall AK, Abid M, Khan MMAA, Ansari MI (2022) Drought stress and sustainable sugarcane production. In: Arora NK, Bouizgarne B (eds) Microbial biotechnology for sustainable agriculture, vol 1. Springer, Singapore, pp 353–368. https://doi.org/10.1007/978-981-16-4843-4_11. ISBN: 978-981-16-4842-7; eISBN: 978-981-16-4843-4
74. Mohammed AS, Kapri A, Goel R (2011) Heavy metal pollution: source, impact, and remedies. In: Khan M, Zaidi A, Goel R, Musarrat J (eds) Handbook of biomanagement of metal contaminated soils. Environmental pollution, vol 20. Springer, Dordrecht, pp 1–28. https://doi.org/10.1007/978-94-007-1914-9_1
75. Naseer M, Zhu Y, Li FM, Yang YM, Wang S, Xiong YC (2022) Nano-enabled improvements of growth and colonization rate in wheat inoculated with arbuscular mycorrhizal fungi. Environ Pollut 295:118724
76. Ostadi A, Javanmard A, Amani Machiani M, Sadeghpour A, Maggi F, Nouraein M, Morshedloo MR, Hano C, Lorenzo JM (2022) Co-application of TiO_2 nanoparticles and arbuscular mycorrhizal fungi improves essential oil quantity and quality of sage (*Salvia officinalis* L.) in drought stress conditions. Plants 11(13):1659
77. Ouziad F, Hildebrandt U, Schmelzer E, Bothe H (2005) Differential gene expressions in arbuscular mycorrhizal-colonized tomato grown under heavy metal stress. J Plant Physiol 162:634–649
78. Pawlowska TE, Blaszkowski J, Rühling Å (1996) The mycorrhizal status of plants colonizing a calamine spoil mound in southern Poland. Mycorrhiza 6:499–505
79. Punamiya P, Datta R, Sarkar D, Barber S, Patel M, Das P (2010) Symbiotic role of *Glomus-mosseae* in phytoextraction of lead in vetiver grass *Chrysopogon zizanioides* L. J Hazard Mater 177:465–474
80. Reed C, Wang R, Zhang Z, Wang H, Chen Y (2023) Soil water deficit reduced root hydraulic conductivity of common reed (*Phragmites australis*). Plants 12(20):3543. https://doi.org/10.3390/plants12203543
81. Regvar M, Groznik N, Goljevšček K, Gogala N (2001) Diversity of arbuscular mycorrhizal fungi form various disturbed ecosystems in Slovenia. Acta Biol Slov 44:27–34
82. Repetto O, Bestel-Corre G, Dumas-Gaudot E, Berta G, Gianinazzi Pearson V, Gianinazzi S (2003) Targeted proteomics to identify cadmium-induced protein modifications in *Glomus mosseae*-inoculated pea roots. New Phytol 157:555–567
83. Rivera-Becerril F, van Tuinen D, Martin-Laurent F, Metwally A, Dietz KJ, Gianinazzi S, Gianinazzi-Pearson V (2005) Molecular changes in *Pisum sativum* L. roots during arbuscular mycorrhiza buffering of cadmium stress. Mycorrhiza 16:51–60
84. Safdar H, Amin A, Shafiq Y, Ali A, Yasin R, Shoukat A, Hussan MUI, Sarwar MI (2019) A review: impact of salinity on plant growth. Nat Sci 17(1):34–40
85. Salomon MJ, Demarmels R, Watts-Williams SJ, McLaughlin MJ, Kafle A, Ketelsen C, Soupir A, Bucking H, Cavagnaro TR, van der Heijden MG (2022) Global evaluation of commercial

arbuscular mycorrhizal inoculants under green house and field conditions. Appl Soil Ecol 169:104225
86. Santos EF, Kondo Santini JM, Paixão AP, Júnior EF, Lavres J, Campos M, dos Reis AR (2017) Physiological highlights of manganese toxicity symptoms in soybean plants. Mn toxicity responses. Plant Physiol Biochem 113:6–19
87. Sarraf M, Vishwakarma K, Kumar V, Arif N, Das S, Johnson R, Hasanuzzaman M (2022) Metal/metalloid-based nanomaterials for plant abiotic stress tolerance: an overview of the mechanisms. Plants 11(3):316
88. Saxena R, Tomar RS, Kumar M (2016) Exploring nano biotechnology to mitigate abiotic stress in crop plants. J Pharm Sci Res 8(9):974
89. Sayed EG, DEsoukey SF, Desouky AF, Farag MF, El-Kholy RI, Azoz SN (2024) Synergistic influence of arbuscular mycorrhizal fungi inoculation with nanoparticle foliar application enhances chili (*Capsicum annuum* L.) antioxidant enzymes, anatomical characteristics, and productivity under cold stress conditions. Plants 13(4):517. https://doi.org/10.3390/plants13040517
90. Seleiman MF, Almutairi KF, Alotaibi M, Shami A, Alhammad BA, Battaglia ML (2020) Nano-fertilization as an emerging fertilization technique: why can modern agriculture benefit from its use? Plants 10:2
91. Seleiman MF, Alotaibi MA, Alhammad BA, Alharbi BM, Refay Y, Badawy SA (2020) Effects of ZnO nanoparticles and biochar of rice straw and cow manure on characteristics of contaminated soil and sunflower productivity, oil quality, and heavy metals uptake. Agronomy 10(6):790
92. Seleiman MF, Santanen A, Mäkelä PS (2020) Recycling sludge on cropland as fertilizer—advantages and risks. Resour Conserv Recycl 155:104647
93. Seyed Sharifi R, Khalilzadeh R, Soltanmoradi S (2017) Effects of mycorrhizal fungi and nano zinc oxide on seed yield, Na^+ and K^+ content of wheat (*Triticum aestivum* L.) under salinity stress. J Crop Nutr Sci 3(3):61–69
94. Shah AA, Usman S, Noreen, Kaleem M, Raja V, Sheikh MAE, Ibrahim Z, Sehar S (2024) Fullerenol nanoparticles and AMF application for optimization of *Brassica napus* L. resilience to lead toxicity through physio-biochemical and antioxidative modulations. Sci Rep 14:30992
95. Siani NG, Fallah S, Pokhrel LR, Rostamnejadi A (2017) Natural amelioration of zinc oxide nanoparticle toxicity in fenugreek (*Trigonella foenum-gracum*) by arbuscular mycorrhizal (*Glomus intraradices*) secretion of glomalin. Plant Physiol Biochem 112:227–238. https://doi.org/10.1016/j.plaphy.2017.01.001
96. Singh MD (2017) Nano-fertilizers is a new way to increase nutrients use efficiency in crop production. Int J Agric Sci 9(7):975–3710
97. Singh A, Margaryan G, Harutyunyan AS, Movsesyan H, Khachatryan H, Rajput VD, Ghazaryan K (2024) Advancing agricultural resilience in Ararat plain, Armenia: utilizing biogenic nanoparticles and biochar under saline environments to optimize food security and foster European trade. Egypt J Soil Sci 64(2):459–483
98. Singh A, Rajput VD, Agrawal S, Ghazaryan K, Minkina T, Al Tawaha ARM, Alexiou A (2024) Nanoparticles mediated salt stress resilience: a holistic exploration of physiological, biochemical, and nano-omics approaches. Rev Environ Contam Toxicol 262(1):19
99. Smith SE, Read DJ (2008) The symbionts forming arbuscular mycorrhizas. In: Smith ES, Read DJ (eds) Mycorrhizal symbiosis. Elsevier, Amsterdam, The Netherlands, pp 13–41
100. Taha RS, Seleiman MF, Shami A, Alhammad BA, Mahdi AH (2021) Integrated application of selenium and silicon enhances growth and anatomical structure, antioxidant defense system and yield of wheat grown in salt-stressed soil. Plants 10(6):1040
101. Tilman D, Balzer C, Hill J, Befort BL (2011) Global food demand and the sustainable intensification of agriculture. Proc Natl Acad Sci 108(50):20260–20264
102. Usman M, Farooq M, Wakeel A, Nawaz A, Cheema SA, Rehman H, Ashraf I, Sanaullah M (2020) Nanotechnology in agriculture: current status, challenges and future opportunities. Sci Total Environ 721:137778. https://doi.org/10.1016/jscitotenv2020137778

103. Varucha M, Solomon S, Mall AK, Prajapati CP, Hassan A, Allah EFE, Ansari MI (2020) Morphological assessment of water stressed sugarcane: a comparison of waterlogged and drought affected crop. Saudi J Biol Sci 27(5):1228–1236. https://doi.org/10.1016/j.sjbs.2020.02.007
104. Wang L, Yang D, Ma F, Wang G, You Y (2022) Recent advances in responses of arbuscular mycorrhizal fungi-plant symbiosis to engineered nanoparticles. Chemosphere 286:131644
105. Wang J, Yuan J, Ren Q, Zhang B, Zhang J, Huang R, Wang GG (2022) Arbuscular mycorrhizal fungi enhanced salt tolerance of *Gleditsia sinensis* by modulating antioxidant activity, ion balance and P/N ratio. Plant Growth Regul 97(1):33–49
106. Wang Z, Yue L, Dhankher OP, Xing B (2020) Nano-enabled improvements of growth and nutritional quality in food plants driven by rhizosphere processes. Environ Int 142:105831. https://doi.org/10.1016/j.envint.2020.105831
107. Wang FY, Liu XQ, Shi ZY, Tong RJ, Adams CA, Shi XJ (2016) Arbuscular mycorrhizae alleviate negative effects of zinc oxide nanoparticle and zinc accumulation in maize plants—a soil microcosm experiment. Chemosphere 147:88–97. https://doi.org/10.1016/j.chemosphere.2015.12.076
108. Zahra N, Kausar A, Abdelghani HTM, Singh S, Vashishth DS, Bachheti A, Bachheti RK, Husen A (2024) Serotonin improves plant growth, foliar functions and antioxidant defence system in Ethiopian mustard (*Brassica carinata* A. Br.). S Afr J Bot 170:1–9. https://doi.org/10.1016/j.sajb.2024.05.002
109. Zahra N, Kausar A, Sharma P, Wabaidur SM, Husen A (2024) Serotonin application to *Guizotia abyssinica* improves growth, physiological functions, antioxidant system and alleviates the deleterious effects induced by salt stress. Plant Biosyst Int J Deal Aspects Plant Biol 1–9. https://doi.org/10.1080/11263504.2024.2384972
110. Zhan F, Li B, Jiang M, Yue X, He Y, Xia Y, Wang Y (2018) Arbuscular mycorrhizal fungi enhance antioxidant defense in the leaves and the retention of heavy metals in the roots of maize. Environ Sci Pollut Res 25:24338–24347
111. Zhang Y, Goss GG (2022) Nanotechnology in agriculture: comparison of the toxicity between conventional and nano-based agrochemicals on non-target aquatic species. J Hazard Mater 439:129559

Changes in Crop Plant Gene Expression in Response to Nanoparticle Exposure Under Abiotic Stress Situation

Kalaivani Nadarajah

Abstract Nanotechnology has revolutionized agricultural practices by introducing nanoparticles (NPs) to address abiotic stressors such as cold, drought, flooding, heat, heavy metals, and salinity, which significantly impact crop productivity. NPs, given their distinct nanoscale dimensions and unique physico-chemical properties, act on plants at the cellular and molecular levels, enhancing stress tolerance. This chapter investigates NP's effect on plant gene expression in response to abiotic stress. The stabilization of cellular structures, the improvement of antioxidant systems, and the modification of stress-related genes are important aspects of stress mitigation in plants. Furthermore, the ability of NPs to influence gene expression in photosynthetic efficiency, secondary metabolite production, and hormonal pathways highlights their multifaceted role in improving plant resilience. While the potential of NPs is undeniable, challenges such as understanding their long-term environmental impact and interactions with soil microbiota remains unresolved. This chapter consolidates the latest findings on NP-mediated gene expression and outlines research gaps, offering insights into their sustainable application for enhancing agricultural productivity amid climate change.

Keywords Nanotechnology · Abiotic stress · Gene expression · Antioxidant systems · Crop resilience · Drought tolerance · Salinity stress · Heavy metal stress · Sustainable agriculture

1 Introduction

Agricultural productivity is fundamentally linked to human welfare, as it provides food, income, and employment for a significant portion of the global population. However, the sector is under increasing pressure from climate change, which

K. Nadarajah (✉)
Department of Biological Sciences and Biotechnology, Faculty of Science and Technology, Universiti Kebangsaan Malaysia (UKM), Bangi, Selangor, Malaysia
e-mail: vani@ukm.edu.my

A. Husen (ed.), *Nanobiotechnology for Abiotic Stress Adaptation and Mitigation in Agricultural Crops*, Smart Nanomaterials Technology,
https://doi.org/10.1007/978-981-96-9298-9_14

brings challenges such as rising temperatures, air pollution, altered rainfall patterns, droughts, floods, and dwindling water reserves. These changes exacerbate abiotic stresses such as heat, heavy metals (HMs), salinity, ultraviolet (UV) radiation, and water-logging resulting in reduced crop yields and impacted global food security [12, 48–51]. Furthermore, a rapidly growing global population and stagnant agricultural performance compounds the urgency for innovative solutions [33, 52, 54–56, 58].

Nanotechnology is a promising avenue to address the above challenges. Nanomaterials (NMs), characterized by dimensions under 100 nm, exhibit unique physicochemical properties such as high surface area-to-volume ratios and enhanced reactivity, making them effective tools in agriculture. Nanoparticles' (NPs) capacity to engage with plants at the cellular and molecular levels has created new avenues for reducing abiotic stress, boosting crop resilience, and increasing yield [4, 53, 57, 59]. NPs such as carbon nanotubes (CNTs), copper oxide (CuO), silicon dioxide (SiO_2), titanium dioxide (TiO_2), and zinc oxide (ZnO) have demonstrated potential in alleviating stress caused by cold, drought, flooding, heat, heavy metals, and salinity, while promoting plant growth and nutrient uptake [101, 103, 104].

Nanotechnology's applications in agriculture are multifaceted, including the development of nanofertilizers, nanopesticides, nanoremediators, and nanosensors [2, 72–74, 97]. Ramírez-Rodríguez et al. [88] reported that urea-doped calcium phosphate nanofertilizers have enabled efficient nutrient delivery, enhancing crop growth and productivity while reducing environmental harm. Similarly, nanohybrids like urea-hydroxyapatite-montmorillonite composites have shown promise in slow nutrient release, improving germination and yield in crops such as tomatoes [22, 23].

One of the critical areas of research involves understanding how NMs impact stress responses at the molecular level in plants. Plants naturally possess sophisticated mechanisms to sense and adapt to abiotic stressors, involving alterations in gene expression, protein activity, and metabolic pathways [123]. Research findings indicate that NPs can modulate the expression of key genes such as aquaporins responsible for water transport and stress-related genes like dehydrins, which play a vital role in plant dehydration responses [117]. These interactions enable plants to better manage oxidative stress, regulate water balance, and maintain metabolic homeostasis under adverse conditions [27, 98, 99].

Notwithstanding the encouraging potential of NPs, there are still issues with comprehending their exact mode of action, long-term consequences on ecosystems, and human well-being. Research gaps include the need for comprehensive studies on NM interactions with soil microbiota, nutrient cycling, and plant-bacteria interactions, as well as the development of regulatory frameworks to ensure their safe and sustainable use [67]. This chapter integrates recent advances in nanotechnology to address abiotic stresses in agriculture, focusing on the physiological, biochemical, and molecular impacts of NMs on plants. By consolidating current knowledge and identifying research gaps, it aims to provide a foundation for future studies to enhance resilient agricultural practices. Researchers in nanotechnology, plant biology, and stress physiology will find this synthesis valuable for tackling the dual challenges of food security and environmental sustainability.

2 Nanoparticles: Classification, Properties, and Applications in Nanotechnology

The rapid advancements in nanotechnology have brought forth an urgent need for an appropriate categorization system for NPs. This system serves as a guideline for scientists and engineers to enhance the practical and secure utilization of these resources. Among the key factors for classification, dimensionality and composition stand out as the most significant criteria. As nanostructured materials proliferated over the past two decades, the development of an accurate and convenient categorization system has become imperative [86].

The dimensionality of NPs forms the primary basis for their classification. According to Chung et al. [15], zero-dimensional (0D) NPs are characterized by dimensions that do not exceed 100 nm in any observable direction. Advances in physical and chemical manufacturing techniques have enabled the precise control of these dimensions, facilitating the production of nanostructures such as quantum dots, nano-lenses, and hollow spheres [69]. One-dimensional (1D) NMs include structures with nanoscale dimensions in one direction, such as nanowires, nanotubes, nanofibers, and nanorods. These materials find extensive use in various applications due to their unique structural and functional properties.

Two-dimensional (2D) NMs exhibit nanoscale dimensions in two directions. Notable examples include nanosheets, nanofilms, and nanolayers [62]. Meanwhile, three-dimensional (3D) NMs are not constrained by any dimension to the nanoscale. Examples of 3D nanostructures include nanoparticle dispersions, bulk powders, arrays of nanowires, and nanotubes. This comprehensive dimensional classification highlights the versatility and broad applicability of NPs across multiple scientific domains.

In addition to dimensionality, composition is another essential criterion for nanoparticle classification. The first category comprises carbon-based NPs, which include fullerenes, graphene, and carbon nanotubes. These materials are renowned for their exceptional strength, reactivity, and wide-ranging applications in fields such as electronics, medicine, and energy storage [86].

The next category encompasses metal-based NPs, which consist of metals such as gold, silver, and titanium, as well as their oxides. These NPs exhibit unique optical properties due to surface plasmon resonance at visible wavelengths, making them highly effective in applications like sensing and imaging [75]. Metallic NPs also include advanced structures such as metal-decorated multi-walled nanotubes, metal nanoclusters, and metallofullerenes, all of which offer innovative possibilities in nanotechnology.

Polymeric NPs represent another important category, characterized by their ability to encapsulate therapeutic substances within their cores or adsorb them onto their surfaces. These NPs are especially relevant in biomedical applications, offering controlled drug delivery and enhanced therapeutic efficacy [114]. Finally, nanocomposites, which are hybrids of NPs and other materials, have garnered significant attention over the past 25 years. Their customizable physico-chemical properties make them suitable for diverse applications, from structural reinforcement to advanced coatings [71].

In conclusion, the development of an accurate and practical classification system for NPs is crucial for advancing nanotechnology and optimizing the use of these materials. By categorizing NPs based on dimensionality and composition, researchers can better understand their properties and harness their potential for applications across various fields. This comprehensive approach not only facilitates innovation but also ensures responsible use of nanotechnology in various industries.

3 The Function of Nanoparticles in Reducing Abiotic Stress

Worldwide, abiotic stressors like cold, drought, flooding, heat, heavy metals, and salinity have a detrimental effect on plant development and agricultural productivity. NPs have become effective instruments for reducing these stressors because of their distinct physico-chemical characteristics. NPs improve stress tolerance in plants through cellular and molecular interactions by enhancing nutrient absorption, photosynthesis, antioxidative responses, and stress-related signaling pathways, their uses in diverse abiotic stressors are examined in depth below.

3.1 The Contribution of Nanoparticles in Reducing Salinity Stress and Improving Plant Function

Excess sodium chloride (NaCl) in soil causes harm to plant development and production by reducing water intake, changing ionic homeostasis, and generating oxidative stress [5, 39, 47, 60, 122, 121]. Over the past two decades, the emergence of NPs as a tool for mitigating abiotic stress has gained substantial attention. NPs not only address the physiological challenges posed by salinity but also improve plant growth and yield through mechanisms including ionic regulation, osmotic adjustment, and activation of antioxidative pathways.

NPs have shown the capacity to reduce the negative impacts of salt stress on a variety of plant species. Foliar treatment of potassium nanoparticles (K NPs) in *Medicago sativa* dramatically reduced electrolyte leakage and lipid peroxidation while increasing antioxidant enzyme activity, including superoxide dismutase (SOD) and catalase (CAT) [25]. Similarly, Ag NPs have shown promise in reducing sodium ion uptake and oxidative stress markers such as malondialdehyde (MDA) in pearl millet, improving photosynthetic efficiency and biomass production [66].

Cerium oxide NPs (CeO_2) enhanced root cell architecture and nutrient uptake in *Brassica napus*, leading to improved photosynthetic performance under salinity stress [13]. In wheat, the utilization of copper NPs (Cu NPs) reduced oxidative stress and improved both plant development and yield under salt stress [85]. Similarly,

Ag NPs pre-applied to wheat seeds enhanced germination and salt tolerance by modulating antioxidant enzyme activities [64]. At the germination stage, Ag NPs have been effective in enhancing osmotic regulation and mitigating salinity effects in *Lathyrus sativus*, improving germination rates, root length, and seedling biomass [43]. ZnO NPs have similarly increased dry weight in *Pisum sativum* under salinity stress [82]. Additionally, mint plants were shielded from oxidative damage in salt stress when treated with FeO_4 NPs [11].

NP use has extended to various crops, demonstrating their versatility in addressing salinity stress. In basil (*Ocimum basilicum*), Ag NPs increased germination rates and reduced stress-induced damage [20]. In cumin (*Cuminum cyminum*), Ag NPs improved salt resistance and growth parameters [21]. The potential of CeO_2 NPs to enhance biomass in salt-stressed canola underscores their applicability to large-scale agricultural systems [91]. NPs represent a transformative approach to mitigating salinity stress in plants by addressing the physiological disruptions caused by excessive NaCl. Their ability to regulate ionic balance, enhance antioxidative pathways, and improve nutrient uptake makes them indispensable tools for sustainable agriculture. As research continues to explore the applications of NPs, their role in combating abiotic stresses and enhancing crop productivity will undoubtedly expand, offering promising solutions for global agricultural challenges.

3.2 The Contribution of Nanoparticles in Reducing Drought Stress and Improving Plant Function

Drought is a severe abiotic stress that reduces crop output by limiting water availability, resulting in decreased cell turgor, impaired photosynthesis, and metabolic disturbances [28, 30, 35, 45, 46]. NPs have emerged as a viable method for mitigating drought stress by increasing water usage efficiency, strengthening antioxidative defense systems, and regulating stress-responsive signaling pathways.

Applying NPs enhances plant resilience and growth in drought-prone land via a number of mechanisms. Si NPs have demonstrated impressive advantages when sprayed on leaves or added to the soil. When Si NPs were applied through foliar spray on pomegranate plants, the levels of antioxidant enzymes, carotenoids, chlorophyll, and phenolic compounds increased dramatically. Additionally, the production of abscisic acid (ABA), which is essential for controlling stomatal activity and conserving water, was up-regulated [26]. Si NPs stabilized photosynthetic pigments, decreased membrane damage, increased relative water content (RWC), and improved water use efficiency in bananas and chickpeas [38, 77].

ZnO NPs have also been effective in addressing drought stress. In tomato plants, ZnO NPs improved antioxidative enzyme activity, including peroxidase (POD) and glutathione reductase (GR), which reduced reactive oxygen species (ROS) accumulation and membrane lipid peroxidation during stress [94]. ZnO NPs also alleviate drought stress and reduced cadmium toxicity in wheat, highlighting their dual protective role [65]. According to Semida et al. [95], FeO_4 NPs improved water uptake and root hydraulic conductivity in wheat, preserving photosynthetic activity and leaf water potential in water-limited environments. According to van Nguyen et al. [84], the application of Cu NPs in maize also improved drought resistance by raising plant biomass, leaf water content, and levels of anthocyanin, chlorophyll, and carotenoids in plants.

ZnO NPs when applied in soybean seeds increased germination percentages in arid settings [93]. Similarly, analcite NPs when applied in wheat under hot, dry conditions promoted germination and growth [81]. Numerous crops have demonstrated the adaptable advantages of silicon dioxide (SiO_2) nanoparticles. For example, SiO_2 NPs reduced membrane damage and superoxide radical production while increasing shoot length, RWC, and antioxidant capacity in barley [19]. Foliar treatment of SiO_2 and TiO_2 NPs in cotton mitigated the adverse impacts of drought by preserving physiological stability [96]. Si NPs enhanced the antioxidant capability and essential oil output of coriander in moderate drought circumstances [24].

NPs confer drought tolerance by modulating molecular and physiological pathways. Si NPs enhance ABA signaling, which regulates stomatal closure, thereby minimizing water loss and maintaining plant hydration [26]. Another important target of NPs is antioxidant defense mechanisms. ZnO and SiO_2 NPs decrease oxidative stress indicators like ROS and lipid peroxidation by increasing the activity of antioxidative enzymes including CAT, SOD, and glutathione reductase (GR) [111, 94].

3.3 The Contribution of Nanoparticles to Reducing Cold Stress and Improving Plant Function

One major abiotic stressor is cold stress, which damages membranes by producing high levels of ROS, dehydrating cells, and reducing enzyme performance [17]. By strengthening antioxidative defenses, maintaining cellular membranes, and increasing photosynthetic efficiency, NPs have demonstrated significant promise in reducing the detrimental impact of cold stress. NPs improve plant resilience to cold stress by modulating physiological and biochemical processes. TiO_2 NPs enhance cold tolerance in chickpea plants. Applications of TiO_2 NPs enhanced antioxidant enzyme activity, including SOD and CAT, strengthened chloroplast membranes, and decreased electrolyte leakage. According to Hasanpour et al. [41], these actions led to decreased oxidative damage and increased photosynthetic rates under cold stress situations.

In sugarcane and rice, the application of Si NPs and ZnO NPs significantly improved cold stress tolerance by increasing photosynthetic pigment content, reducing ROS accumulation, and enhancing the activities of antioxidant enzymes. These physiological improvements contributed to better growth, increased biomass, and improved yield in plants exposed to low temperatures [3]. The application of NPs offers a promising approach to mitigate the damaging effects of cold stress in plants. By enhancing antioxidative defense systems, stabilizing cellular structures, and maintaining photosynthetic efficiency, NPs not only protect plants from cold-induced damage but also promote growth and productivity under abiotic stress conditions.

3.4 The Contribution of Nanoparticles to Reducing Heavy Metal Stress and Improving Plant Function

Heavy metals (HMs) such as cadmium (Cd), mercury (Hg), and lead (Pb) disrupt cellular homeostasis, photosynthesis, and nutrient uptake. These toxic elements disrupt metabolic activities and generate excessive ROS, leading to oxidative stress. NPs have emerged as an effective tool to mitigate HM stress by chelating metals, modifying soil properties, and activating plant defense mechanisms. The utilization of NPs in plants under HM stress demonstrates remarkable benefits, including reduced metal toxicity, enhanced photosynthesis, and improved plant development.

Fe_3O_4 NPs immobilize Cd in the soil, reducing its bioavailability and uptake by plants [70]. Similarly, hydroxyapatite NPs improve soil pH and adsorb HM, mitigating their toxic effects [66]. Further, Si NPs induce the synthesis of defense compounds like root exudates, phytochelatins, and organic acids, which chelate HMs and limit their translocation to aerial tissues in the plant. Si NPs increased antioxidant enzyme activity, including SOD and ascorbate peroxidase (APX), hence reducing ROS damage in pea seedlings under chromium (Cr) stress [69]. In maize, Si NPs activated antioxidant defense mechanisms to mitigate aluminum (Al) toxicity [106].

In Chinese cabbage under cadmium stress, foliar treatments with selenium nanoparticles (Se NPs) increased plant biomass, plant height, chlorophyll content, and antioxidant enzymes levels while lowering malondialdehyde (MDA) content and cadmium accumulation [124]. Similarly, Si NPs reduced Cd toxicity in rice by enhancing physiological and biochemical responses in alleviating stress [10]. Further studies have shown that combined application of NPs and other amendments has synergistic effects in mitigating HM stress. For instance, co-application of ZnO NPs and soil biochar was more effective in alleviating Cd stress than individual treatments [127]. Similarly, Fe NPs combined with biochar reduced Cd uptake and improved antioxidant enzyme activities in rice [118]. In tomato plants, soil application of Fe_3O_4 NPs reduced Cd accumulation and increased nutrient uptake, improving overall plant health under stress [87].

NPs modulate molecular and physiological processes to alleviate HM stress. Soil application of TiO_2 NPs improved photosynthetic efficiency and physiological parameters in soybean plants under Cd stress, highlighting their role in mitigating oxidative damage [4]. In addition, Si NPs boosted both photochemical and non-photochemical quenching, which are essential for preserving photosystem II efficiency, and decreasing oxidative stress indicators in maize under arsenic (As) stress [108]. The application of NPs is an effective strategy to mitigate HM stress in plants. In addition to increasing plant resistance to HMs, NPs aid in the cleanup of contaminated environments by lowering metal bioavailability, triggering antioxidant pathways, and enhancing physiological stability. The incorporation of NPs into sustainable agricultural techniques holds promise for major improvements in HM stress management and food security.

3.5 The Contribution of Nanoparticles to Reducing Flooding Stress and Improving Plant Function

Flooding creates hypoxic conditions, disrupting respiration and energy production in plant roots. Nanoparticles enhance oxygen availability and promote metabolic adaptations to flooding stress. Ag NPs improved glycolysis and amino acid synthesis pathways in soybean under flood stress. They also enhanced the formation of root aerenchyma, which facilitates oxygen transport, leading to better growth and yield [107]. Aluminum oxide nanoparticles (Al_2O_3 NPs) increased ROS scavenging capacity and improved root development, enabling plants to better cope with flooding-induced stress [1].

3.6 The Contribution of Nanoparticles to Reducing Heat Stress and Improving Plant Function

Heat stress accelerates transpiration, disrupts protein structure, and impairs photosynthesis. NPs alleviate these effects by stabilizing proteins, enhancing antioxidative responses, and improving water retention. Selenium nanoparticles (Se NPs) improved heat tolerance in sorghum by reducing lipid peroxidation of membranes and enhancing the activity of SOD and CAT within the cell. These reactions improved photosynthetic and water use efficiency under high temperatures [13]. Zn and Si nanoparticles in wheat reduced ROS levels, stabilized chlorophyll function, and enhanced heat-shock protein (HSP) production, which are critical in maintaining cellular homeostasis during heat stress [66]. Figure 1 provides a simple illustration of how nanoparticles affect the biochemical and physiological processes in plants through the regulation of gene expression in these processes.

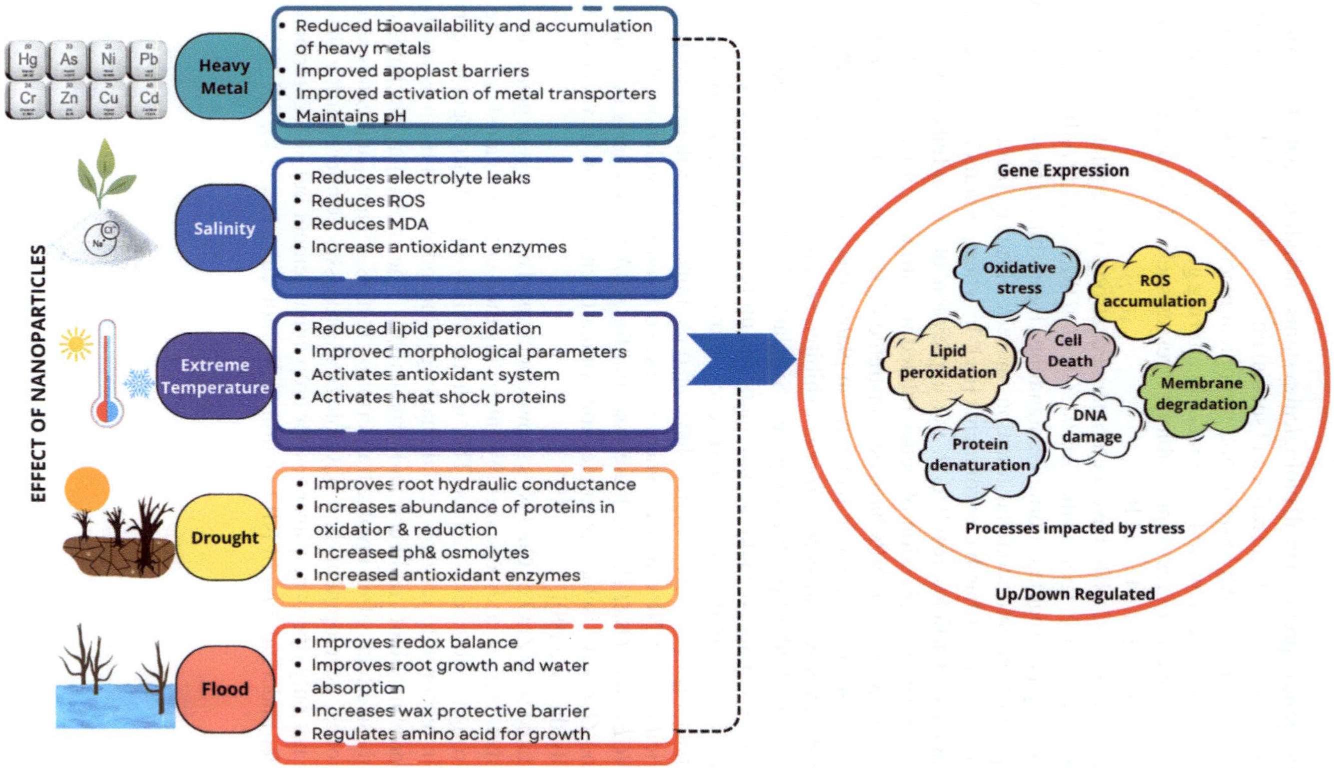

Fig. 1 Nanoparticles enhance plant tolerance to abiotic stresses by modulating key physiological and biochemical processes, such as reducing ROS, stabilizing membranes, and boosting antioxidant activity through the up- and down-regulation of specific genes

4 Nanoparticle and Its Influence on Gene Expression

NPs can affect gene expression in all biological systems, including plants. Due to their size, high surface area-to-volume ratio, and special physico-chemical characteristics, NPs can interact with cellular systems and biological macromolecules in ways that are not possible with conventional materials. By altering gene expression pathways, NPs can make plants more resistant to abiotic stresses such salt, drought, flooding, and HM toxicity, to name a few [116].

Under drought stress, NPs have demonstrated strong up-regulation of genes related to stress resilience, such as the *GmRD20A, GmDREB2,* and *GmWRKY* families [89]. These genes regulate hormone signaling, lignin biosynthesis, secondary metabolite production, and stomatal function, thereby ensuring plant stability and water conservation under harsh environments [61]. In wheat, treatments with ceria-based or ZnO nanoparticles resulted in the expression of stress-linked proteins, while down-regulating expression of genes related to nutrient storage and glucose metabolism [89]. Moreover, NPs have demonstrated their capacity to influence the antioxidant systems and metabolic pathways in plants. Incidentally, under drought stress, CS NP-treated *Catharanthus roseus* exhibited enhanced antioxidant activity and expression of alkaloid production genes. Similarly, Si NPs when administered under salt stress enhanced plant development by up-regulating genes (*CrMPK3* and *STR*) involved in photosynthetic efficiency, nitrogen assimilation, and stress-responsive pathways [32]. Researchers have also examined these molecular reactions and elicitation of NPs in rice, where stress led to an up-regulation of genes such as *OsNCED* and *OsZEP*, which are essential for ABA biosynthesis in drought response, and *OsHMA3*, which promotes ion compartmentalization in vacuoles to lessen salinity damage [90].

Plants treated with ZnO or CuO NPs showed increased expression of genes associated with metal stress in shoots, thereby enhancing their defense against combined threats of drought and heavy metals. This unique ability of NPs to regulate gene expression in response to environmental challenges has positioned them as pivotal tools in improving plant resilience and productivity. The molecular mechanisms underlying these effects often involve transcription factors, kinases, and genes regulating hormones, development, and physiological processes. Understanding these interactions not only opens new avenues for crop enhancement but also contributes to our broader understanding of NP function at the molecular level [40, 90].

4.1 Nanoparticle Influence on Gene Expression During Drought Stress

Recent advances in nanotechnology have demonstrated that NPs can mitigate drought stress effects by modulating gene expression to enhance drought tolerance. By interacting with cellular pathways, NPs activate stress-responsive genes, activate antioxidant defense systems, and facilitate hormonal adjustments that bolster plant resilience [14, 89]. Research has demonstrated that the application of nanoparticles (NPs), such as ZnO, CuO, and ceria-based NPs, can significantly enhance drought tolerance by up-regulating critical drought-responsive genes. In soybean (*Glycine max*), genes like *GmRD20A* (responsive to dehydration), *GmDREB2* (a key transcription factor in the drought-responsive element-binding protein family), *GmMYB118*, and *GmMYB174* (involved in drought-related signaling pathways) are up-regulated upon NP treatment. This up-regulation results in improved water-use efficiency and stress adaptation [90].

Additionally, NP treatment enhances the expression of genes such as *GmWRKY27, GmWRKY118,* and *GmWRKY174* (*WRKY* gene family), which regulate critical processes including ABA production, lignin biosynthesis, and the formation of secondary metabolites. These genes are integral in drought adaptation which facilitates bolstering of structural integrity and promoting stress-induced metabolic adjustments [113]. In addition to modulating stress-related genes, NPs enhance the plant's antioxidative systems, which are crucial for weathering drought stress. Ceria-based NPs in kidney beans promote the expression of proteins associated with stress resistance while down-regulating those linked to nutrient storage and glucose metabolism, enabling plants to prioritize survival mechanisms under water-limited conditions [78]. In addition, TiO_2 and ceria-based NPs also demonstrate the ability to influence root architecture and water uptake by targeting drought-related genes in plant roots such as members of the AP2/EREBP gene family (*DREB1B/CBF1*, *DREB1C/CFB2*, *DREB1D/CBF4*, and *DREB2A*) [110]. Wheat plants treated with ZnO and CuO NPs exhibited intensified expression of drought-resilient genes in root tissues, enhancing their capacity to access water in dry conditions [40, 92]. In addition to these genes, drought results in up-regulation of antioxidative and osmolyte regulating genes to improve physiological stress response and improved water utilization efficiency during stress [94]. Moreover, CS NPs improve drought tolerance in *Catharanthus roseus* by boosting antioxidant activity (*SOD* and *CAT* genes) and activating genes involved in secondary metabolite production (alkaloid production), such as strictosidine synthase (*STR*) and tryptophan decarboxylase (*TDC*) [32, 90].

4.2 Nanoparticle Influence on Gene Expression During Flooding Stress

Submergence stress, which is frequently induced by protracted flooding, alters oxygen availability and metabolic processes in plants, resulting in lower growth, survival, and productivity. Nanoparticles (NPs) have developed as a new method for modulating plant responses to submergence stress by affecting the expression of key stress-responsive genes. These molecular adjustments enable plants to adapt and survive during hypoxic conditions associated with submergence stress [34]. Research has highlighted the potential role of NPs in enhancing the expression of genes involved in anaerobic metabolism and ethylene signaling, which are vital for submergence tolerance [4]. Silicon (Si) NPs up-regulate genes like *ADH1, ADH2* (alcohol dehydrogenase 1 and 2), and *PDC* (pyruvate decarboxylase), which are essential for anaerobic fermentation processes in plants during low-oxygen conditions [44, 109]. Silver (Ag) and aluminum oxide (Al_2O_3) NPs on the other hand down-regulate *PDC* and *ADH1*, reducing the reliance on anaerobic metabolism under flooding. NPs also enhance antioxidant defenses by increasing the expression of genes like *SOD, CAT,* and *APX* mitigating oxidative damage caused by ROS [6, 80, 83]. Additionally, dehydroascorbate reductase (*DHAR*), glutathione reductase (*GR*), and glutathione S-transferase (*GST*) are other antioxidant-related genes that contribute to ROS scavenging and cellular protection under submergence stress [102]. Further genes involved in lipid biosynthesis, such as Beta-Ketoacyl Reductase 1 (*BKR1*), are also modulated by NPs to strengthen physical barriers and support recovery post submergence stress.

NPs also influence hormonal and photosynthetic pathways critical for submergence tolerance. CeO_2 NPs regulate ethylene biosynthesis by modulating 1-Aminocyclopropane-1-Carboxylate Synthase (*ACS1*) and 1-Aminocyclopropane-1-Carboxylate Oxidase (*ACO1*), which balances ethylene levels to improve stress tolerance [36]. Furthermore, Si NPs enhance photosynthetic efficiency by up-regulating genes like ribulose-1,5-bisphosphate carboxylase/oxygenase large subunit (*rbcL*) and chlorophyll a/b-binding protein (*CAB*), aiding in recovery and productivity [100]. Additionally, the regulation of ABA-responsive genes is crucial during submergence stress to maintain cellular homeostasis and survival [4]. These metabolic shifts allow plants to maintain energy production and avoid cellular damage under submerged environments.

4.3 Nanoparticle Influence on Gene Expression During Salinity Stress

Salinity and osmotic stress disrupt osmotic balance, ionic homeostasis, and metabolic pathways leading to oxidative damage, nutrient deficiencies, and impaired physiological functions. Under osmotic stress, NPs modulate the ABA signaling pathway, which plays a crucial role in salinity stress adaptation. Ag and Si NPs up-regulate genes critical in ABA signaling which include ABA response element binding protein (*AREB*), cystein-rich receptor-like protein kinase 42-Like (*CRK1*), 9-Cis-epoxycarotenoid dioxygenase 3 (*NCED3*), and zeaxanthin epoxidase (*ZEP*). These genes regulate stomatal closure, osmotic adjustment, and proline biosynthesis by eliciting sucrose non-fermenting 1-related protein kinase (*SnRK2*), which phosphorylates AREB. Meanwhile, genes such as dwarf and delayed flowering 2 (*DDF2*), *TAS14*, and zinc finger homeodomain transcription factor (*ZFHD1*) are selectively down-regulated by NPs, further enhancing ABA-dependent responses while suppressing gibberellic acid pathways that alleviate osmotic stress [8, 109].

Salinity stress exacerbates oxidative damage by promoting excessive sodium (Na^+) accumulation and ROS generation. Si NPs mitigate salinity effects by up-regulating heavy metal ATPase 3 (*OsHMA3*), which facilitates Na^+ sequestration into vacuoles, maintaining ionic homeostasis. NPs such as ZnO and TiO_2 enhance antioxidant defenses by increasing the expression of *SOD, CAT*, and APX, enabling plants to scavenge ROS effectively and avoid cellular damage [36, 100]. Moreover, *CRK1*, a ROS-sensing gene, helps maintain redox balance, enhancing overall plant resilience.

In addition, NPs modulate secondary metabolite production and protect protein stability under stress. CeO_2 NPs up-regulate genes encoding heat shock proteins (*HSPs*) and aquaporins, vital for protein folding and water transport [6]. Further, CS NPs activate genes linked to secondary metabolite production and stress resistance. In tomato plants, CS NPs down-regulate salinity stress genes such as mitogen-activated protein kinase 3 (*MAPK3*) and respiratory burst oxidase homolog 1 (*RBOH1*), while enhancing growth-related processes [7, 8]. NPs also influence photosynthetic efficiency and energy metabolism. Si NPs up-regulate genes like *CAB* and *RbcL* to maintain photosynthesis under stress.

4.4 Nanoparticle Influence on Gene Expression During Heat Stress

Heat stress significantly impairs plant growth and agricultural productivity by disrupting photosynthesis, protein stability, hormonal balance, and cellular integrity. NPs such as ZnO, TiO_2, Si, and CeO_2 have demonstrated the ability to protect plants from heat-induced damage by altering cellular structures and boosting stress tolerance mechanisms. ZnO NPs modify the ultrastructure of chloroplasts, mitochondria, and cell walls in *Medicago sativa* (alfalfa) seedlings, minimizing heat-induced damage and maintaining plant growth [63]. This effect is supported by evidence that ZnO NPs can mitigate the reduction of heat-responsive genes such as *TGS* in *Arabidopsis thaliana* seedlings exposed to elevated temperatures [115]. In wheat seedlings, Si NPs enhanced expression of aquaporin genes *TaPIP1* and *TaNIP2*, resulting in the increase of relative water content under heat stress [119].

Heat shock proteins (HSPs) are critical for maintaining protein folding and preventing aggregation during heat stress. According to Khodakovskaya et al. [68], multiwall carbon nanotubes have been demonstrated to enhance the gene expression of heat shock proteins like *HSP90.* Additionally, CeO_2 nanoparticle-exposed maize plants produced excessive amounts of H_2O_2 and up-regulated *HSP70* [125, 126]. Further, heat stress induces excessive ROS production, leading to oxidative damage. NPs such as ZnO and TiO_2 activate antioxidant-related genes enhancing ROS scavenging and cellular integrity [29].

NPs also influence hormonal pathways critical for heat stress adaptation. CS NPs stimulate the expression of *NCED* which improves stomatal regulation and reduces water loss. Similarly, NPs modulate ethylene-related genes like *ACO1*, enhancing thermotolerance by balancing growth and stress responses [7, 8, 36, 128]. Photosynthetic efficiency, often impaired under heat stress, is also bolstered by Si NPs, which up-regulates genes such as *rbcL* and CAB, resulting in stable photosynthetic activity and chlorophyll function [100]. Furthermore, NPs enhance the expression of genes involved in cell wall and membrane stability, such as expansin (*EXP*) and cytochrome P450 (*CYP*), supporting structural integrity during heat stress [100].

Secondary metabolites contribute to stress adaptation by functioning as antioxidants and osmoprotectants. Ag and CeO_2 NPs enhance the expression of phenylalanine ammonia-lyase (*PAL*) and chalcone synthase (*CHS*), which play pivotal roles in the synthesis of phenolic compounds, flavonoids, and other secondary metabolites [7, 8]. These compounds mitigate oxidative damage and support osmotic balance, further improving plant tolerance to heat stress.

4.5 Nanoparticle Influence in Gene Expression During Heavy Metal Stress

NPs influence several genes in metal transport, detoxification, antioxidant defense, and structural integrity in response to heavy metal (HM) stress. ZnO NPs regulate metal transporter genes in rice seedlings by impeding the molecular expression of *IRT1, IRT2, YSL2,* and *YSL15*, which are associated with Fe and Cd absorption [40]. In addition, ZnO NPs also enhanced heat tolerance mechanisms that up-regulated antioxidant enzymes and osmoprotectants [92]. Further, Si NPs up-regulated *OsLsi1* (Si uptake) and *OsHMA3* (Cd transport into vacuoles) while down-regulating *OsLCT1* and *OsNramp5* (Cd uptake and transport genes) [69]. Si NPs suppressed the expression of genes like *LCT1* and *NRAMP5*, which mediates Cd movement from root to shoot [31].

NPs also enhanced the biosynthesis of chelators and detoxification compounds. Ag NPs up-regulate *PCS* (phytochelatin synthase) and *MT* (metallothionein), improving the ability to chelate and sequester HM [6]. Meanwhile, sulfur NPs decreased the expression of *BnGST, BnPOD*, and *BnAPX* (antioxidant genes) under

Hg stress [7]. These responses not only mitigate heavy metal toxicity but also strengthen the plants' defense mechanisms.

Heavy metal (HM) accumulation triggers the buildup of ROS, causing oxidative stress. NPs like ZnO and CeO_2 enhance the expression of antioxidant-related genes, such as *SOD, CAT,* and *GPX* [79]. These NPs mitigate ROS accumulation, protecting cellular components from oxidative damage. Si NPs demonstrate significant benefits to photosynthetic efficiency by enhancing the expression of *CAB* and *rbcL*, ensuring photosynthetic stability under stress [36].

Structural integrity and hormonal regulation also play crucial roles in plant resilience to heavy metal stress. NPs promote structural stability by up-regulating genes like *EXP* (expansin), which maintains cell wall integrity and prevents structural damage caused by heavy metals [36]. Additionally, CS NPs were found to stimulate *NCED*, along with ethylene and jasmonic acid pathways involving *ACO1* and *LOX* (lipoxygenase) [8]. These hormonal changes improve stress perception and adaptive responses in plants.

NPs also boost secondary metabolite production (flavonoids and phenolics) which act as metal chelators and antioxidants. Ag NPs and CeO_2 NPs increased the expression of *PAL* and *CHS* genes involved in the synthesis of these metabolites [42, 76]. These secondary metabolites enhance detoxification and alleviate stress caused by HMs. Practical applications of NPs have shown significant promise, particularly in rice plants. Overexpression of *OsVIT1* and *OsCAX4* resulted in Cd retention within vacuoles, reducing its toxic effects [7]. Si NPs modulated similar genes to enhance tolerance mechanisms, such as reducing Cd translocation to shoots and improving vacuolar sequestration.

4.6 *Nanoparticle Influence on Gene Expression During Cold Stress*

Common physiological impacts of cold stress include reduced transpiration, decreased CO_2 intake, inhibited chlorophyll synthesis and gaseous exchange. Nanoparticles (NPs) mitigate cold stress by modulating the expression of stress-responsive genes, enhancing antioxidant systems, and promoting cold acclimation mechanisms. TiO_2 NPs led to differential expression of genes related to cellular defense, chromatin modification, cell signaling, and transcriptional regulation in *Cicer arietinum* L. under cold stress [57]. This included up-regulation of genes coding for the large and small subunits of *RUBISCO, CAB,* and phosphoenolpyruvate carboxylase (*PEPC*). These changes enhanced photosynthesis, altered energy metabolism, and reduced H_2O_2 concentrations [59]. Similarly, Si NPs and CS NPs have been linked to the up-regulation of cold-responsive (*COR*) genes such as C-repeat binding factor (*CBF*) and *COR15A*. These genes play a critical role in cellular stability, membrane fluidity, and freezing tolerance in plants. ZnO NPs up-regulate antioxidative defense-related genes in rice, including *OsCATA, OsCATB. OsCu/*

ZnSOD1, OsCu/ZnSOD2, OsCu/ZnSOD3, OsPRX11, OsPRX65, and *OsPRX89.* Additionally, ZnO NPs also induced chilling-response transcription factors such as *OsMYB4, OsMYB30, OsNAC5, OsWRKY76, OsWRKY94,* and *OsbZIP52* [105]. Cold induces oxidative stress due to the imbalance between ROS production and ROS scavenging. ZnO and CeO_2 NPs boost the expression of antioxidant genes, mitigating ROS accumulation and protecting cellular components from oxidative damage during cold stress [6]. ABA and ethylene signaling pathways are also vital in cold stress responses. Gold nanoparticles (Au-NPs) used in wheat improve cold tolerance by up-regulating the expression of *RuBisCo* and *Wcor15*, which are critical for photosynthesis and cold stress adaptation. These changes maintain energy production and protect plants from photo-inhibition during cold stress [112]. Additionally, NPs enhance the expression of structural and metabolic genes to support cold stress tolerance. Ag NPs and Si NPs stimulate genes involved in cell wall modifications, such as *EXP* and xyloglucan endotransglucosylase/hydrolase (*XTH*), which improve cell wall elasticity and flexibility. Genes related to osmolyte biosynthesis, such as pyrroline-5-carboxylate synthase (*P5CS*) are also up-regulated to maintain osmotic balance and mitigate stress [7, 8]. Table 1 provides simplified compilation of the up- and down-regulated genes in different plant systems and the nanoparticles that have been used to elicit these responses.

5 Concluding Remarks

By modifying gene expression and strengthening stress tolerance systems, nanoparticles have the potential to significantly reduce the negative impacts of abiotic stresses on crops. Across a variety of crops, their capacity to enhance physiological stability, nutrient absorption, and stress-related signaling pathways has shown promise. To maximize their use, resolve safety issues, and assess their long-term ecological and environmental impact, more research is necessary. Sustainable crop management may be possible with the use of nanotechnology in precision agriculture, such as creating NP-based targeted delivery systems for genes that confer stress resilience. In order to guarantee global food security, future research should concentrate on multifunctional NPs, their function in improving soil health, and their use in climate-resilient farming systems.

Table 1 Genes up and down-regulated by nanoparticles in plants

Abiotic stress	Nanoparticles	Plant species	Up-regulated genes	Down-regulated genes	Key references
Drought	ZnO, CuO, TiO_2, CeO_2	*Arabidopsis thaliana, Glycine max, Catharanthus roseus*	*GmRD20A, GmDREB2, GmWRKY27, OsNCED, OsHMA3, AP2/EREBP* (*DREB1B, DREB1C, DREB2A*)	Genes related to glucose metabolism	[32, 40, 89, 90, 110]
Salinity	SiO_2, TiO_2, ZnO	*Oryza sativa, Triticum aestivum, Pisum sativum, Solanum lycopersicum*	*OsHMA3, SOD, CAT, APX, CAB, RbcL, STR, CrMPK3*	*MAPK3, RBOH1, TAS14, ZFHD1*	[7, 36, 100, 94]
Heat	ZnO, TiO_2, Si, CeO_2	*Medicago sativa, Zea mays, Triticum aestivum*	*HSP70, HSP90, NCED, ACO1, rbcL, CAB, TaPIP1, TaNIP2, PAL, CHS*	Genes related to gibberellin pathways	[7, 36, 68, 100, 119]
Cold	TiO_2, ZnO, SiO_2	*Cicer arietinum, Hordeum vulgare, Saccharum officinarum*	*SOD, CAT, APX, CAB, RbcL, CAB, aquaporins* (*OsLsi1*)	Genes involved in ROS signaling	[3, 32, 36, 41]
Heavy metals	Fe_3O_4, Si, CeO_2	*Brassica napus, Oryza sativa, Pisum sativum*	*PCS, MT, OsHMA3, OsLsi1, SOD, APX, PAL, CYP, CRK1*	*OsLCT1, OsNramp5*	[6, 10, 66, 124, 127]
Flooding	Si, Ag, Al_2O_3, CeO_2	*Glycine max, Triticum aestivum*	*ADH1, ADH2, PDC, CAB, RbcL, ACS1, ACO1, DHAR, GR, GST, BKR*	*PDC, ADH1*	[4, 44, 80, 100, 109]

References

1. Abd El-Mageed TA, El-Sherif AMA, Abd El-Mageed SA, Abdou NM (2019) A novel compost alleviate drought stress for sugar beet production grown in Cd-contaminated saline soil. Agric Water Manag 226:105831. https://doi.org/10.1016/j.agwat.2019.105831
2. Abd-Elsalam KA (2024) Special issue: agricultural nanotechnology. Plants (Basel) 13(4):489. https://doi.org/10.3390/plants13040489
3. Afshari M, Pazoki A, Sadeghipour O (2021) Foliar-applied silicon and its nanoparticles stimulates physio-chemical changes to improve growth, yield and active constituents of coriander (*Coriandrum sativum* L.) essential oil under different irrigation regimes. Silicon 13:4177–4188. https://doi.org/10.1007/s12633-021-01101-8
4. Al-Khayri JM, Rashmi R, Surya Ulhas R, Sudheer WN, Banadka A, Nagella P, Aldaej MI, Rezk AA-S, Shehata WF, Almaghasla MI (2023) The role of nanoparticles in response of plants to abiotic stress at physiological, biochemical, and molecular levels. Plants (Basel) 12(2):292. https://doi.org/10.3390/plants12020292
5. Alharbi K, Al-Osaimi AA, Alghamdi BA (2022) Sodium chloride (NaCl)-induced physiological alteration and oxidative stress generation in *Pisum sativum* (L.): a toxicity assessment. ACS Omega 7(24):20819–20832. https://doi.org/10.1021/acsomega.2c01427
6. Ali E, El-Shehawi AM, Ibrahim O, Abdul-Hafeez E, Moussa M, Hassan F (2021) A vital role of chitosan nanoparticles in improvisation the drought stress tolerance in *Catharanthus roseus* (L.) through biochemical and gene expression modulation. Plant Physiol Biochem 161:166–175. https://doi.org/10.1016/j.plaphy.2021.02.008
7. Almutairi ZM (2016a) Influence of silver nano-particles on the salt resistance of tomato (*Solanum lycopersicum*) during germination. Int J Agric Biol 18:449–457. https://doi.org/10.17957/ijab/15.0114
8. Almutairi ZM (2016b) Effect of nano-silicon application on the expression of salt tolerance genes in germinating tomato (*Solanum lycopersicum* L.) seedlings under salt stress. Plant Omics 9:106–114
9. Amini S, Maali-Amiri R, Mohammadi R, Shahandashti S-SK (2017) CDNA-AFLP analysis of transcripts induced in chickpea plants by TiO_2 nanoparticles during cold stress. Plant Physiol Biochem 111:39–49
10. Ashraf H, Ghouri F, Liang J, Xia W, Zheng Z, Shahid MQ, Fu X (2024) Silicon dioxide nanoparticles-based amelioration of Cd toxicity by regulating antioxidant activity and photosynthetic parameters in a line developed from wild rice. Plants (Basel) 13(12):1715. https://doi.org/10.3390/plants13121715
11. Askary M, Talebi SM, Amini F, Bangan ADB (2017) Effects of iron nanoparticles on *Mentha piperita* L. under salinity stress. Biologija 63:65–67. https://doi.org/10.6001/biologija.v63i1.3476
12. Badawy SA, Zayed BA, Bassiouni SMA, Mahdi AHA, Majrashi A, Ali EF, Seleiman MF (2021) Influence of nano silicon and nano selenium on root characters, growth, ion selectivity, yield, and yield components of rice (*Oryza sativa* L.) under salinity conditions. Plants 10:1657. https://doi.org/10.3390/plants10081657
13. Batool T, Ali S, Seleiman MF, Naveed NH, Ali A, Ahmed K, Abid M, Rizwan M, Shahid MR, Alotaibi M, Al-Ashkar I, Mubushar M (2020) Plant growth promoting rhizobacteria alleviates drought stress in potato iresponse to suppressive oxidative stress and antioxidant enzymes activities. Sci Rep 10:16975. https://doi.org/10.1038/s41598-020-73489-z
14. Bekkam R, Chitdeshwari T, Mohanapriya G (2024) Fascinating role of nanosilica in mitigating drought and nutrient stress—a review. Plant Stress 14:100672
15. Chung C, Kim Y-K, Shin D, Ryoo S-R, Hong BH, Min D-H (2013) Biomedical applications of graphene and graphene oxide. Acc Chem Res 46:2211–2224. https://doi.org/10.1021/ar300159f
16. Cong W, Miao Y, Xu L, Zhang Y, Yuan C, Wang J, Zhuang T, Lin X, Jiang L, Wang N, Ma J, Sanguinet KA, Liu B, Rutsgi S, Ou X (2019) Transgenerational memory of gene expression changes induced by heavy metal stress in rice (*Oryza sativa* L.). BMC Plant Biol 19:282

17. Cruz de Carvalho MH (2008) Drought stress and reactive oxygen species: production, scavenging and signaling. Plant Signal Behav 3:156–165. https://doi.org/10.4161/psb.3.3.5536
18. Cui J, Liu T, Li F, Yi J, Liu C, Yu H (2017) Silica nanoparticles alleviate cadmium toxicity in rice cells: mechanisms and size effects. Environ Pollut 228:363–369
19. Daler S, Kaya O, Canturk S, Korkmaz N, Kiliç T, Karadağ A, Hatterman-Valenti H (2024) Silicon nanoparticles (SiO_2 NPs) boost drought tolerance in grapevines by enhancing some morphological, physiological, and biochemical traits. Plant Mol Biol Report. https://doi.org/10.1007/s11105-024-01520-y
20. Darvishzadeh F, Najatzadeh F, Iranbakhsh AR (2015) Effect of silver nanoparticles on salinity tolerance of basil plant in germination stages under laboratory conditions. New Cell Mol Biotechnol J 20:63–70
21. Ekhtiyari R, Moraghebi F (2012) Effect of nanosilver particles on salinity tolerance of cumin (*Cuminum cyminum* L.). J Plant Biotechnol 25:99–107
22. El-Saadony MT, Abd El-Hack ME, Taha AE, Fouda M, Ajarem JS, Maodaa SN, Allam AA, Elshaer N (2020) Ecofriendly synthesis and insecticidal application of copper nanoparticles against the storage pest *Tribolium castaneum*. Nanomaterials 10:587. https://doi.org/10.3390/nano10030587
23. El-Saadony MT, Almoshadak AS, Shafi ME, Albaqami NM, Saad AM, El-Tahan AM, Desoky EM, Elnahal ASM, Almakas A, Abd El-Mageed TA, Taha AE, Elrys AS, Helmy AM (2021) Vital roles of sustainable nano-fertilizers in improving plant quality and quantity—an updated review. Saudi J Biol Sci 28:7349–7359. https://doi.org/10.1016/j.sjbs.2021.08.032
24. El-Saadony MT, Saad AM, Soliman SM, Salem Heba M, Desoky El-Sayed M, Babalghith Ahmad O, El-Tahan Amira M, Ibrahim Omar M, Ebrahim Alia AM, Abd El-Mageed Taia A, Elrys Ahmed S, Elbadawi Alaa A, El-Tarabily Khaled A, AbuQamar Synan F (2022) Role of nanoparticles in enhancing crop tolerance to abiotic stress: a comprehensive review. J Plant Biotechnol 13:946717. https://doi.org/10.3389/fpls.2022.946717
25. El-Sharkawy MS, El-Beshsbeshy TR, Mahmoud EK, Abdelkader NI, Al-Shal RM, Missaoui AM (2017) Response of alfalfa under salt stress to the application of potassium sulfate nanoparticles. Am J Plant Sci 8:1751–1773
26. El-Zohri M, Al-Wadaani NA, Bafeel SO (2021) Foliar sprayed green zinc oxide nanoparticles mitigate drought-induced oxidative stress in tomato. Plants 10:2400. https://doi.org/10.3390/plants10112400
27. Elsakhawy T, Omara AED, Alshaal T, El-Ramady H (2018) Nanomaterials and plant abiotic stress in agroecosystems. Environ Biodiver Soil Secur 2:73–94. https://doi.org/10.21608/jenvbs.2018.3897.1030
28. Embiale A, Hussein M, Husen A, Sahile S, Mohammed K (2016) Differential sensitivity of *Pisum sativum* L. cultivars to water-deficit stress: changes in growth, water status, chlorophyll fluorescence and gas exchange attributes. J Agron 15:45–57
29. Fallah S, Elham Y-T, Jose RP-V (2024) Interaction of nanoparticles and reactive oxygen species and their impact on macromolecules and plant production. Plant Nanobiol 10:2024
30. Faraji J, Sepehri A (2020) Exogenous nitric oxide improves the protective effects of TiO_2 nanoparticles on growth, antioxidant system, and photosynthetic performance of wheat seedlings under drought stress. J Soil Sci Plant Nutr 20:703–714. https://doi.org/10.1007/s42729-019-00158-0
31. Feng SJ, Che J, Yamaji N, Fang SR, Feng MJ (2017). Silicon reduces cadmium accumulation by suppressing expression of transporter genes involved in cadmium uptake and translocation in rice. J Exp Bot 68(20):5641–5651. https://doi.org/10.1093/jxb/erx364
32. Fouad A, Hegazy AE, Azab E, Khojah E, Kapiel T (2021) Boosting of antioxidants and alkaloids in *Catharanthus roseus* suspension cultures using silver nanoparticles with expression of CrMPK3 and STR genes. Plants (Basel) 10(10):2202. https://doi.org/10.3390/plants10102202
33. Fróna D, Szenderák J, Harangi-Rákos M (2019) The challenge of feeding the world. Sustainability 11(20):5816. https://doi.org/10.3390/su11205816

34. Fukao T, Barrera-Figueroa BE, Juntawong P, Peña-Castro JM (2019) Submergence and waterlogging stress in plants: a review highlighting research opportunities and understudied aspects. Front Plant Sci 10:340. https://doi.org/10.3389/fpls.2019.00340
35. Getnet Z, Husen A, Fetene M, Yemata G (2015) Growth, water status, physiological, biochemical and yield response of stay green sorghum {*Sorghum bicolor* (L.) Moench} varieties-a field trial under drought-prone area in Amhara regional state, Ethiopia. J Agron 14:188–202. https://doi.org/10.3923/ja.2015.188.202
36. Guerriero G, Sutera FM, Torabi-Pour N, Renaut J, Hausman J-F, Berni R, Pennington HC, Welsh M, Dehsorkhi A, Zancan LR, Suzanne S-S (2021) Phyto-courier, a silicon particle-based nano-biostimulant: evidence from *Cannabis sativa* exposed to salinity. ACS Nano 15:3061–3069. https://doi.org/10.1021/acsnano.0c09488
37. Guha T, Barman S, Mukherjee A, Kundu R (2020) Nano-scale zero valent iron modulates Fe/Cd transporters and immobilizes soil Cd for production of Cd free rice. Chemosphere 260:127533
38. Gunes A, Pilbeam DJ, Inal A, Bagci EG, Coban S (2007) Influence of silicon on antioxidant mechanisms and lipid peroxidation in chickpea (*Cicer arietinum* L.) cultivars under drought stress. J Plant Interact 2:105–113. https://doi.org/10.1080/17429140701529399
39. Hafeez MB, Ghaffar A, Zahra N, Ahmad N, Al-Qahtani WH, Sharma P, Husen A, Li J (2024) Exploring the efficacy of plant growth promoters in nutrient acquisition and antioxidant defense responses of late sown wheat under saline conditions. S Afr J Bot 171:812–824. https://doi.org/10.1016/j.sajb.2024.06.042
40. Hanif S, Rabia J, Mumtaz C, Misbah ZK, Snovia F, Muhammad Z (2024) Harnessing the potential of zinc oxide nanoparticles and their derivatives as nanofertilizers: trends and perspectives. Plant Nano Biol 10:100110
41. Hasanpour H, Maali-Amir R, Zeinali H (2020) Effect of TiO_2 nanoparticles on metabolic limitations to photosynthesis under cold in chickpea. Russ J Plant Physiol 62:779–787
42. Hassanvand A, Saadatmand S, Lari Yazdi H, Iranbakhsh A (2021) Biosynthesis of nanosilver and its effect on key genes of flavonoids and physicochemical properties of *Viola tricolor* L. Iran J Sci Technol Trans Sci 45:805–819. https://doi.org/10.1007/s40995-021-01091-7
43. Hojjat SS (2019) Effect of interaction between Ag nanoparticles and salinity on germination stages of *Lathyrus sativus* L. J Environ Soil Sci 2:186–191. https://doi.org/10.32474/oajess.2019.02.000132
44. Hsu SY, Morris R, Cheng F (2021) Signaling pathways regulated by silica nanoparticles. Molecules 26(5):1398. https://doi.org/10.3390/molecules26051398
45. Husen A (2010) Growth characteristics, physiological and metabolic responses of teak (*Tectona grandis* Linn. f.) clones differing in rejuvenation capacity subjected to drought stress. Silvae Genet 59:124–136. https://doi.org/10.1515/sg-2010-0015
46. Husen A, Iqbal M, Aref MI (2014) Growth, water status and leaf characteristics of *Brassica carinata* under drought stress and rehydration conditions. Braz J Bot 37:217–227. https://doi.org/10.1007/s40415-014-0066-1
47. Husen A, Iqbal M, Khanum N, Aref IM, Sohrab SS, Meshresa G (2019) Modulation of salt-stress tolerance of niger (*Guizotia abyssinica*), an oilseed plant, by application of salicylic acid. J Environ Biol 40:94–104
48. Husen A (2021a) Plant performance under environmental stress. Springer, Cham, pp 1–15. https://doi.org/10.1007/978-3-030-78521-5_1
49. Husen A (2021b) Harsh environment and plant resilience. Springer, Cham, pp 1–23. https://doi.org/10.1007/978-3-030-65912-7_1
50. Husen A (2022) Environmental pollution and medicinal plants. CRC Press, Boca Raton, FL. https://doi.org/10.1201/9781003178866
51. Husen A (2023) Plants and their interaction to environmental pollution (Damage detection, adaptation, tolerance, physiological and molecular responses). Elsevier Inc., Cambridge, MA. https://doi.org/10.1016/C2020-0-03319-0
52. Husen A (2024a) Essential minerals in plant-soil system. In: Coordination, signalling, and interaction under adverse situations. Elsevier Inc., Cambridge, MA. https://doi.org/10.1016/C2022-0-02437-5

53. Husen A (2024b) Plant response to gold nanoparticles. Springer Nature, Singapore. https://doi.org/10.1007/978-981-97-7174-5
54. Husen A (2024c) Biostimulants in plant protection and performance. Elsevier Inc., Cambridge, MA. https://doi.org/10.1016/C2022-0-02207-8
55. Husen A (2025a) Roles of osmolytes in changing environment. Elsevier Inc., Cambridge
56. Husen A (2025b) Agricultural crop improvement: plant and soil relationships. Taylor & Francis Group, LLC, Boca Raton. https://doi.org/10.1201/9781032630366
57. Husen A (2025c) Plant response to silver nanoparticles. Springer Nature, Singapore. https://doi.org/10.1007/978-981-97-7352-7
58. Husen A (2025d) Role of antioxidants in mitigating plant stress. Elsevier Inc., Cambridge, MA
59. Husen A (2025e) Emerging carbon nanomaterials for sustainable agricultural practices. In: Synthesis, plant growth, performance, production and protection. Springer Nature, Singapore
60. Husen A, Iqbal M, Sohrab SS, Ansari MKA (2018) Salicylic acid alleviates salinity-caused damage to foliar functions, plant growth and antioxidant system in Ethiopian mustard (*Brassica carinata* A. Br.). Agric Food Secur 7(44):1–14
61. Jan R, Asaf S, Numan M, Lubna KK-M (2021) Plant secondary metabolite biosynthesis and transcriptional regulation in response to biotic and abiotic stress conditions. Agronomy 11(5):968. https://doi.org/10.3390/agronomy11050968
62. Joudeh N, Linke D (2022) Nanoparticle classification, physicochemical properties, characterization, and applications: a comprehensive review for biologists. J Nanobiotechnol 20:262. https://doi.org/10.1186/s12951-022-01477-8
63. Kareem HA, Mahmood UH, Muhammad Z, Annie I, Noman S, Sana S, Niu J, Milan S, Zhao C, Guo Z, Wang Q (2022) Nanosized zinc oxide (n-ZnO) particles pretreatment to alfalfa seedlings alleviate heat-induced morpho-physiological and ultrastructural damages. Environ Pollut 303:119069
64. Kashyap PL, Xiang X, Heiden P (2015) Chitosan nanoparticle based delivery systems for sustainable agriculture. Int J Biol Macromol 77:36–51. https://doi.org/10.1016/j.ijbiomac.2015.02.039
65. Khan ZS, Rizwan M, Hafeez M, Ali S, Javed MR, Adrees M (2019) The accumulation of cadmium in wheat (*Triticum aestivum*) as influenced by zinc oxide nanoparticles and soil moisture conditions. Environ Sci Pollut Res 26:19859–19870. https://doi.org/10.1007/s11356-019-05333-5
66. Khan I, Seleiman MF, Chattha MU, Jalal RS, Mahmood F, Hassan FA, Izzet W, Alhammad BA, Ali EF, Roy R, Uslu OS, Hassan MU (2021) Enhancing antioxidant defense system of mung bean with a salicylic acid exogenous application to mitigate cadmium toxicity. Not Bot Horti Agrobo Cluj-Napoca 49:12303
67. Khanna K, Kaur KS, Neha H, Kaur H, Ohri P, Bhardwaj R, Yousaf B, Rinklebe J, Ahmad P (2021) Enthralling the impact of engineered nanoparticles on soil microbiome: a concentric approach towards environmental risks and cogitation. Ecotoxicol Environ Saf 222:112459
68. Khodakovskaya MV, de Silva K, Nedosekin DA, Dervishi E, Biris AS, Shashkov EV (2011) Complex genetic, photothermal, and photoacoustic analysis of nanoparticle-plant interactions. Proc Natl Acad Sci USA 108:1028–1033
69. Kim D-Y, Kadam A, Shinde S, Saratale RG, Patra J, Ghodake G (2018) Recent developments in nanotechnology transforming the agricultural sector: a transition replete with opportunities. J Sci Food Agric 98:849–864
70. Konate A, He X, Zhang Z, Ma Y, Zhang P, Alugongo GM, Rui Y (2017) Magnetic (Fe_3O_4) nanoparticles reduce heavy metals uptake and mitigate their toxicity in wheat seedling. Sustainability 9:790. https://doi.org/10.3390/su9050790
71. Kubiak M (2014) Dendrimers-fascinating nanoparticles in the application in medicine. Chemik 68:146–150
72. Kumar A, Choudhary A, Kaur H, Guha S, Mehta S, Husen A (2022) Potential applications of engineered nanoparticles in plant disease management: a critical update. Chemosphere 295(133798):1–21. https://doi.org/10.1016/j.chemosphere.2022.133798

73. Kumar A, Choudhary A, Kaur H, Mehta S, Husen A (2021a) Smart nanomaterial and nanocomposite with advanced agrochemical activities. Nanoscale Res Lett 16(156):1–26. https://doi.org/10.1186/s11671-021-03612-0
74. Kumar A, Choudhary A, Kaur H, Mehta S, Husen A (2021b) Metal-based nanoparticles, sensors and their multifaceted application in food packaging. J Nanobiotechnol 19(256):1–25. https://doi.org/10.1186/s12951-021-00996-0
75. Kumar S, Dilbaghi N, Saharan R, Bhanjana G (2012) Nanotechnology as emerging tool for enhancing solubility of poorly water-soluble drugs. Bionanoscience 2:227–250. https://doi.org/10.1007/s12668-012-0060-7
76. Lala S (2024) Nanoparticles as elicitors and harvesters of economically important secondary metabolites in plants. In: Mérillon JM, Ramawat KG (eds) Plant specialized metabolites. Reference series in phytochemistry. Springer, Cham. https://doi.org/10.1007/978-3-031-30037-0_45-1
77. Mahmoud LM, Dutt M, Shalan AM, El-Kady ME, El-Boray MS, Shabana YM, Grosser JW (2020) Silicon nanoparticles mitigate oxidative stress of in vitroderived banana (*Musa acuminata* 'Grand Nain') under simulated water deficit or salinity stress. S Afr J Bot 132:155–163. https://doi.org/10.1016/j.sajb.2020.04.027
78. Majumdar S, Almeida IC, Arigi EA, Choi H, Verberkmoes NC, TrujilloReyes J, Flores-Margez JP, White JC, Peralta-Videa JR, Gardea-Torresdey JL (2015) Environmental effects of nanoceria on seed production of common bean (*Phaseolus vulgaris*): a proteomic analysis. Environ Sci Technol 49:13283–13293. https://doi.org/10.1021/acs.est.5b03452
79. Mansoor S, Ali A, Kour N, Bornhorst J, AlHarbi K, Rinklebe J, Moneim DAE, Ahmad P, Chung YS (2023) Heavy metal induced oxidative stress mitigation and ROS scavenging in plants. Plants (Basel) 12(16):3003. https://doi.org/10.3390/plants12163003
80. Maqsood MF, Shahbaz M, Khalid F, Rasheed Y, Asif K, Naz N, Zulfiqar U, Zulfiqar F, Moosa A, Alamer KH, Attia H (2023) Biogenic nanoparticles application in agriculture for ROS mitigation and abiotic stress tolerance: a review. Plant Stress 10:100281
81. Melash AA, Bogale AA, Bytyqi B, Nyandi MS, Ábrahám ÉB (2023) Nutrient management: as a panacea to improve the caryopsis quality and yield potential of durum wheat (*Triticum turgidum* L.) under the changing climatic conditions. Front Plant Sci 14:1232675. https://doi.org/10.3389/fpls.2023.1232675
82. Mustafa G, Chaudhari SK, Manzoor M, Batool S, Hatami M, Hasan M (2024) Zinc oxide nanoparticles mediated salinity stress mitigation in *Pisum sativum*: a physio-biochemical perspective. BMC Plant Biol 24:835. https://doi.org/10.1186/s12870-024-05554-y
83. Nadarajah KK (2020) ROS homeostasis in abiotic stress tolerance in plants. Int J Mol Sci 21(15):5208. https://doi.org/10.3390/ijms21155208
84. Nguyen Van D, Nguyen HM, Le NT, Nguyen AT, Dinh NTT, Hoang SA, Ha CV (2022) Copper nanoparticle application enhances plant growth and grain yield in maize under drought stress conditions. J Plant Growth Regul 41:364–375. https://doi.org/10.1007/s00344-021-10301-w
85. Noman M, Shahid M, Ahmed T, Tahir M, Naqqash T, Muhammad S, Song F, Abid HMA, Aslam Z (2020) Green copper nanoparticles from a native *Klebsiella pneumoniae* strain alleviated oxidative stress impairment of wheat plants by reducing the chromium bioavailability and increasing the growth. Ecotoxicol Environ Saf 192:110303. https://doi.org/10.1016/j.ecoenv.2020.110303
86. Prajitha N, Athira S, Mohanan P (2019) Bio-interactions and risks of engineered nanoparticles. Environ Res 172:98–108. https://doi.org/10.1016/j.envres.2019.02.003
87. Rahmatizadeh R, Javad Arvin SM, Jamei R, Mozaffari H, Nejhad FR (2019) Response of tomato plants to interaction effects of magnetic (Fe_3O_4) nanoparticles and cadmium stress. J Plant Interact 14:474–481. https://doi.org/10.1080/17429145.2019.1626922
88. Ramírez-Rodríguez GB, Miguel-Rojas C, Montanha GS, Carmona FJ, Dal Sasso G, Sillero JC, Skov Pedersen J, Masciocchi N, Guagliardi A, Pérez-de-Luque A, Delgado-López JM (2020) Reducing nitrogen dosage in triticum durum plants with urea-doped nanofertilizers. Nanomater 10(6):1043. https://doi.org/10.3390/nano10061043

89. Rasheed A, Li H, Tahir MM, Mahmood A, Nawaz M, Shah AN, Aslam MT, Negm S, Moustafa M, Hassan MU, Wu Z (2022) The role of nanoparticles in plant biochemical, physiological, and molecular responses under drought stress: a review. Front Plant Sci 13:976179. https://doi.org/10.3389/fpls.2022.976179
90. Rehman A, Khan S, Sun F, Peng Z, Feng K, Wang N, Jia Y, Pan Z, He S, Wang L, Qayyum A, Du X, Li H (2024) Exploring the nano-wonders: Unveiling the role of nanoparticles in enhancing salinity and drought tolerance in plants. Front Plant Sci 14:1324176. https://doi.org/10.3389/fpls.2023.1324176
91. Rossi L, Zhang W, Lombardini L, Ma X (2016) The impact of cerium oxide nanoparticles on the salt stress responses of *Brassica napus* L. Environ Pollut 219:28–36. https://doi.org/10.1016/j.envpol.2016.09.060
92. Sadati Raeisi SY, Jahanbakhsh Godehkahriz S, Ebadi A, Sedghi M (2022) Zinc oxide nanoparticles enhance drought tolerance in wheat via physio-biochemical changes and stress genes expression. Iran J Biotechnol 20(1):e3027. https://doi.org/10.30498/ijb.2021.280711.3027
93. Sedghi M, Hadi M, Toluie SG (2013) Effect of nano zinc oxide on the germination parameters of soybean seeds under drought stress. Ann West Univ Timisoara Ser Biol 16:73
94. Seleiman MF, Al-Suhaibani N, Ali N, Akmal M, Alotaibi M, Refay Y, Dindaroglu T, Abdul-Wajid HH, Battaglia ML (2021) Drought stress impacts on plants and different approaches to alleviate its adverse effects. Plants (Basel) 10(2):259. https://doi.org/10.3390/plants10020259. PMID: 33525688; PMCID: PMC7911879
95. Semida WM, Abdelkhalik A, Mohamed GF, Abd El-Mageed TA, Abd ElMageed SA, Rady MM, Ali EF (2021) Foliar application of zinc oxide nanoparticles promotes drought stress tolerance in eggplant (*Solanum melongena* L.). Plants 10:421. https://doi.org/10.3390/plants10020421
96. Shallan MA, Hassan H, Namich AA, Ibrahim AA (2016) Biochemical and physiological effects of TiO_2 and SiO_2 nanoparticles on cotton plant under drought stress. Res J Pharm Biol Chem Sci 7:1540–1551
97. Sharma P, Pandey V, Sharma MMM, Patra A, Singh B, Mehta S, Husen A (2021) A review on biosensors and nanosensors application in agroecosystems. Nanoscale Res Lett 16(136):1–24. https://doi.org/10.1186/s11671-021-03593-0
98. Sharma MMM, Kapoor D, Loyal A, Kumar R, Sharma P, Husen A (2024) Mitigation of drought stress in plants using gold nanoparticles. In: Husen A (eds) Plant response to gold nanoparticles. Smart nanomaterials technology. Springer, Singapore. https://doi.org/10.1007/978-981-97-7174-5_10
99. Sharma MMM, Kapoor D, Loyal A, Kumar R, Sharma P, Husen A (2025) Plant response to silver nanoparticles in terms of growth, development, production, and protection: an overview. In: Husen A (eds) Plant response to silver nanoparticles. Smart nanomaterials technology. Springer, Singapore. https://doi.org/10.1007/978-981-97-7352-7_1
100. Siddiqui H, Ahmed KBM, Sami F, Hayat S (2020) Silicon nanoparticles and plants: current knowledge and future perspectives. In: Sustainable agriculture reviews: nanotechnology for plant growth and development, vol 41. Springer, Cham, pp 129–142
101. Singh J, Lee BK (2016) Influence of nano-TiO_2 particles on the bioaccumulation of Cd in soybean plants (*Glycine max*): a possible mechanism for the removal of Cd from the contaminated soil. J Environ Manage 170:88–96. https://doi.org/10.1016/j.jenvman.2016.01.015
102. Singh D, Sillu D, Kumar A, Agnihotri S (2021) Dual nanozyme characteristics of iron oxide nanoparticles alleviate salinity stress and promote the growth of an agroforestry tree, *Eucalyptus tereticornis* Sm. Environ Sci Nano 8:1308–1325. https://doi.org/10.1039/D1EN00040C
103. Singh S, Husen A (2019) Role of nanomaterials in the mitigation of abiotic stress in plants. In: Husen A, Iqbal M (eds) N*anomaterials* and plant potential. Springer International Publishing AG, Cham, pp 441–471. https://doi.org/10.1007/978-3-030-05569-1_18
104. Singh S, Husen A (2020) Behavior of agricultural crops in relation to nanomaterials under adverse environmental conditions. In: Husen, A, Jawaid M (eds) Nanomaterials for agriculture

and forestry applications. Elsevier Inc., Cambridge, MA, pp 219–256. https://doi.org/10.1016/B978-0-12-817852-2.00009-3
105. Song Y, Jiang M, Zhang H, Li R (2021) Zinc oxide nanoparticles alleviate chilling stress in rice (*Oryza sativa* L.) by regulating antioxidative system and chilling response transcription factors. Molecules 26:2196
106. de Sousa A, Saleh AM, Habeeb TH, Hassan YM, Zrieq R, Wadaan MAM, Hozzein WN, Selim S, Matos M, Abd Elgawad H (2019) Silicon dioxide nanoparticles ameliorate the phytotoxic hazards of aluminum in maize grown on acidic soil. Sci Total Environ 693:133636. https://doi.org/10.1016/j.scitotenv.2019.133636
107. Taha R, Seleiman MF, Alotaibi M, Alhammad BA, Rady MM, Mahdi AHA (2020) Exogenous potassium treatments elevate salt tolerance and performances of *Glycine max* L. by boosting antioxidant defense system under actual saline field conditions. Agronomy 10:1741. https://doi.org/10.3390/agronomy10111741
108. Tripathi DK, Singh S, Singh VP, Prasad SM, Chauhan DK, Dubey NK (2016) Silicon nanoparticles more efficiently alleviate arsenate toxicity than silicon in maize cultiver and hybrid differing in arsenate tolerance. Front Environ Sci 4:46. https://doi.org/10.3389/fenvs.2016.00046
109. Tripathi DK, Vishwakarma K, Singh VP, Prakash V, Sharma S, Muneer S, Nikolic M, Deshmukh R, Vaculik M, Corpas FJ (2021) Silicon crosstalk with reactive oxygen species, phytohormones and other signaling molecules. J Hazard Mater 408:124820. https://doi.org/10.1016/j.jhazmat.2020.124820
110. Tumburu L, Andersen CP, Rygiewicz PT, Reichman JR (2017) Molecular and physiological responses to titanium dioxide and cerium oxide nanoparticles in Arabidopsis. Environ Toxicol Chem 36(1):71–82. https://doi.org/10.1002/etc.3500
111. Turgeon R (2010) The puzzle of phloem pressure. Plant Physiol 154:578–581. https://doi.org/10.1104/pp.110.161679
112. Venzhik Y, Deryabin A, Zhukova K (2024) Au-based nanoparticles enhance low temperature tolerance in wheat by regulating some physiological parameters and gene expression. Plants 13(9):1261. https://doi.org/10.3390/plants13091261
113. Wang S, Wang F, Gao S (2015) Foliar application with nano-silicon alleviates Cd toxicity in rice seedlings. Environ Sci Pollut Res 22:2837–2845. https://doi.org/10.1007/s11356-014-3525-0
114. Weir A, Westerhoff P, Fabricius L, Hristovski K, Von Goetz N (2012) Titanium dioxide nanoparticles in food and personal care products. Environ Sci Technol 46:2242–2250. https://doi.org/10.1021/es204168d
115. Wu J, Wang T (2020) Synergistic effect of zinc oxide nanoparticles and heat stress on the alleviation of transcriptional gene silencing in *Arabidopsis thaliana.* Bull Environ Contam Toxicol 104(1):49–56. https://doi.org/10.1007/s00128-019-02749-0
116. Xuan Y, Zhang W, Zhu X, Zhang S (2023) An updated overview of some factors that influence the biological effects of nanoparticles. Front Bioeng Biotechnol 11:1254861. https://doi.org/10.3389/fbioe.2023.1254861
117. Yepes-Molina L, Bárzana G, Carvajal M (2020) Controversial regulation of gene expression and protein transduction of aquaporins under drought and salinity stress. Plants 9(12):1662. https://doi.org/10.3390/plants9121662
118. Yin D, Wang X, Peng B, Tan C, Ma LQ (2017) Effect of biochar and Fe-biochar on Cd and As mobility and transfer in soil-rice system. Chemosphere 186:928–937
119. Younis AA, Khattab H, Emam MM (2020) Impacts of silicon and silicon nanoparticles on leaf ultrastructure and TaPIP1 and TaNIP2 gene expressions in heat stressed wheat seedlings. Biol Plant 64:343–352. https://doi.org/10.32615/bp.2020.030
120. Yuan H, Liu Q, Guo Z, Fu J, Sun Y, Gu C, Xing B, Dhankher OP (2021) Sulfur nanoparticles improved plant growth and reduced mercury toxicity via mitigating the oxidative stress in *Brassica napus* L. J Clean Prod 318:128589

121. Zahra N, Kausar A, Sharma P, Wabaidur SM, Husen A (2024) Serotonin application to *Guizotia abyssinica* improves growth, physiological functions, antioxidant system and alleviates the deleterious effects induced by salt stress. Plant Biosyst Int J Deal Aspects Plant Biol. https://doi.org/10.1080/11263504.2024.2384972
122. Zahra N, Kausar A, Abdelghani HTM, Singh S, Vashishth DS, Bachheti A, Bachheti RK, Husen A (2024) Serotonin improves plant growth, foliar functions and antioxidant defence system in Ethiopian mustard (*Brassica carinata* A. Br.). South Afr J Bot 170:1–9 https://doi.org/10.1016/j.sajb.2024.05.002
123. Zhang Y, Xu J, Li R, Ge Y, Li Y, Li R (2023) Plants' response to abiotic stress: mechanisms and strategies. Int J Mol Sci 24(13):10915. https://doi.org/10.3390/ijms241310915
124. Zhang S (2019) Mechanism of migration and transformation of nano selenium and mitigates cadmium stress in plants. Master's thesis. Shandong University, Jinan
125. Zhao L, Peng B, Hernandez-Viezcas JA, Rico C, Sun Y, Peralta-Videa JR (2012) Stress response and tolerance of *Zea mays* to CeO_2 nanoparticles: cross talk among H_2O_2, heat shock protein and lipid peroxidation. ACS Nano 6:9615–9622
126. Zhao L, Peralta-Videa JR, Ren M, Varela-Ramirez A, Li C, Hernandez-Viezcas JA, Aguilera RJ, Gardea-Torresdey JL (2012) Transport of Zn in a sandy loam soil treated with ZnO NPs and uptake by corn plants: electron microprobe and confocal microscopy studies. Chem Eng J 184:1–8. https://doi.org/10.1016/j.cej.2012.01.041
127. Zhou P, Adeel M, Shakoor N, Guo M, Hao Y, Azeem I, Li M, Liu M, Rui Y (2020) Application of nanoparticles alleviates heavy metals stress and promotes plant growth: an overview. Nanomaterials (Basel) 11(1):26. https://doi.org/10.3390/nano11010026
128. Zhu J-K (2002) Salt and drought stress signal transduction in plants. Annu Rev Plant Biol 53:247–273

Proteomic Insights into the Effects of Nanoparticle Exposure on Crop Plants Under Abiotic Stress Conditions

H. Shayista, S. Niranjan Raj, S. Anjani, K. Manju, Ravikumara, T. K. Pavan, and Syed Baker

Abstract Nanoparticles (NPs) have emerged as promising tools in modern agriculture, offering potential solutions to enhance plant growth, nutrient uptake, pest control, and stress tolerance. Due to their unique properties, such as large surface area and high reactivity, NPs are capable of interacting at the molecular level, providing mechanisms to mitigate abiotic stresses, including environmental contaminants, water scarcity, and natural calamities, which jeopardize global crop production. Despite their potential, the molecular mechanisms underlying NP-plant interactions, particularly under stress conditions, remain poorly understood, highlighting the need for further investigation. Proteomics, which involves the study of proteins and their functions, provides a powerful approach to investigating the molecular responses of plants to NP exposure. Advanced mass spectrometry (MS)-based techniques enable the identification and quantification of proteins, as well as the analysis of post-translational modifications (PTMs) that regulate stress-responsive pathways. Techniques such as SILAC, iTRAQ, and label-free quantification offer precise measurements of protein abundance changes in response to NP treatments. Furthermore, proteomics-based approaches, including multiple reaction monitoring (MRM), enable the validation of key proteins involved in stress adaptation. The integration of proteomics with other omics technologies offers a comprehensive understanding of NP-induced molecular changes and unveils complex regulatory networks involved in stress tolerance. Although challenges, such as the complexity of plant proteomes and low-abundance protein detection, persist, advancements in MS sensitivity and computational proteomics are addressing these obstacles. This chapter examines the combined potential of NPs and proteomics to enhance crop resilience, focusing on key proteins and signalling pathways involved in stress tolerance, with the goal of providing sustainable agricultural solutions in the face of increasing abiotic stresses.

H. Shayista
Department of Biotechnology, JSS Science and Technology University, Mysore, Karnataka, India

S. Niranjan Raj · S. Anjani · K. Manju · Ravikumara · T. K. Pavan · S. Baker (✉)
Department of Studies in Microbiology, Karnataka State Open University, Mukthagangotri, Mysuru, Karnataka, India
e-mail: syedbaker3@gmail.com

A. Husen (ed.), *Nanobiotechnology for Abiotic Stress Adaptation and Mitigation in Agricultural Crops*, Smart Nanomaterials Technology,
https://doi.org/10.1007/978-981-96-9298-9_15

Keywords Proteomics · Nanoparticle exposure · Abiotic stress · Stress response mechanisms · Oxidative stress

1 Introduction

Nanoparticles (NPs) are regarded as modern practices of the century which are employed across various sectors to offer innovative solutions. Their unique behaviour to attenuate particular activity can be attributed to their inherent properties like large surface area, high reactivity, surface charges chemical composition, solubility, and reactivity, enable precise interactions with plant systems at the molecular level [76]. This has led to their increasing applications, recent advances in agriculture, where NPs are reported to improve uptake of nutrient, pest control, and stress tolerance. Among the most significant challenges facing global agriculture are abiotic stressors, which has tampered the crop production. These stresses can be in the form of environmental contaminants, low water availability, and natural calamities. These environmental stressors disrupt cellular homeostasis, impaired photosynthesis, and disrupt metabolism, ultimately threatening global agriculture production [47, 48, 51, 53, 54, 106]. The use of NPs to mitigate the impacts of detrimental effects of abiotic stresses has gained significant attention in recent years. NPs can enhance stress tolerance, promote growth, and optimize resource utilization in plants [32, 50, 52, 55]. However, the intricate mechanisms underlying NP-plant interactions, particularly under stress conditions, remain poorly understood. There is great need to address these issues and demand more such studies pertaining to NPs in agriculture.

To encounter various challenges faced by the farming community, different domains need to collaborate to attenuate profound activities. One such area with the scope of involvement includes proteomics and NPs. Proteomics is study of proteins and functions which offer great success. Proteomics and NPs can address various applications [37]. Studies highlight NPs influence on protein expression, modification, and function is crucial for deciphering their impact on plant physiology and stress tolerance mechanisms.

Proteomic studies utilize advanced analytical techniques, primarily mass spectrometry (MS)-based approaches, to systematically identify and quantify proteins within complex biological samples [11]. These techniques enable comprehensive profiling of the proteome, including alterations in protein availability, translational stages of their modifications called post-translational modifications (PTMs), and protein–protein interactions. In the context of NP exposure, proteomics can reveal alterations in key proteins involved in stress signalling pathways, antioxidant defence systems, and metabolic networks [17]. NPs are reported to modulate the expression of various proteins, scavenging activity enzymes and proteins involved in osmoregulation, thus enhancing plant resilience [105].

Quantitative proteomics, particularly label-free quantification and isotope-labelling methods (e.g., SILAC and iTRAQ), provide precise measurements of protein abundance changes in response to NP treatment under stress conditions. These techniques allow for comparative analysis between treated and control samples, enabling barcoding differentially expressed associated with pivotal roles in stress adaptation. Additionally, proteomic-based approaches such as multiple reaction monitoring (MRM) can be employed to validate key proteins involved in NP-induced stress responses, offering insights into specific regulatory mechanisms [115].

Furthermore, proteomic studies facilitate the investigation of PTMs, which are crucial for dynamic regulation of protein activity and stability. NPs can induce various PTMs, including phosphorylation, ubiquitination, and glycosylation, influencing stress-responsive signalling cascades [67]. For example, phosphorylation of transcription factors and signalling proteins can activate downstream stress defence pathways, while ubiquitination may modulate protein turnover under stress conditions. Advanced MS techniques, such as phosphoproteomics, enable the identification and quantification of these modifications, shedding light on the signalling networks modulated by NPs [67, 68].

Integrating proteomics strategies with advance techniques like omics provides a holistic view of NP-induced molecular changes [1]. This system biology approach can elucidate the crosstalk between different regulatory layers, revealing how NPs orchestrate a coordinated response to abiotic stress. Bioinformatics tools and pathway analysis further enhance the interpretation of proteomic data, enabling the important regulatory advances which highlights signalling pathways involved in NP-mediated stress tolerance [61].

Despite the significant potential of proteomics in understanding NP-plant interactions, several challenges remain. These include the complexity of plant proteomes, low-abundance protein detection, and the need for robust data analysis pipelines. Advancements in MS sensitivity, data-independent acquisition (DIA) techniques, and computational proteomics are addressing these challenges, paving the way for more comprehensive and accurate proteomic studies.

Thus, proteomics is an indispensable approach for deciphering the process working out at the molecular level and the interactions of NPs with plants. By providing detailed insights into protein dynamics and regulatory networks, proteomics also enhances NP-Phyto component interaction which in turn improves crop production. As proteomic technologies continue to evolve, their integration with nanotechnology holds immense potential.

This book chapter evaluates the role of NPs and proteomics together to combat plant production and their ability to withstand abiotic stress. It seeks to elucidate the molecular mechanisms underlying NP-induced stress tolerance by exploring changes in protein expression, post-translational modifications, and regulatory networks. By integrating findings from recent proteomic studies, this review aims to highlight key proteins and signalling pathways involved in NP-mediated stress adaptation. Furthermore, it will discuss the potential of advanced proteomic technologies in unravelling complex NP-plant interactions and identifying critical knowledge gaps that warrant further investigation. Ultimately, this review aspires to contribute to the strategic design and application of NP for enhancing crop resilience, productivity, and sustainable agricultural practices.

2 Nanoparticles in Agriculture

2.1 Types of Nanoparticles Used in Crop Plants

The use of NPs in agriculture is rapidly transforming conventional farming practices, offering innovative strategies to enhance crop growth, productivity, and stress resilience [8]. With the escalating challenges posed by climate change, soil degradation, and the need for sustainable food production, NPs present a promising frontier for modern agriculture. Their nanoscale dimensions and unique physicochemical properties enable them to interact with plant systems at the molecular level, leading to enhanced nutrient uptake, improved photosynthetic efficiency, and increased resistance to environmental stresses [10]. As a result, the integration of NPs in agricultural practices is gaining momentum, driven by their potential to optimize resource utilization and mitigate the adverse effects of abiotic stress factors as well as addressing drug resistant crisis [111].

A wide range of NPs has been explored for use in crop plants, each designed to perform specific functions that support plant growth and productivity as shown in Fig. 1. These NPs vary not only in their chemical composition but also in their structural and functional properties, influencing their interactions with plant cells. From enhancing nutrient delivery to providing protection against pathogens and environmental stresses, the applications of NPs in agriculture are diverse and dynamic [9]. Moreover, the development of eco-friendly and biodegradable NPs is further expanding their potential for sustainable agricultural practices.

Table 1 highlights various types of NPs used in crop plants, examining their unique attributes and mechanisms of action. By exploring the different categories, including metal and metal oxide NPs, carbon-based materials, polymeric carriers, and biologically synthesized NPs, it aims to provide a comprehensive overview of their roles in agriculture. Understanding these diverse nanoparticle systems and their interactions with plant physiology is essential for advancing nanotechnology applications in sustainable crop production.

2.2 Mechanisms of Nanoparticle Uptake and Transport in Plants

The entry and transport of NPs into plants are complex processes determined by the physicochemical properties of the NPs, plant type, and environmental factors. Knowledge of these processes is important to maximize the application of nanotechnology in agriculture with least potential risks. NPs enter plants mainly through the root system, where they cross the root cortex and pericycle to reach the xylem for subsequent transportation. The charge, surface properties, and size of NPs play major roles in determining their ability to be taken up with high efficiency through roots [64, 113]. NPs also absorb through the leaves, though this is less well understood

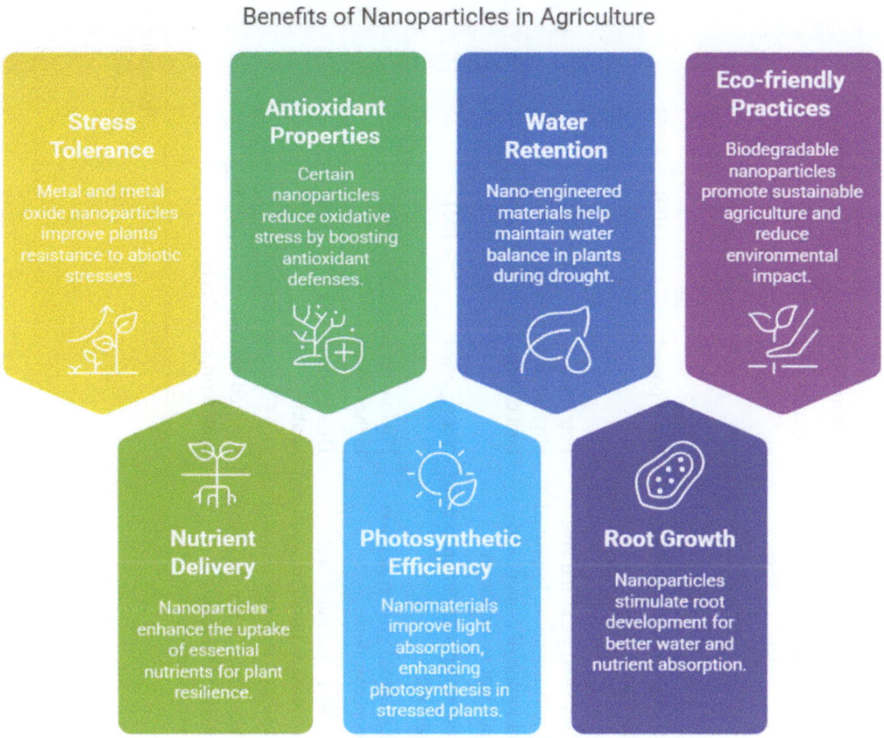

Fig. 1 Benefits of nanoparticles in agriculture

than uptake through the roots. Foliar uptake comprises NPs' penetration into stomata or through cuticular pathways, the efficiency of which depends on the type of NP as well as on the leaf surface features of the plant [20, 102].

After entering the plant, NPs are mostly transported by the xylem, which allows them to move from roots to aerial plant parts, such as leaves and other above-ground tissues [95, 113]. Certain NPs may also be transported by the phloem, allowing movement from leaves to roots and other plant parts. Bidirectional transport is essential for the redistribution of NPs in the plant system [26]. The deposition of NPs in various parts of the plant depends on their size, form, and the physiological state of the plant. Metal-based NPs, for instance, tend to accumulate in roots, while polymer-based NPs tend to be evenly distributed in the plant [20].

Recent imaging methods, including transmission electron microscopy (TEM), have shown that NPs are capable of localizing in the different subcellular compartments, like chloroplasts and mitochondria. Subcellular localization might influence cellular processes and determine the physiological response and general well-being of the plant [28]. While uptake and transportation of NPs by plants present interesting avenues in plant agriculture, e.g., the facilitation of better nutrient supply and disease/pest control, risk exposure also has potential ascribable to NPs' exposure

Table 1 Nanoparticles used in crop plants, examining their unique attributes and mechanisms of action

Type of nanoparticle	Function and benefits	Crops studied	Physiological impact	Potential risks	Key references
Zinc oxide (ZnO) NPs	Nanofertilizer enhancing seed germination and enzyme activity; Improves nutrient absorption and distribution	Rice	Increased grain yield and zinc content	Not reported	[70]
(ZnO) NPs	Seed priming and foliar application enhance growth parameters	Pearl Millet	Higher growth parameters and grain yield	Not reported	[109]
(ZnO) NPs	Enhances growth parameters, biochemical attributes, and antioxidant enzyme activity	Wheat	Increased yield and zinc biofortification; Improved chlorophyll content and nutrient assimilation	Potential phytotoxicity at high concentrations	[84]
(ZnO) NPs	Enhances stem thickness and zinc biofortification	Lettuce	Improved nutrient uptake and growth	Not reported	[22]
(ZnO) NPs	Increases seed germination and seedling vigour	Maize	Enhanced growth at lower concentrations	Phytotoxic effects at higher doses	[93]
(ZnO) NPs	Analyses impact on soil ecosystems and microbial processes	–	Affects soil microbiota; Influences mineralization and nitrogen fixation	Disruption of soil health	[108]
Green synthesis of ZnONPs	Eco-friendly alternative to reduce soil toxicity	–	Maintains beneficial effects while minimizing environmental impact	Limited data on long-term soil effects	[40, 89]

(continued)

Table 1 (continued)

Type of nanoparticle	Function and benefits	Crops studied	Physiological impact	Potential risks	Key references
Titanium dioxide (TiO_2) NPs	Enhances plant growth and physiological parameters; Improves photosynthesis, nutrient uptake, and antioxidant enzyme production	*Swietenia macrophylla*	Improved seed germination, survival rates, root and shoot development, and leaf growth; Increased chlorophyll content and DNA replication	Potential genotoxicity at high concentrations	[117]
(TiO_2) NPs	Enhances plant resilience to biotic and abiotic stresses; Functions as nanofertilizer ensuring efficient nutrient delivery	–	Improved physiological responses to environmental challenges	Adverse effects at high concentrations	[18]
(TiO_2) NPs	Used in combination with bio-fertilizers for enhanced growth; Facilitates better interaction with plant surfaces	–	Improved nutrient absorption and stress resilience	Potential environmental accumulation	[94, 103]
Silver (Ag) NPs	Antimicrobial properties reduce disease incidence by up to 50%; Activates stress-responsive genes, enhancing plant defence mechanisms	–	Reduced disease incidence; Improved plant health and productivity	Nanoparticle accumulation in the food chain; Toxicity to non-target organisms	[39]
(Ag) NPs	Biosynthesized using *Chaenomeles* leaf extracts; Demonstrates effective antifungal and antibacterial activities	–	Enhanced antimicrobial action; Disrupts microbial cell membranes; Induces oxidative stress	Potential environmental and health risks	[25]

(continued)

Table 1 (continued)

Type of nanoparticle	Function and benefits	Crops studied	Physiological impact	Potential risks	Key references
(Ag) NPs	Acts as signalling molecules, activating stress-responsive genes	–	Enhances plant innate defence mechanisms	Requires regulatory frameworks for safe usage	[39]
(Ag) NPs	Inhibits growth of pathogens like *Fusarium oxysporum* and *Fusarium culmorum*	–	Effective disease control and productivity enhancement	Risk of environmental toxicity and bioaccumulation	[30, 98]
Iron oxide (FeO) NPs	Enhance growth and productivity in leguminous plants under stress conditions Improve symbiotic interactions, photosynthesis, and nitrogen fixation	Garden cress	Improved shoot length and chlorophyll concentration Enhanced photosynthetic activity	Phytotoxicity at high concentrations	[79]
(FeO) NPs	Enhance growth and productivity in leguminous plants under stress conditions Improve symbiotic interactions, photosynthesis, and nitrogen fixation	*Dracocephalum kotschyi*	Increased leaf length, dry weight, and chlorophyll content	Phytotoxicity at high concentrations	[69]

(continued)

Table 1 (continued)

Type of nanoparticle	Function and benefits	Crops studied	Physiological impact	Potential risks	Key references
(FeO) NPs	Enhance growth and productivity in leguminous plants under stress conditions Improve symbiotic interactions, photosynthesis, and nitrogen fixation	Soybean, Chickpea	Enhanced nitrogenase enzyme activity, root nodule formation, and nutrient uptake	Phytotoxicity at high concentrations	[104]
(FeO) NPs	Enhance growth and productivity in leguminous plants under stress conditions Improve symbiotic interactions, photosynthesis, and nitrogen fixation	Maize	Improved salinity stress tolerance Enhanced antioxidant enzyme activities	Phytotoxicity at high concentrations	[81]
(FeO) NPs	Enhance growth and productivity in leguminous plants under stress conditions Improve symbiotic interactions, photosynthesis, and nitrogen fixation	*Vigna radiata*	Improved growth under heavy metal stress	Phytotoxicity at high concentrations	[75]

(continued)

Table 1 (continued)

Type of nanoparticle	Function and benefits	Crops studied	Physiological impact	Potential risks	Key references
Other metal oxide nanoparticles (MONPs) (CuO, CeO_2, MgO)	Enhance nutrient uptake and bioavailability Improve stress tolerance and antioxidant systems Facilitate symbiotic interactions for better nitrogen fixation Reduce need for traditional fertilizers and pesticides	Not crop-specific (general agricultural use)	Increased nutrient absorption and stress resilience Enhanced defence mechanisms and antioxidant activity	Negative effects at high concentrations (e.g., reduced photosynthesis, altered gene expression) Potential toxicity to non-target organisms Nanoparticle accumulation in food chain	[27, 39, 65, 73, 92, 119]
Graphene oxide (GO)	Enhances nutrient delivery Slow and controlled release of nutrients Effective crop protection against diseases Improves stress tolerance	Various crops (general use)	Enhanced nutrient uptake efficiency Improved plant growth and stress tolerance	Potential toxicity to non-target organisms Possible impact on soil microbiota	[14, 91]
Carbon nanotubes (CNTs)	Promotes plant growth Enhances water absorption, root growth, and nutrient uptake Used in nanofertilizers and as pesticide carriers	Various crops (general use)	Increased biomass production Enhanced seed germination and water uptake	Toxicity concerns for non-target organisms Long-term environmental impact	[62, 77]

(continued)

Table 1 (continued)

Type of nanoparticle	Function and benefits	Crops studied	Physiological impact	Potential risks	Key references
Graphene oxide (GO)	Prevents fungal infections Reduces bacterial complexity in rhizosphere Helps manage antibiotic resistance genes	Various crops (general use)	Fungal infection prevention Improved microbial health in rhizosphere	Potential impact on beneficial soil microbes	[122]
Carbon nanotubes (CNTs)	Enhances seed germination under stress conditions - Improves water uptake and nutrient availability	Crops in arid regions	Improved seed germination under salinity stress	Potential toxicity to soil organisms and water systems	[38, 120]
Polymeric Nanoparticles (Biopolymer-based)	Encapsulate natural compounds for controlled release Enhance crop yields and resilience to abiotic stress	Various crops (general use)	Enhanced plant resilience under adverse conditions Minimized reliance on chemical fertilizers and pesticides	Environmental risks of nanoplastics accumulation	[78]
Polymeric nanoparticles (biopolymer-based)	Slow-release mechanism for nutrients Eco-friendly synthesis Effective carriers for herbicides, fungicides, and insecticides	Various crops (general use)	Increased nutrient use efficiency Enhanced efficacy of agrochemicals	Potential for nanoplastics contamination in soil	[4, 90]
Polymeric nanoparticles (Biodegradable plastics)	Reduces environmental footprint Supports eco-friendly farming practices	Various crops (general use)	Reduced dependency on traditional agrochemicals Increased crop productivity	Accumulation of nanoplastics in crops and soil	[7, 31]

(continued)

Table 1 (continued)

Type of nanoparticle	Function and benefits	Crops studied	Physiological impact	Potential risks	Key references
Polymeric nanoparticles [Polyvinyl chloride (PVC), Polyethylene (PE)]	Used in agrochemical carriers Nanoplastics accumulation in crops and soil	Various crops (general use)	Effective delivery of agrochemicals	Toxicity of nanoplastics to soil and crops Environmental contamination	[121]
Actinomycetes (e.g., *Streptomyces* spp.)	Synthesize metal NPs such as AgNPs and CuONPs Potent antimicrobial properties Enhance plant protection against pathogens	Actinomycetes (e.g., *Streptomyces* spp.)	Improved disease resistance Enhanced crop protection	Ecological disruption Impact on soil microbiota	[99, 100]
Actinomycetes (e.g., *Streptomyces* spp.)	Biogenic synthesis of NPs Activation of cryptic biosynthetic gene clusters for increased NP yield	Actinomycetes (e.g., *Streptomyces spp.*)	Enhanced nanoparticle synthesis efficiency	Scaling challenges for large-scale production	[100]
Plant endophytes (e.g., fungi and bacteria)	Synthesize NPs such as ZnONPs Improve antimicrobial and antioxidant properties Enhance plant growth and resistance to diseases	Endophytic fungi and bacteria	Enhanced plant growth Improved resistance to diseases	Impact on host plant health and soil ecosystem	[19, 29]

(continued)

Table 1 (continued)

Type of nanoparticle	Function and benefits	Crops studied	Physiological impact	Potential risks	Key references
Plant endophytes (e.g., fungi and bacteria)	Biogenic synthesis of nanoparticles Provide sustainable alternatives to conventional agrichemicals	Endophytic fungi and bacteria	Reduced reliance on chemical inputs Enhanced plant stress tolerance	Complex microbial interactions may impact plant–microbe dynamics	[35, 85]
Plant endophytes (e.g., fungi and bacteria)	Produce bioactive metabolites for nanoparticle synthesis Customized NPs for specific agricultural applications	Endophytic fungi and bacteria	Improved plant fitness and stress tolerance	Challenges in controlling nanoparticle properties	[87, 112]

to them. As an example, their interaction with the plant system will facilitate ROS formation and production that have the ability to bring oxidative stress as well as cause damage at the cellular level [63]. Additionally, the long-term environmental effect and the possibility of NPs entering the food chain are issues that need to be studied further [26, 95]. These dynamics need to be understood in order to create safe and efficient NP-based agricultural technologies.

3 Proteomic Strategies in Plant Stress Studies

Proteomic methods in plant stress biology have emerged as central to the understanding of plant response to environmental stresses as shown in Fig. 2. Proteomics, the high-throughput analysis of proteins, gives thorough information on the cellular processes and mechanisms plants utilize to survive against stresses. Stress-tolerant crop development is a major requirement for food security in the context of climate change, and proteomics plays a key role in its understanding. Through the examination of the proteome, which is the entire complement of proteins produced by a genome, cell, tissue, or organism, scientists can isolate stress-responsive proteins and determine their functions in stress tolerance. Such information is critical to crop improvement initiatives that target increasing resistance to abiotic stressors like drought, salinity, and temperature extremes [110].

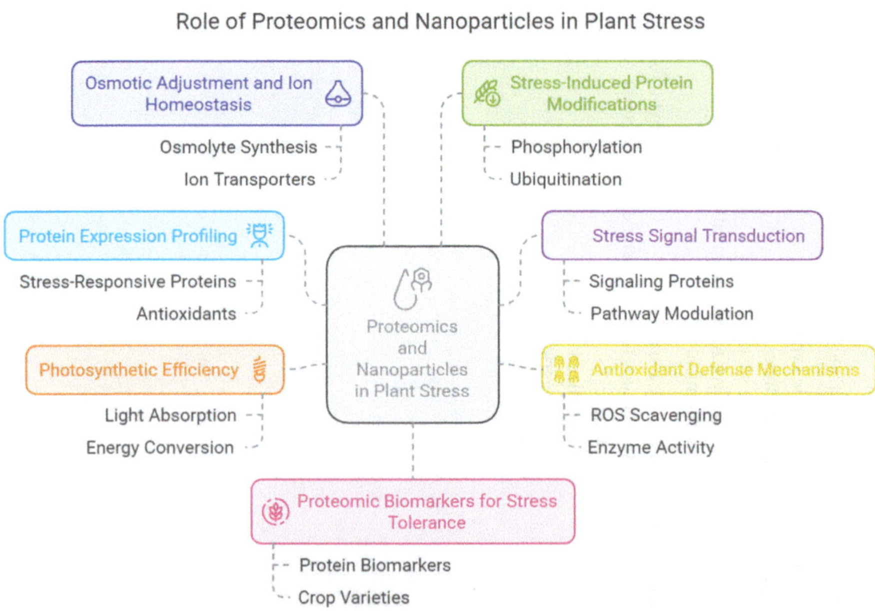

Fig. 2 Role of proteomics and nanoparticles in plant stress

Proteomics is especially useful in plant stress studies since proteins are the major effectors of biological responses and activities towards environmental changes. In contrast to genomics or transcriptomics, which indicate genetic potential or mRNA expression, proteomics immediately reflects the functional state of the cell. It uncovers shifts in protein quantity, post-translational adjustment, and protein–protein communications, all essential for insight into the molecular dynamics of stress accommodation [88, 110]. These understandings can be translated into biotechnological intervention towards increasing crop stress tolerance, such as stress-inducible gene manipulation of responsive proteins or crop improvement schemes towards selecting genotypes tolerant of stresses.

Various sophisticated methods are utilized in plant proteomics to analyse protein expression patterns under stress conditions as shown in Fig. 3. Two-dimensional gel electrophoresis (2D-GE) is one of the conventional techniques utilized to separate proteins according to their isoelectric point and molecular weight so that researchers can observe differential protein expression under different stress conditions. While 2D-GE gives a general overview of the proteome, it has its own set of shortcomings such as low sensitivity towards the detection of low-abundance proteins and difficulty in separating highly hydrophobic or membrane proteins [43]. Mass spectrometry (LC–MS/MS) is used extensively to mitigate these shortcomings because it is highly sensitive, specific, and can analyse complex mixtures of proteins. LC–MS/MS is able to detect and quantify proteins with high precision and also identify post-translational modifications, giving a better insight into protein regulation under stress [68].

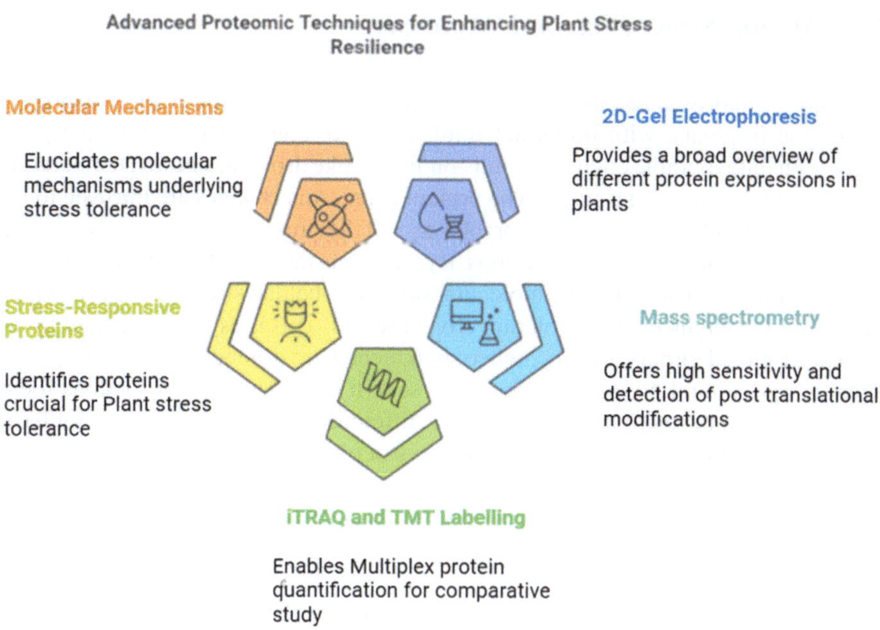

Fig. 3 Advanced proteomic techniques for enhancing plant stress resilience

Labelling methods like Isobaric Tags for Relative and Absolute Quantification (iTRAQ) and Tandem Mass Tags (TMT) are also frequently employed in plant proteomics. iTRAQ enables the simultaneous quantification of proteins across several samples, and thus it is very suitable for comparative proteomic analyses with the aim of detecting differentially abundant proteins under stress [71]. TMT, which works like iTRAQ, allows multiplexed protein quantitation and the comparison of protein expression levels between samples. These methods play a crucial role in giving information on the molecular mechanisms of stress responses and allow researchers to identify key proteins playing a role in stress tolerance [12].

In spite of the tremendous improvement provided by proteomics, understanding mechanisms of stress tolerance is beyond proteomics and depends on integration of proteomics with other omics technologies like genomics, transcriptomics, and metabolomics. Integration is valuable in explaining sophisticated interactions between metabolites, genes, and proteins and presents the whole picture of plant reactions towards environmental stress factors [118]. But translating proteomic results into useful applications, e.g., generating stress-resistant crop lines, remains problematic. Some of the issues include the intricate nature of protein networks, heterogeneity of proteomic data, and the requirement for technological progress to improve sensitivity and throughput. Research and technology advancement are necessary to overcome these limitations and achieve the full potential of proteomics in plant stress studies.

3.1 Abiotic Stress Proteomic Analysis

Proteomics analysis under abiotic stress is found to reveal dramatic changes in protein expression in plants, with important implications in understanding how the plant copes with unfavourable environmental conditions like drought, salinity, temperature extremes, and heavy metal stress. The protein expression changes are intrinsic to the strategies of the plant for avoidance or tolerance of the stress, helping it survive as well as produce even under hostile conditions. By analysing these changes at the proteomic level, scientists can decipher the molecular processes behind stress responses, which can be used to engineer stress-tolerant crop varieties. This is especially useful in the context of global climate change, where heightened environmental stressors pose a threat to agricultural productivity and food security. The subsequent sections focus on the change in protein expression under different abiotic stresses with emphasis on pivotal proteins and the involved pathways in plant adaptation.

3.1.1 Drought Stress

Drought stress considerably affects plant physiology through the modification of water intake, photosynthesis, and homeostasis of cells [34, 41, 45]. It has been discovered through proteomics that plants undergo changes in expression of proteins dealing with photosynthesis, stress signalling, and metabolism in response to drought. In chickpea (*Cicer arietinum*), for instance, FtsH family proteases display differential expression when under drought stress. These proteases are involved in chloroplast formation and the regulation of photosynthetic control by controlling damaged protein degradation and thylakoid membrane stability [24]. This process guarantees the efficiency of the photosynthetic machinery under water deficiency, leading to drought tolerance.

In addition, proteins involved in the abscisic acid (ABA) signalling pathway are highly expressed under drought stress. ABA is a central phytohormone that plays a role in stomatal closure, osmotic adjustment, and gene expression under water stress. Proteomic research has shown a number of ABA-responsive proteins that increase the capacity of the plant to retain water by decreasing transpiration rates [58]. Additionally, universal stress proteins (USPs) are strongly induced in cotton (*Gossypium hirsutum*) during drought. These proteins are linked with cellular defence processes, such as reactive oxygen species (ROS) detoxification and osmoprotection, hence protecting cellular structures against dehydration stress-induced damage [74]. Such synchronized expression of drought-responsive proteins is indicative of the intricate regulatory networks that facilitate plants to adapt to water-deficit conditions.

3.1.2 Salinity Stress

Salinity stress is an important challenge for plants as it interferes with ionic homeostasis and causes ion toxicity, osmotic stress, and oxidative damage [44, 46, 56, 57, 124, 123]. Proteomic analysis has shown that plants react to salinity stress by changing the expression of antioxidant defence, ion transport, and osmoprotectant production proteins. For example, proteins involved in ROS detoxification, including superoxide dismutase, catalase, and glutathione peroxidase, are induced to counteract oxidative stress caused by excessive salt levels [67]. These antioxidant proteins are crucial for the maintenance of cellular redox homeostasis, preservation of membrane integrity, and inhibition of oxidative damage to proteins, lipids, and DNA. In addition, salinity stress induces the synthesis of osmoprotectants such as proline, glycine betaine, and sugars that support osmotic balance of cells and protect proteins and membranes [53]. Proteomic research has shown new protein markers associated with salt tolerance, which include ion transporters such as Na^+/H^+ antiporters responsible for ionic homeostasis through sequestration of excess sodium ions into vacuoles [67]. This pathway reduces ion toxicity and maintains peak cellular function when subjected to salinity. In addition, salt-induced protein-small protein interactions combined with regulatory gene responses underscore complex molecular mechanisms driving salt stress signalling. Elucidating these through proteomics holds key importance to engineer salt tolerant crops that produce acceptable yields from saline soils.

3.1.3 Temperature Stress

Temperature extremes, such as heat and cold stress, have a deep impact on plant growth and development by influencing membrane fluidity, enzyme activity, and metabolic processes [49]. Proteomic research has demonstrated that plants react to temperature fluctuations by expressing heat shock proteins (HSPs) and other stress-related proteins that contribute to cellular homeostasis and protect cellular structures. HSPs function as molecular chaperones, stabilizing unfolded or misfolded proteins and preventing protein aggregation under heat stress [3]. These chaperones facilitate the refolding of denatured proteins and aid in protein degradation, ensuring proteostasis under thermal stress.

In addition to HSPs, temperature stress alters the expression of proteins involved in metabolic pathways, including those related to carbohydrate metabolism, energy production, and secondary metabolite biosynthesis. Proteomic studies have shown that cold stress upregulates antifreeze proteins and dehydrins, which stabilize membranes and inhibit ice crystal growth to safeguard cellular structures [72]. In contrast, heat stress upregulates the production of ROS detoxifying proteins to neutralize oxidative damage induced by high temperatures. These heat- and cold-stress-induced proteins are responsible for maintaining cellular homeostasis and plant survival during extreme thermal stresses.

3.1.4 Heavy Metal Stress

Heavy metal stress significantly endangers plant growth and yield by triggering oxidative injury, interrupting metabolic activity, and disrupting nutrient absorption [49]. Proteomics research has elucidated major changes in protein expression towards detoxification and metal ion balance upon heavy metal stress. For instance, metallothioneins and phytochelatins are induced by heavy metals like cadmium, lead, and arsenic to sequester and detoxify the metals [74]. The heavy metal ions are bound into stable complexes by these metal-binding proteins, thereby inhibiting interactions with cellular components and reducing toxicity.

In addition, heavy metal stress also triggers the synthesis of antioxidant defence proteins to mitigate metal-induced ROS generation. Enzymes like glutathione S-transferases, peroxidases, and superoxide dismutases are crucial in ROS scavenging and the prevention of oxidative damage to cellular structures. Proteomic studies have further shown the upregulation of metal transport and homeostasis proteins, such as ATP-binding cassette transporters and metal ion transporters, which control the uptake, sequestration, and efflux of metals [110]. Such regulation allows plants to survive in heavy metal-contaminated environments by keeping metal ion homeostasis and reducing toxicity.

3.1.5 Integrated Proteomics Strategy in Stress Research

Proteomic analysis gives a holistic view of plant responses to abiotic stresses by uncovering intricate regulatory networks and protein interactions. Plant responses to environmental stressors are not, however, independent but tend to interact with one another, such as in the combined impact of drought and salinity or heavy metal stress. This dialogue may make the stress response of the plant complex, and integrated strategies involving proteomics along with other omics technologies like genomics, transcriptomics, and metabolomics would be required. These integrative studies may deconvolute crosstalk among various signalling pathways so that it gives an overall understanding of stress tolerance mechanisms [74]. Translating these results to translational biotechnology applications to develop stress-tolerant crop varieties is still difficult, even with the development of proteomics. Such challenges involve the intricate nature of protein networks, proteomic data heterogeneity, and technological innovations necessary to increase sensitivity and throughput. Sustained research and advancing technology are vital to maximize the potential of proteomics in explaining plant stress response and enhancing crop resistance under varied environmental conditions.

4 Molecular Mechanisms of Nanoparticle-Induced Proteomic Alterations

NPs induce intricate molecular processes that modulate cellular functions through interactions with the biological system, resulting in profound proteomic alterations. One of the major mechanisms by which NPs modulate cellular functions is through the activation of diverse signalling cascades, such as those related to oxidative stress, hormone signalling, and immune systems. Many of these cascades are via PTMs like phosphorylation and ubiquitination that can modulate the regulation of stress-related genes and protein stability with a downstream effect on cellular responses to NP exposure.

4.1 Signalling Pathways Triggered by NPs

ROS is the primary mediator of the signalling pathways triggered by NPs. NPs, especially metal-based, produce ROS as signalling molecules. In plants, aluminium oxide nanoparticles (Al_2O_3 NPs) strengthen the ascorbate/glutathione (AsA/GSH) pathway, one of the critical antioxidant defence mechanisms, thereby enhancing ROS scavenging ability and overcoming oxidative stress [82]. In the same manner, in human cells, silver nanoparticles (AgNPs) trigger serine/threonine protein kinase (PAK) and mitogen-activated protein kinase (MAPK) signalling pathways, which are important for cellular stress responses such as inflammation, apoptosis, and cell proliferation [116]. Triggering of these pathways by NPs assists in the regulation of the expression of genes related to antioxidant defence, detoxification, and other stress-related processes.

Hormone signalling is also a vital pathway affected by NPs. NPs have been reported to affect plant hormone signalling pathways, e.g., auxin and cytokinin signalling, which play a role in growth regulation and stress acclimation. For example, gold nanoparticles (AuNPs) modulate auxin signalling in Arabidopsis thaliana and increase stress tolerance by overexpressing osmotic stress-related genes [60]. This hormone signalling modulation can influence plant growth, development, and abiotic stress response, leading to enhanced stress tolerance.

4.2 Crosstalk Between NP-Induced Signalling and Stress-Responsive Pathways

NPs affect the crosstalk between different signalling and stress-responsive pathways. Immune and inflammatory responses initiated by NPs are mostly deregulated under NP exposure. Research has demonstrated that NPs such as silver and ZnONPs induce the activation of transcription factors such as zinc finger proteins that control stress responses and immune signalling [23]. These proteins are essential in triggering defence mechanisms, including the synthesis of antimicrobial peptides and second metabolites in plants. NPs also affect immune responses by modifying the production of cytokines and other mediators of immunity, leading to inflammatory responses that can affect cellular health and function.

Metabolic pathways are also an area where NPs have an important function. In plants, NPs such as CuNPs are capable of regulating core metabolic processes, such as glycolysis and the tricarboxylic acid (TCA) cycle, used for energy generation and adaptation to cellular stress. Treatment with CuNPs in salt-stressed wheat plants promotes gene expression during these metabolic processes, thus increasing stress tolerance [82]. This metabolic reprogramming enables plants to maximize energy generation and redox balance, which enables cellular adaptation to unfavourable environmental conditions. Likewise, in animal cells, NPs may modulate metabolic processes through the regulation of critical enzymes that are involved in cellular respiration and biosynthesis.

4.3 Role of Post-translational Modifications (PTMs)

PTMs play a central role in the regulation of protein function upon NP-induced stress. Phosphorylation is among the most prevalent PTMs that participate in the activation and regulation of signalling pathways. For example, AgNPs cause the phosphorylation of proteins in the MAPK pathway, including MAPK3 and MAPK6, which are implicated in stress signalling and antioxidant defence regulation [116]. Phosphorylation processes trigger downstream transcription factors, such as MYB and WRKY proteins, that control the expression of genes involved in cellular protection, stress adaptation, and antioxidant defence. This is essential for coordinating the cellular response to oxidative damage induced by NP exposure, allowing for appropriate cellular adaptation to stress.

Ubiquitination is another significant PTM that controls protein stability, function, and turnover. While less explored in the context of NP exposure, the ubiquitin–proteasome system (UPS) is particularly important in regulating cellular responses to stress through the mediation of degradation of damaged proteins and regulatory protein stability. NPs, including gold and copper NPs, are reported to modulate the function of E3 ligases that catalyse selective ubiquitination of target proteins [23]. This UPS modulation can control the stability of stress proteins and transcription factors and, in this way, can change the capacity of the plant or cell to respond to NP-induced stress.

Besides the described signalling pathways and PTMs, biotransformation of NPs and protein corona formation are significant features that affect NPs' biological impacts. Protein corona results when proteins present in the vicinity of the biological environment adsorb onto the NPs' surface, modifying their physicochemical properties and biointeractions. This interaction can modify proteins' conformation, resulting in changes in function and interactions with cellular machinery [80]. Protein corona also modulates NP entry into cells and tissues, dictating their localization and potential toxicity.

Nanoparticle-induced oxidative damage is another important issue. NPs, especially metal-based NPs such as silver and copper, produce ROS that can cause harm to cellular structures, such as lipids, proteins, and nucleic acids. Oxidative damage resulting from this process can inhibit the function of enzymes and result in cellular dysfunction. For instance, it has been shown that AgNPs suppress the action of antioxidant enzymes like SOD and CAT through binding at their active sites and structural changes [16]. This suppression has the ability to increase the oxidative stress response at the cellular level, promoting more damage and potentially initiating apoptotic processes. In addition, NP generation of ROS has the capacity to destroy the mitochondrial respiratory chain and impair energy metabolism and cellular function.

Consequently, NPs engage biological systems in a multifaceted process, which modulates an extensive range of signalling pathways and cellular functions. The activation of ROS-associated and hormone signalling pathways is critical for the mediation of cellular responses to NP-induced stress. Crosstalk among these pathways and immune, metabolic, and stress response mechanisms further amplify the cellular capacity to adapt to NP exposure. Post-translational modifications including phosphorylation and ubiquitination play pivotal roles in controlling such responses, sustaining cellular survival and function under conditions of stress. Nevertheless, risks posed by NP-induced oxidative stress and biotransformation make it essential that NP interaction at the molecular level is better understood. This information is critical to enabling the optimal exploitation of NPs in different uses, as well as reducing their environmental and health hazards.

5 Proteomic Responses of Exposure to Nanoparticles Under Abiotic Stress

5.1 *Proteomic Responses Under Drought Stress*

Drought stress is one of the serious threats to crop productivity by disrupting diverse physiological and metabolic processes of the plant such as photosynthesis, osmotic balance, and cell homeostasis. There is evidence showing that NPs are capable of modifying proteomic responses in crops in drought situations and increasing the crop's resilience towards water scarcity. NPs affect the protein expression related to stress responses, conferring enhanced drought tolerance. For example, metal-based NPs, ZnO and TiO_2, were reported to modulate the expression of proteins involved in oxidative stress and photosynthesis, which play significant roles in the preservation of cellular processes during drought [114, 125]. Particularly, such NPs augment the level of photosystem II (PSII) proteins and thus stabilize the photosynthetic machinery and uphold photosynthetic performance under drought stress. In addition, NPs regulate the expression of proteins associated with oxidative stress, such as superoxide dismutase (SOD), catalase (CAT), and peroxidases, that are crucial for scavenging reactive oxygen species (ROS) and safeguarding cellular structures from oxidative damage [125].

NPs also impact the levels of stress-related proteins that are essential for cellular homeostasis during drought conditions. For instance, the presence of NPs results in the upregulation of HSPs, which function as molecular chaperones, inhibiting protein denaturation and aggregation under stress conditions [125]. In addition, NPs increase the expression of aquaporins (AQPs), which are transmembrane proteins that allow the transport of water and hence enable plants to balance osmotic pressure under drought stress. This regulation of AQPs augments uptake and holding of water, leading to improved tolerance to drought [125]. Moreover, NPs enhance the functioning of antioxidant enzymes, including glutathione peroxidase (GPX) and ascorbate peroxidase (APX), which play crucial roles in preventing oxidative damage through the regulation of redox balance at cellular levels. Such enhanced antioxidant activity lessens cellular damage and maintains metabolic processes under drought stress [96].

NPs also affect hormonal regulation and gene expression on drought stress responses. They can modulate the concentrations of major phytohormones like abscisic acid (ABA), which is a central component of drought stress signalling. Through the enhancement of ABA accumulation, NPs enhance stomatal closure, conserving water while inducing ABA-responsive genes associated with the synthesis of protective proteins and osmolytes [96, 125]. Also, NPs regulate the biosynthesis and signalling network of other phytohormones such as salicylic acid (SA) and jasmonic acid (JA) that increase antioxidative defence systems. NPs additionally trigger drought-responsive genes that control the synthesis of proteins and stress acclimation. This comprises the upregulation of genes coding for late embryogenesis abundant (LEA) proteins and osmolytes such as proline and glycine betaine, which protect cellular proteins and membranes from dehydration stress [125]. In addition, NPs modulate the expression of transcription factors including Dehydration-Responsive Element Binding (DREB) and NAM, ATAF, and CUC (NAC) proteins that control downstream stress-responsive genes and improve drought tolerance.

Although the use of NPs in agriculture is highly promising in terms of improving drought tolerance, their risks must be carefully considered. The interaction of NPs with plant systems may result in phytotoxicity, such as genotoxic effects, membrane damage, and nutrient imbalance, which may adversely affect plant growth and development if not properly controlled [15, 114]. In addition, the deposition of NPs in soil and water ecosystems is an environmental concern, which may impact beneficial soil microorganisms and aquatic life. Thus, continued research is important to maximize NP formulations and application techniques so that their advantages are achieved with the least possible side effects. With the knowledge of intricate interactions between NPs and plant proteomes under drought stress, sustainable agricultural management can be designed to increase crop productivity and tolerance to environmental stresses.

5.2 Salinity Stress Proteomic Responses

It has been the recent research that has called attention to the ability of NPs to regulate crops' proteomic environment under salinity stress, allowing them to be more resistant and productive. NPs have the ability to modulate the expression and activity of salt-stress-inducible proteins, facilitating ion homeostasis, osmotic adaptation, and redox stress regulation. These interactions are multifaceted and comprise several pathways, and hence it is crucial to know the underlying proteomic mechanisms for formulating effective measures to counteract the negative impacts of salinity on crop yield and quality.

NPs have the ability to modify the expression of stress-responsive proteins significantly, thus modulating the plant's capacity to adapt to salinity stress. For instance, calcium nanoparticles (CaNPs) have been reported to alter protein profiles in tomato plants under high salinity, promoting growth and survival. This promotion is mainly attributed to the upregulation of stress-responsive proteins such as HSPs and dehydrins, which are crucial in maintaining cellular homeostasis and defending cellular structures against salt-induced damage [2]. HSPs function as molecular chaperones, inhibiting protein denaturation and aggregation under salt stress, whereas dehydrins stabilize cellular membranes and promote cellular osmotic balance. Besides CaNPs, ZnO, and SiO_2 NPs have also been found to regulate the amount of proteins playing roles in ion transport and osmotic regulation, thus increasing salt tolerance [2].

Exposure of crops to NPs in the conditions of salinity stress also results in the stimulation of antioxidant enzymes, which are important for preventing oxidative stress caused by excessive salt levels. Salinity stress tends to cause overproduction of reactive oxygen species (ROS), including hydrogen peroxide (H_2O_2) and superoxide radicals (O_2^-), that have the capability to damage the cells and impair the metabolic processes. NPs regulate the expression of antioxidant defence system proteins, thus activating the activity of major antioxidant enzymes such as superoxide dismutase (SOD), catalase (CAT), and ascorbate peroxidase (APX) [128]. For example, treatment with silicon nanoparticles (SiNPs) was found to enhance SOD and CAT activity in rice plants under salt stress, thus alleviating ROS buildup and guarding cellular constituents against oxidative damage. Likewise, CuNPs have also been found to escalate the expression of glutathione peroxidase (GPX) and thioredoxin, which play vital roles in redox balance and warding off oxidative stress [82].

Apart from augmenting antioxidant defence, NPs induce the production of stress proteins that are essential for sustaining cellular homeostasis under salinity stress. These include late embryogenesis abundant (LEA) proteins, osmoprotectants, and ion transporters, which enable plants to adapt to high salinity levels. LEA proteins maintain cell structures, including proteins and membranes, in dehydration stress, whereas osmoprotectants like glycine betaine and proline manage osmotic balance inside cells [82]. NPs improve the biosynthesis of these osmoprotectants by regulating the expression of stress enzymes, for example, proline dehydrogenase and choline monooxygenase. In addition, NPs affect the function of ion transporters, such as sodium/hydrogen exchangers (NHXs) and high-affinity potassium transporters (HKTs), which play an important role in the maintenance of ion homeostasis under salinity stress [82]. Through regulation of the expression and activity of these transporters, NPs assist in compartmentalization of toxic sodium ions (Na^+) in vacuoles, thus decreasing Na^+ toxicity in the cytoplasm and maintaining potassium (K^+) levels necessary for cellular processes.

NPs are also important in the regulation of ion transport mechanisms, which are crucial for sustaining ion homeostasis when the plant is subjected to salt stress. For instance, copper nanoparticles (CuNPs) have been found to make ion transport processes easier, promoting salt tolerance in wheat by regulating the functions of Na^+/H^+ antiporters and K^+ channels [82]. These antiporters facilitate sequestration of Na^+ within vacuoles, mitigating cytoplasmic Na^+ toxicity and a high K^+/Na^+ ratio necessary for enzyme function and metabolic processes. Moreover, NPs modulate the expression of ion transport protein members like salt overly sensitive (SOS) proteins and NHX antiporters that are important in sustaining ion balance when subjected to high salinity [82]. This regulation assists plants to be in osmotic balance and cellular turgor, advancing salt tolerance.

In addition, NPs enhance the accumulation of osmolytes, which are low molecular weight substances that assist with osmotic adjustment by equilibrating the osmotic pressure within plant cells. This is important for cell turgor and function under conditions of high salinity. For example, CaNP application has been demonstrated to increase proline and glycine betaine accumulation in tomato plants subjected to salt stress, thus optimizing osmotic balance and saving cellular structures from osmotic damage [2]. Besides, NPs regulate the biosynthesis of other compatible solutes such as soluble sugars and polyols that are involved in osmotic adaptation and ROS scavenging. NPs also help in redox homeostasis by regulating the redox status of cells, which is required for ion equilibrium and osmotic adjustment. This includes the control of redox-regulated proteins and pathways like thioredoxin and glutaredoxin systems that help in cellular acclimation to salinity stress [114, 128].

Although the use of NPs provides potential means to increase plant tolerance to salinity stress, their possible risks should be taken into account. The physicochemical characteristics of NPs, such as size, shape, concentration, and surface charge, can determine their interaction with plant systems and may result in phytotoxicity and environmental issues [15]. Large amounts of NPs induce genotoxicity, membrane damage, and alterations in nutrient availability, affecting plant growth and development adversely. Additionally, NPs' incorporation into soil and aquatic environments has ramifications for the environmental setting, influencing beneficial microorganisms in the soil and aquatic life forms. Hence, a harmonized strategy making optimal use of NPs and limiting their risks becomes pivotal for sustainable agriculture. Optimizing NP formulation and application procedures can help to increase salt tolerance in plants without increasing environmental and health hazards.

5.3 Proteomic Response to Temperature Stress

Temperature stress, both heat and cold stress, greatly impacts crop productivity by modifying several physiological and biochemical processes that are involved in growth and development. Exposure to NPs has been demonstrated to impact the expression of specific proteins like HSPs and metabolic enzymes that are pivotal for thermotolerance as well as cold acclimation mechanisms in plants. These molecular responses, which are usually linked to complex proteomic modifications, increase the tolerance of plants against extreme temperatures and environmental stress [13].

HSPs are key to the plant's heat stress response, acting as molecular chaperones that stabilize proteins and cell membranes, inhibit protein aggregation, and assist in protein refolding. HSPs are universally expressed in organisms under conditions of stress, such as heat, cold, and other abiotic stresses, and play a critical role in cell protection against the lethal action of heat-induced protein denaturation [83]. In plants, HSPs also function to preserve cellular membrane integrity and prevent oxidative damage by eliminating ROS, which are frequently produced during heat stress. Induction of HSPs during heat stress is important in maintaining cellular function and preventing plant death at higher temperatures [83].

During heat stress, plants experience significant proteomic alterations in the rates of protein turnover related to redox signalling, stress defence pathways, and metabolism. Dynamic proteomic adjustment is vitally important to cellular homeostasis because it enables plants to reconcile oxidative damage and metabolic activity while under stress. During *Arabidopsis thaliana* heat stress, strong changes in expression of redox regulation and carbohydrate metabolism proteins show an integrated mechanism for stabilizing the cellular medium [36]. Heat stress during grape cells causes changes in phenylpropanoid pathways and sugar metabolism, reflecting a diversion of metabolic reactions to sustain the plant functions in elevated temperatures [36].

Conversely, cold stress also influences metabolic processes but differently. Cold temperatures increase the stiffness of cellular membranes, which can impair crucial processes like photosynthesis and respiration. Cold stress triggers the production of certain sugars and signalling compounds, including proline, which are critical for the defence of plant cells against freezing injury. In addition, cold stress initiates regulation of cold-responsive transcription factors and downstream genes participating in osmotic regulation and cryoprotection [59]. Cold-induced proteomic changes in *Nicotiana tabacum* lead to upregulation of genes associated with photosynthesis and flavonoid biosynthesis, two processes vital for cold tolerance and optimal metabolic function at low temperatures [107].

The function of NPs in the modulation of thermotolerance and cold acclimation in plants is interesting, as NPs have the ability to regulate the expression of HSPs and other stress-related proteins. In tomato plants, for instance, RS2Z proteins are implicated in regulating the expression of HSPs and other heat shock transcription factors that lead to acquired thermotolerance. The regulation is essential for increasing the ability of the plant to endure heat stress. The use of NPs may augment this pathway by affecting the production of the crucial proteins that are engaged in thermotolerance, indicating that NPs provide a new strategy to advance plant heat stress resistance [97]. Likewise, NPs could enhance cold acclimation by regulating cold-inducible genes and the accumulation of protective metabolites, including sugars and amino acids, that facilitate cell integrity preservation during cold stress [59].

Proteomic profiling of temperature-stressed plants shows drastic changes in the rate of protein turnover, particularly in stress response proteins, metabolic proteins, and cell signalling pathway proteins. Moderate heat stress in *Arabidopsis*, for instance, increased protein turnover to 1.4 times, stressing the role of proteomic plasticity in compensatory ability to temperature change [36]. This enhanced turnover is a manifestation of the adaptive ability of the plant to reorganize its proteome in response to environmental change.

Transporters, being critical in regulating ion homeostasis and solute transport across cell membranes in the case of compatible solutes like sucrose and proline, are central to the proteomic response against temperature stress. The transporters maintain osmotic balance within the cell during heat and cold stress by enabling the transport of solutes across the cell membrane. Their functioning is vital to acclimatize to thermal stress because they govern the movement of ions and molecules responsible for stabilizing cellular functions and metabolism [13]. Transporters, therefore, help preserve homeostasis, which is crucial for plant life and strength in the face of thermal stress. Although the effect of NPs on plant temperature stress responses is promising, it is important to take into account the possible environmental and ecological effects of NP application in agriculture. NPs interact with plant systems in intricate manners, and their use in agricultural systems could have unforeseen effects, including nanoparticle toxicity or bioaccumulation. More research is required to gain a clearer understanding of how NPs influence temperature stress response in plants and their long-term impacts on plant health and the environment. The development of environmentally friendly approaches for harnessing NPs to promote crop resistance while keeping their environmental impact low is crucial for pushing agricultural productivity against the backdrop of climate change.

5.4 Proteomic Responses to Heavy Metal Stress

Heavy metal stress is an important environmental issue, which inhibits plant growth and productivity. The phytotoxicity of heavy metals like cadmium, lead, and mercury causes disruptions in normal plant physiology by inducing ROS and inhibiting critical metabolic processes. In order to adapt to such stress, plants have developed different defence mechanisms, including the utilization of NPs and proteomic adjustments linked with metal transporters and detoxification enzymes. NPs can perform a double role in plant response to heavy metals, either mitigating or exacerbating toxicity, depending on their interactions with plant systems. This action is intimately connected with remodeling of the plant proteome, which assists in coping with the stress by the metal transport and detoxification systems.

NPs may assist in reducing heavy metal toxicity by their specific physicochemical properties. Some NPs, especially metal oxides including CeO_2, TiO_2, and Mn_3O_4, can reduce oxidative stress in plants. These NPs function by adsorbing and sequestering harmful metal ions from the surrounding medium, thus decreasing their bioavailability to the plant. These NPs can also increase the activity of antioxidant enzymes like SOD, CAT, and ascorbate peroxidase (APX), which are pivotal in neutralizing ROS that are created by heavy metal exposure [33]. By enhancing the antioxidative defence system, NPs enable plants to preserve cellular integrity and minimize oxidative stress-induced damage.

The interaction between NPs and plants is not always positive. The possibility of NPs releasing metal ions into the plant system sometimes enhances heavy metal toxicity. This is when NPs interact with plant cellular constituents, causing enhanced metal ion accumulation or modifying the metal uptake pathways. These unintended consequences may reinforce stress responses, perturbing metabolic processes and inflicting more harm to plant tissues [33]. Hence, though NPs can provide possible advantages in alleviating heavy metal stress, their use needs to be meticulously contemplated to avert detrimental effects.

Plants utilize various metal transporters to manage the uptake, translocation, and compartmentalization of heavy metals. They play an important function to counter metal toxicity by modulating the transit of metals to the vacuole, in which they become deposited, or through effluxation of metals exceeding the required limits from plant cells. These roles are achieved with the involvement of ABC transporters, natural resistance-associated macrophage protein (NRAMP), and heavy metal ATPase (HMA) [42]. These proteins reduce the poisonous impact of metal ions on plant growth and metabolism by sequestering metals within vacuoles or transporting them out of the cell [66].

Other than metal transporters, plants also induce heavy metal stress responses by upregulating detoxifying enzymes. Plant antioxidant defence is composed of SOD, CAT, and APX enzymes that neutralize ROS produced due to metal-induced oxidative stress. These enzymes catalyse reactions that transform toxic superoxide radicals into less toxic compounds, which guard cells against oxidative damage [6, 86]. The enhanced activity of these enzymes assists plants in maintaining cellular homeostasis and minimizing the harmful effects of heavy metal exposure.

The other essential part of the plant's defence mechanism against heavy metal stress is the sequestration and chelation of metals by proteins such as phytochelatins and metallothioneins. The proteins sequester the heavy metals, making it possible for them to be compartmentalized in plant cells and blocking their toxic impact on cellular components [66]. Sequestration of the metals into vacuoles or metallothionein and phytochelatin binding effectively detoxifies the metals, making them less accessible to engage with critical cellular constituents.

Progress in omics technologies such as genomics, transcriptomics, and metabolomics has generated significant information regarding the molecular processes that plants use to adapt to heavy metal stress. Transcriptomics, for instance, have found the upregulation of genes responsible for metal transport and antioxidation pathways under lead stress exposure [126]. The findings indicate how plants use certain genes to fight oxidative damage and the efflux of toxic metals.

Metabolomics has also contributed to the understanding of the metabolic reorganizations that take place in plants under heavy metal stress. Metabolomics enables the characterization of changes in primary and secondary metabolites, showing how plants modulate their metabolic pathways to cope with the toxicity of metals [5]. Through the identification of the major metabolites participating in stress responses, metabolomics offers insights into possible strategies for reducing the toxic effects of heavy metals in plants.

In general, NPs provide promising avenues for reducing heavy metal stress in plants, especially through increased antioxidant defence and detoxification of metal ions. Their potential to exacerbate toxicity through metal release or interaction with cellular constituents must be carefully evaluated, though. Proteomic alterations, particularly metal transporters and detoxification enzymes, are key to plant survival under heavy metal stress. Knowledge of these mechanisms by using omics technologies may contribute to the development of more tolerant plant genotypes and successful remediation protocols. However, a holistic approach taking into account the positive as well as the potential adverse effects of NPs is critical for heavy metal toxicity management in plants in a sustainable manner.

6 Future Outlook

6.1 Proteomic Technologies Developments

New proteomic technologies, including dual-fraction mass spectrometry, are opening doors for more extensive studies of NP-plant interactions. This new technique enables analysis of both NP-bound and free proteins, giving a more holistic understanding of the molecular processes through which NPs influence plant physiology. Through the recognition of protein expression and structural conformation changes, these technologies demystify the intricacies by which NPs affect plant development, nutrient uptake, stress processes, and metabolism. Such an advancement is capable of unveiling mild but critical variations in plant proteomes that would otherwise go undetected, enhancing our potential to evaluate the functional effects of NP exposure on a molecular level [101].

6.2 Integrative Omics Approaches

In order to have a better understanding of the impact of NPs on plants, combining proteomics with other omics technologies, like transcriptomics and metabolomics, is an exciting future direction. By integrating proteomic information with gene expression data (from transcriptomics) and metabolite formation data (from metabolomics), scientists can construct a complete map of NP-induced alterations at different biological levels. This integrative strategy will assist in identifying major molecular pathways impacted by NPs, providing greater understanding of how the nanomaterials modify plant function. For example, modifications in gene expression associated with nutrient metabolism or stress response can be linked to alterations in protein content and the biosynthesis of certain metabolites, ultimately providing a greater understanding of how NPs affect plant health and productivity. The convergence of these omics platforms can have a crucial role in driving the formulation of strategies for enhancing crop resilience and productivity through targeted application of NPs [127].

6.3 Implications for Sustainable Agriculture

A greater insight into NP-induced proteomic alterations in plants has far-reaching implications for sustainable agriculture. Nanotechnology has the capability to transform agricultural practices by facilitating the creation of novel NP-based agricultural technologies, including nanofertilizers. These nanofertilizers would be able to increase the bioavailability of key nutrients, enhancing the efficiency of nutrient uptake in crops and supporting healthy growth, particularly in poor soils. NPs can also increase stress tolerance in plants, rendering them more resilient to environmental stresses like drought, salinity, or temperature variations. With improved nutrient absorption and stress resistance, crops may have enhanced yield, leading to food security and sustainable agriculture. As the agricultural sector fights to increase yields in order to take care of a growing population, these technologies would be an essential element in bringing more efficient and environmentally friendly farming [21].

6.4 Possible Risks and Safety Evaluation

Although the potential benefits of NPs to agriculture are promising, their potential risks to the environment and human health should not be underestimated. With the increasing use of NPs in agricultural uses, it is imperative to assess their long-term impact on plant ecosystems, soil quality, and whether they have the ability to become part of the food chain. Experiments need to evaluate how NPs adsorb into plants, how they interact with microbes in the soil, and whether or not they disrupt the ecological harmony of agricultural ecosystems. Safety evaluations also need to be performed in order to ensure that exposure to NP does not lead to unwanted effects, such as toxicity in non-target species or environmental pollution. Longitudinal studies dedicated to the assessment of the fate of NPs in the environment and their ecotoxicological implications will be indispensable to provide safety and guarantee sustainable integration of nanotechnology into agronomy [26].

7 Conclusion

In summary, although nanotechnology has enormous prospect for boosting productivity and sustainability of agriculture, serious knowledge gaps in the NP-plant interactions persist, especially how NP concentration and exposure time shape plant proteomes. The absence of routinely standardized protocols for proteomic analysis additionally hinders attempts at comparison and reproduction of results across studies. Recent improvements in proteomic technologies, including dual-fraction mass spectrometry, and incorporation of multi-omics strategies, provide encouraging paths towards a more holistic understanding of NPs' impact on plant biology. These advances, combined with an increased comprehension of the molecular bases of NP-induced changes in the plant proteome, can inform the creation of NP-based agricultural technologies to enhance crop resilience, nutrient acquisition, and productivity. Concurrently, environmental and health hazards of NP application should be examined in detail for ensuring their safe use in agriculture. Overcoming these challenges and adopting an integrative strategy for NP research will be central to realizing the full potential of nanotechnology for sustainable agriculture.

References

1. Abdelkader Y, Perez-Davalos L, LeDuc R, Zahedi RP, Labouta HI (2023) Omics approaches for the assessment of biological responses to nanoparticles. Adv Drug Deliv Rev 200:114992
2. Abeed AHA, Al-Huqail AA, Albalawi S, Alghamdi SA, Ali B, Alghanem SMS, Al-Haithloul HAS, Amro A, Tammam SA, El-Mahdy MT (2023) Calcium nanoparticles mitigate severe salt stress in Solanum lycopersicon by instigating the antioxidant defense system and renovating the protein profile. South African J Bot 161:36–52
3. Alugoju P, Tencomnao T (2023) Medicinal plants proteomics in response to abiotic stresses. In: Medicinal plants: their response to abiotic stress, pp 35–107
4. Anitha S, Lakshmi BS, Manojkumar S, Gnanadesigan M, Chandrasekaran R, Kapildev G (2024) Biopolymers-based nanofertilizers. In: Nanofertilizer synthesis: methods and types, pp 109–128
5. Anjitha KS, Sarath NG, Sameena PP, Janeeshma E, Shackira AM, Puthur JT (2023) Plant response to heavy metal stress toxicity: the role of metabolomics and other omics tools. Funct Plant Biol 50(12):965–982
6. Ansari Mohd. KA, Iqbal M, Ahmad M, Munir M, Gaffar SA, Chaachouay N (2024) Heavy metal stress and cellular antioxidant systems of plants: a review. Agri Rev
7. Astner AF, Hayes DG, O'Neill HM, Evans BR, Pingali SV, Urban VS, Young TM (2022) Forming micro-and nano-plastics from agricultural plastic films for employment in fundamental research studies. J Vis Exp 185
8. Baker S, Volova T, Prudnikova SV, Satish S, Prasad MN (2017) Nanoagroparticles emerging trends and future prospect in modern agriculture system. Environ Toxicol Pharmacol 53:10–17
9. Baker S, Prudnikova SV, Sukhanova AA, Perianova OV, Nayaka SC, Garkusha TA, Shidlovskiy IP, Potkina NK, Khohlova OY, Kuzmin AK, Ranjini HK (2023) Nanoparticle embedded cellulosic film: Synergistic influence on antibacterial activity and sustainable agronomic evaluation. Biocatal Agri Biotechnol 52:102828
10. Baker S, Satish S, Prasad N, Chouhan RS (2019) Nano-agromaterials: influence on plant growth and crop protection. In: Industrial applications of nanomaterials, pp 341–363
11. Bantscheff M, Schirle M, Sweetman G, Rick J, Kuster B (2007) Quantitative mass spectrometry in proteomics: a critical review. Anal Bioanal Chem 389(4):1017–1031
12. Bawa G, Liu Z, Zhou Y, Fan S, Ma Q, Tissue DT, Sun X (2022) Cotton proteomics: dissecting the stress response mechanisms in cotton. Front Plant Sci 13:1035801
13. Bhattacharyya S (2022) Mechanism of temperature stress acclimation and the role of transporters in plants. In: Plant perspectives to global climate changes: developing climate-resilient plants, pp 413–457
14. Bhushan I, Mehta M, Sharma M, Chopra C, Chandra R, Noor ISM, Yahya MZA, Tripathi A, Yadav AK (2024) Role of nanomaterials in modern agriculture. Zastita Materijala
15. Biswas A, Pal S (2024) Plant-nano interactions: a new insight of nano-phytotoxicity. Plant Physiol Biochem 210:108646
16. Cai X, Liu X, Jiang J, Gao M, Wang W, Zheng H, Xu S, Li R, Cai X, Liu X, Jiang J, Gao M, Wang W, Zheng H, Xu S, Li R (2020) Molecular mechanisms, characterization methods, and utilities of nanoparticle biotransformation in nanosafety assessments. Small 16(36):1907663
17. Cameron SJ, Sheng J, Hosseinian F, Willmore WG (2022) Nanoparticle effects on stress response pathways and nanoparticle–protein interactions. Int J Mol Sci 23(14):7962
18. Chandoliya R, Sharma S, Sharma V, Joshi R, Sivanesan I (2024) Titanium dioxide nanoparticle: a comprehensive review on synthesis, applications and toxicity. Plants 13(21):2964
19. Chauhan A, Anand J, Parkash V, Rai N (2023) Biogenic synthesis: a sustainable approach for nanoparticles synthesis mediated by fungi. Inorgan Nano-Metal Chem 53(5):460–473
20. Chen Y, He E, Peijnenburg WJGM, Jiang X, Qiu H (2024) Differential leaf-to-root movement, trophic transfer, and tissue-specific biodistribution of metal-based and polymer-based nanoparticles when present singly and in mixture. Environ Sci Technol

21. Daniel AI, Hüsselmann L, Shittu OK, Gokul A, Keyster M, Klein A (2024) Application of nanotechnology and proteomic tools in crop development towards sustainable agriculture. J Crop Sci Biotechnol 27(3):359–379
22. de Francisco M, Fernandes-Silva P, Durães L, Romeiro A, Álvarez-Torrellas S, Almendros P (2024) Influence of the application of different zinc oxide nanoparticles on a lettuce crop grown in an acidic Mediterranean soil. Horticulturae 10(7):681
23. del Giudice G, Serra A, Saarimäki LA, Kotsis K, Rouse I, Colibaba SA, Jagiello K, Mikolajczyk A, Fratello M, Papadiamantis AG, Sanabria N, Annala ME, Morikka J, Kinaret PAS, Voyiatzis E, Melagraki G, Afantitis A, Tämm K, Puzyn T, et al (2023) An ancestral molecular response to nanomaterial particulates. Nat Nanotechnol 18(8):957–966
24. Dey S, Malviya R, Pandey A, Banavath HN, Muthamilarasan M, Gayen D (2023) Identification and expression analysis of the FtsH protein family in chickpea in response to drought stress
25. Didur OO, Khromykh NO, Drehval OA, Sklyar TV, Dzhagan VM, Mazur NV, Skoryk M, Kurahina NV (2024) Influence of silver nanoparticles synthesized from Chaenomeles leaf extracts on pathogenic microorganisms *Klebsiella pneumoniae*, *Staphylococcus aureus*, and *Fusarium culmorum*. Biosyst Divers 32(3):380–388
26. Djanaguiraman M, Anbazhagan V, Dhankher OP, Prasad PVV (2024) Uptake, translocation, toxicity, and impact of nanoparticles on plant physiological processes. Plants 13(22):3137
27. Dubourg G, Pavlović Z, Bajac B, Kukkar M, Finčur N, Novaković Z, Radović M (2024) Advancement of metal oxide nanomaterials on agri-food fronts. Sci Total Environ 928:172048
28. Ďúranová H, Kšiňan S, Kuželová L, Šimora V, Ďurišová Ľ, Olexíková L, Ernst D, Kolenčík M (2024) Nanoparticle-plant interactions: physico-chemical characteristics, application strategies, and transmission electron microscopy-based ultrastructural insights, with a focus on stereological research. Chemosphere 363:142772
29. Eid AM, Fouda A, Niedbała G, Hassan SED, Salem SS, Abdo AM, Hetta HF, Shaheen TI (2020) Endophytic *Streptomyces laurentii* mediated green synthesis of Ag-NPs with antibacterial and anticancer properties for developing functional textile fabric properties. Antibiotics 9(10):641
30. Eker F, Duman H, Akdaşçi E, Witkowska AM, Bechelany M, Karav S (2024) Silver nanoparticles in therapeutics and beyond: a review of mechanism insights and applications. Nanomaterials 14(20):1618
31. El Amerany F, Aboudamia FZ, Janah I, Taourirte M, Rhazi M (2022) Polymer-based nanoparticles (NPs): a promising approach for crop productivity. In: Plant and nanoparticles, pp 119–154
32. El-Saadony MT, Saad AM, Soliman SM, Salem HM, Desoky ESM, Babalghith AO, El-Tahan AM, Ibrahim OM, Ebrahim AAM, Abd El-Mageed TA, Elrys AS, Elbadawi AA, El-Tarabily KA, AbuQamar SF (2022) Role of nanoparticles in enhancing crop tolerance to abiotic stress: a comprehensive review. Front Plant Sci 13:946717
33. Emamverdian A, Barker J, Pehlivan N, Ghorbani A (2024) Role of nanomaterials for alleviating heavy metal(oid) toxicity in plants. In: Nanotechnology for abiotic stress tolerance and management in crop plants, pp 289–306
34. Embiale A, Hussein M, Husen A, Sahile S, Mohammed K (2016) Differential sensitivity of *Pisum sativum* L. cultivars to water-deficit stress: changes in growth, water status, chlorophyll fluorescence and gas exchange attributes. J Agron 15:45–57
35. Fadiji AE, Mortimer PE, Xu J, Ebenso EE, Babalola OO, Fadiji AE, Mortimer PE, Xu J, Ebenso EE, Babalola OO (2022) Biosynthesis of nanoparticles using endophytes: a novel approach for enhancing plant growth and sustainable agriculture. Sustainability 14(17):10839
36. Fan K-T, Xu Y (2024) Dynamic changes in proteomics during the response of Arabidopsis seedlings to heat stress. AgriRxiv
37. Farooq A, Khan I, Shehzad J, Hasan M, Mustafa G (2024) Proteomic insights to decipher nanoparticle uptake, translocation, and intercellular mechanisms in plants. Environ Sci Pollut Res 31(12):18313–18339
38. Fonseca JDS, Wojciechowska E, Kulesza J, Barros BS (2024) Carbon nanomaterials in seed priming: current possibilities. ACS Omega 9(45):44891–44906

39. Francis DV, Abdalla AK, Mahakham W, Sarmah AK, Ahmed ZFR (2024) Interaction of plants and metal nanoparticles: exploring its molecular mechanisms for sustainable agriculture and crop improvement. Environ Int 190:108859
40. Gandhi RR, Koche DK (2024) An insight of zinc oxide nanoparticles (ZnO NPS): green synthesis, characteristics and agricultural applications. Biosci Biotechnol Res Asia 21(3):863–876
41. Getnet Z, Husen A, Fetene M, Yemata G (2015) Growth, water status, physiological, biochemical and yield response of stay green sorghum {*Sorghum bicolor* (L.) Moench} varieties-a field trial under drought-prone area in Amhara regional state, Ethiopia. J Agron 14:188–202. https://doi.org/10.3923/ja.2015.188.202
42. Ghuge SA, Nikalje GC, Kadam US, Suprasanna P, Hong JC (2023) Comprehensive mechanisms of heavy metal toxicity in plants, detoxification, and remediation. J Hazard Mater 450:131039
43. Gupta S, Kaur R, Sharma T, Bhardwaj A, Sharma S, Sohal JS, Singh SV (2023) Multi-omics approaches for understanding stressor-induced physiological changes in plants: an updated overview. Physiol Mol Plant Pathol 126:102047
44. Hafeez MB, Ghaffar A, Zahra N, Ahmad N, Al-Qahtani WH, Sharma P, Husen A, Li J (2024) Exploring the efficacy of plant growth promoters in nutrient acquisition and antioxidant defense responses of late sown wheat under saline conditions. S Afr J Bot 171:812–824. https://doi.org/10.1016/j.sajb.2024.06.042
45. Husen A (2010) Growth characteristics, physiological and metabolic responses of teak (*Tectona grandis* Linn. f.) clones differing in rejuvenation capacity subjected to drought stress. Silvae Genet 59:124–136. https://doi.org/10.1515/sg-2010-0015
46. Husen A, Iqbal M, Khanum N, Aref IM, Sohrab SS, Meshresa G (2019) Modulation of salt-stress tolerance of niger (*Guizotia abyssinica*), an oilseed plant, by application of salicylic acid. J Environ Biol 40:94–104
47. Husen A (2021) Plant performance under environmental stress. In: Hormones, biostimulants and sustainable plant growth management. Springer Nature, Cham. https://doi.org/10.1007/978-3-030-78521-5
48. Husen A (2021) Harsh environment and plant resilience. In: Molecular and functional aspects. Springer Nature, Cham. https://doi.org/10.1007/978-3-030-65912-7
49. Husen A (2021) The harsh environment and resilient plants: an overview. In: Husen A (ed) Harsh environment and plant resilience. Springer International Publishing AG, Cham, pp 1–23. https://doi.org/10.1007/978-3-030-65912-7_1
50. Husen A (2024) Plant response to gold nanoparticles. Springer Nature Singapore Pte Ltd., Singapore https://doi.org/10.1007/978-981-97-7174-5
51. Husen A (2025) Agricultural crop improvement: plant and soil relationships. Taylor & Francis Group, LLC, Boca Raton. https://doi.org/10.1201/9781032630366
52. Husen A (2025) Plant response to silver nanoparticles. Springer Nature Singapore Pte Ltd., Singapore https://doi.org/10.1007/978-981-97-7352-7
53. Husen A (2025) Role of osmolytes in changing environment. Elsevier Inc., Cambridge, MA
54. Husen A (2025) Role of antioxidants in mitigating plant stress. Elsevier Inc., Cambridge, MA
55. Husen A (2025) Emerging carbon nanomaterials for sustainable agricultural practices. In: Synthesis, plant growth, performance, production and protection. Springer Nature Singapore Pte Ltd., Singapore
56. Husen A, Iqbal M, Sohrab SS, Ansari MKA (2018) Salicylic acid alleviates salinity-caused damage to foliar functions, plant growth and antioxidant system in Ethiopian mustard (*Brassica carinata* A. Br.). Agri Food Secur 7(44):1–14
57. Hussein M, Embiale A, Husen A, Aref IM, Iqbal M (2017) Salinity-induced modulation of plant growth and photosynthetic parameters in faba bean (*Vicia faba*) cultivars. Pak J Bot 49:867–877
58. Jamshidi Goharrizi K, Karami S, Hamblin MR, Momeni MM, Basaki T, Mostafaei Dehnavi M, Nazari M (2022) Transcriptomic and proteomic profile approaches toward drought and salinity stresses. Biol Plantarum 66(1):255–271. https://doi.org/10.32615/bp.2022.035

59. Jan S, Rustgi S, Barmukh R, Shikari AB, Leske B, Bekuma A, Sharma D, Ma W, Kumar U, Kumar U, Bohra A, Varshney RK, Mir RR (2024) Advances and opportunities in unraveling cold-tolerance mechanisms in the world's primary staple food crops. Plant Genome 17(1):e20402
60. Jha S, Pudake RN (2016) Molecular mechanism of plant–nanoparticle interactions. In: Plant nanotechnology: principles and practices, pp 155–181
61. Joshi S, Joshi R (2024) Comparative transcriptome profiling unveils the role of differentially charged gold nanoparticle in growth, metabolic regulation and defence response of micropropagated *Nardostachys jatamansi* Wall. ex DC. S Afr J Bot 167:585–596
62. Jyotsna T, Durga Venkat Hemu Y, Shanthi B, Swathi D (2024) Nanotechnology in agriculture. In: Futuristic trends in agriculture engineering and food sciences, volume 3, book 17, pp 125–130
63. Kandhol N, Singh VP, Pandey S, Sharma S, Zhao L, Corpas FJ, Chen ZH, White JC, Tripathi DK (2024) Nanoscale materials and NO-ROS homeostasis in plants: trilateral dynamics. Trends Plant Sci 29(12):1310–1318
64. Karthiga P, Kaushik R, Rama Krishna K, Srivignesh S (2024) Nanoparticles uptake, utilization and phytotoxicity in plants. In: Futuristic trends in chemical material sciences and nano technology Volume 3 Book 4, pp 8–20
65. Kaur S, Garg T, Joshi A, Awasthi A, Kumar V, Kumar A (2024) Potential effects of metal oxide nanoparticles on leguminous plants: practical implications and future perspectives. Sci Hortic 331:113146
66. Kaur H, Kohli SK, Khanna K, Dhiman S, Kour J, Bhardwaj T, Bhardwaj R (2022) Deciphering the role of metal binding proteins and metal transporters for remediation of toxic metals in plants. In: Bioremediation of toxic metal(loid)s, pp 257–272
67. Kausar R, Komatsu S (2022) Proteomic approaches to uncover salt stress response mechanisms in crops. Int J Mol Sci 24(1):518
68. Kausar R, Wang X, Komatsu S (2022) Crop proteomics under abiotic stress: from data to insights. Plants 11(21):2877
69. Khanizadeh P, Mumivand H, Morshedloo MR, Maggi F (2024) Application of Fe2O3 nanoparticles improves the growth, antioxidant power, flavonoid content, and essential oil yield and composition of *Dracocephalum kotschyi* Boiss. Front Plant Sci 15:1475284
70. Kheyri N, Kheyri N (2024) Impact of zinc oxide nanoparticles application on rice yield and quality
71. Kosová K, Prášil IT, Klíma M, Nesvadba Z, Vítámvás P, Ovesná J (2023) Proteomics of wheat and barley cereals in response to environmental stresses: current state and future challenges. J Proteomics 282:104923
72. Kumar J, Jain K, Kumar RR, Mohanty A, Kumar A, Ranjan T (2022) Proteomic approaches in physiological studies of plant abiotic stress. In: Omics analysis of plants under abiotic stress, pp 19–53
73. Kumari A, Gupta AK, Sharma S, Jadon VS, Sharma V, Chun SC, Sivanesan I (2024) Nanoparticles as a tool for alleviating plant stress: mechanisms, implications, and challenges. Plants 13(11):1528
74. Li Y, Zheng A, Li Z, Wang H, Wang J, Dong Z, Yao L, Han X, Wei F (2024) Characterization and gene expression analysis reveal universal stress proteins respond to abiotic stress in *Gossypium hirsutum*. BMC Genom 25(1):1–13
75. Madhogaria B, Banerjee S, Chakraborty S, Dhak P, Kundu A (2024) Synergistic impacts of plant growth promoting *Pseudomonas geniculata* TIU16A3 and green-synthesized iron-oxide nanoparticles from *Eichhornia crassipes* for the amelioration of heavy metal stress in *Vigna radiata* L.
76. Manju K, Ranjini HK, Raj SN, Nayaka SC, Lavanya SN, Chouhan RS, Prasad MN, Satish S, Harini BP, Baker S (2023) Nanoagrosomes: future prospects in the management of drug resistance for sustainable agriculture. Plant Nano Biol 4:100039
77. Mgadi K, Ndaba B, Roopnarain A, Rama H, Adeleke R (2024) Nanoparticle applications in agriculture: overview and response of plant-associated microorganisms. Front Microbiol 15:1354440

78. Mondéjar-López M, García-Simarro MP, Navarro-Simarro P, Gómez-Gómez L, Ahrazem O, Niza E (2024) A review on the encapsulation of "eco-friendly" compounds in natural polymer-based nanoparticles as next generation nano-agrochemicals for sustainable agriculture and crop management. Int J Biol Macromol 280:136030
79. Mošenoka A, Kokina I, Plaksenkova I, Jermaļonoka M, Sledevskis E, Krasovska M (2024) Effects of metal oxide nanoparticles on the growth and genotoxicity of garden cress (*Lepidium sativum* L.). Agronomy 14(10):2324
80. Mukhopadhyay A, Basu S, Singha S, Patra HK (2018) Inner-view of nanomaterial incited protein conformational changes: insights into designable interaction. Research
81. Mukhtiar A, Zia MA, Alawadi HFN, Naqve M, Seleiman MF, Mahmood A, Majeed MI, Hafeez MB, Khan BA, Khan N (2024) Role of iron oxide nanoparticles in maize (*Zea mays* L.) to enhance salinity stress tolerance. Not Bot Horti Agrobot Cluj-Napoca 52(3):13695–13695
82. Mustafa G, Komatsu S (2022) Proteomics of plant-nanoparticle interaction mechanism. In: Plant and nanoparticles, pp 67–84
83. Nayak L, Lal MK, Tiwari RK, Kumar R, Lal P, Das R, Behera B, Thakur P, Altaf MA, Kumar A (2023) A balancing act: exploring the interplay between HSPs and osmoprotectants in temperature stress responses. S Afr J Bot 162:64–71
84. Nazir MA, Hasan M, Mustafa G, Tariq T, Ahmed MM, Golzari Dehno R, Ghorbanpour M (2024) Zinc oxide nano-fertilizer differentially effect on morphological and physiological identity of redox-enzymes and biochemical attributes in wheat (*Triticum aestivum* L.). Sci Rep 14(1)
85. Nehru L, Kandasamy GD, Sekar V, Alshehri MA, Panneerselvam C, Alasmari A, Kathirvel P (2023) Green synthesis of ZnO-NPs using endophytic fungal extract of *Xylaria arbuscula* from *Blumea axillaris* and its biological applications. Artif Cells Nanomed Biotechnol 51(1):318–333
86. Ningombam L, Hazarika BN, Yumkhaibam T, Heisnam P, Singh YD (2024) Heavy metal priming plant stress tolerance deciphering through physiological, biochemical, molecular and omics mechanism. S Afr J Bot 168:16–25
87. Nishad JH, Singh A, Gautam VS, Kumar D, Kumar J, Kharwar RN (2019) Endophytic fungi: a cryptic fountainhead for biodiversity, functional metabolites, host stress tolerance, and myco-mediated nanoparticles (Nps) synthesis, pp 1–29
88. Noor Mohamed Ibrahim PMS, Basri S (2023) Current understanding of proteomics in plants under drought stress conditions. In: Genomics, transcriptomics, proteomics and metabolomics of crop plants, pp 277–301. https://doi.org/10.1016/B978-0-323-95989-6.00006-1
89. Parmar K (2024) Green synthesis of zinc oxide nanoparticles and its application as a green fertilizer in agriculture. Holist Approach Environ 14(3):109–113
90. Pascoli M, Lopes-Oliveira PJ, Fraceto LF, Seabra AB, Oliveira HC (2018) State of the art of polymeric nanoparticles as carrier systems with agricultural applications: a minireview. Energy Ecol Environ 3(3):137–148
91. Pathak J, Singh DK, Singh PR, Kumari N, Jaiswal J, Gupta A, Sinha RP (2024) Application of nanoparticles in agriculture: nano-based fertilizers, pesticides, herbicides, and nanobiosensors. In: Molecular impacts of nanoparticles on plants and algae, pp 305–331
92. Pietrzak M, Skiba E, Wolf WM (2024) Root-applied cerium oxide nanoparticles and their specific effects on plants: a review. Int J Mol Sci 25(7):4018
93. Prajapati BJ, Patel KC, Kumar D, Patel SH, Patel VJ (2024) Impact assessment of zinc oxide nanoparticles (ZnO-NPS) on seed germination and seedling development of maize. Int J Plant Soil Sci 36(9):810–820
94. Raheman SB, Joseph M, Hemalatha M, Bhuvaneswari J, Srinivasan S, Leninraja D (2024) Nano-fertilizers: the future of nutrient approaches for cereals. Indian J Agri Sci 94(11):1155–1164
95. Rajput VD, Singh A, Tomar B, Minkina T, Movsesyan HS, Elshikh MS, Chena SM, Singh RK, Ghazaryan K (2024) Nanoparticle-mediated approaches in agriculture addressing abiotic stress from soil to plant cells

96. Rehman A, Khan S, Sun F, Peng Z, Feng K, Wang N, Jia Y, Pan Z, He S, Wang L, Qayyum A, Du X, Li H (2023) Exploring the nano-wonders: unveiling the role of Nanoparticles in enhancing salinity and drought tolerance in plants. Front Plant Sci 14:1324176
97. Richard R, Rosenkranz E (2022) The involvement of RS2Z proteins in tomato heat stress response and thermotolerance
98. Rodrigues AS, Batista JGS, Rodrigues MÁV, Thipe VC, Minarini LAR, Lopes PS, Lugão AB (2024) Advances in silver nanoparticles: a comprehensive review on their potential as antimicrobial agents and their mechanisms of action elucidated by proteomics. Front Microbiol 15:1440065
99. Samiappan SC, Natarajan SD, Munisamy P, Pandiyan R, Rangaraj S, Kumaresan M, Ravichandran M (2024) Synthesis and applications of biopharmaceuticals from whole-cell actinobacterial isolates. In: Whole-cell biocatalysis: next-generation technology for green synthesis of pharmaceutical, chemicals, and biofuels, pp 221–266
100. Schniete JK, Fernández-Martínez LT (2024) Natural product discovery in soil actinomycetes: unlocking their potential within an ecological context. Curr Opin Microbiol 79:102487
101. Schvartz M, Saudrais F, Boulard Y, Renault JP, Henry C, Chédin S, Pin S, Aude JC (2024) Dual fractions proteomic analysis of silica nanoparticle interactions with protein extracts. Materials 17(19):4909
102. Sembada AA, Lenggoro IW (2024) Transport of nanoparticles into plants and their detection methods. Nanomaterials 14(2):131
103. Sethi SK, Dalei SR, Oko AO, Kumar A, Kindo A (2024) Effect of nano-fertilizer on crop plant growth and its effect on rhizospheric microorganism and its application. In: Advances in chemical and materials engineering, pp 177–202
104. Shahroz M, Ashiq M, Infal M, Riaz A, Chaudhry Z, Hassan MN, Rana A, Arif A, Aslam R, Ulah Q, Rana BA, Acad G, Biosci JA (2024) Nanoparticles mediated modulation of nitrogen fixation in legumes: a biochemical perspective. Glob Acad J Agri Biosci 6(5):119–133
105. Singh R, Choudhary P, Kumar S, Daima HK (2024) Mechanistic approaches for crosstalk between nanomaterials and plants: plant immunomodulation, defense mechanisms, stress resilience, toxicity, and perspectives. Environ Sci Nano 11(6):2324–2351
106. Singh A, Mehta S, Yadav S, Nagar G, Ghosh R, Roy A, Chakraborty A, Singh IK (2022) How to cope with the challenges of environmental stresses in the era of global climate change: an update on ROS stave off in plants. Int J Mol Sci 23(4):1995
107. Song X, Wang H, Wang Y, Zeng Q, Zheng X (2024) Metabolomics combined with physiology and transcriptomics reveal how *Nicotiana tabacum* leaves respond to cold stress. Plant Physiol Biochem 208:108464
108. Strekalovskaya EI, Perfileva AI, Krutovsky KV (2024) Zinc oxide nanoparticles in the "soil–bacterial community–plant" system: impact on the stability of soil ecosystems. Agronomy 14(7):1588
109. Sumersinh G, Chuodhary R, Dudwal BL, Shivran AC, Meena BR, Mittal GK, Kumhar BL, Choudhary R, Gunjan (2024) Stimulatory effect of zno nanoparticles as a nanofertilizer on growth and yield of pearl millet (*Pennisetum glaucum*). Int J Plant Soil Sci 36(8):1068–1073
110. Suvarna, Yashaswini R, Sagar SPP, Kuchanur PH, Raghavendra VC, Prasad BK, Amaregouda A, Patil A (2024) Plant proteome in response to abiotic stresses. In: Abiotic stress in crop plants—ecophysiological responses and molecular approaches
111. Syed B, Nagendra Prasad MN, Mohan Kumar K, Satish S (2018) Bioconjugated nano-bactericidal complex for potent activity against human and phytopathogens with concern of global drug resistant crisis. Sci Total Environ 637–638:274–281
112. Tanvir R, Sheikh AA, Javeed A (2019) Endophytic Actinomycetes in the Biosynthesis of Bioactive Metabolites: Chemical Diversity and the Role of Medicinal Plants. Stud Nat Prod Chem 60:399–424
113. Thokchom S, Sekhar G, Bharteey PK, Rai S (2024) Engineered nanoparticles for plant-originated environmental consequence. In: Environmental nexus for resource management, pp 363–378

114. Tripathi S, Sharma S, Suri S, Tiwari K, Tripathi DK, Sharma S (2024) Insights into physiological and molecular responses of plants under metal-nanoparticle stresses. In: Molecular and physiological insights into plant stress tolerance and applications in agriculture—Part 2, pp 147–173
115. Van Leene C, Van Moortel L, De Bosscher K, Gevaert K (2024) Exploring protein conformations with limited proteolysis coupled to mass spectrometry. Trends Biochem Sci 50(2):143-155
116. Verano-Braga T, Miethling-Graff R, Wojdyla K, Rogowska-Wrzesinska A, Brewer JR, Erdmann H, Kjeldsen F (2014) Insights into the cellular response triggered by silver nanoparticles using quantitative proteomics. ACS Nano 8(3):2161–2175
117. Verma DK, Thakor AB, Patel S (2024) Titanium dioxide (TiO_2) nanoparticles improved morpho-physiological characters and DNA content in *Swietenia macrophylla* an IUCN red listed endangered plant species. Curr Bot 138–150
118. Waseem M, Aslam MM (2024) Integrated approaches of "omics" for the improvement of stress tolerance in plants. In: Improving stress resilience in plants: physiological and biochemical basis and utilization in breeding, pp 321–329
119. Wroblewski C, Barbhuiya RI, Kaur G, Nair GR, Elsayed A, Singh A (2024) A comprehensive insight into metal oxide nanoparticle synthesis, associated regulations, and application in the agri-food sector. Can J Chem Eng
120. Yan N, Cao J, Wang J, Zou X, Yu X, Zhang X, Si T (2024) Seed priming with graphene oxide improves salinity tolerance and increases productivity of peanut through modulating multiple physiological processes. J Nanobiotechnol 22(1):1–18
121. Ye Q, Wu Y, Liu W, Ma X, He D, Wang Y, Li J, Wu W (2024) Identification and quantification of nanoplastics in different crops using pyrolysis gas chromatography-mass spectrometry. Chemosphere 354:141689–141689
122. Yin G, Luo Y, Jia W, Jiang X, Jiang Y, Wu X, Song S, Shen H, Shen J (2024) Graphene oxide-based antifungal pesticide delivery system for tobacco fungal disease (tobacco target spot) control. Langmuir 40(35):18598–18609
123. Zahra N, Kausar A, Sharma P, Wabaidur SM, Husen A (2024) Serotonin application to *Guizotia abyssinica* improves growth, physiological functions, antioxidant system and alleviates the deleterious effects induced by salt stress. Plant Biosyst Int J Deal Aspects Plant Biol. https://doi.org/10.1080/11263504.2024.2384972
124. Zahra N, Kausar A, Abdelghani HTM, Singh S, Vashishth DS, Bachheti A, Bachheti RK, Husen A (2024) Serotonin improves plant growth, foliar functions and antioxidant defence system in Ethiopian mustard (*Brassica carinata* A. Br.). South Afri J Bot 170:1–9 https://doi.org/10.1016/j.sajb.2024.05.002
125. Zhao W, Wu Z, Amde M, Zhu G, Wei Y, Zhou P, Zhang Q, Song M, Tan Z, Zhang P, Rui Y, Lynch I (2023) Nanoenabled enhancement of plant tolerance to heat and drought stress on molecular response. J Agric Food Chem 71(51):20405–20418
126. Zhu C, Yu J, Cao S, Wu X, Meng W, Hou X (2022) Transcriptomics-based analysis of genes related to lead stress and their expression in the roots of *Pogonatherum crinitum*. Front Plant Sci 13:1066329
127. Zhuang M, Qiao C, Han L, Bi Y, Cao M, Wang S, Guo L, Pang R, Xie H (2024) Multi-omics analyses reveal the responses of wheat (*Triticum aestivum* L.) and rhizosphere bacterial community to nano(micro)plastics stress. J Nanobiotechnol 22(1):1–18
128. Zia-ur-Rehman M, Anayatullah S, Irfan E, Hussain SM, Rizwan M, Sohail MI, Jafir M, Ahmad T, Usman M, Alharby HF (2023) Nanoparticles assisted regulation of oxidative stress and antioxidant enzyme system in plants under salt stress: a review. Chemosphere 314:137649

Specific Physiological and Molecular Response of Nanoparticle Exposure to Crop Plants

Manoj Kumar Srinivasan, Siva Sankari Thirugnanam, Subasri Pandian, Harish Krishnan, Kamalesh Balakumar Venkatesan, and Sathish-Kumar Kamaraj

Abstract Nanoparticles (NPs) have emerged as transformative tools in agriculture, offering innovative solutions to enhance plant growth, stress resilience, and productivity. Their distinct physicochemical properties enable efficient uptake, translocation, and accumulation within plant tissues, providing targeted strategies to mitigate environmental and physiological stressors such as drought, salinity, high temperatures, and heavy metal toxicity. This chapter underscores the multifaceted roles of NPs in enhancing crop performance, emphasizing their ability to improve water retention, regulate osmotic balance, and activate antioxidant defense mechanisms. At the molecular level, nanoparticle exposure triggers the modulation of stress-responsive genes, activation of key signaling pathways, and synthesis of protective metabolites, fostering improved plant adaptability. Additionally, NPs facilitate nutrient uptake and alleviate deficiencies or toxicities, further supporting sustainable crop health. However, challenges persist in ensuring the safe, sustainable, and effective application of NPs, addressing phytotoxicity risks, and evaluating long-term environmental impacts. Advancing research in biocompatible NP synthesis, elucidating plant-NPs

M. K. Srinivasan · S. S. Thirugnanam · S. Pandian · H. Krishnan · K. B. Venkatesan
Department of Biochemistry and Biotechnology, Faculty of Science, Annamalai University, Tamilnadu, India
e-mail: manojsrinivasan1002@gmail.com

S. S. Thirugnanam
e-mail: stssankari@gmail.com

S. Pandian
e-mail: subasrilaksh35@gmail.com

H. Krishnan
e-mail: harishyuvaraj2001@gmail.com

K. B. Venkatesan
e-mail: kamalvengat2000@gmail.com

S.-K. Kamaraj (✉)
Instituto Politécnico Nacional (IPN)-Centro de Investigación en Ciencia Aplicada y Tecnología Avanzada (CICATA), Unidad Altamira, Mexico City, Mexico
e-mail: sathish.bot@gmail.com

A. Husen (ed.), *Nanobiotechnology for Abiotic Stress Adaptation and Mitigation in Agricultural Crops*, Smart Nanomaterials Technology,
https://doi.org/10.1007/978-981-96-9298-9_16

interactions at the molecular scale, and integrating nanotechnology into precision agriculture will be pivotal for harnessing the full potential of NPs in sustainable crop production.

Keywords Nanoparticles · Plant crop · Drought · Salinity · Heavy metal · Nutrition

1 Introduction

The agricultural sector confronts significant challenges from a variety of environmental factors that reduce crop output. Stress factors such as soil salinity, severe temperatures, drought, nutritional shortages, and heavy metal and air pollution drive plants to undertake biochemical, physiological, or molecular adaptations in order to survive [60-66, 83, 160]. While these adaptations enable plants to cope with adverse environments, they also impose a substantial energy cost, which negatively affects overall crop yield and quality. This growing challenge necessitates innovative approaches to sustain and enhance agricultural productivity under increasingly unpredictable environmental conditions. Greenhouse cultivation has emerged as an effective strategy to manage adverse environmental conditions, mitigate the impact of abiotic stress, and improve crop productivity [24]. However, even in controlled environments, abiotic stresses like salinity, temperature extremes, and nutrient imbalances continue to affect plant growth and development. In recent years, nanotechnology has emerged as a transformative science with vast potential in agriculture, providing innovative solutions to improve crop performance under stress conditions [130, 142, 143].

Nanoparticles (NPs), defined as particles with dimensions below 100 nm, have garnered significant attention in agricultural research. While initial studies focused primarily on assessing the potential risks and toxicity of NPs to plants [34, 115, 152], a growing body of evidence now highlights the beneficial effects of NPs on various crops. These benefits range from enhanced germination and growth to improved quality and stress tolerance, suggesting the potential of NPs as a new frontier in agricultural innovation.

Several studies have highlighted the beneficial effects of NPs on plant germination and growth. For instance, increased germination rates have been observed in crops like *Solanum lycopersicum L.* (tomato) and *Allium cepa L.* (onion) following the application of NPs [8, 134]. Furthermore, NPs have been shown to promote greater growth in crops like tomato [57, 75], strawberry [107], and rice [98, 141]. The application of NPs has also been associated with improved fruit quality in tomatoes and strawberries, and even extended vase life in ornamental plants [7, 22, 113]. One of the promising applications of NPs in agriculture lies in their ability to enhance crop resilience under abiotic stress conditions. A number of studies have shown that NPs can help alleviate the negative effects of stress factors, such as salinity [11, 123], drought [107], extreme temperatures [9, 56, 104], heavy metal contamination

[51], and nutrient deficiency. These favorable benefits are related to NPs' potential to produce physiological, morphological, biochemical, and genetic changes in plants, allowing them to survive and adapt to different stress situations [106, 152].

NPs achieve their good impacts through a variety of techniques. They can, for example, increase nutritional availability and absorption, boost the manufacture of stress-protective molecules, and regulate the activity of stress-related enzymes. Additionally, NPs can influence gene expression, leading to the activation of stress-responsive pathways. These multifaceted effects make NPs a promising tool to enhance crop resilience and productivity, particularly under conditions of abiotic stress. Beyond stress mitigation, NPs have also been shown to improve the overall efficiency of agricultural inputs [157]. For example, NPs can be used to develop slow-release fertilizers and pesticides, reducing environmental contamination and enhancing the effectiveness of these inputs. Additionally, the ability of NPs to penetrate plant cells and tissues allows for targeted delivery of nutrients and bioactive compounds, further improving plant health and productivity [121].

Despite the growing body of evidence supporting the potential benefits of NPs in agriculture, challenges remain [128]. The variability in nanoparticle formulations, application methods, and plant responses underscores the need for further research to optimize their use. Additionally, concerns about the long-term environmental and ecological impacts of NPs must be addressed through comprehensive risk assessments and the development of sustainable practices. As global food security continues to face threats from climate change and environmental degradation, the integration of nanotechnology into agriculture represents a promising avenue for innovation. By harnessing the unique properties of NPs, it is possible to develop new strategies to transcend the constraints of conventional agriculture practices and enhance the resilience of crops to stress [174].

2 Uptake, Translocation, and Accumulation of Nanoparticles in Plants

Soil, water, and air are among the environmental matrices that expose terrestrial plants directly to NPs. Nanoparticles are known to absorb, transport, and bioaccumulate within plant systems, with the potential to migrate along the food chain. Plants can be exposed to NPs through a variety of methods, including in vitro culture, spraying, hydroponics, isolated cells, soil, protoplast incubation, and biolistics [111]. To investigate the interactions between nanomaterials and plant systems, seedlings are often immersed in a nanomaterial solution or grown in a medium or soil enriched with NPs.

Nanoparticle absorption through the roots allows for their transfer to the shoots via xylem tissue, assuming the NPs can migrate. However, tiny NPs with large surfaces frequently agglomerate in aqueous solutions, which reduces their bioavailability.

Furthermore, root tips and hairs generate mucilage made up of hydrated polysaccharides, which might help NPs aggregate and plug cell membrane pores, limiting nanoparticle entrance into plant systems [111].

Figure 1 represents the uptake and translocation of NPs through root. The roots of plants are submerged in a nutritional solution or inert materials like gravel or perlite when using the hydroponics method. Although it establishes a regulated setting, it limits the ability to assess the impact of soil microbes and the interplay between nutrients and nanomaterials [13, 32].

Plant cell walls are semipermeable, with pores typically smaller than a few nanometers, blocking particles larger than this size. However, studies indicate that NPs can penetrate plant cells by either sliding through or bypassing the cell wall, potentially creating new channels.

Many plant researchers are currently focusing on nanomaterial exposure through foliar sprays. This approach is designed to model interactions between air NPs and plant leaves, as well as to examine the effects of synthetic NPs. Nanoparticles applied to leaf surfaces penetrate the plant system through stomatal apertures or trichome bases and are subsequently transported to various tissues [96, 154, 166].

A variety of variables effect nanoparticle (NP) uptake and mobility in terrestrial plants. While some studies have revealed their absorption and mobility in plants [28, 77] discovered that they are much less mobile in soil. Mucilage, root exudates, hyphae, and bacterial populations can all have a substantial impact on nanoparticle mobility in the soil. This change is mostly owing to sluggish desorption kinetics or sequestration mechanisms [139, 175].

NPs infiltrate plant tissues through wounds and root linkages. NPs pass through plant cell walls and membranes before reaching the leaves [152]. NPs may pass through leaf cuticles, stomata, trichomes, hydathodes, and the cell cytoplasm [135]. Once inside, they can adhere to numerous cytoplasmic organelles, possibly interfering with essential metabolic operations [172]. Moreover, NPs can travel across the seed coat's intercellular gaps [152].

The accumulation, translocation, and absorption of NPs in plants depend on factors such as plant species, NP size, type, stability, and interactions with roots, soil, and microorganisms. Characteristics like size, surface charge, and morphology, as well as environmental conditions like soil pH, organic matter, and microbial activity, significantly influence the absorption of metal and metal oxide-based NPs by plant roots. For instance, NPs such as metal and metal oxide-based types can exhibit varying levels of root uptake based on their surface functionalization and interactions with soil components [5, 102]. Silver nanoparticles (Ag NPs) have been found to significantly boost their content in lettuce roots and shoots [33]. Furthermore, root microorganisms have an important role in the mobility of NPs delivered by foliar sprays, soil application, or root application. For example, mycorrhizae have been shown to inhibit the absorption of Ag NPs in *Trifolium repens* roots [42], but microorganisms have been shown to greatly increase the uptake of selenium NPs (Se NPs) [35].

Mucilage improves NP absorption and microbial development in soil by acidifying the rhizosphere, which increases the uptake of metal- and carbon-based NPs [99, 128]. NPs adsorption is influenced by different factors, including the technique of

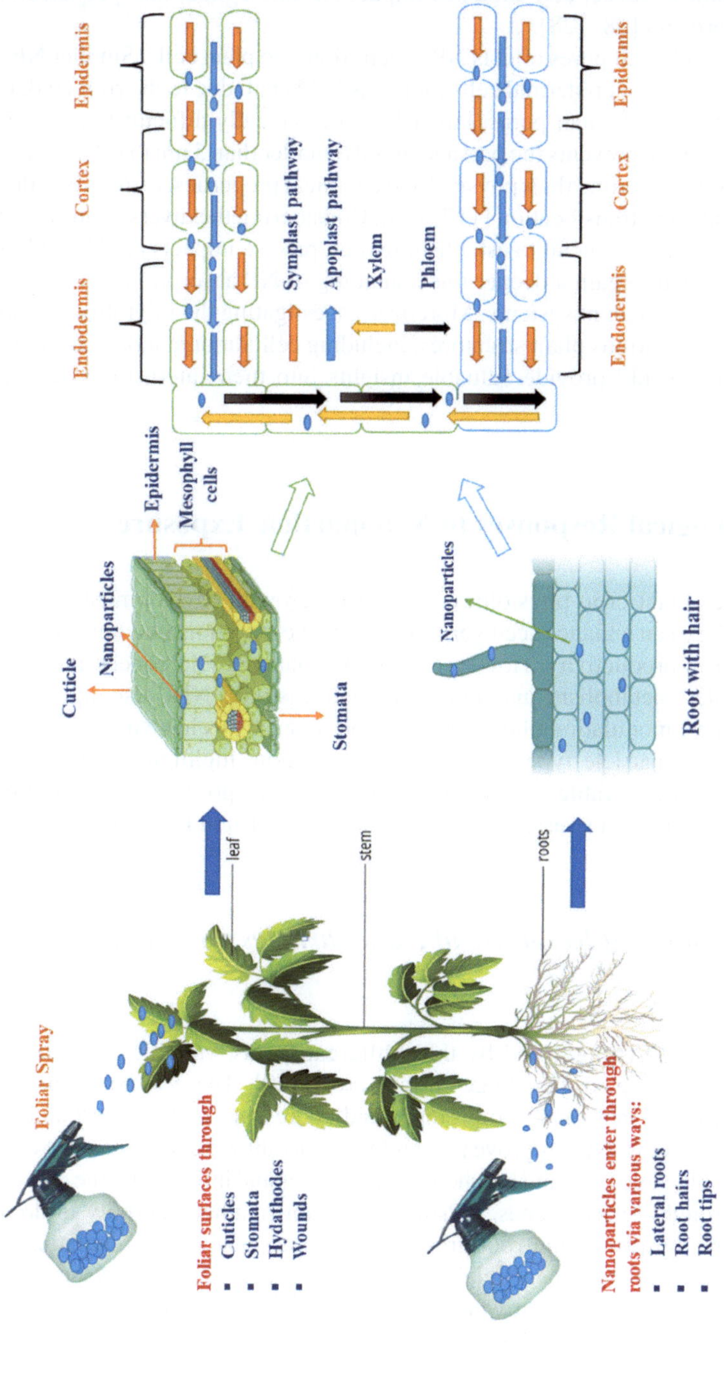

Fig. 1 Schematic representation of nanoparticle uptake via various routes and their translocation pathways within different parts of plants

application, meteorological conditions, size, and concentration. Additionally, leaf shape, trichomes, waxes, and exudates impact NP absorption, and pesticides may modify this process [88, 128].

The size of cell wall pores affects NP penetration into plant cells. Smaller NPs can directly penetrate root epidermal cells more easily [5] but, due to the root epidermis' semipermeability and small pores, larger NPs are typically difficult to enter. Similarly, the leaf cuticle prevents the passage of NPs smaller than 5 nm [5]. The structural qualities of NPs, including their physical and chemical properties, primarily influence their absorption and translocation [117, 118]. Considering the diverse nature of NPs, laboratory research is crucial to establishing their precise impacts [171]. Additional research is needed to gain a deeper understanding of NP absorption, accumulation, and transport within plant systems. Moreover, investigating the migration and localization of NPs in various plant structures, including cellular organelles, and tracking their pathways would provide valuable insights into their interactions with plant biology.

3 Physiological Responses to Nanoparticle Exposure

NPs influence plant crop physiology by altering growth, development, and stress responses. They can enhance seed germination, root elongation, and nutrient uptake, improving crop productivity. However, excessive nanoparticle concentrations may disrupt cellular metabolism and induce oxidative stress. NPs also interact with phytohormones, modulating photosynthesis and defense mechanisms. Their impact depends on nanoparticle type, size, and concentration, highlighting the need for balanced application. Table 1 shows the physiological responses of NPs on drought stress, temperature, salt stress, and heavy metal and nutrition imbalance.

3.1 Physiological Response of Nanoparticles on Drought Stress

Nanoparticles are distinguished by their different forms, variable pore sizes, and high reactivity owing to their huge surface area [129]. These features make NPs a viable option for increasing crop output and addressing both present and future agricultural concerns. NPs improve plant tolerance to abiotic conditions like drought by activating physiological and biochemical responses and influencing the expression pattern of drought-related genes. NPs enhance photosynthetic activity in drought-stressed plants by promoting root growth, activating aquaporins, regulating water metabolism, accumulating beneficial solutes, and maintaining ionic balance, thereby alleviating osmotic stress from water scarcity. They also reduce leaf water loss by increasing abscisic acid (ABA) levels, which induces stomatal closure, and mitigate

Table 1 Physiological responses to nanoparticle exposure

Nanoparticles	Effects	Key references
Physiological response on NPs on drought stress		
TiO_2 NPs	Enhances plant photosynthesis, increases Rubisco activase activity, chlorophyll formation, and photosynthetic rate, promotes root growth, improves seed yield	[32, 47, 105]
ZnO NPs	Enhances root growth, lateral root formation, water absorption, and reduces oxidative stress	[81, 82, 167]
CuO NPs	Induces root hair proliferation, lignification of cell walls, increases free radical accumulation, activates antioxidant enzymes, enhances proline levels	[92, 110, 122]
CNTs	Increases seed germination, antioxidant content, root and shoot growth in wheat	[92]
FNPs	Enhances water absorption in plant tissues, binds intracellular water, increases proline content, ameliorates oxidative stress by activating antioxidant systems	[20, 156]
Si NPs	Ameliorates drought stress by improving growth and biochemical characteristics, activates antioxidant enzymes	[122, 138]
Ag NPs	Modulates antioxidant enzyme activity, alleviates oxidative stress	[2, 137]
Physiological response of NPs on temperature stress		
Multiwall carbon nanotubes	Upregulate heat shock proteins (e.g., HSP90) and assist in heat stress resistance	[132]
CeO_2 NPs	Induce excessive H_2O_2 generation and upregulate HSP70 in maize	[159]
Biological selenium NPs	Improve plant productivity, growth, photosynthesis, and gas exchange at elevated temperatures	[37]

(continued)

Table 1 (continued)

Nanoparticles	Effects	Key references
Nano-ZnO NPs	Increase chlorophyll activity, gas exchange, enzymatic balance, pod size, and total grain yield in mung bean	[78]
ZnO and TiO_2 NPs	Improve membrane stability and antioxidant defense mechanisms under heat stress in wheat seedlings	[149]
ZnO NPs	Regulate osmotic potential, reduce thylakoid damage, and activate antioxidant defense in rice	[145]
$nSiO_2$, nSe, nZnO, GNRs	Mitigate cold stress in sugarcane by improving chlorophyll content, photosynthetic rate, and gas exchange	[38]
Various NPs	Help maintain ion concentration, membrane stability, osmotic regulation, and antioxidant enzyme production	[116, 160]
ZnO, TiO_2 NPs	Improve defense against heat stress by promoting antioxidant enzyme production (e.g., SOD, GPX, H_2O_2 reduction)	[56, 153]
TiO_2 NPs	Reduce H_2O_2 content, improve photosynthetic activity, and enhance PSII activity under heat stress	[30]
Physiological response of NPs on salt stress		
Mn NPs	Enhances photosynthetic electron transport rates and oxygen evolution. Maintains optimal photosynthesis rates under abiotic stress	[168]
Cu NPs	Reduces the negative effects of salinity on water relations and photosynthesis in maize plants	[71]
Nano-SiO_2	Increases leaf $K+$ concentration and enhances soybean seedling growth under salt stress	[39]
Cu NPs	Enhances tomato plant growth performance and $Na+/K+$ ratio after salt stress	[114]

(continued)

Table 1 (continued)

Nanoparticles	Effects	Key references
Fe_2O_3 NPs	Reduces sensitivity to salinity stress in Trachyspermum ammi plants	[114]
Mn NPs	Enhances growth of pepper plants under salinity stress	[168]
Ag NPs	Increases antioxidant enzyme activity (SOD, CAT, GPX) and reduces Na + /K + ratio in Pearl millet under salinity stress	[81, 82, 164]
TiO_2 NPs	Decreases H_2O_2 concentration and increases antioxidant content in Dracocephalum moldavica under salinity stress	[49]
Co, Fe, Ce NPs	Act like catalase (CAT) and peroxidase (POD) enzymes, enhancing ROS scavenging in plants	[164]
Physiological response of NPs on heavy metals		
FeO NPs	Alleviated Cd toxicity in wheat; enhanced photosynthetic pigments, dry biomass, and morphological parameters	[69]
Nano-TiO_2	Reduced Cd content and oxidative injuries in coriander; improved agronomic traits and photosynthesis in soybean	[126, 140]
Si NPs	Increased biomass of summer savory; reduced Cd-induced stress; improved antioxidant defense systems	[100]
Graphite Carbon Nitride	Mitigated Cd toxicity in rice; increased plant biomass and reduced Cd-induced oxidative stress	[54]
ZnO NPs	Decreased Cd accumulation; enhanced SOD and POD activity in wheat and brassica	[69, 159]
Cu NPs	Reduced Cd toxicity in brassica; increased antioxidant enzyme activities (SOD, POD, CAT) and plant biomass	[159]

(continued)

Table 1 (continued)

Nanoparticles	Effects	Key references
Fe_2O_3 NPs	Restricted As uptake in *Vigna radiata*; enhanced total antioxidant capacity (SOD, CAT) and reduced oxidative stress	[131]
CuO NPs	Reduced As accumulation in rice; increased plant biomass and antioxidant activity	[162]
Si NPs	Alleviated Pb stress in coriander; increased antioxidant activity and reduced Pb toxicity	[41]
Nano-hydroxyapatite	Increased biomass of ryegrass affected by Pb toxicity; reduced Pb stress	[74]
Citrate-coated Magnetite NPs	Reduced Cd and Cr toxicity; enhanced root length and alleviated heavy metal stress in wheat	[94]
Physiological response of NPs on nutrients imbalance		
Zn and Mg hydroxyapatite nanohybrids	Slow release of urea molecules in soil, enhancing nitrogen nutrient uptake and wheat crop yield	[136]
Urea/MIL-100(Fe)/silica heterojunction nanomaterials	Sustained release of nitrogen, increasing nitrogen use efficiency by 34.7% in potted rice	[163]
Urea-coated hydroxyapatite NPs (UHA)	Slow release of urea, improving nutrient uptake and efficiency during rice growth	[17]
Nano U-NPK (multinutrient nanofertilizer)	Slow release of essential nutrients (N, P, K), reduces nitrogen supply by 40%	[119]
Potassium-based NPs (K_2SiO_3-NP, $K_{18}Mo_8O_{33}$-NPs)	Improved potassium utilization efficiency, 40% higher in soybean growth due to slow-release effect	[161]

oxidative stress by decreasing ROS while stimulating antioxidant defenses. Figure 2 shows the role of NPs on drought stress tolerance in plant crop [120].

Nanotechnology can develop the plant photosynthesis efficiency by modifying enzyme activity in the C3 cycle and adjusting photosynthetic pigments, which are essential for plant growth and development [93]. While NPs promote plant germination and growth, their effectiveness depends on the dosage applied and the type of host plant. For instance, in drought-stressed sorghum plants, foliar spraying with nanowax significantly boosted seed yield compared to water spraying. Similarly, titanium dioxide (TiO_2) NPs have been found to notably influence crop morphology, biochemistry, and physiological traits [32]. Exogenous NP treatment to cucumber plants throughout their growth period increased rubisco activase activity, chlorophyll production, and photosynthetic rate, resulting in increased plant dry mass [47]. Furthermore, foliar application of NPs was shown to increase soybean seed production by enhancing photosynthetic efficiency [91].

The effectiveness of TiO_2 depends on ambient factors, plant type, and application dose. Mohammadi et al. [105] found that TiO_2 increased dragonhead plant

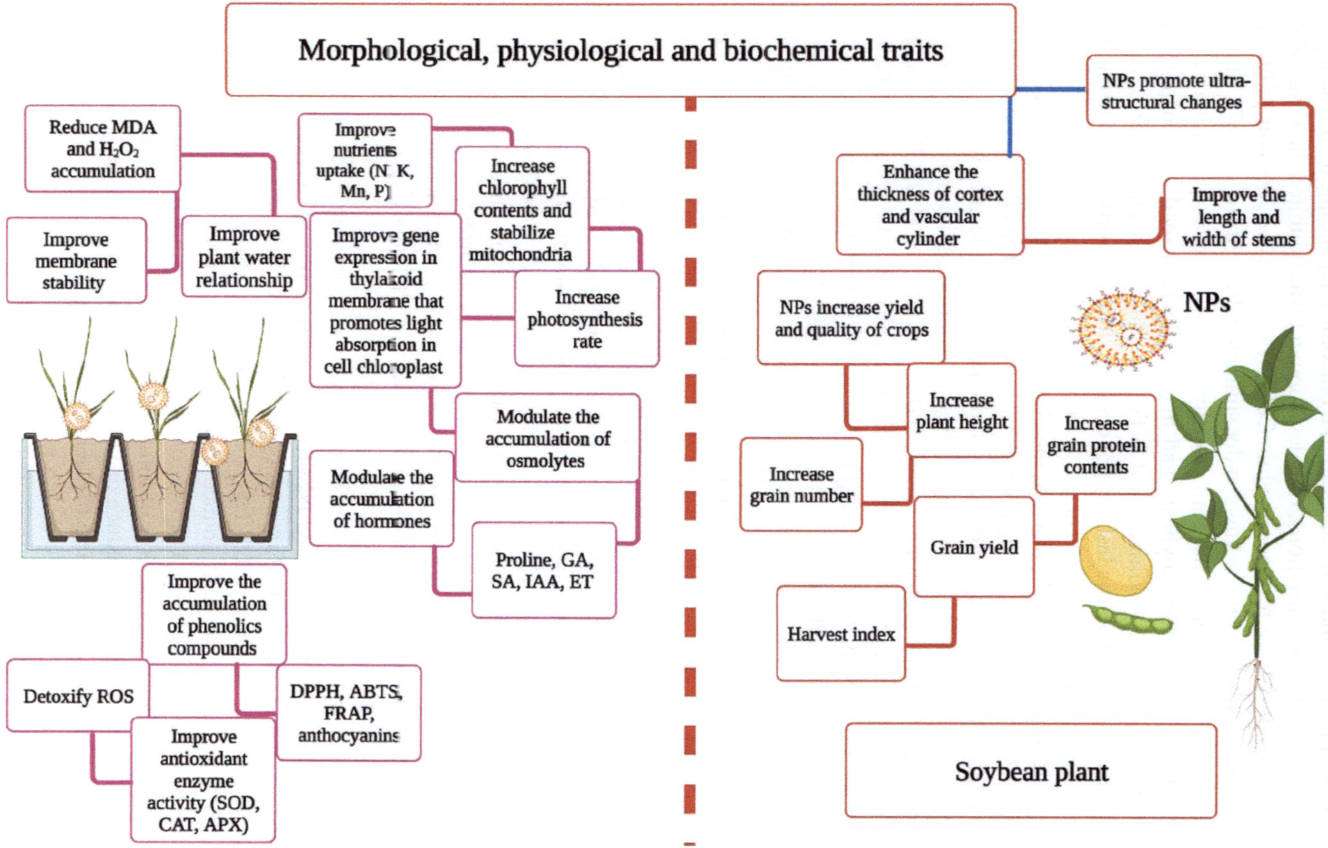

Fig. 2 Role of nanoparticles in drought tolerance in plant crop. They help reduce MDA accumulation, stabilize cellular membranes, stimulate the expression of stress-responsive proteins, enhance nutrient and water uptake, boost photosynthesis, and ultimately improve grain yield and harvest index [120]

growth and essential oil production during water stress. Furthermore, a fertilizer mixture including ZnO and CuO had variable outcomes depending on the nanoparticle dose. At certain concentrations, root development was inhibited, however, ZnO NPs enhanced lateral root formation, while CuO NPs stimulated root hair proliferation and elongation in wheat seedlings under simulated drought stress [167]. Studies have shown that NP therapies have noteworthy results. For example, using carbon nanotubes (CNTs) in barley, soybean, and maize enhanced seed germination and antioxidant content [92]. Additionally, CNTs stimulated root and shoot development in wheat plants. Over the last several decades, major efforts have been undertaken to minimize drought stress's negative impact on plant quality and production. Fullerenol nanoparticles (FNPs) having the chemical formula $C_{60}(OH)_{24}$ have also been proposed to ameliorate drought stress by serving as intracellular water stores. FNPs may enter plant root and leaf tissues and bind water molecules within cellular compartments. FNPs' water-absorbing ability helps offer additional water supply, reducing the consequences of water scarcity [20, 156]. Borišev et al. [20] found that nanofullerenol treatment through foliar application enhanced the water metabolism in drought-stressed plants by increasing proline levels in both roots and leaves [20]. This suggests that FNPs might act as intracellular water binders, forming stores to help in drought adaption. Ag NPs have received the most attention among NPs due to their wide range of applications [2].

NPs can target plant organelles, influencing Superoxide Dismutase (SOD), Catalase (CAT), and Peroxidase (POD) [27, 137]. For instance, TiO_2 NP treatment enhances SOD activity in stressed plants [48]. Elemental oxide NPs improve plant drought resilience in agriculture. Silica (Si) NPs are widely used to alleviate the negative effects of different abiotic stresses, such as drought [138]. Treatments with silica and ZnO NPs enhance crop growth, physiology, and biochemistry [129]. Si NPs reduce drought stress in wheat [122], while ZnO NPs alleviate both salinity and drought stress [82]. However, excessive nanoparticle application can induce oxidative stress, leading to antioxidant enzyme production, cell cycle arrest, and programmed cell death [72]. TiO_2 NP treatment boosts antioxidant enzyme activity, like POD and CAT, reducing malondialdehyde (MDA) and free radicals (H_2O_2, O_2^-), thus enhancing the plant's antioxidant defense during drought [3].

Plants treated to CuO NPs showed an increase in anthocyanin levels and proline content during drought stress. CuO NPs treated to wheat roots produced more free radicals, showing that the plant responded to NP-induced reactive oxygen species (ROS) bursts. The increased ROS levels during drought stress, possibly mediated by an increase in abscisic acid, induce transcriptional changes that enhance tolerance mechanisms. The enhanced activity of antioxidant enzymes in drought-stressed plants highlights the protective role of foliar-applied nanofullerenols. Depending on the dose, these NPs appear to reduce drought-induced oxidative stress [92]. While FNPs show promise in mitigating the negative impacts of drought by increasing antioxidant responses, further research is needed to determine their specific mode of action and the underlying physiological processes in plants.

The study by Krishnan et al. [85] highlights the growth-promoting effects of CaO NPs at an optimal concentration (50 mg L^{-1}) on *Trigonella foenum-graecum*

L. seedlings, improving growth parameters such as root and stem length, leaf area, and photosynthetic pigment content. The enhanced enzymatic (SOD, CAT) and non-enzymatic antioxidant activities, along with reduced MDA levels, suggest improved plant stress tolerance. These findings emphasize the potential of CaO NPs as eco-friendly growth enhancers in agriculture. However, higher concentrations (100 mg L^{-1}) showed stagnation or limited benefits, underlining the importance of concentration optimization.

3.2 Physiological Response of Nanoparticles on Temperature Stress

NPs enhance plant resistance to abiotic stressors, particularly temperature stress, by improving growth and hydration [1, 31]. At low concentrations, NPs activate antioxidant defense, while higher concentrations can cause oxidative stress. NPs like MWCNTs promote heat shock protein (HSP) production, aiding in heat stress resistance [132]. CeO_2 NPs in maize increase H_2O_2 levels and upregulate HSP70. Additionally, NPs help regulate stomatal opening, reducing heat stress impacts [159].

NPs aid plants in managing cold and heat stress. El-Saadony et al. [37] demonstrated that selenium NPs increased the growth and photosynthesis under heat stress while Kareem et al. [78] found that nano-ZnO NPs enhanced gas exchange and chlorophyll activity in mung beans, resulting in larger pods and higher grain yield.

ZnO and TiO_2 NPs also improved membrane stability and boosted antioxidant defenses in heat-stressed wheat seedlings [149]. ZnO NPs regulated osmotic potential, reduced thylakoid damage, and activated antioxidant defenses, resulting in increased plant yield. Moreover, ZnO NPs were linked to enhanced cold tolerance, photosynthetic activity, reduced oxidative stress, and improved root and shoot growth in rice [145].

NPs like SiO_2, Se, ZnO, and Gold nanorods (GNRs) helped reduce cold stress in sugarcane by increasing chlorophyll content, enhancing photosynthesis, reducing oxidative stress, and improving gas exchange [38]. NPs assist plants maintain ion concentrations, which improves membrane stability and osmotic control. This, in turn, enhances water and nutrient transfer during stress, increasing vegetative yield. NP concentration is critical, since high levels can cause toxicity and harm plant health [116, 160]. Heat stress has a detrimental impact on plants by reducing growth, development, physiological processes, and total output. High temperatures cause an excess of ROS, resulting in oxidative stress [103]. Enzymes, which are required for a variety of metabolic processes, are extremely sensitive to temperature variations. Heat stress can cause enzymes to become uncoupled from their metabolic functions, leading to the generation of ROS [29].

Enhancing plant defense systems against heat stress is a significant role of NPs. It has been discovered that ZnO and TiO_2 NPs enhance plant defense by lowering H_2O_2 levels and boosting the synthesis of SOD and GPX. According to [56], for example,

TiO_2 NP treatment enhanced photosynthetic activity and reduced H_2O_2 levels. Zn and Fe NPs boosted antioxidant enzyme activity under high-temperature stress [153]. Moreover, plants under mild heat stress have been demonstrated to exhibit increased photosystem II (PSII) activity in response to TiO_2 NPs [30].

3.3 Physiological Response of Nanoparticles on Salt Stress

Plants primarily absorb NPs through their roots and leaves. Khan et al. [82] found that once NPs enter plants, they influence morphology, physiology, biochemistry, and molecular processes. The relationship between NPs and plants can be advantageous or disadvantageous, depending on the type of NPs and the plant species involved. The size, reactivity, and chemical composition of NPs significantly influence plant responses [112, 152]. NPs enhance salt tolerance and boost plant resilience in harsh environmental conditions, and enhance the plant growth and development [178].

Salinity stress has a significant impact on photosynthesis, although the effects differ depending on plant species, salt concentration, and environmental circumstances [58]. NP foliar application has been shown to enhance the synthesis of light-harvesting complexes, thereby increasing light absorption and photosynthesis. Studies have found that NPs can also boost chlorophyll content in plants [5]. According to salinity stress study, most NPs promote photosynthesis in salt-stressed plants by increasing the concentration of photosynthetic pigments. Figure 3 shows the role and mechanism of NPs on salinity stress tolerance in plants crop.

Manganese (Mn) is crucial for photosynthesis and oxygen evolution in cellular compartments. Mn NPs enhance photosynthesis, especially under stress [168]. Mn supplementation improves chlorophyll, membrane stability, and nitrate reductase activity in *Vigna radiata* under salt stress [133]. Similarly, copper (Cu) supplementation helps maize plants mitigate salt stress effects on photosynthesis and water relations [71].

NPs significantly influence nutrient absorption, transport, and distribution in plants [83]. Salt stress often disrupts the essential high potassium-to-sodium (K^+/Na^+) ratio for plant tolerance. NPs can boost plant growth by improving osmotic potential. For example, nano-SiO_2 has been shown to enhance the growth of soybean seedling under salt stress by increasing leaf K^+ concentration [39]. Similarly, Cu NPs enhance tomato growth and improve the K^+/Na^+ ratio under salt stress [114]. Additionally, Fe_2O_3 NPs reduce the sensitivity of *Trachyspermum ammi* to salt stress, while Mn NPs support pepper plant growth under similar conditions [168].

Plants generate antioxidant enzymes to prevent ROS damage caused by abiotic stress, such as salt, which has been well established. NPs can assist plants control oxidative stress by increasing antioxidant enzyme levels [50, 108]. Fe, Co and Ce NPs, for example, mimic CAT activity, while Ce, Mn, Cu, and Fe NPs resemble POD activity. [164] found that ROS-responsive nanostructured carbon composites (ROS-NSCC) enhanced ROS scavenging in Arabidopsis treated with cerium. Priming

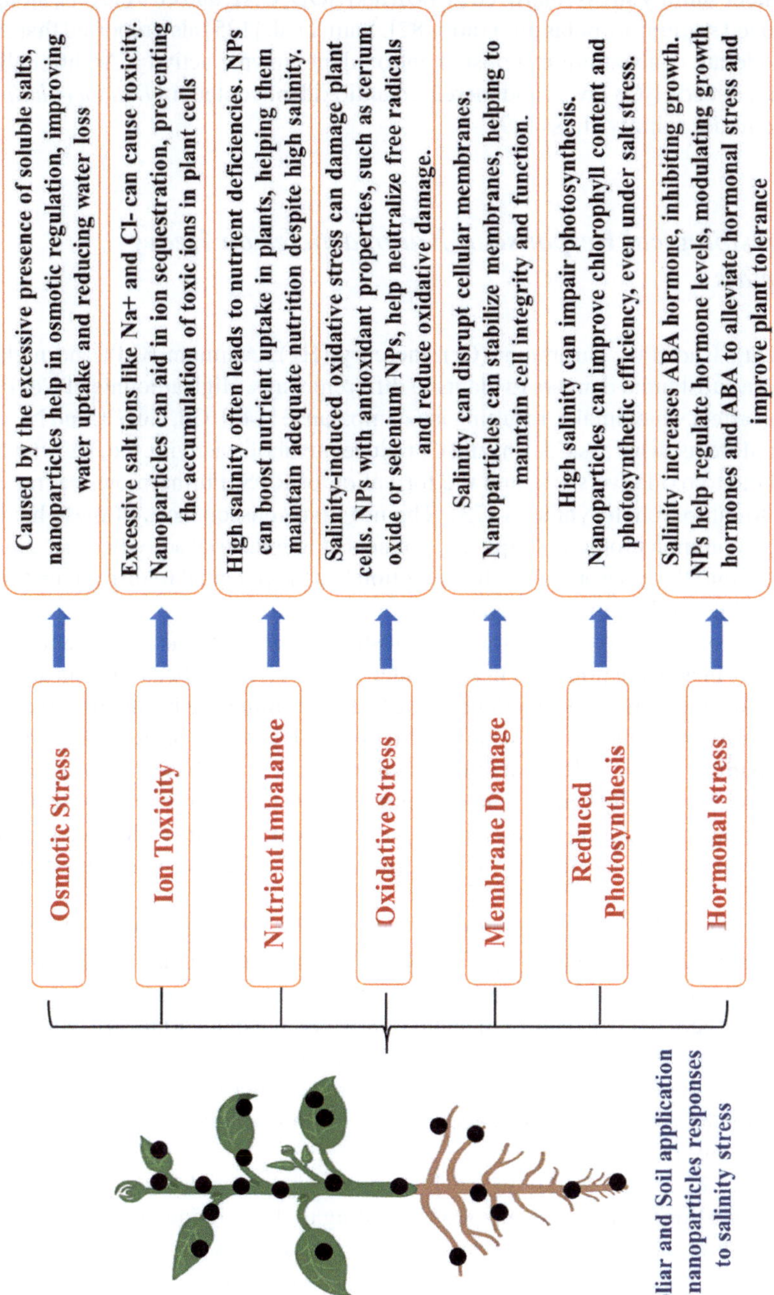

Fig. 3 Role and mechanisms of nanoparticles on salinity stress tolerance in plant crops

Pennisetum glaucum L. (pearl millet) with Ag NPs at different concentrations (10–30 mM) under salinity stress improved growth and SOD, CAT, and GPX levels, alongside a reduced sodium-to-potassium ratio [82]. Sami et al. [125] also reported that Ag NPs at moderate concentrations boosted antioxidant enzyme activity. Additionally, TiO_2 NPs reduced H_2O_2 levels and increased antioxidant content in *Dracocephalum moldavica* under salinity stress [49].

3.4 Physiological Response of Nanoparticles on Heavy Metals

Copper (Cu), lead (Pb), chromium (Cr), mercury (Hg), cadmium (Cd), and nickel (Ni) all pose substantial dangers to plants in different ways. High amounts of Cd, Pb, and Cu can affect water intake, resulting in an imbalance [148]. Cd, Cu, Cr, and Ni are linked to chlorosis, whereas Cd, Zn, and Cr reduce metabolic activity. Cu, Hg, and Ni all contribute to oxidative stress and the formation of ROS. Furthermore, cadmium is known to inhibit photosynthesis [73]. The negative consequences of these heavy metals are a common problem, frequently connected to industrial activities that cause their release into the environment. This infection has the potential to disrupt the food chain, lowering production and food quality.

NPs are gaining traction across various industries, from medicine to cosmetics and apparel. With increasing pollution in soil, water, and air, NPs are becoming an attractive option for environmental remediation due to their minimal environmental impact. In comparison to bioremediation, which is sluggish and reliant on bacteria, and chemical remediation, which is dependent on reaction kinetics, NP-based remediation is efficient, environmentally benign, and does not generate toxic byproducts. Nanotechnology's versatility and long-term competitiveness are driving its widespread use in sectors like agriculture, particularly in the form of nanofertilizers and nanopesticides [40].

NPs enhance chloroplast pigments, photosynthesis, and membrane stability in plants under heavy metal stress [4]. FeO NPs, for example, mitigate Cd toxicity in wheat by improving growth, photosynthetic activity, and biomass [69]. Similarly, nano-TiO_2 reduces Cd concentration and oxidative damage in coriander, enhancing agronomic features [126]. Nano-TiO_2 also boosts photosynthesis and growth in soybean [140], while Si NPs alleviate Cd stress in summer savory [100]. Graphite carbon nitride has been shown to reduce Cd effects in rice, increasing biomass and reducing toxicity [54].

NPs reduce MDA and H_2O_2 levels while enhancing antioxidant enzyme activity [12]. They also limit heavy metal uptake by binding to them, reducing their mobility and bioavailability. Because of their large surface area and tiny size, NPs can interact with molecules and aid in reducing heavy metal stress in plants.

TiO_2 NPs at varying concentrations were used to mitigate Cd toxicity in maize, with higher concentrations showing a reduction in toxicity by increasing SOD and glutathione levels and upregulating metabolic pathways [89]. Similarly, [68] demonstrated that ZnO NPs decreased Cd toxicity in wheat by lowering Cd concentrations and boosting SOD and POD activity. Singh and Lee [140] explored the potential of nano-TiO_2 in reducing cadmium (Cd) absorption in soybeans. [159] reduced Cd toxicity in brassica plants by using Cu NPs, which boosted SOD, CAT, POD activities, and plant biomass.

ZnO NPs in rice reduce arsenic (As) and Cd accumulation, alleviating phytotoxicity [95]. FeO NPs activate antioxidant enzymes, sequester arsenic in vacuoles and cell walls, and promote chelating agent accumulation [18]. In *Vigna radiata*, Fe_2O_3 NPs lower arsenic uptake, boost antioxidant capacity, and enhance SOD and CAT activity while reducing guaiacol peroxidase [131]. CuO NPs also reduce arsenic levels in rice, improving biomass and antioxidant activity [162].

Phytoremediation of lead (Pb) is vital due to its toxicity and challenges in phytoextraction. Fatemi et al. [41] found that silicon nanoparticles (Si NPs) helped mitigate Pb toxicity in coriander by improving antioxidant enzyme activity, despite Pb stress reducing plant biomass and vitamin C levels while increasing flavonoid content. Similarly, [74] showed that nano-hydroxyapatite enhanced ryegrass biomass under Pb stress.

Cr phytotoxicity slows plant growth by affecting photosynthetic pigments and chlorophyll fluorescence [151]. The usage of Si NPs has been found to increase antioxidant activity, decrease Cr buildup, minimize oxidative stress, and boost plant defense systems and nutritional levels. López-Luna et al. [94] studied the effects of citrate-coated magnetite NPs on wheat and found that root length rose dramatically while Cd and Cr accumulation reduced, resulting in a reduction in toxicity.

3.5 Physiological Response of Nanoparticles on Nutrients Imbalance

NPs have great potential to enhance nutrient absorption in plants by improving nutrient usage efficiency through directed distribution and sustained release [36, 144, 147]. They act as nutrient transporters and can modify soil microbes to boost plant nutrient uptake [21, 80]. Additionally, NPs enable precise distribution of nutrients within the plant, enhancing overall nutrient consumption efficiency [43].

NPs serve as effective slow-release agents in soil, significantly enhancing nutrient use efficiency. For instance, a study on indigenous wheat irrigated with a nanohybrid demonstrated the gradual release of urea molecules from Zn and Mg hydroxyapatite-doped nanohybrids over two weeks. This nanocompound, which utilized only half the nitrogen dose, sustained wheat crop productivity while improving nitrogen uptake [136]. Recent studies highlight a core–shell nanocarrier structure consisting of urea-loaded metal–organic frameworks (MOFs) and silica, designed for sustained nitrogen release. In potted rice, treatment with urea/MIL-100(Fe)/silica heterojunction nanomaterials improved nitrogen utilization efficiency by 34.7% compared to conventional urea applications [163].

Urea-coated hydroxyapatite nanoparticles (UHA) release nutrients more slowly than normal urea, resulting in better nutrient delivery during the rice growth cycle. Nano U-NPK, a multinutrient nanofertilizer, can lower nitrogen availability in plants by up to 40% [119]. Potassium-based nanoparticles (K_2SiO_3-NPs and K18Mo8O33-NPs) are about 40% more effective than regular potassium fertilizers in enhancing soybean growth, thanks to their slow-release effect, which prevents the harmful impact of rapid high-dose potassium application [161].

4 Molecular Responses to Nanoparticles

When plants are subjected to abiotic stress, they respond in a complex manner, with the early phase dominated by stress phytohormone stimulation, ROS accumulation, kinase cascades, and ion channel activation. These events influence both functional electrophysiology and oscillatory signaling [15, 44, 87]. Following the first phase, the next stage involves stress- and species-dependent regulation of gene networks. These networks' downstream objectives frequently include defense systems and stress-induced damage repair [16].

Some stress-induced proteins help to neutralize ROS and activate signaling cascades such as mitogen-activated protein (MAP) kinase and salt excessively sensitive (SOS) kinase [44, 127]. Recent studies have also discovered functional results of these regulatory networks, including ion and water transport, uptake, and transcription factor regulation for secondary response components [19, 124]. NPs applied to plants under abiotic stress help mitigate stress by triggering various chemical mechanisms. NPs such as Al_2O_3, ZnO, Si, multi-walled carbon nanotubes (MWCNTs), and TiO_2 have been extensively studied for their beneficial effects on plants. Zhao et al. [173] reported that silver and copper oxide could activate stress responses and defense systems in plants, thereby enhancing their resilience to abiotic stress. Table 2 shows the molecular responses of NPs on drought stress, temperature, salt stress and heavy metal.

4.1 Molecular Responses of Nanoparticles on Drought Stress

Plants adapt to drought stress by reducing water loss through stomatal closure and enhancing root water uptake via morphological and anatomical changes. Furthermore, molecular understanding of these adaptations, particularly regarding the role of NPs in improving drought tolerance, remains limited. Garcia-Sanchez et al. [46] found that drought stress affected 16 out of 351 genes in *Arabidopsis thaliana* treated with TiO_2, Ag NPs, and MWCNTs. While the number of coding genes influenced by

Table 2 Molecular responses of NPs on plant crops

Nanoparticles	Effects	Key reference
Molecular responses of NPs on drought stress		
TiO_2 NPs, Ag NPs, MWCNTs	16 genes responsive to drought stress identified from 351 differentially regulated genes in *Arabidopsis thaliana*; potential involvement of non-coding elements like miRNAs in transcriptional regulation	[46]
Molecular responses of NPs on heat stress		
ZnO NPs	Modified ultrastructure of chloroplasts, mitochondria, and cell wall, preventing heat-induced damage and promoting better plant growth in alfalfa seedlings. Improved heat stress-induced reduction of TGS-GUS genes in *Arabidopsis thaliana*	[79, 165]
Si NPs	Increased overexpression of TaPIP1 and TaNIP2 aquaporin genes in wheat seedlings, improving relative water content under heat stress	[169]
Molecular responses of NPs on salt stress		
Ag NPs	Upregulate genes for AREB, CRK1, and NCED3. Downregulate genes for ABA and TASI4, ZFHD1, Dwarf and Delayed Flowing 2 (DDF2), and MAPK variants. Activate MAPK2	[6, 14, 177]
Si NPs	Upregulate genes for AREB, CRK1, and NCED3. Downregulate genes for ABA and TASI4, ZFHD1, Dwarf and Delayed Flowing 2 (DDF2), and MAPK variants. Repress MAPK2. Inhibit ABA signaling and activate APX2 expression, triggering ROS and H_2O_2 production	[6, 14, 23]
Both Ag and Si NPs	Activate ABA signaling through SnRK2 pathway; regulate water and salt balance in plants via stomatal closure. Activate ROS signaling through RBOH1 phosphorylation. Upregulate P5CS1 for proline biosynthesis and osmoregulation	[6, 53, 76, 90]
Ag NPs	Downregulate ZFHD1, an ABA-independent stress response protein	[150]
Si NPs	Downregulate DDF2, which alleviates osmotic stress via reduced gibberellic acid biosynthesis	[97]
Ag and Si NPs	Activate SOS, MAPK, and ROS signaling pathways. SOS pathway regulates salt tolerance and ion homeostasis, while MAPK and ROS pathways regulate oxidative stress	[6, 101]
Molecular responses of NPs on heavy metal stress		
ZnO NPs	Upregulation of antioxidant enzymes and osmoprotectants, improved heat tolerance mechanisms in mung bean	[78]

(continued)

Table 2 (continued)

Nanoparticles	Effects	Key reference
ZnO NPs	Decreased molecular expression of IRT1, IRT2, YSL2, and YSL15 genes involved in Fe and Cd absorption in rice seedlings. Overexpression of OsVIT1 and OsCAX4 genes resulted in Cd retention in vacuoles	[52]
Si NPs	Upregulation of OsLsi1 (Si uptake) and OsHMA3 (Cd transport) genes, downregulation of OsLCT1 and OsNramp5 (Cd uptake and transport) genes in rice	[26]
ZnO NPs	Genomic modifications, novel DNA bands or absence of typical bands in RAPD pattern under Cd and Pb stress in plants	[155]
Si NPs	Suppression of LCT1 and NRAMP5 expression, stimulation of LSI1 and HMA3 genes for Si absorption and Cd transport into vacuoles in rice	[25]
Sulfur NPs	Decreased expression of antioxidant genes (BnGST, BnPOD, BnAPX) under Hg stress	[170]

drought stress was small, non-coding elements such as microRNAs (miRNAs) may play a significant role, regulating key stress-related genes through transcriptional suppression [45, 146].

4.2 Molecular Responses of Nanoparticles on Temperature Stress

NPs influence genes and transcription factors that regulate heat stress responses [79]. For example, ZnO NP pretreatment in alfalfa seedlings prevents heat-induced damage by altering chloroplast, mitochondria, and cell wall ultrastructure, enhancing plant growth. In Arabidopsis thaliana, ZnO NPs reduced heat stress-induced suppression of TGS-GUS genes [165]. Additionally, [169] demonstrated that silicon (Si) increased aquaporin gene expression (TaPIP1 and TaNIP2) in wheat seedlings, improving their relative water content under heat stress.

4.3 Molecular Responses of Nanoparticles on Salt Stress

Through osmotic and oxidative stress, excessive concentration of salt ions can be hazardous to plants and result in nutritional imbalances [14]. When Ag and Si NPs were treated under osmotic stress in a research on *Solanum lycopersicum L.,* an unusual transcriptome response was observed. Genes including 9-Cis-Epioxycarotinoid Dioxygenase 3 (NCED3), Cystein-Rich Receptor-Like Protein

Kinase 42-Like (CRK1), and ABA Response Element Binding Protein (AREB) were upregulated. In contrast, there was a downregulation of the genes encoding ABA, Zinc Finger Homeodomain Transcription Factor (ZFHD1), Dwarf and Delayed Flowering 2 (DDF2), Environmental Stress-Induced Protein (TASI4), and MAPK variants. These differentially regulated genes play key roles in the ABA signaling pathway, which is critical for maintaining water and salt balance in plants by promoting stomatal closure through SnRK2 activation [6, 177]. Activated SnRK2 phosphorylates AREB transcription factors, in turn, enhance the expression of ABA-responsive genes, creating a positive feedback loop that strengthens the stress response [86]. Additionally, SnRK2 phosphorylates RBOH1, an NADPH oxidase, which activates the ROS signaling pathway, further contributing to the plant's response to stress [76].

In osmotic stress, NP-mediated SnRK activation increases ABA signaling via AREB rather than RBOH. During drought stress, Si NPs can block ABA signaling and enhance APX2 production, which triggers a chloroplast electron transport chain leading to ROS and H_2O_2 production [23]. AREB1, a key transcription factor, regulates plant development and osmotic stress responses by increasing P5CS1 expression, which is involved in proline biosynthesis and osmoregulation [59, 90]. Proline also functions as an antioxidant and metal ion chelator [10]. Notably, NPs can directly upregulate P5CS1 [6].

Muñoz-Mayor et al. [109] demonstrated that NCED3 and TAS14 activate the ABA pathway during osmotic and drought stress, respectively. While ABA-dependent stress response proteins like TAS14 and NCED3 are significantly upregulated, Ag NPs downregulate the gene ZFHD1, which encodes an ABA-independent stress response protein [150]. This suggests that NPs selectively activate ABA-dependent response pathways under osmotic stress. Furthermore, NPs seem to enhance ABA signaling more than other stress-related phytohormones. For instance, NPs often downregulate DDF2, a protein that inhibits gibberellic acid synthesis to reduce osmotic stress [97]. Reduced oxidative stress is a critical result of ABA signaling [55]. Additionally, CRK1, a gene upregulated by NPs, plays a role in detecting ROS and regulating redox balance through extracellular protein domains [70].

During osmotic stress, NPs alter crucial signaling pathways including SOS, MAPK, and ROS. The MAPK and ROS pathways regulate oxidative stress, whereas SOS pathway controls salt tolerance and ion balance [101]. The effect of NPs on various pathways varies with their nature. For instance, Ag NPs activate MAPK2 under salt stress, whereas Si NPs inhibit it [6].

4.4 Molecular Responses of Nanoparticles on Heavy Metal Stress

Recent studies show that NPs enhance molecular and signaling pathways in plants, boosting their resilience to heavy metal stress [176]. Kareem et al. [78] found that ZnO NPs improved heat tolerance in mung beans by upregulating antioxidant enzymes and osmoprotectants. In rice seedlings exposed to ZnVI (100 mg/L), the molecular expression of genes like IRT1, IRT2, YSL2, and YSL15, which regulate iron (Fe) and cadmium (Cd) absorption, was significantly reduced. Additionally, overexpression of the OsVIT1 and OsCAX4 genes enhanced Cd retention in vacuoles [52].

Si NPs have been demonstrated to increase the activity of genes associated with Cd transport (OsHMA3) and silicon uptake (OsLsi1) in rice plants, while suppressing the activity of genes associated with Cd absorption and transport, such as OsLCT1 and OsNramp5 [26]. When subjected to Cd and Pb stress, ZnO NPs induced considerable genomic changes in plants, including the creation of novel DNA bands and the lack of conventional RAPD bands [155].

Si NPs also reduce the expression of genes involved in Cd movement from the root to the shoot, like LCT1 and NRAMP5, while enhancing the expression of genes such as LSI1 and HMA3, which aid in Cd sequestration within vacuoles [25]. Additionally, sulfur NPs were found to lower the expression of antioxidant genes like BnGST, BnPOD, and BnAPX, which are generally increased during Hg stress [170].

5 Challenges and Future Directions

The rapid growth of nanotechnology has led to an increase in the global production of NPs for various applications, including in agriculture. This expansion has resulted in the unintended release of NPs into the environment, with a significant portion leaching into the soil. In the soil, NPs interact with plant roots and may be incorporated into the entire plant system. Depending on factors such as NP concentration, properties (e.g., size, shape, zeta potential, and stability), and the load applied, these NPs can persist in soils for extended periods, potentially causing long-term effects.

Current research indicates that NPs can influence seed germination, root elongation, plant redox potential, and overall growth and development. However, the outcomes—whether positive, negative, or neutral—are influenced by several factors, including plant species, environmental conditions, and the specific characteristics of the NPs. To maximize the benefits of NPs while minimizing potential risks, it is essential to carefully optimize their concentration, application methods, and timing. Moreover, plants in agricultural fields are often exposed to multiple stressors, such as salinity and drought, which complicates the study and application of NPs. The limited understanding of the molecular mechanisms, metabolic roles, and transporter proteins related to various NPs further restricts their effective use.

To address these challenges, several avenues for future research are suggested as follows.

Mechanistic Understanding: Explore the role of NPs in modulating plant physiology under salinity stress, focusing on their impact on host defense mechanisms at the genomic, metabolomics, and proteomic levels.

Comparative Studies: Conduct comparative research on interactions between different NPs and plants to uncover the mechanisms governing salinity stress responses.

Metabolic Roles: Explore the molecular and metabolic roles of NPs, especially Mn, Zn, Si, and Fe, in nutrient-mediated stress tolerance.

Optimal Application: Determine the optimal concentrations, timing, and methods for NP application, particularly under salinity and other concurrent stress conditions.

Simultaneous Stress Studies: Evaluate the effects of NPs on plants facing combined stresses, such as salinity and drought, to mimic real-world agricultural scenarios.

NP Transporters: Identify transporters or channel proteins responsible for NP uptake in plants, beyond the limited research on Si transporters.

Subcellular Distribution: Investigate the distribution of NPs within cellular organelles, including the nucleus, to elucidate their biological roles in stress tolerance.

Proteomics and Genetic Engineering: Study the impacts of NPs on plant proteomics and investigate genetic engineering approaches for enhanced stress tolerance.

Microorganism Interactions: Examine the combined effects of NPs and salinity-tolerant microorganisms on plants, including the underlying mechanisms of their interactions.

Environmental impacts: Assess the translocation, accumulation, and secondary effects of NPs on soil, air, water, and ecosystems to ensure sustainable use.

6 Conclusion

NPs hold considerable promise for enhancing plant health and resilience under salinity stress, offering a sustainable solution for low-input agriculture. However, a thorough understanding of their mechanisms, environmental impacts, and long-term feasibility is essential. Large-scale, well-designed, and long-term field trials are crucial to validating laboratory findings and assessing the economic viability of various NP sources. By addressing these challenges and pursuing the recommended research avenues, we can unlock the full potential of NPs to revolutionize agriculture while ensuring environmental sustainability.

References

1. Aghdam MTB, Mohammadi H, Ghorbanpour M (2016) Effects of nanoparticulate anatase titanium dioxide on physiological and biochemical performance of *Linum usitatissimum* (Linaceae) under well-watered and drought stress conditions. Braz J Bot 39:139–146

2. Ahmad I, Akhtar MS (2019) Use of nanoparticles in alleviating salt stress. Salt Stress Microbes Plant Interact Causes Solut 1:199–215
3. Ahmad J, Bashir H, Bagheri R, Baig A, Al-Huqail A, Ibrahim MM, Qureshi MI (2017) Drought and salinity induced changes in ecophysiology and proteomic profile of Parthenium hysterophorus. PLoS ONE 12(9):e0185118
4. Ahmed T, Noman M, Ijaz M, Ali S, Rizwan M, Ijaz U, Hameed A, Ahmad U, Wang Y, Sun G, Li B (2021) Current trends and future prospective in nanoremediation of heavy metals contaminated soils: a way forward towards sustainable agriculture. Ecotoxicol Environ Saf 227:112888
5. Ali S, Mehmood A, Khan N (2021) Uptake, translocation, and consequences of nanomaterials on plant growth and stress adaptation. J Nanomater 2021(1):6677616
6. Almutairi ZM (2016) Influence of silver nano-particles on the salt resistance of tomato (*Solanum lycopersicum*) during germination. Int J Agric Biol 18(2):449–457
7. Amingad V, Sreenivas KN, Fakrudin B, Seetharamu GK, Shankarappa TH, Venugopalan R (2017) Comparison of silver nanoparticles and other metal nanoparticles on postharvest attributes and bacterial load in cut roses var. Taj Mahal. Int J Pure Appl Biosci 5(6):579–584
8. Anandaraj K, Natarajan N (2017) Effect of nanoparticles for seed quality enhancement in onion [Allium cepa (Linn) cv. CO (On)] 5. Int J Curr Microbiol Appl Sci 6(11):3714
9. Asadpour S, Madani H, Mohammadi GN, Heravan IM, Abad HHS (2016) Control of high temperature stress in maize crop by nano-silicon and nano-zinc foliar application
10. Ashraf MFMR, Foolad MR (2007) Roles of glycine betaine and proline in improving plant abiotic stress resistance. Environ Exp Bot 59(2):206–216
11. Askary M, Talebi SM, Amini F, Bangan ADB (2017) Effects of iron nanoparticles on *Mentha piperita* L. under salinity stress. Biologija 63(1)
12. Azeez NA, Dash SS, Gummadi SN, Deepa VS (2021) Nano-remediation of toxic heavy metal contamination: hexavalent chromium [Cr (VI)]. Chemosphere 266:129204
13. Ballikaya P, Brunner I, Cocozza C, Grolimund D, Kaegi R, Murazzi ME, Schaub M, Schönbeck LC, Sinnet B, Cherubini (2023) First evidence of nanoparticle uptake through leaves and roots in beech (*Fagus sylvatica* L.) and pine (*Pinus sylvestris* L.). Tree Physiol 43(2), 262–276
14. Bano A, Fatima M (2009) Salt tolerance in Zea mays (L). following inoculation with Rhizobium and Pseudomonas. Biol Fertil Soils 45:405–413
15. Bari R, Jones JD (2009) Role of plant hormones in plant defence responses. Plant Mol Biol 69:473–488
16. Ben Rejeb I, Pastor V, Mauch-Mani B (2014) Plant responses to simultaneous biotic and abiotic stress: molecular mechanisms. Plants 3(4):458–475
17. Bhavani P, Prakash SS, Harinikumar KM, Thimmegowda MN, Benherlal PS, Yoganand SB (2020) Performance of slow release hydroxyapatite coated urea nanofertilizer on aerobic paddy. Int J Curr Microbiol App Sci. 9(11):1320–1330
18. Bidi H, Fallah H, Niknejad Y, Tari DB (2021) Iron oxide nanoparticles alleviate arsenic phytotoxicity in rice by improving iron uptake, oxidative stress tolerance and diminishing arsenic accumulation. Plant Physiol Biochem 163:348–357
19. Blumwald E (2000) Sodium transport and salt tolerance in plants. Curr Opin Cell Biol 12(4):431–434
20. Borišev M, Borišev I, Župunski M, Arsenov D, Pajević S, Ćurčić Ž, Vasin J, Djordjevic A (2016) Drought impact is alleviated in sugar beets (*Beta vulgaris* L.) by foliar application of fullerenol nanoparticles. PLoS One 11(11):e0166248
21. Bortoletto-Santos R, Plotegher F, Majaron VF, da Silva MG, Polito WL, Ribeiro C (2020) Polyurethane nanocomposites can increase the release control in granulated fertilizers by controlling nutrient diffusion. Appl Clay Sci 199:105874
22. Byczyńska A (2017) Improvement of postharvest quality of cut tulip 'White Parrot 'by nano silver. World Scientific News 83:224–228
23. Chang CCC, Ball L, Fryer MJ, Baker NR, Karpinski S, Mullineaux M (2004) Induction of ASCORBATE PEROXIDASE 2 expression in wounded Arabidopsis leaves does not involve

known wound-signalling pathways but is associated with changes in photosynthesis. Plant J 38(3):499–511
24. Clark M, Tilman D (2017) Comparative analysis of environmental impacts of agricultural production systems, agricultural input efficiency, and food choice. Environ Res Lett 12(6):064016
25. Cong W, Miao Y, Xu L, Zhang Y, Yuan C, Wang J, Zhuang T, Lin X, Jiang L, Wang N, Ma J (2019) Transgenerational memory of gene expression changes induced by heavy metal stress in rice (Oryza sativa L.). BMC Plant Biol 19:1–14
26. Cui J, Liu T, Li F, Yi J, Liu C, Yu H (2017) Silica nanoparticles alleviate cadmium toxicity in rice cells: mechanisms and size effects. Environ Pollut 228:363–369
27. Cunningham FJ, Goh NS, Demirer GS, Matos JL, Landry MP (2018) Nanoparticle-mediated delivery towards advancing plant genetic engineering. Trends Biotechnol 36(9):882–897
28. Darlington TK, Neigh AM, Spencer MT, Guyen OT, Oldenburg SJ (2009) Nanoparticle characteristics affecting environmental fate and transport through soil. Environ Toxicol Chem Int J 28(6):1191–1199
29. Debnath S, Chakraborty M, Kumawat R (2022) Physiological responses of plants under high temperature stress. Biot Res Today 4:347–349
30. Dias MC, Santos C, Pinto G, Silva AM, Silva S (2019) Titanium dioxide nanoparticles impaired both photochemical and non-photochemical phases of photosynthesis in wheat. Protoplasma 256:69–78
31. Dimkpa CO, Bindraban S (2017) Nanofertilizers: new products for the industry? J Agric Food Chem 66(26):6462–6473
32. Djanaguiraman M, Belliraj N, Bossmann SH, Prasad V (2018) High-temperature stress alleviation by selenium nanoparticle treatment in grain sorghum. ACS Omega 3(3):2479–2491
33. Doolette CL, McLaughlin MJ, Kirby JK, Navarro DA (2015) Bioavailability of silver and silver sulfide nanoparticles to lettuce (*Lactuca sativa*): effect of agricultural amendments on plant uptake. J Hazard Mater 300:788–795
34. Du W, Tan W, Peralta-Videa JR, Gardea-Torresdey JL, Ji R, Yin Y, Guo H (2017) Interaction of metal oxide nanoparticles with higher terrestrial plants: physiological and biochemical aspects. Plant Physiol Biochem 110:210–225
35. Durán P, Acuña JJ, Jorquera MA, Azcón R, Borie F, Cornejo, Mora ML (2013) Enhanced selenium content in wheat grain by co-inoculation of selenobacteria and arbuscular mycorrhizal fungi: a preliminary study as a potential Se biofortification strategy. J CeR Sci 57(3):275–280
36. Dutta S, Pal S, Panwar P, Sharma RK, Bhutia L (2022) Biopolymeric nanocarriers for nutrient delivery and crop biofortification. ACS Omega 7(30):25909–25920
37. El-Saadony MT, Saad AM, Najjar AA, Alzahrani SO, Alkhatib FM, Shafi ME, Selem E, Desoky ES, Fouda SE, El-Tahan AM, Hassan MA (2021) The use of biological selenium nanoparticles to suppress *Triticum aestivum* L. crown and root rot diseases induced by Fusarium species and improve yield under drought and heat stress. Saudi J Biol Sci 28(8):4461–4471
38. Elsheery NI, Helaly MN, El-Hoseiny HM, Alam-Eldein SM (2020) Zinc oxide and silicone nanoparticles to improve the resistance mechanism and annual productivity of salt-stressed mango trees. Agronomy 10(4):558
39. Farhangi-Abriz S, Torabian S (2018) Nano-silicon alters antioxidant activities of soybean seedlings under salt toxicity. Protoplasma 255:953–962
40. Farooqi ZH, Akram MW, Begum R, Wu W, Irfan A (2021) Inorganic nanoparticles for reduction of hexavalent chromium: physicochemical aspects. J Hazard Mater 402:123535
41. Fatemi H, Esmaiel Pour B, Rizwan M (2021) Foliar application of silicon nanoparticles affected the growth, vitamin C, flavonoid, and antioxidant enzyme activities of coriander (*Coriandrum sativum* L.) plants grown in lead (Pb)-spiked soil. Environ Sci Pollut Res 28:1417–1425
42. Feng Y, Cui X, He S, Dong G, Chen M, Wang J, Lin X (2013) The role of metal nanoparticles in influencing arbuscular mycorrhizal fungi effects on plant growth. Environ Sci Technol 47(16):9496–9504

43. Fincheira P, Tortella G, Seabra AB, Quiroz A, Diez MC, Rubilar O (2021) Nanotechnology advances for sustainable agriculture: current knowledge and prospects in plant growth modulation and nutrition. Planta 254:66. https://doi.org/10.1007/s00425-021-03714-0
44. Fraire-Velázquez S, Rodríguez-Guerra R, Sánchez-Calderón L (2011) Abiotic and biotic stress response crosstalk in plants. Abiotic Stress Response Plants—Physiological Biochem Genet Perspect 3–26
45. Frazier TP, Burklew CE, Zhang B (2014) Titanium dioxide nanoparticles affect the growth and microRNA expression of tobacco (*Nicotiana tabacum*). Funct Integr Genomics 14:75–83
46. García-Sánchez S, Bernales I, Cristobal S (2015) Early response to nanoparticles in the Arabidopsis transcriptome compromises plant defence and root-hair development through salicylic acid signalling. BMC Genomics 16:1–17
47. Ghani MI, Saleem S, Rather SA, Rehmani MS, Alamri S, Rajput VD, Kalaji HM, Saleem N, Sial TA, Liu M (2022) Foliar application of zinc oxide nanoparticles: an effective strategy to mitigate drought stress in cucumber seedling by modulating antioxidant defense system and osmolytes accumulation. Chemosphere 289:133202
48. Ghasemlou F, Amiri H, Karamian R, Mirzaie-asl A (2019) Alleviation of the effects of on drought stress *Verbascum nudicuale* by methyl jasmonate and titanium dioxide nanoparticles. Iran J Plant Physiol 9(4):2911–2920
49. Gohari G, Mohammadi A, Akbari A, Panahirad S, Dadpour MR, Fotopoulos V, Kimura S (2020) Titanium dioxide nanoparticles (TiO_2 NPs) promote growth and ameliorate salinity stress effects on essential oil profile and biochemical attributes of *Dracocephalum moldavica*. Sci Rep 10(1):912
50. González-García Y, Cárdenas-Álvarez C, Cadenas-Pliego G, Benavides-Mendoza A, Cabrera-De-La-Fuente M, Sandoval-Rangel A, Valdés-Reyna J, Juárez-Maldonado A (2021) Effect of three nanoparticles (Se, Si and Cu) on the bioactive compounds of bell pepper fruits under saline stress. Plants 10(2):217
51. Gowayed SM, Kadasa NM (2016) Effect of zinc oxide nanoparticles on antioxidative system of faba bean (Vicia faba L.) seedling expose to cadmium. Life Sci. J 13:18–27
52. Guha T, Barman S, Mukherjee A, Kundu R (2020) Nano-scale zero valent iron modulates Fe/Cd transporters and immobilizes soil Cd for production of Cd free rice. Chemosphere 260:127533
53. Guiltinan MJ, Marcotte WR Jr, Quatrano RS (1990) A plant leucine zipper protein that recognizes an abscisic acid response element. Science 250(4978):267–271
54. Hao Y, Lv R, Ma C, Adeel M, Zhao Z, Rao Y, Rui Y (2021) Graphitic carbon nitride (gC 3 N 4) alleviates cadmium-induced phytotoxicity to rice (*Oryza sativa* L.). Environ Sci Pollut Res 28:21276–21284
55. Hao Y, Yu F, Lv R, Ma C, Zhang Z, Rui Y, Liu L, Cao W, Xing B (2016) Carbon nanotubes filled with different ferromagnetic alloys affect the growth and development of rice seedlings by changing the C: N ratio and plant hormones concentrations. PLoS ONE 11(6):e0157264
56. Hasanpour H, Maali-Amir R, Zeinali H (2015) Effect of TiO_2 nanoparticles on metabolic limitations to photosynthesis under cold in chickpea. Russ J Plant Physiol 62:779–787
57. Hernández HH, Benavides-Mendoza A, Ortega-Ortiz H, Hernández-Fuentes AD, Juárez-Maldonado A (2017) Cu Nanoparticles in chitosan-PVA hydrogels as promoters of growth, productivity and fruit quality in tomato. Emir J Food Agric 29(8):573–580
58. Hnilickova H, Kraus K, Vachova P, Hnilicka F (2021) Salinity stress affects photosynthesis, malondialdehyde formation, and proline content in *Portulaca oleracea* L. Plants 10(5):845
59. Hong Z, Lakkineni K, Zhang Z, Verma DPS (2000) Removal of feedback inhibition of Δ1-pyrroline-5-carboxylate synthetase results in increased proline accumulation and protection of plants from osmotic stress. Plant Physiol 122(4):1129–1136
60. Husen A (2021a) The harsh environment and resilient plants: an overview. In: Husen A (ed) Harsh environment and plant resilience. Springer, Cham; Gewerbestrasse 11:6330 1–23. https://doi.org/10.1007/978-3-030-65912-7_1
61. Husen A (2021b) Morpho-anatomical, physiological, biochemical and molecular responses of plants to air pollution. In: Husen A (ed) Harsh environment and plant resilience. Springer, Cham. Gewerbestrasse 11:6330 203–234. https://doi.org/10.1007/978-3-030-65912-7_9

62. Husen A (2021c) Cross talk between autophagy and hormones for abiotic stress tolerance in plants. In: Husen A (ed) Plant performance under environmental stress. Springer, Cham. Gewerbestrasse 11:6330 1–15. https://doi.org/10.1007/978-3-030-78521-5_1
63. Husen A (2022) Environmental pollution and medicinal plants. CRC Press, Boca Raton. https://doi.org/10.1201/9781003178866
64. Husen A (2023) Plants and their interaction to environmental pollution (damage detection, adaptation, tolerance, physiological and molecular responses). Elsevier Inc., Cambridge. https://doi.org/10.1016/C2020-0-03319-0
65. Husen A (2025a) Agricultural crop improvement: plant and soil relationships. Taylor & Francis Group, NW. https://doi.org/10.1201/9781032630366
66. Husen A (2025b) Roles of osmolytes in changing environment. Elsevier Inc., Cambridge
67. Husen A (2025c) Role of antioxidants in mitigating plant stress. Elsevier Inc., Cambridge
68. Hussain A, Ali S, Rizwan M, ur Rehman MZ, Javed MR, Imran M, Chatha SAS, Nazir R (2018) Zinc oxide nanoparticles alter the wheat physiological response and reduce the cadmium uptake by plants. Environ Pollut 242:1518–1526
69. Hussain A, Ali S, Rizwan M, ur Rehman MZ, Qayyum MF, Wang H, Rinklebe J (2019). Responses of wheat (*Triticum aestivum*) plants grown in a Cd contaminated soil to the application of iron oxide nanoparticles. Ecotoxicol Environ Saf 173:156–164
70. Idänheimo N, Gauthier A, Salojärvi J, Siligato R, Brosché M, Kollist H, Mähönen AP, Kangasjärvi J, Wrzaczek M (2014) The Arabidopsis thaliana cysteine-rich receptor-like kinases CRK6 and CRK7 protect against apoplastic oxidative stress. Biochem Biophys Res Commun 445(2):457–462
71. Iqbal MN, Rasheed R, Ashraf MY, Ashraf MA, Hussain I (2018) Exogenously applied zinc and copper mitigate salinity effect in maize (*Zea mays* L.) by improving key physiological and biochemical attributes. Environ Sci Pollut Res 25:23883–23896
72. Jalil SU, Ansari MI (2019) Nanoparticles and abiotic stress tolerance in plants: synthesis, action, and signaling mechanisms. Plant Signal Mol 549–561
73. Jin C, Fan J, Liu R, Sun R (2015) Single and joint toxicity of sulfamonomethoxine and cadmium on three agricultural crops. Soil Sediment Contam Int J 24(4):454–470
74. Jin Y, Liu W, Li XL, Shen SG, Liang SX, Liu C, Shan L (2016) Nano-hydroxyapatite immobilized lead and enhanced plant growth of ryegrass in a contaminated soil. Ecol Eng 95:25–29
75. Juárez-Maldonado A, Ortega-Ortíz H, Pérez-Labrada F, Cadenas-Pliego G, Benavides-Mendoza A (2016) Cu nanoparticles absorbed on chitosan hydrogels positively alter morphological, production, and quality characteristics of tomato. J Appl Bot Food Qual 89
76. Kadota Y, Sklenar J, Derbyshire P, Stransfeld L, Asai S, Ntoukakis V, Jones JD, Shirasu K, Menke F, Jones A, Zipfel C (2014) Direct regulation of the NADPH oxidase RBOHD by the PRR-associated kinase BIK1 during plant immunity. Mol Cell 54(1):43–55
77. Kapoor D, Singh MP (2021) Nanoparticles: sources and toxicity. In: Plant responses to nanomaterials: recent interventions, and physiological and biochemical responses, pp 217–232
78. Kareem HA, Saleem MF, Saleem S, Rather SA, Wani SH, Siddiqui MH, Alamri S, Kumar R, Gaikwad NB, Guo Z, Niu J (2022) Zinc oxide nanoparticles interplay with physiological and biochemical attributes in terminal heat stress alleviation in mungbean (*Vigna radiata* L.). Front Plant Sci 13:842349
79. Khalid MF, Iqbal Khan R, Jawaid MZ, Shafqat W, Hussain S, Ahmed T, Rizwan M, Ercisli S, Pop OL, Alina Marc R (2022) Nanoparticles: the plant saviour under abiotic stresses. Nanomaterials 12(21):3915
80. Khan ST, Adil SF, Shaik MR, Alkhathlan HZ, Khan M, Khan M (2021) Engineered nanomaterials in soil: their impact on soil microbiome and plant health. Plants 11(1):109
81. Khan I, Raza MA, Awan SA, Shah GA, Rizwan M, Ali B, Tariq R, Hassan MJ, Alyemeni MN, Brestic M, Zhang X (2020) Amelioration of salt induced toxicity in pearl millet by seed priming with silver nanoparticles (AgNPs): The oxidative damage, antioxidant enzymes and ions uptake are major determinants of salt tolerant capacity. Plant Physiol Biochem 156:221–232

82. Khan ZS, Rizwan M, Hafeez M, Ali S, Adrees M, Qayyum MF, Khalid S, ur Rehman MZ, Sarwar MA (2020) Effects of silicon nanoparticles on growth and physiology of wheat in cadmium contaminated soil under different soil moisture levels. Environ Sci Pollut Res27:4958–4968
83. Kopittke M, Lombi E, Wang P, Schjoerring JK, Husted S (2019) Nanomaterials as fertilizers for improving plant mineral nutrition and environmental outcomes. Environ Sci Nano 6(12):3513–3524
84. Kosová K, Vítámvás P, Urban MO, Prášil IT, Renaut J (2018) Plant abiotic stress proteomics: the major factors determining alterations in cellular proteome. Front Plant Sci 9:122
85. Krishnan H, SRINIVASAN MK, Pratima BJ, Premnath BJ, Venkatesan KB, Namasivayam N, Panneerselvam C, Alshehri MA, Sayed S, Kamaraj SK (2023) Synthesis and characterization of calcium oxide nanoparticles and evaluating their diverse effects on plant growth and development of *Trigonella foenum graecum* L. Pak J Bot 57:2
86. Kulik A, Wawer I, Krzywińska E, Bucholc M, Dobrowolska G (2011) SnRK2 protein kinases—key regulators of plant response to abiotic stresses. Omics J Integr Biol 15(12):859–872
87. Laloi C, Apel K, Danon A (2004) Reactive oxygen signalling: the latest news. Curr Opin Plant Biol 7(3):323–328
88. Larue C, Castillo-Michel H, Sobanska S, Cécillon L, Bureau S, Barthès V, Ouerdane L, Carrière M, Sarret G (2014) Foliar exposure of the crop Lactuca sativa to silver nanoparticles: evidence for internalization and changes in Ag speciation. J Hazard Mater 264:98–106
89. Lian J, Zhao L, Wu J, Xiong H, Bao Y, Zeb A, Tang J, Liu W (2020) Foliar spray of TiO_2 nanoparticles prevails over root application in reducing Cd accumulation and mitigating Cd-induced phytotoxicity in maize (*Zea mays* L.). Chemosphere 239:124794
90. Liao Y, Zou HF, Wei W, Hao YJ, Tian AG, Huang J, Liu YF, Zhang JS, Chen SY (2008) Soybean GmbZIP44, GmbZIP62 and GmbZIP78 genes function as negative regulator of ABA signaling and confer salt and freezing tolerance in transgenic Arabidopsis. Planta 228:225–240
91. Linh TM, Mai NC, Hoe T, Lien LQ, Ban NK, Hien LTT, Chau NH, Van NT (2020) Metal-based nanoparticles enhance drought tolerance in soybean. J Nanomater 2020(1):4056563
92. Liu FY, Xiong FX, Fan YK, Li J, Wang HZ, Xing GM, Yan FM, Tai FJ, He R (2016) Facile and scalable fabrication engineering of fullerenol nanoparticles by improved alkaline-oxidation approach and its antioxidant potential in maize. J Nanopart Res 18:1–13
93. Lowry GV, Avellan A, Gilbertson LM (2019) Opportunities and challenges for nanotechnology in the agri-tech revolution. Nat Nanotechnol 14(6):517–522
94. López-Luna J, Silva-Silva MJ, Martinez-Vargas S, Mijangos-Ricardez OF, González-Chávez MC, Solís-Domínguez FA, Cuevas-Díaz MC (2016) Magnetite nanoparticle (NP) uptake by wheat plants and its effect on cadmium and chromium toxicological behavior. Sci Total Environ 565:941–950
95. Ma X, Sharifan H, Dou F, Sun W (2020) Simultaneous reduction of arsenic (As) and cadmium (Cd) accumulation in rice by zinc oxide nanoparticles. Chem Eng J 384:123802
96. Ma C, White JC, Dhankher OP, Xing B (2015) Metal-based nanotoxicity and detoxification pathways in higher plants. Environ Sci Technol 49(12):7109–7122
97. Magome H, Yamaguchi S, Hanada A, Kamiya Y, Oda K (2008) The DDF1 transcriptional activator upregulates expression of a gibberellin-deactivating gene, GA2ox7, under high-salinity stress in Arabidopsis. Plant J 56(4):613–626
98. Mankad M, Fougat RS, Patel A, Mankad P, Patil G, Subhash N (2017) Assessment of physiological and biochemical changes in rice seedlings exposed to bulk and nano iron particles. Int J Pure Appl Biosci5:150–159
99. Mckenzie BM, Mullins CE, Tisdall JM, Bengough AG (2013) Root–soil friction: quantification provides evidence for measurable benefits for manipulation of root-tip traits. Plant Cell Environ 36(6):1085–1092
100. Memari-Tabrizi EF, Yousefpour-Dokhanieh A, Babashpour-Asl M (2021) Foliar-applied silicon nanoparticles mitigate cadmium stress through physio-chemical changes to improve growth, antioxidant capacity, and essential oil profile of summer savory (*Satureja hortensis* L.). Plant Physiol Biochem 165:71–79

101. Miller G, Shulaev V, Mittler R (2008) Reactive oxygen signaling and abiotic stress. Physiol Plant 133(3):481–489
102. Mittal D, Kaur G, Singh P, Yadav K, Ali SA (2020) Nanoparticle-based sustainable agriculture and food science: recent advances and future outlook. Front Nanotechnol 2:579954
103. Mittler R, Zandalinas SI, Fichman Y, Van Breusegem F (2022) Reactive oxygen species signalling in plant stress responses. Nat Rev Mol Cell Biol 23(10):663–679
104. Mohammadi R, Maali-Amiri R, Mantri NL (2014) Effect of TiO_2 nanoparticles on oxidative damage and antioxidant defense systems in chickpea seedlings during cold stress. Russ J Plant Physiol 61:768–775
105. Mohammadi H, Esmailpour M, Gheranpaye A (2016) Effects of TiO_2 nanoparticles and water-deficit stress on morpho-physiological characteristics of dragonhead (*Dracocephalum moldavica* L.) plants. Acta Agriculturae Slovenica 107(2):385–396
106. Monica RC, Cremonini R (2009) Nanoparticles and higher plants. Caryologia 62(2):161–165
107. Mozafari AA, Havas F, Ghaderi N (2018) Application of iron nanoparticles and salicylic acid in in vitro culture of strawberries (*Fragaria× ananassa* Duch.) to cope with drought stress. Plant Cell Tissue Organ Culture (PCTOC) 132:511–523
108. Mushtaq A, Khan Z, Khan S, Rizwan S, Jabeen U, Bashir F, Ismail T, Anjum S, Masood A (2020) Effect of silicon on antioxidant enzymes of wheat (*Triticum aestivum* L.) grown under salt stress. SILICON 12:2783–2788
109. Muñoz-Mayor A, Pineda B, Garcia-Abellán JO, Antón T, Garcia-Sogo B, Sanchez-Bel P, Flores FB, Atarés A, Angosto T, Pintor-Toro JA, Moreno V (2012) Overexpression of dehydrin tas14 gene improves the osmotic stress imposed by drought and salinity in tomato. J Plant Physiol 169(5):459–468
110. Nair MG, Chung IM (2015) Study on the correlation between copper oxide nanoparticles induced growth suppression and enhanced lignification in Indian mustard (*Brassica juncea* L.). Ecotoxicol Environ Saf 113:302–313
111. Nazari M (2021) Plant mucilage components and their functions in the rhizosphere. Rhizosphere 18:100344
112. Paramo LA, Feregrino-Pérez AA, Guevara R, Mendoza S, Esquivel K (2020) Nanoparticles in agroindustry: applications, toxicity, challenges, and trends. Nanomaterials 10(9):1654
113. Park DY, Naing AH, Ai TN, Han JS, Kang IK, Kim CK (2017) Synergistic effect of nano-sliver with sucrose on extending vase life of the carnation CV Edun. Front Plant Sci 14(8):1601
114. Pérez-Labrada F, López-Vargas ER, Ortega-Ortiz H, Cadenas-Pliego G, Benavides-Mendoza A, Juárez-Maldonado A (2019) Responses of tomato plants under saline stress to foliar application of copper nanoparticles. Plants 8(6):151
115. Rajput VD, Minkina T, Suskova S, Mandzhieva S, Tsitsuashvili V, Chapligin V, Fedorenko A (2018) Effects of copper nanoparticles (CuO NPs) on crop plants: a mini review. BioNanoScience 8:36–42
116. Rajput VD, Minkina T, Kumari A, Harish, Singh VK, Verma KK, Mandzhieva S, Sushkova S, Srivastava S, Keswani C (2021) Coping with the challenges of abiotic stress in plants: new dimensions in the field application of nanoparticles. Plants 10(6):1221
117. Raliya R, Franke C, Chavalmane S, Nair R, Reed N, Biswas (2016) Quantitative understanding of nanoparticle uptake in watermelon plants. Front Plant Sci7:1288
118. Raliya R, Tarafdar JC, Biswas (2016) Enhancing the mobilization of native phosphorus in the mung bean rhizosphere using ZnO nanoparticles synthesized by soil fungi. J Agric Food Chem 64(16):3111–3118
119. Ramírez-Rodríguez GB, Dal Sasso G, Carmona FJ, Miguel-Rojas C, Pérez-de-Luque A, Masciocchi N, Guagliardi A, Delgado-López JM (2020) Engineering biomimetic calcium phosphate nanoparticles: a green synthesis of slow-release multinutrient (NPK) nanofertilizers. ACS Appl Bio Mater 3(3):1344–1353
120. Rasheed A, Li H, Tahir MM, Mahmood A, Nawaz M, Shah AN, Aslam MT, Negm S, Moustafa M, Hassan MU, Wu Z (2022) The role of nanoparticles in plant biochemical, physiological, and molecular responses under drought stress: a review. Front Plant Sci 13:976179

121. Rivero-Montejo SDJ, Vargas-Hernandez M, Torres-Pacheco I (2021) Nanoparticles as novel elicitors to improve bioactive compounds in plants. Agriculture 11(2):134
122. Rizwan M, Ali S, Ali B, Adrees M, Arshad M, Hussain A, ur Rehman MZ, Waris AA (2019) Zinc and iron oxide nanoparticles improved the plant growth and reduced the oxidative stress and cadmium concentration in wheat. Chemosphere 214:269–277
123. Rossi L, Zhang W, Ma X (2017) Cerium oxide nanoparticles alter the salt stress tolerance of *Brassica napus* L. by modifying the formation of root apoplastic barriers. Environ Pollut 229:132–138
124. Saibo NJ, Lourenço T, Oliveira MM (2009) Transcription factors and regulation of photosynthetic and related metabolism under environmental stresses. Ann Bot 103(4):609–623
125. Sami F, Siddiqui H, Hayat S (2020) Impact of silver nanoparticles on plant physiology: a critical review. Sustain Agric Rev 41 Nanotechnol Plant Growth Dev 111–127
126. Sardar R, Ahmed S, Yasin NA (2022) Titanium dioxide nanoparticles mitigate cadmium toxicity in *Coriandrum sativum* L. through modulating antioxidant system, stress markers and reducing cadmium uptake. Environ Pollut 292:118373
127. Scandalios JG (2005) Oxidative stress: molecular perception and transduction of signals triggering antioxidant gene defenses. Braz J Med Biol Res 38:995–1014
128. Schwab F, Zhai G, Kern M, Turner A, Schnoor JL, Wiesner MR (2016) Barriers, pathways and processes for uptake, translocation and accumulation of nanomaterials in plants–critical review. Nanotoxicology 10(3):257–278
129. Seleiman MF, Almutairi KF, Alotaibi M, Shami A, Alhammad BA, Battaglia ML (2020) Nano-fertilization as an emerging fertilization technique: why can modern agriculture benefit from its use? Plants 10(1):2
130. Servin AD, White JC (2016) Nanotechnology in agriculture: next steps for understanding engineered nanoparticle exposure and risk. NanoImpact 1:9–12
131. Shabnam N, Kim M, Kim H (2019) Iron (III) oxide nanoparticles alleviate arsenic induced stunting in *Vigna radiata*. Ecotoxicol Environ Saf 183:109496
132. Shafqat W, Jaskani MJ, Maqbool R, Chattha WS, Ali Z, Naqvi SA, Haider MS, Khan IA, Vincent CI (2021) Heat shock protein and aquaporin expression enhance water conserving behavior of citrus under water deficits and high temperature conditions. Environ Exp Bot 181:104270
133. Shahi S, Srivastava M (2018) Influence of foliar application of manganese on growth, pigment content, and nitrate reductase activity of *Vigna radiata* (L.) R. Wilczek under salinity. J Plant Nutr 41(11):1397–1404
134. Shankramma K, Yallappa S, Shivanna MB, Manjanna J (2016) Fe 2 O 3 magnetic nanoparticles to enhance S. lycopersicum (tomato) plant growth and their biomineralization. Appl Nanosci 6:983–990
135. Sharif F, Westerhoff, Herckes P (2013) Sorption of trace organics and engineered nanomaterials onto wetland plant material. Environ Sci: Process Impacts15(1):267–274
136. Sharma B, Shrivastava M, Afonso LO, Soni U, Cahill DM (2022) Zinc-and magnesium-doped hydroxyapatite nanoparticles modified with urea as smart nitrogen fertilizers. ACS Appl Nano Mater 5(5):7288–7299
137. Siddiqi KS, Husen A (2016) Engineered gold nanoparticles and plant adaptation potential. Nanoscale Res Lett 11:1–10
138. Siddiqui H, Ahmed KBM, Sami F, Hayat S (2020) Silicon nanoparticles and plants: current knowledge and future perspectives. Sustain Agric Rev 41 Nanotechnol Plant Growth Dev 129–142
139. Simonin M, Martins JM, Uzu G, Spadini L, Navel A, Richaume A (2021) Low mobility of CuO and TiO_2 nanoparticles in agricultural soils of contrasting texture and organic matter content. Sci Total Environ 783:146952
140. Singh J, Lee BK (2016) Influence of nano-TiO_2 particles on the bioaccumulation of Cd in soybean plants (Glycine max): a possible mechanism for the removal of Cd from the contaminated soil. J Environ Manage 170:88–96

141. Singh A, Prasad SM, Singh S (2018) Impact of nano ZnO on metabolic attributes and fluorescence kinetics of rice seedlings. Environ Nanotechnol Monit Manag 9:42–49
142. Singh S, Husen A (2019) Role of nanomaterials in the mitigation of abiotic stress in plants. In: Husen A, Iqbal M (eds) N*anomaterials* and plant potential. Springer, Cham. Gewerbestrasse 11(6330):441–471. https://doi.org/10.1007/978-3-030-05569-1_18
143. Singh S, Husen A (2020) Behavior of agricultural crops in relation to nanomaterials under adverse environmental conditions. In: Husen A, Jawaid M (eds) Nanomaterials for agriculture and forestry applications. Elsevier Inc. Cambridge. https://doi.org/10.1016/B978-0-12-817852-2.00009-3
144. Solanki P, Bhargava A, Chhipa H, Jain N, Panwar J (2015) Nano-fertilizers and their smart delivery system. Nanotechnologies Food Agric 81–101
145. Song Y, Jiang M, Zhang H, Li R (2021) Zinc oxide nanoparticles alleviate chilling stress in rice (*Oryza Sativa* L.) by regulating antioxidative system and chilling response transcription factors. Molecules 26(8):2196
146. Sunkar R (2010) MicroRNAs with macro-effects on plant stress responses. In: Seminars in cell & developmental biology, vol 21, no 8. Academic Press, pp 805–811
147. Taware AS, Rathod B, Katariya AP, Tagad CK, Wagh S, Sonar JP, Deshmukh SU, Kanagare AB (2024) + Technological advancement in the development of nano fertilizers for sustainable agriculture. J Soil Sci Plant Nutr 1–17
148. Thakur M, Praveen S, Divte R, Mitra R, Kumar M, Gupta CK, Kalidindi U, Bansal R, Roy S, Anand A, Singh B (2022) Metal tolerance in plants: molecular and physicochemical interface determines the "not so heavy effect" of heavy metals. Chemosphere 287:131957
149. Thakur S, Asthir B, Kaur G, Kalia A, Sharma A (2021) Zinc oxide and titanium dioxide nanoparticles influence heat stress tolerance mediated by antioxidant defense system in wheat. CeR Res Commun 1–12
150. Tran LSP, Nakashima K, Sakuma Y, Osakabe Y, Qin F, Simpson SD, Maruyama K, Fujita Y, Shinozaki K, Yamaguchi-Shinozaki K (2007) Co-expression of the stress-inducible zinc finger homeodomain ZFHD1 and NAC transcription factors enhances expression of the ERD1 gene in Arabidopsis. Plant J 49(1):46–63
151. Tripathi DK, Singh VP, Prasad SM, Chauhan DK, Dubey NK (2015) Silicon nanoparticles (SiNp) alleviate chromium (VI) phytotoxicity in *Pisum sativum* (L.) seedlings. Plant Physiol Biochem 96:189–198
152. Tripathi DK, Singh S, Singh S, Pandey R, Singh VP, Sharma NC, Prasad SM, Dubey NK, Chauhan DK (2017) An overview on manufactured nanoparticles in plants: uptake, translocation, accumulation and phytotoxicity. Plant Physiol Biochem 110:2–12
153. Tului V, Janmohammadi M, Abbasi A, Vahdati-Khajeh S, Nouraein M (2021) Influence of iron, zinc and bimetallic Zn-Fe nanoparticles on growth and biochemical characteristics in chickpea (*Cicer arietinum*) cultivars. Poljoprivreda i Sumarstvo 67(2):179–193
154. Uzu G, Sobanska S, Sarret G, Muñoz M, Dumat C (2010) Foliar lead uptake by lettuce exposed to atmospheric fallouts. Environ Sci Technol 44(3):1036–1042
155. Venkatachalam P, Jayaraj M, Manikandan R, Geetha N, Rene ER, Sharma NC, Sahi SV (2017) Zinc oxide nanoparticles (ZnONPs) alleviate heavy metal-induced toxicity in *Leucaena leucocephala* seedlings: a physiochemical analysis. Plant Physiol Biochem 110:59–69
156. Verma SK, Das AK, Gantait S, Kumar V, Gurel E (2019) Applications of carbon nanomaterials in the plant system: a perspective view on the pros and cons. Sci Total Environ 667:485–499
157. Verma SK, Das AK, Patel MK, Shah A, Kumar V, Gantait S (2018) Engineered nanomaterials for plant growth and development: a perspective analysis. Sci Total Environ 630:1413–1435
158. Vishwakarma K, Upadhyay N, Kumar N, Yadav G, Singh J, Mishra RK, Kumar V, Verma R, Upadhyay RG, Pandey M, Sharma S (2017) Abscisic acid signaling and abiotic stress tolerance in plants: a review on current knowledge and future prospects. Front Plant Sci 8:161
159. Wang S, Fu Y, Zheng S, Xu Y, Sun Y (2022) Phytotoxicity and accumulation of copper-based nanoparticles in brassica under cadmium stress. Nanomaterials 12(9):1497
160. Wang P, Lombi E, Zhao FJ, Kopittke M (2016) Nanotechnology: a new opportunity in plant sciences. Trends Plant Sci 21(8):699–712

161. Wang Q, Liao Y, Zhao W, Yi T, Jiang Y, Zhu G, Sun Y, Wang Q, Huang L, Chen F, Zhang P (2024) Potassium-based nanomaterials significantly enhance nutrient utilization efficiency and promote high crop yields. Environ Sci Nano
162. Wang Y, Lin Y, Xu Y, Yin Y, Guo H, Du W (2019) Divergence in response of lettuce (var. ramosa Hort.) to copper oxide nanoparticles/microparticles as potential agricultural fertilizer. Environ Pollut Bioavailab 31(1):80–84
163. Wu K, Li W, Ma F, Gan F, Zhou J, Zeng H, Du C (2024) Urea-loaded core-shell MOF/silica nanocarriers for continuous nitrogen release to crops. ACS Appl Nano Materi 7(10):11645–11654
164. Wu H, Shabala L, Shabala S, Giraldo JP (2018) Hydroxyl radical scavenging by cerium oxide nanoparticles improves Arabidopsis salinity tolerance by enhancing leaf mesophyll potassium retention. Environ Sci Nano 5(7):1567–1583
165. Wu J, Wang T (2020) Synergistic effect of zinc oxide nanoparticles and heat stress on the alleviation of transcriptional gene silencing in Arabidopsis thaliana. Bull Environ Contam Toxicol 104(1):49–56
166. Wu H, Li Z (2022) Nano-enabled agriculture: How do nanoparticles cross barriers in plants? Plant Commun 3(6)
167. Yang KY, Doxey S, McLean JE, Britt D, Watson A, Al Qassy D, Jacobson A, Anderson AJ (2018) Remodeling of root morphology by CuO and ZnO nanoparticles: effects on drought tolerance for plants colonized by a beneficial pseudomonad. Botany 96(3):175–186
168. Ye Y, Cota-Ruiz K, Hernández-Viezcas JA, Valdes C, Medina-Velo IA, Turley RS, Peralta-Videa JR, Gardea-Torresdey JL (2020) Manganese nanoparticles control salinity-modulated molecular responses in *Capsicum annuum* L. through priming: a sustainable approach for agriculture. ACS Sustain Chem Eng 8(3):1427–1436
169. Younis AA, Khattab H, Emam MM (2020) Impacts of silicon and silicon nanoparticles on leaf ultrastructure and TaPIP1 and TaNIP2 gene expressions in heat stressed wheat seedlings. Biologia Plantarum 64(1)
170. Yuan H, Liu Q, Guo Z, Fu J, Sun Y, Gu C, Xing B, Dhankher OP (2021) Sulfur nanoparticles improved plant growth and reduced mercury toxicity via mitigating the oxidative stress in *Brassica napus* L. J Clean Prod 318:128589
171. Zhang Z, Kong F, Vardhanabhuti B, Mustapha A, Lin M (2012) Detection of engineered silver nanoparticle contamination in pears. J Agric Food Chem 60(43):10762–10767
172. Zhang LW, Monteiro-Riviere NA (2009) Mechanisms of quantum dot nanoparticle cellular uptake. Toxicol Sci 110(1):138–155
173. Zhao L, Bai T, Wei H, Gardea-Torresdey JL, Keller A, White JC (2022) Nanobiotechnology-based strategies for enhanced crop stress resilience. Nature Food 3(10):829–836
174. Zhao L, Lu L, Wang A, Zhang H, Huang M, Wu H, Xing B, Wang Z, Ji R (2020) Nano-biotechnology in agriculture: use of nanomaterials to promote plant growth and stress tolerance. J Agric Food Chem 68(7):1935–1947
175. Zhao L, Peralta-Videa JR, Varela-Ramirez A, Castillo-Michel H, Li C, Zhang J, Aguilera RJ, Keller AA, Gardea-Torresdey JL (2012) Effect of surface coating and organic matter on the uptake of CeO_2 NPs by corn plants grown in soil: insight into the uptake mechanism. J Hazard Mater 225:131–138
176. Zhou P, Adeel M, Shakoor N, Guo M, Hao Y, Azeem I, Li M, Liu M, Rui Y (2020) Application of nanoparticles alleviates heavy metals stress and promotes plant growth: an overview. Nanomaterials 11(1):26
177. Zhu JK (2002) Salt and drought stress signal transduction in plants. Annu Rev Plant Biol 53(1):247–273
178. Zulfiqar F, Ashraf M (2021) Nanoparticles potentially mediate salt stress tolerance in plants. Plant Physiol Biochem 160:257–268

MIX
Papier aus verantwortungsvollen Quellen
Paper from responsible sources
FSC® C105338

If you have any concerns about our products,
you can contact us on
ProductSafety@springernature.com

In case Publisher is established outside the EU,
the EU authorized representative is:
Springer Nature Customer Service Center GmbH
Europaplatz 3, 69115 Heidelberg, Germany

Printed by Libri Plureos GmbH
in Hamburg, Germany